T0281348

Undergraduate Texts in Mathematics

Undergraduate Texts in Mathematics

Undergraduate Texts in Mathematics are generally aimed at third- and fourth-year undergraduate mathematics students at North American universities. These texts strive to provide students and teachers with new perspectives and novel approaches. The books include motivation that guides the reader to an appreciation of interrelations among different aspects of the subject. They feature examples that illustrate key concepts as well as exercises that strengthen understanding.

For further volumes:
http://www.springer.com/series/666

Peter D. Lax • Maria Shea Terrell

Calculus With Applications

Second Edition

Peter D. Lax
Courant Institute of Mathematical Sciences
New York University
New York, NY, USA

Maria Shea Terrell
Department of Mathematics
Cornell University
Ithaca, NY, USA

ISSN 0172-6056
ISBN 978-1-4939-3688-5 ISBN 978-1-4614-7946-8 (eBook)
DOI 10.1007/978-1-4614-7946-8
Springer New York Heidelberg Dordrecht London

Mathematics Subject Classification: 00-01

Softcover reprint of the hardcover 2nd edition 2014

Printed on acid-free paper

Springer is part of Springer Science+Business Media (www.springer.com)

Preface

Our purpose in writing a calculus text has been to help students learn at first hand that mathematics is the language in which scientific ideas can be precisely formulated, that science is a source of mathematical ideas that profoundly shape the development of mathematics, and that mathematics can furnish brilliant answers to important scientific problems. This book is a thorough revision of the text *Calculus with Applications and Computing* by Lax, Burstein, and Lax. The original text was predicated on a number of innovative ideas, and it included some new and nontraditional material. This revision is written in the same spirit. It is fair to ask what new subject matter or new ideas could possibly be introduced into so old a topic as calculus. The answer is that science and mathematics are growing by leaps and bounds on the research frontier, so what we teach in high school, college, and graduate school must not be allowed to fall too far behind. As mathematicians and educators, our goal must be to simplify the teaching of old topics to make room for new ones.

To achieve that goal, we present the language of mathematics as natural and comprehensible, a language students can learn to use. Throughout the text we offer proofs of all the important theorems to help students understand their meaning; our aim is to foster understanding, not "rigor." We have greatly increased the number of worked examples and homework problems. We have made some significant changes in the organization of the material; the familiar transcendental functions are introduced before the derivative and the integral. The word "computing" was dropped from the title because today, in contrast to 1976, it is generally agreed that computing is an integral part of calculus and that it poses interesting challenges. These are illustrated in this text in Sects. 4.4, 5.3, and 10.4, and by all of Chap. 8. But the mathematics that enables us to discuss issues that arise in computing when we round off inputs or approximate a function by a sequence of functions, i.e., uniform continuity and uniform convergence, remains. We have worked hard in this revision to show that uniform convergence and continuity are more natural and useful than pointwise convergence and continuity. The initial feedback from students who have used the text is that they "get it."

This text is intended for a two-semester course in the calculus of a single variable. Only knowledge of high-school precalculus is expected.

Chapter 1 discusses numbers, approximating numbers, and limits of sequences of numbers. Chapter 2 presents the basic facts about continuous functions and describes the classical functions: polynomials, trigonometric functions, exponentials, and logarithms. It introduces limits of sequences of functions, in particular power series.

In Chapter 3, the derivative is defined and the basic rules of differentiation are presented. The derivatives of polynomials, the exponential function, the logarithm, and trigonometric functions are calculated. Chapter 4 describes the basic theory of differentiation, higher derivatives, Taylor polynomials and Taylor's theorem, and approximating derivatives by difference quotients. Chapter 5 describes how the derivative enters the laws of science, mainly physics, and how calculus is used to deduce consequences of these laws.

Chapter 6 introduces, through examples of distance, mass, and area, the notion of the integral, and the approximate integrals leading to its definition. The relation between differentiation and integration is proved and illustrated. In Chapter 7, integration by parts and change of variable in integrals are presented, and the integral of the uniform limit of a sequence of functions is shown to be the limit of the integrals of the sequence of functions. Chapter 8 is about the approximation of integrals; Simpson's rule is derived and compared with other numerical approximations of integrals.

Chapter 9 shows how many of the concepts of calculus can be extended to complex-valued functions of a real variable. It also introduces the exponential of complex numbers. Chapter 10 applies calculus to the differential equations governing vibrating strings, changing populations, and chemical reactions. It also includes a very brief introduction to Euler's method. Chapter 11 is about the theory of probability, formulated in the language of calculus.

The material in this book has been used successfully at Cornell in a one-semester calculus II course for students interested in majoring in mathematics or science. The students typically have credit for one semester of calculus from high school. Chapters 1, 2, and 4 have been used to present sequences and series of numbers, power series, Taylor polynomials, and Taylor's theorem. Chapters 6–8 have been used to present the definite integral, application of integration to volumes and accumulation problems, methods of integration, and approximation of integrals. There has been adequate time left in the term then to present Chapter 9, on complex numbers and functions, and to see how complex functions and calculus are used to model vibrations in the first section of Chapter 10.

We are grateful to the many colleagues and students in the mathematical community who have supported our efforts to write this book. The first edition of this book was written in collaboration with Samuel Burstein. We thank him for allowing us to draw on his work. We wish to thank John Guckenheimer for his encouragement and advice on this project. We thank Matt Guay, John Meluso, and Wyatt Deviau, who while they were undergraduates at Cornell, carefully read early drafts of the manuscript, and whose perceptive comments helped us keep our student audience in mind. We also wish to thank Patricia McGrath, a teacher at Maloney High School in Meriden, Connecticut, for her thoughtful review and suggestions, and Thomas

Kern and Chenxi Wu, graduate students at Cornell who assisted in teaching calculus II with earlier drafts of the text, for their help in writing solutions to some of the homework problems. Many thanks go to the students at Cornell who used early drafts of this book in fall 2011 and 2012. Thank you all for inspiring us to work on this project, and to make it better.

This current edition would have been impossible without the support of Bob Terrell, Maria's husband and long-time mathematics teacher at Cornell. From TEX-ing the manuscript to making the figures, to suggesting changes and improvements, at every step along the way we owe Bob more than we can say.

Peter Lax thanks his colleagues at the Courant Institute, with whom he has discussed over 50 years the challenge of teaching calculus.

New York, NY Peter Lax
Ithaca, NY Maria Terrell

Contents

Chapter 1
Numbers and Limits

Abstract This chapter introduces basic concepts and properties of numbers that are necessary prerequisites for defining the calculus concepts of limit, derivative, and integral.

1.1 Inequalities

One cannot exaggerate the importance in calculus of inequalities between numbers. Inequalities are at the heart of the basic notion of convergence, an idea central to calculus. Inequalities can be used to prove the equality of two numbers by showing that one is neither less than nor greater than the other. For example, Archimedes showed that the area of a circle was neither less than nor greater than the area of a triangle with base the circumference and height the radius of the circle.

A different use of inequalities is descriptive. Sets of numbers described by inequalities can be visualized on the number line.

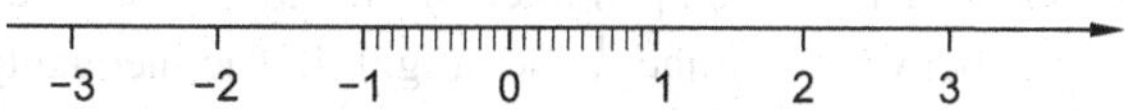

Fig. 1.1 The number line

To say that a is less than b, denoted by $a < b$, means that $b - a$ is positive. On the number line in Fig. 1.1, a would lie to the left of b. Inequalities are often used to describe intervals of numbers. The numbers that satisfy $a < x < b$ are the numbers between a and b, not including the endpoints a and b. This is an example of an *open* interval, which is indicated by round brackets, (a,b).

To say that a is less than or equal to b, denoted by $a \le b$, means that $b - a$ is not negative. The numbers that satisfy $a \le x \le b$ are the numbers between a and b, including the endpoints a and b. This is an example of a *closed* interval, which is indicated by square brackets, $[a,b]$. Intervals that include one endpoint but not

P.D. Lax and M.S. Terrell, *Calculus With Applications*, Undergraduate Texts in Mathematics, DOI 10.1007/978-1-4614-7946-8_1, © Springer Science+Business Media New York 2014

Fig. 1.2 *Left*: the open interval (a,b). *Center*: the half open interval $(a,b]$. *Right*: the closed interval $[a,b]$

the other are called *half-open* or *half-closed*. For example, the interval $a < x \le b$ is denoted by $(a,b]$ (Fig. 1.2).

The absolute value $|a|$ of a number a is the distance of a from 0; for a positive, then, $|a| = a$, while for a negative, $|a| = -a$. The absolute value of a difference, $|a-b|$, can be interpreted as the distance between a and b on the number line, or as the length of the interval between a and b (Fig. 1.3).

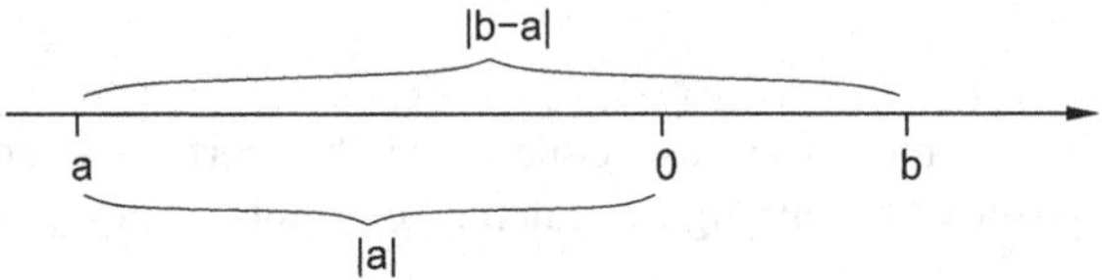

Fig. 1.3 Distances are measured using absolute value

The inequality

$$|a-b| < \varepsilon$$

can be interpreted as stating that the distance between a and b on the number line is less than ε. It also means that the difference between a and b is no more than ε and no less than $-\varepsilon$:

$$-\varepsilon < a-b < \varepsilon. \tag{1.1}$$

In Problem 1.9, we ask you to use some of the properties of inequalities stated in Sect. 1.1a to obtain inequality (1.1).

Example 1.1. The inequality $|x-5| < \frac{1}{2}$ describes the numbers x whose distance from 5 is less than $\frac{1}{2}$. This is the open interval $(4.5, 5.5)$. It also tells us that the difference $x-5$ is between $-\frac{1}{2}$ and $\frac{1}{2}$. See Fig. 1.4. The inequality $|x-5| \le \frac{1}{2}$ describes the closed interval $[4.5, 5.5]$.

Fig. 1.4 *Left*: the numbers specified by the inequality $|x-5| < \frac{1}{2}$ in Example 1.1. *Right*: the difference $x-5$ is between $-\frac{1}{2}$ and $\frac{1}{2}$

The inequality $|\pi - 3.141| \le \dfrac{1}{10^3}$ can be interpreted as a statement about the precision of 3.141 as an approximation of π. It tells us that 3.141 is within $\dfrac{1}{10^3}$ of π, and that π is in an interval centered at 3.141 of length $\dfrac{2}{10^3}$.

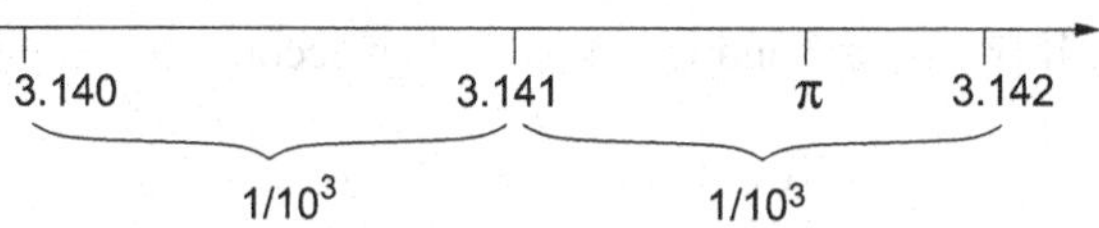

Fig. 1.5 Approximations to π

We can imagine smaller intervals contained inside the larger one in Fig. 1.5, which surround π more closely. Later in this chapter we will see that one way to determine a number is by trapping it within progressively tighter intervals. This process is described by the nested interval theorem in Sect. 1.3c.

We use (a,∞) to denote the set of numbers that are greater than a, and $[a,\infty)$ to denote the set of numbers that are greater than or equal to a. Similarly, $(-\infty,a)$ denotes the set of numbers less than a, and $(-\infty,a]$ denotes those less than or equal to a. See Fig. 1.6.

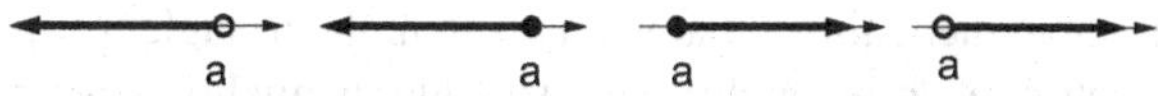

Fig. 1.6 The intervals $(-\infty,a)$, $(-\infty,a]$, $[a,\infty)$, and (a,∞) are shown from left to right

Example 1.2. The inequality $|x-5| \geq \frac{1}{2}$ describes the numbers whose distance from 5 is greater than or equal to $\frac{1}{2}$. These are the numbers that are in $(-\infty,4.5]$ or in $[5.5,\infty)$. See Fig. 1.7.

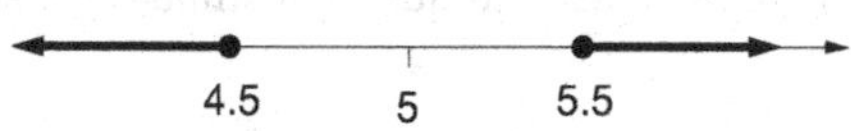

Fig. 1.7 The numbers specified by the inequality in Example 1.2

1.1a Rules for Inequalities

Next we review some rules for handling inequalities.

(a) *Trichotomy:* For any numbers a and b, either $a < b$ or $a = b$ or $b < a$.
(b) *Transitivity:* If $a < b$ and $b < c$, then $a < c$.
(c) *Addition:* If $a < b$ and $c < d$, then $a+c < b+c$ and $a+c < b+d$.
(d) *Multiplication:* If $a < b$ and p is positive, then $pa < pb$, but if $a < b$ and n is negative, then $nb < na$.
(e) *Reciprocal:* If a and b are positive numbers and $a < b$, then $\dfrac{1}{b} < \dfrac{1}{a}$.

The rules for inequalities can be used algebraically to simplify inequalities or to derive new inequalities from old ones. With the exception of trichotomy, these rules are still true if $<$ is replaced by $\leq$. In Problem 1.8 we ask you to use trichotomy to show that if $a \leq b$ and $b \leq a$, then $a = b$.

Example 1.3. If $|x-3|<2$ and $|y-4|<6$, then according to the inequality rule on addition,

$$|x-3|+|y-4|<2+6.$$

Example 1.4. If $0<a<b$, then according to inequality rule on multiplication,

$$a^2<ab \text{ and } ab<b^2.$$

Then by the transitivity rule, $a^2<b^2$.

1.1b The Triangle Inequality

There are two notable inequalities that we use often, the triangle inequality, and the arithmetic–geometric mean inequality. The triangle inequality is as important as it is simple:

$$|a+b|\le|a|+|b|.$$

Try substituting in a few numbers. What does it say, for example when $a=-3$ and $b=1$? It is easy to convince yourself that when a and b are of the same sign, or one of them is zero, equality holds. If a and b have opposite signs, inequality holds.

The triangle inequality can be used to quickly estimate the accuracy of a sum of approximations.

Example 1.5. Using

$$|\pi-3.141|<10^{-3} \text{ and } |\sqrt{2}-1.414|<10^{-3},$$

the inequality addition rule gives $|\pi-3.141|+|\sqrt{2}-1.414|<10^{-3}+10^{-3}$. The triangle inequality then tells us that

$$\begin{aligned}|(\pi+\sqrt{2})-4.555| &= |(\pi-3.141)+(\sqrt{2}-1.414)|\\ &\le|\pi-3.141|+|\sqrt{2}-1.414|\le 2\times10^{-3}.\end{aligned}$$

That is, knowing $\sqrt{2}$ and π within 10^{-3}, we know their sum within 2×10^{-3}.

Another use of the triangle inequality is to relate distances between numbers on the number line. The inequality says that the distance between x and z is less than or equal to the sum of the distance between x and y and the distance between y and z. That is,

$$|z-x|=|(z-y)+(y-x)|\le|z-y|+|y-x|.$$

In Fig. 1.8 we illustrate two cases: in which y is between x and z, and in which it is not.

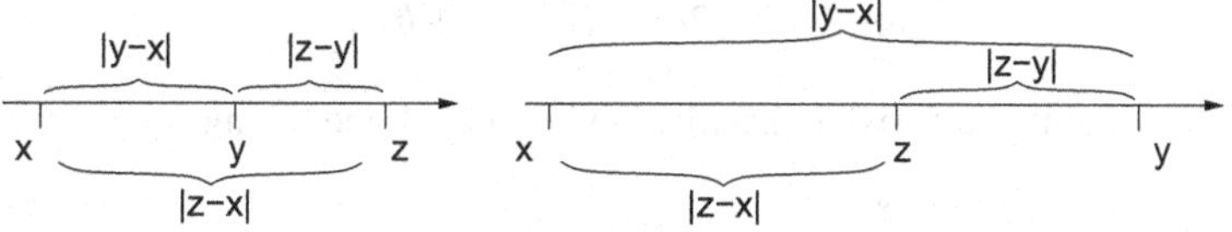

Fig. 1.8 Distances related by the triangle inequality

1.1c The Arithmetic–Geometric Mean Inequality

Next we explore an important but less familiar inequality.

Theorem 1.1. The arithmetic–geometric mean inequality. *The geometric mean of two positive numbers is less than their arithmetic mean:*

$$\sqrt{ab} \le \frac{a+b}{2},$$

with equality only in the case $a = b$.

We refer to this as the "A-G" inequality. The word "mean" is used in the following sense:

(a) The mean lies between the smaller and the larger of the two numbers a and b.
(b) When a and b are equal, their mean is equal to a and b.

You can check that each side of the inequality is a mean in this sense.

A visual proof: Figure 1.9 provides a visual proof that $4ab \le (a+b)^2$. The A-G inequality follows once you divide by 4 and take the square root.

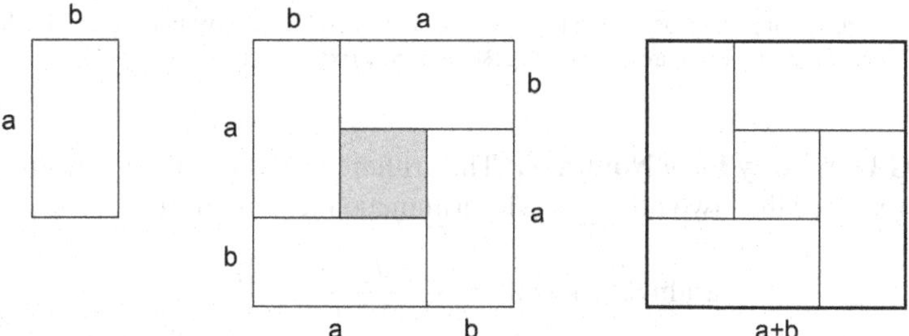

Fig. 1.9 A visual proof that $4ab \le (a+b)^2$, by comparing areas

An algebraic proof: Since the square of any number is positive or zero, it follows that

$$0 \le (a-b)^2 = a^2 - 2ab + b^2,$$

with equality holding only when $a = b$. By adding $4ab$ to both sides, we get

$$4ab \le a^2 + 2ab + b^2 = (a+b)^2,$$

the same inequality we derived visually. Dividing by 4 and taking square roots, we get

$$\sqrt{ab} \le \frac{a+b}{2},$$

with equality holding only when $a = b$.

Example 1.6. The A-G inequality can be used to prove that among all rectangles with the same perimeter, the square has the largest area. See Fig. 1.10. *Proof:* Denote the lengths of the sides of the rectangle by W and L. Its area is WL. The lengths of the sides of the square with the same perimeter are $\frac{W+L}{2}$, and its area is $\left(\frac{W+L}{2}\right)^2$. The inequality

$$WL \le \left(\frac{W+L}{2}\right)^2$$

follows from squaring both sides of the A-G inequality.

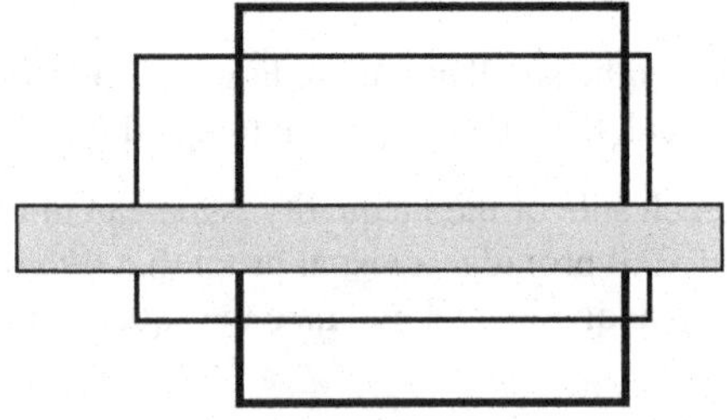

Fig. 1.10 Three rectangles measuring 6 by 6, 8 by 4, and 11 by 1. All have perimeter 24. The areas are 36, 32, and 11, and the square has the largest area. See Example 1.6

The A-G Inequality for n Numbers. The arithmetic and geometric means can be defined for more than two numbers. The arithmetic mean of $a_1, a_2, \ldots, a_n$ is

$$\text{arithmetic mean} = \frac{a_1 + a_2 + \cdots + a_n}{n}.$$

The geometric mean of n positive numbers is defined as the nth root of their product:

$$\text{geometric mean} = (a_1 a_2 \cdots a_n)^{1/n}.$$

For n numbers, the A-G inequality is

$$(a_1 a_2 \cdots a_n)^{1/n} \le \frac{a_1 + a_2 + \cdots + a_n}{n},$$

with equality holding only when $a_1 = a_2 = \cdots = a_n$. As in the case of a rectangle, the A-G inequality for three numbers can be interpreted geometrically: Consider the volume of a box that measures a_1 by a_2 by a_3. Then the inequality states that among all boxes with a given edge sum, the cube has the largest volume.

The proof of the case for n numbers is outlined in Problem 1.17. The key to the proof is understanding how to use the result for two numbers to derive it for four numbers. Curiously, the result for $n = 4$ can then be used to prove the result for $n = 3$. The general proof proceeds in a similar manner. Use the result for $n = 4$ to get the result for $n = 8$, and then use the result for $n = 8$ to get the result for $n = 5$, 6, and 7, and so forth.

Here is the proof for $n = 4$. Let c, d, e, f be four positive numbers. Denote by a the arithmetic mean of c and d, and denote by b the arithmetic mean of e and f:

$$a = \frac{c+d}{2}, \quad b = \frac{e+f}{2}.$$

By the A-G inequality for two numbers, applied three times, we get

$$\sqrt{cd} \le a, \quad \sqrt{ef} \le b, \tag{1.2}$$

and

$$\sqrt{ab} \le \frac{a+b}{2}. \tag{1.3}$$

Combining inequalities (1.2) and (1.3) gives

$$(cdef)^{1/4} \le \frac{a+b}{2}. \tag{1.4}$$

Since

$$\frac{a+b}{2} = \frac{\frac{c+d}{2} + \frac{e+f}{2}}{2} = \frac{c+d+e+f}{4},$$

we can rewrite inequality (1.4) as

$$(cdef)^{1/4} \le \frac{c+d+e+f}{4},$$

with equality holding only when $a = b$ and when $c = d$ and $e = f$. This completes the argument for four numbers. Next we see how to use the result for four to prove the result for three numbers.

We start with the observation that if a, b, and c are any three numbers, and m is their arithmetic mean,

$$m = \frac{a+b+c}{3}, \tag{1.5}$$

then m is also the arithmetic mean of the four numbers a, b, c, and m:

$$m = \frac{a+b+c+m}{4}.$$

To see this, multiply Eq. (1.5) by 3 and add m to both sides. We get $4m = a+b+c+m$. Dividing by 4 gives the result we claimed. Now apply the A-G inequality to the four numbers a, b, c, and m. We get

$$(abcm)^{1/4} \le m.$$

Raise both sides to the fourth power. We get $abcm \le m^4$. Divide both sides by m and then take the cube root of both sides; we get the desired inequality

$$(abc)^{1/3} \le m = \frac{a+b+c}{3}.$$

This completes the argument for $n = 2$, 3, and 4. The rest of the proof proceeds similarly.

Problems

1.1. Find the numbers that satisfy each inequality, and sketch the solution on a number line.

(a) $|x-3| \le 4$
(b) $|x+50| \le 2$
(c) $1 < |y-7|$
(d) $|3-x| < 4$

1.2. Find the numbers that satisfy each inequality, and sketch the solution on a number line.

(a) $|x-4| < 2$
(b) $|x+4| \le 3$
(c) $|y-9| \ge 2$
(d) $|4-x| < 2$

1.3. Use inequalities to describe the numbers *not* in the interval $[-3,3]$ in two ways:

(a) using an absolute value inequality
(b) using one or more simple inequalities.

1.4. Find the arithmetic mean $A(a,b)$ and geometric mean $G(a,b)$ of the pairs $(a,b) = (5,5)$, $(3,7)$, $(1,9)$. Sketch a square corresponding to each case, as in the geometric proof. Interpret the pairs as dimensions of a rectangle. Find the perimeter and area of each.

1.5. Find the geometric mean of 2, 4, and 8. Verify that it is less than the arithmetic mean.

1.6. Which inequalities are true for all numbers a and b satisfying $0 < a < b < 1$?

(a) $ab > 1$
(b) $\frac{1}{a} < \frac{1}{b}$
(c) $\frac{1}{b} > 1$
(d) $a + b < 1$
(e) $a^2 < 1$
(f) $a^2 + b^2 < 1$
(g) $a^2 + b^2 > 1$
(h) $\frac{1}{a} > b$

1.7. You know from algebra that when x and y are positive numbers,

$$(\sqrt{x} - \sqrt{y})(\sqrt{x} + \sqrt{y}) = x - y.$$

(a) Suppose $x > y > 5$. Show that $\sqrt{x} - \sqrt{y} \le \frac{1}{4}(x - y)$.
(b) Suppose y is within 0.02 of x. Use the inequality in part (a) to estimate how close $\sqrt{y}$ is to $\sqrt{x}$.

1.8. Use the trichotomy rule to show that if $a \le b$ and $b \le a$, then $a = b$.

1.9. Suppose $|b - a| < \varepsilon$. Explain why each of the following items is true.

(a) $0 \le (b - a) < \varepsilon$ or $0 \le -(b - a) < \varepsilon$
(b) $-\varepsilon < b - a < \varepsilon$
(c) $a - \varepsilon < b < a + \varepsilon$
(d) $-\varepsilon < a - b < \varepsilon$
(e) $b - \varepsilon < a < b + \varepsilon$

1.10.(a) A rectangular enclosure is to be constructed with 16 m of fence. What is the largest possible area of the enclosure?
(b) If instead of four fenced sides, one side is provided by a large barn wall, what is the largest possible area of the enclosure?

1.11. A shipping company limits the sum of the three dimensions of a rectangular box to 5 m. What are the dimensions of the box that contains the largest possible volume?

1.12. Two pieces of string are measured to within 0.001 m of their true length. The first measures 4.325 m and the second measures 5.579 m. A good estimate for the total length of string is 9.904 m. How accurate is that estimate?

1.13. In this problem we see how the A-G inequality can be used to derive various inequalities. Let x be positive.

(a) Write the A-G inequality for the numbers 1, 1, x, to show that $x^{1/3} \le \frac{x+2}{3}$.

(b) Similarly, show that $x^{1/n} \le \frac{x+n-1}{n}$ for every positive integer n.

(c) By letting $x = n$ in the inequality in (b), we get

$$n^{1/n} \le \frac{2n-1}{n}.$$

Explain how it follows that $n^{1/n}$ is always less than 2.

1.14. The *harmonic mean* is defined for positive numbers a and b by

$$H(a,b) = \frac{2}{\frac{1}{a}+\frac{1}{b}}.$$

(a) For the cases $(a,b) = (2,3)$ and $(3,3)$, verify that

$$H(a,b) \le G(a,b) \le A(a,b), \tag{1.6}$$

i.e., $\dfrac{2}{\frac{1}{a}+\frac{1}{b}} \le \sqrt{ab} \le \dfrac{a+b}{2}$.

(b) On a trip, a driver goes the first 100 miles at 40 mph, and the second 100 miles at 60 mph. Show that the average speed is the harmonic mean of 40 and 60.

(c) Deduce $H(a,b) \le G(a,b)$ from $G(\frac{1}{a},\frac{1}{b}) \le A(\frac{1}{a},\frac{1}{b})$.

(d) A battery supplies the same voltage V to each of two resistors in parallel in Fig. 1.11. The current I splits as $I = I_1 + I_2$, so that Ohm's law $V = I_1R_1 = I_2R_2$ holds for each resistor. Show that the value R to be used in $V = IR$ is one-half the harmonic mean of R_1 and R_2.

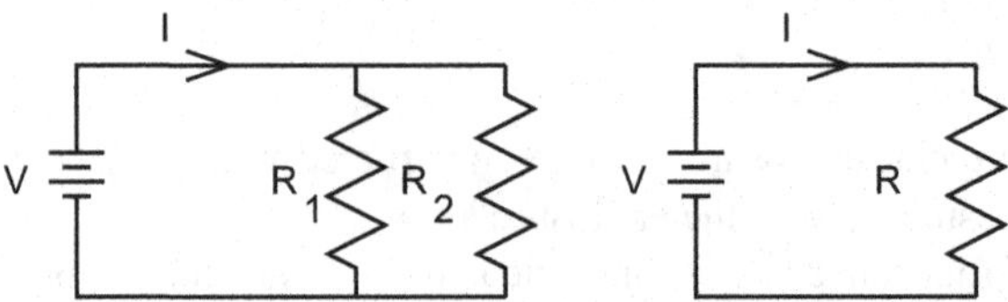

Fig. 1.11 Two resistors in parallel with a battery, and an equivalent circuit with only one resistor. See Problem 1.14

1.15. The product of the numbers 1 through n is the factorial $n! = (1)(2)(3)\cdots(n)$. Their sum is

$$1+2+3+\cdots+n = \frac{1}{2}n(n+1).$$

(a) Show that $(n!)^{1/n} \le n$.

(b) Use the A-G inequality to derive the better result that $(n!)^{1/n} \le \dfrac{n+1}{2}$.

1.16. If we want to know how much the product ab varies when we allow a and b to vary independently, there is a clever algebra trick that helps in this:

$$ab - a_0b_0 = ab - ab_0 + ab_0 - a_0b_0.$$

(a) Show that
$$|ab - a_0 b_0| \le |a||b - b_0| + |b_0||a - a_0|.$$

(b) Suppose a and b are in the interval $[0, 10]$, and that a_0 is within 0.001 of a and b_0 is within 0.001 of b. How close is $a_0 b_0$ to ab?

1.17. Here you may finish the proof of the A-G inequality.

(a) Prove the A-G inequality for eight numbers by using twice the A-G mean inequality for four numbers, and combine it with the A-G inequality for two numbers.
(b) Show that if a, b, c, d, and e are any five numbers, and m is their arithmetic mean, then the arithmetic mean of the eight numbers a, b, c, d, e, m, m, and m is again m. Use this and the A-G inequality for eight numbers to prove the A-G inequality for five numbers.
(c) Prove the general case of the A-G inequality by generalizing (a) and (b).

1.18. Another important inequality is due to the French mathematician Cauchy and the German mathematician Schwarz: Let $a_1, a_2, \ldots, a_n$ and $b_1, b_2, \ldots, b_n$ be two sets of numbers. Then

$$a_1 b_1 + \cdots + a_n b_n \le \sqrt{a_1^2 + \cdots + a_n^2}\sqrt{b_1^2 + \cdots + b_n^2}.$$

Verify each of these steps of the proof:

(a) The roots of the polynomial $p(x) = Px^2 + 2Qx + R$ are $\dfrac{-Q \pm \sqrt{Q^2 - PR}}{P}$.
(b) Show that if $p(x)$ does not take negative values, then $p(x)$ has at most one real root. Show that in this case, $Q^2 \le PR$.
(c) Take $p(x) = (a_1 x + b_1)^2 + \cdots + (a_n x + b_n)^2$. Show that

$$P = a_1^2 + \cdots + a_n^2, \qquad Q = a_1 b_1 + \cdots + a_n b_n, \qquad \text{and} \quad R = b_1^2 + \cdots + b_n^2.$$

(d) Since $p(x)$ defined above is a sum of squares, it does not take negative values. Therefore, $Q^2 \le PR$. Deduce from this the Cauchy–Schwarz inequality.
(e) Determine the condition for equality to occur.

1.2 Numbers and the Least Upper Bound Theorem

1.2a Numbers as Infinite Decimals

There are two familiar ways of looking at numbers: as infinite decimals and as points on a number line. The integers divide the number line into infinitely many intervals of unit length. If we include the left endpoint of each interval but not the right, we can cover the number line with nonoverlapping intervals such that each number a

belongs to exactly one of them, $n \le a < n+1$. Each interval can be subdivided into ten subintervals of length $\frac{1}{10}$. As before, if we agree to count the left endpoint but not the right as part of each interval, the intervals do not overlap. Our number a belongs to exactly one of these ten subintervals, say to

$$n + \frac{\alpha_1}{10} \le a < n + \frac{\alpha_1 + 1}{10}.$$

This determines the first decimal digit α_1 of a. For example, Fig. 1.12 illustrates how to find the first decimal digit of a number a between 2 and 3.

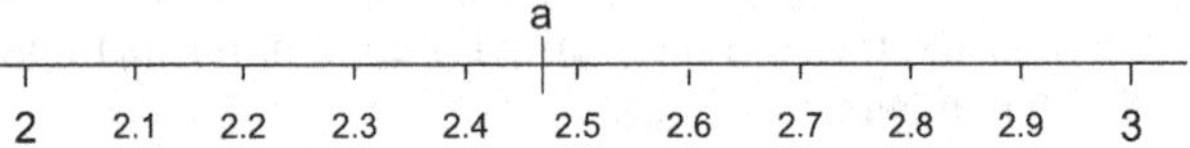

Fig. 1.12 a is in the interval $[2.4, 2.5)$, so $\alpha_1 = 4$

The second decimal digit α_2 is determined similarly, by subdividing the interval $[2.4, 2.5)$ into ten equal subintervals, and so on. Figure 1.13 illustrates the example $\alpha_2 = 7$.

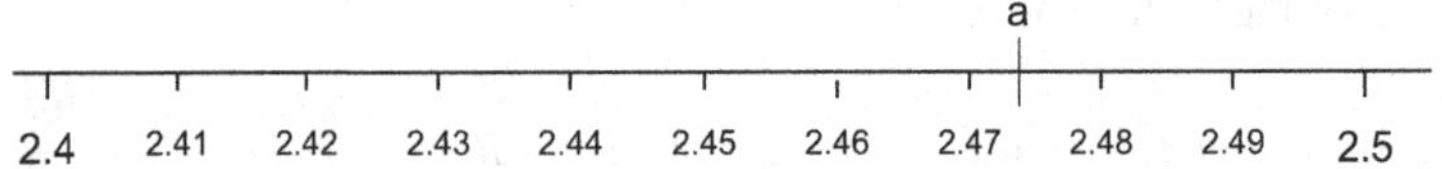

Fig. 1.13 a is in the interval $[2.47, 2.48)$, so $\alpha_2 = 7$

Thus using the representation of a as a point on the number line and the procedure just described, we can find α_k in $a = n.\alpha_1\alpha_2\ldots\alpha_k\ldots$ by determining the appropriate interval in the kth step of this process. Conversely, once we have the decimal representation of a number, we can identify its location on the number line.

Example 1.7. Examining the decimal representation $\frac{31}{39} = 0.7948717\ldots$, we see that

$$0.79487 \le \frac{31}{39} < 0.79488.$$

Repeated Nines in Decimals. The method we described for representing numbers as infinite decimals does not result in decimal fractions that end with infinitely many nines. Nevertheless, such decimals come up when we do arithmetic with infinite decimals. For instance, take the sum

$$\begin{array}{r} 1/3 = 0.333333333\cdots \\ +\ 2/3 = 0.666666666\cdots \\ \hline 1 = 0.999999999\cdots \end{array}$$

Similarly, every infinite decimal ending with all nines is equal to a finite decimal, such as

$$0.39529999999\cdots = 0.3953.$$

Decimals and Ordering. The importance of the infinite decimal representation of numbers lies in the ease with which numbers can be compared. For example, which of the numbers

$$\frac{17}{20}, \quad \frac{31}{39}, \quad \frac{45}{53}, \quad \frac{74}{87}$$

is the largest? To compare them as fractions, we would have to bring them to a common denominator. If we represent the numbers as decimals,

$$\frac{17}{20} = 0.85000\ldots$$
$$\frac{31}{39} = 0.79487\ldots$$
$$\frac{45}{53} = 0.84905\ldots$$
$$\frac{74}{87} = 0.85057\ldots$$

we can tell which number is larger by examining their integer parts and decimal digits, place by place. Then clearly,

$$\frac{31}{39} < \frac{45}{53} < \frac{17}{20} < \frac{74}{87}.$$

1.2b The Least Upper Bound Theorem

The same process we used for comparing four numbers can be used to find the largest number in any finite set of numbers that are represented as decimals. Can we apply a similar procedure to find the largest number in an infinite set S of numbers? Clearly, the set S of positive integers has no largest element. Suppose we rule out sets that contain arbitrarily large numbers and assume that all numbers in S are less than some number k. Such a number k is called an *upper bound* of S.

Definition 1.1. A number k is called an *upper bound* for a set S of numbers if

$$x \le k$$

for every x in S, and we say that S is *bounded above* by k. Analogously, k is called a *lower bound* for S if $k \le x$ for every x in S, and we say that S is *bounded below* by k.

Imagine pegs in the number line at all points of the set S. Let k be an upper bound for S that is to the right of every point of S. Put the point of your pencil at k and move it as far to the left as the pegs will let it go (Fig. 1.14). The point where the pencil gets stuck is also an upper bound of S. There can be no smaller upper bound, for if there were, we could have slid the pencil further to the left. It is the *least upper bound* of S.[1]

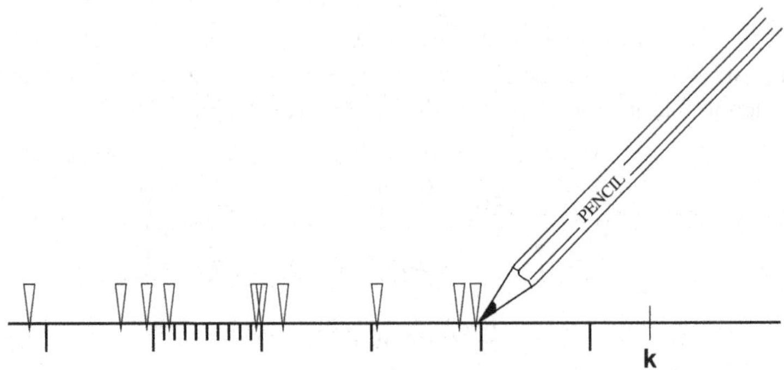

Fig. 1.14 The least upper bound of a bounded set of numbers

This result is so important it deserves restatement and a special name:

Theorem 1.2. The least upper bound theorem. *Every set S of numbers that is bounded above has a least upper bound.*

Proof. We prove the theorem when S is an infinite set of numbers between 0 and 1. The proof of the general case is similar. Examine the first decimal digits of the numbers in S and keep only those with the largest first digit. We call the remaining numbers *eligible* after the first step. Examine the second digits of the numbers that were eligible after the first step and keep only those with the largest second digit. Those are the numbers that are eligible after the second step. Define the number s by setting its jth digit equal to the jth digit of any number that remains eligible after j steps. By construction, s is greater than or equal to every number in S, i.e., s is an upper bound of S.

Next we show that every number that is smaller than s is not an upper bound of S, i.e., s is the smallest, or least, upper bound of S. Let $m = 0.m_1m_2m_3\dots m_n\dots$ be any number smaller than $s = 0.s_1s_2s_3\dots s_n\dots$. Denote by j the first position in which the digits of s and m differ. That means that for $n < j$, $s_n = m_n$. Since m is smaller than s, $m_j < s_j$. At the jth step in our construction of s there was at least one number x in

[1] The story is told that R.L. Moore, a famous mathematician in Texas, asked a student to give a proof or find a counterexample to the statement "Every bounded set of numbers has a largest element." The student came up with a counterexample: the set consisting of the numbers 1 and 2; it has a *larger* element, but no *largest*.

S that agreed with s up through the jth decimal digit. By comparing decimal digits, we see that m is less than x. So m is not an upper bound for S. Since no number less than s is an upper bound for S, s is the least upper bound of S. □

An analogous theorem is true for lower bounds:

Theorem 1.3. The greatest lower bound theorem. *Every set of numbers that is bounded below has a greatest lower bound.*

The least upper bound of set S is also known as the *supremum* of S, and the greatest lower bound as the *infimum* of S, abbreviated as $\sup S$ and $\inf S$ respectively.

The least upper bound theorem is one of the workhorses for proving things in calculus. Here is an example.

Existence of Square Roots. If we think of positive numbers geometrically as representing lengths of intervals and areas of geometric figures such as squares, then it is clear that every positive number p has a square root. It is the length of the edge of a square with area p. We now think of numbers as infinite decimals. We can use the least upper bound theorem to prove that a positive number has a square root. Let us do this for a particular positive number, say

$$p = 5.1.$$

A calculator produces the approximation $\sqrt{5.1} \approx 2.2583$. By squaring, we see that $(2.2583)^2 = 5.09991889$. Let S be the set of numbers a with $a^2 < 5.1$. Then S is not empty, because as we just saw, 2.2583 is in S, and so are 1 and 2 and many other numbers. Also, S is bounded above, for example by 3, because numbers larger than 3 cannot be in S; their squares are too large. The least upper bound theorem says that the set S has a least upper bound; call it r.

We show that $r^2 = 5.1$ by eliminating the possibility that $r^2 > 5.1$ or $r^2 < 5.1$. By squaring,

$$\left(r + \frac{1}{n}\right)^2 = r^2 + \frac{1}{n}\left(2r + \frac{1}{n}\right),$$

we see that the square of a number slightly bigger than r is more than r^2, but not much more when n is sufficiently large. Also,

$$\left(r - \frac{1}{n}\right)^2 = r^2 - \frac{1}{n}\left(2r - \frac{1}{n}\right)$$

shows that the square of a number slightly less than r is less than r^2, but not much less when n is sufficiently large. So, if r^2 is more than 5.1, there is a smaller number of the form $r - \frac{1}{n}$ whose square is also more than 5.1, so r is not the least upper bound of S, a contradiction. If r^2 is less than 5.1, then there is a larger number of the form $r+\frac{1}{n}$ whose square is also less than 5.1, so r is not an upper bound at all, a contradiction. The only other possibility is that $r^2 = 5.1$, and $r = \sqrt{5.1}$.

1.2c Rounding

As a practical matter, comparing two infinite decimal numbers involves rounding. If two decimal numbers with the same integer part have n digits in common, then they differ by less than 10^{-n}. The converse is not true: two numbers can differ by less than 10^{-n} but have no digits in common. For example, the numbers 0.300000 and 0.299999 differ by 10^{-6} but have no digits in common. The operation of *rounding* makes it clear by how much two numbers in decimal form differ.

Rounding a number a to m decimal digits starts with finding the decimal interval of length 10^{-m} that contains a. Then *a rounded down* to m digits is the left endpoint of this interval. Similarly, *a rounded up* to m digits is the right endpoint of the interval. Another way to round a up to m digits is to round a down to m digits and then add 10^{-m}. For example, $\frac{31}{39} = 0.7948717949\ldots$ rounded down to three digits is 0.794, and $\frac{31}{39}$ rounded up to three digits is 0.795.

When calculating, we frequently round numbers up or down. If after rounding, two numbers appear equal, how far apart might they be? Here are two observations about the distance between two numbers a and b and their roundings:

Theorem 1.4. *If a and b are two numbers given in decimal form and if one of the two roundings of a to m digits agrees with one of the two roundings of b to m digits, then $|a-b| < 2\cdot 10^{-m}$.*

Proof. If a and b rounded down to m digits agree, then a and b are in the same interval of width 10^{-m}, and the difference between them is less than 10^{-m}. In the case that one of these numbers rounded up to m digits agrees with the other number rounded down to m digits, a and b lie in adjacent intervals of length 10^{-m}, and hence a and b differ by less than 2×10^{-m}. □

Similarly, if we know how close a and b are, we can conclude something about their roundings:

Theorem 1.5. *If the distance between a and b is less than 10^{-m}, then one of the roundings of a to m digits agrees with one of the roundings of b to m digits.*

Proof. The interval between a rounded down and a rounded up to m digits contains a and is 10^{-m} wide. Similarly, the interval between b rounded down and b rounded up to m digits contains b and is 10^{-m} wide. Since a and b differ by less than 10^{-m}, these two intervals are either identical or adjacent. In either case, they have at least one endpoint in common, so one of the roundings of a must agree with one of the roundings of b. □

Rounding and Calculation Errors. There are infinitely many real numbers, but calculators and computers have finite capacities to represent them. So numbers are stored by rounding. Calculations of basic arithmetic operations are a source of error due to rounding. Here is an example.

In Archimedes' work *Measurement of a Circle*, he approximated π by computing the perimeters of inscribed and circumscribed regular polygons with n sides. There are recurrence formulas for these estimates. Let p_1 be the perimeter of a regular hexagon inscribed in a unit circle. The length of each side of the hexagon is $s_1 = 1$. Then $p_1 = 6s_1 = 6$. Let p_2 be the perimeter of the regular 12-gon. The length of each side s_2 can be expressed in terms of s_1 using the Pythagorean theorem. We have in Fig. 1.15,

$$D = \frac{1}{2}s_1, \quad C = s_2, \quad \text{and} \quad B = 1 - A.$$

By the Pythagorean theorem, $A = \sqrt{1-D^2}$ and $C = \sqrt{B^2+D^2}$. Combining these, we find that

$$s_2 = \sqrt{\left(1 - \sqrt{1 - \left(\frac{1}{2}s_1\right)^2}\right)^2 + \left(\frac{1}{2}s_1\right)^2} = \sqrt{2 - 2\sqrt{1 - \left(\frac{1}{2}s_1\right)^2}}.$$

The same formula can be used to express the side s_n of the polygon of $3(2^n)$ sides in terms of s_{n-1}. The perimeter $p_n = 3(2^n)s_n$ approximates the circumference of the unit circle, 2π. The table in Fig. 1.15 shows that the formula appears to work well through $n = 16$, but after that something goes wrong, as you certainly see by line 29. This is an example of the catastrophic effect of round-off error.

As we will see, many of the key concepts of calculus rely on differences of numbers that are nearly equal, sums of many numbers near zero, or quotients of very small numbers. This example shows that it is unwise to naively implement an algorithm in a computer program without considering the effects of rounding.

Problems

1.19. What would you choose for m in $|\sqrt{3} - 1.7| < 10^{-m}$, and why?

1.20. Find the least upper bound and the greatest lower bound of each of the following sets. Or if it is not possible, explain why.

(a) the interval $(8, 10)$.
(b) the interval $(8, 10]$.

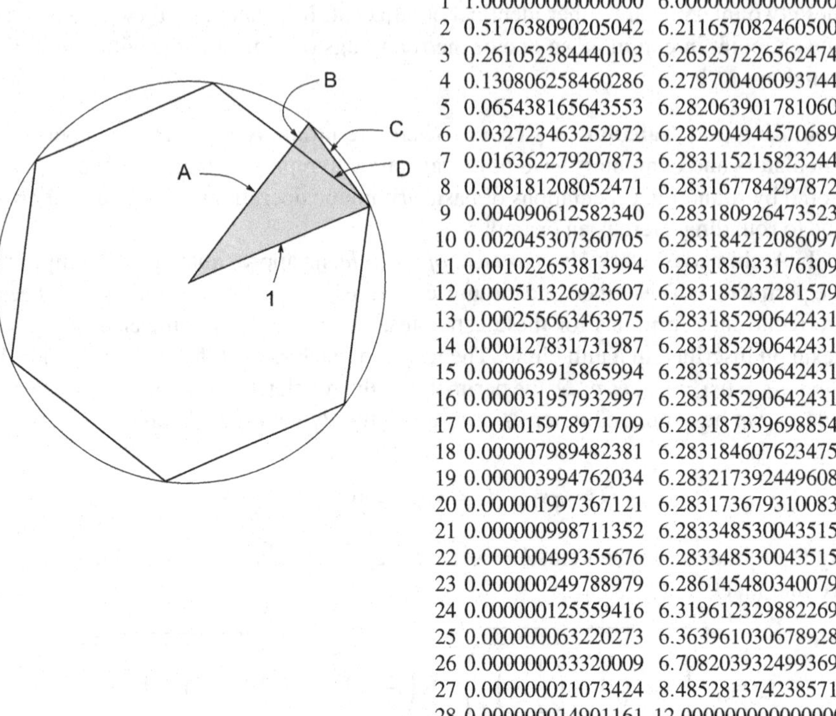

n	s_n	p_n
1	1.000000000000000	6.000000000000000
2	0.517638090205042	6.211657082460500
3	0.261052384440103	6.265257226562474
4	0.130806258460286	6.278700406093744
5	0.065438165643553	6.282063901781060
6	0.032723463252972	6.282904944570689
7	0.016362279207873	6.283115215823244
8	0.008181208052471	6.283167784297872
9	0.004090612582340	6.283180926473523
10	0.002045307360705	6.283184212086097
11	0.001022653813994	6.283185033176309
12	0.000511326923607	6.283185237281579
13	0.000255663463975	6.283185290642431
14	0.000127831731987	6.283185290642431
15	0.000063915865994	6.283185290642431
16	0.000031957932997	6.283185290642431
17	0.000015978971709	6.283187339698854
18	0.000007989482381	6.283184607623475
19	0.000003994762034	6.283217392449608
20	0.000001997367121	6.283173679310083
21	0.000000998711352	6.283348530043515
22	0.000000499355676	6.283348530043515
23	0.000000249788979	6.286145480340079
24	0.000000125559416	6.319612329882269
25	0.000000063220273	6.363961030678928
26	0.000000033320009	6.708203932499369
27	0.000000021073424	8.485281374238571
28	0.000000014901161	12.000000000000000
29	0.000000000000000	0.000000000000000

Fig. 1.15 *Left*: the regular hexagon and part of the 12-gon inscribed in the *circle*. *Right*: calculated values for the edge lengths s_n and perimeters p_n of the inscribed $3(2^n)$-gon. Note that as n increases, the exact value of p_n approaches $2\pi = 6.2831853071795\ldots$

(c) the nonpositive integers.
(d) the set of four numbers $\frac{30}{279}, \frac{29}{263}, \frac{59}{525}, \frac{1}{9}$.
(e) the set $1, \frac{1}{2}, \frac{1}{3}, \frac{1}{4}, \ldots$.

1.21. Take the unit square, and by connecting the midpoints of opposite sides, divide it into $2^2 = 4$ subsquares, each of side 2^{-1}. Repeat this division for each subsquare, obtaining $2^4 = 16$ squares whose sides have length 2^{-2}. Continue this process so that after n steps, there are 2^{2n} squares, each having sides of length 2^{-n}. See Fig. 1.16. With the lower left corner as center and radius 1, inscribe a unit quarter circle into the square. Denote by a_n the total area of those squares that at the nth step of the process, lie entirely inside the quarter circle. For example, $a_1 = 0$, $a_2 = \frac{1}{4}$, $a_3 = \frac{1}{2}$.

(a) Is the set S of numbers $a_1, a_2, a_3, \ldots, a_n, \ldots$ bounded above? If so, find an upper bound.
(b) Does S have a least upper bound? If so, what number do you think the least upper bound is?

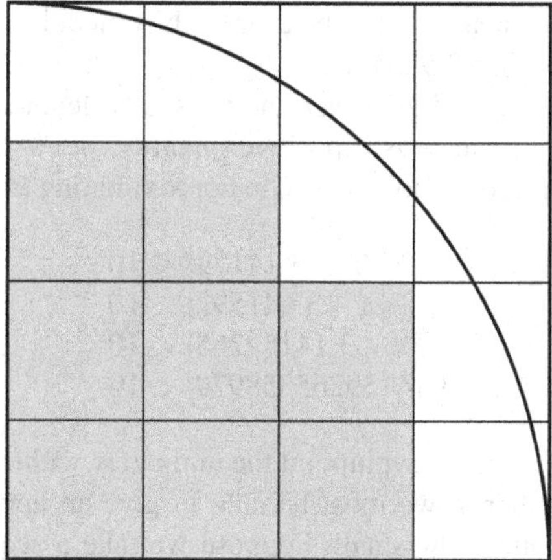

Fig. 1.16 The square in Problem 1.21. Area a_3 is shaded

1.22. Use rounding to add these two numbers so that the error in the sum is not more than 10^{-9}:

$$\begin{array}{r} 0.12345678987654321045678987654321 01 \\ +\ 9.11111122211122211187654321045678 92 \\ \hline \end{array}$$

1.23. Tell how, in principle, to add the two numbers having the indicated pattern of decimals:

$$\begin{array}{r} 0.101100111000111100001111100000\cdots \\ +\ 0.898989898989898989898989898989\cdots \\ \hline \end{array}$$

Does your explanation involve rounding?

1.24. Show that the least upper bound of a set S is *unique*. That is, if x_1 and x_2 both are least upper bounds of S, then $x_1 = x_2$.

Hint: Recall that given any numbers a and b, exactly one of the following holds: $a < b$, $a > b$, or $a = b$.

1.3 Sequences and Their Limits

In Sect. 1.2a we described numbers as infinite decimals. That is a very good theoretical description, but not a practical one. How long would it take to write down infinitely many decimal digits of a number, and where would we write them?

For an alternative practical description of numbers, we borrow from engineering the idea of *tolerance*. When engineers specify the size of an object to be used in a design, they give its magnitude, say 3 m. But they also realize that nothing built by human beings is exact, so they specify the error they can tolerate, say 1 mm, and still use the object. This means that to be usable, the object has to be no larger than 3.001 m and no smaller than 2.999 m.

This tolerable error is called *tolerance* and is usually denoted by the Greek letter ε. By its very nature, tolerance is a positive quantity, i.e., $\varepsilon > 0$. Here are some examples of tolerable errors, or tolerances, in approximating π:

$$\begin{aligned} |\pi - 3.14159| &< 10^{-5} \\ |\pi - 3.141592| &< 10^{-6} \\ |\pi - 3.14159265| &< 10^{-8} \\ |\pi - 3.14159265358979| &< 10^{-14} \end{aligned}$$

Notice that the smaller tolerances pinpoint the number π within a smaller interval.

To determine a number a, we must be able to give an approximation to a for any tolerance ε, no matter how small. Suppose we take a sequence of tolerances ε_n tending to zero, and suppose that for each n, we have an approximation a_n that is within ε_n of a. The approximations a_n form an infinite sequence of numbers a_1, a_2, a_3, ... that tend to a in the sense that the difference between a_n and a tends to zero as n grows larger and larger. This leads to the general concept of the *limit of a sequence*.

Definition 1.2. A list of numbers is called a *sequence*. The numbers are called the terms of the sequence. We say that an infinite sequence a_1, a_2, a_3, ..., a_n, ... *converges* to the number a (is *convergent*) if given any tolerance $\varepsilon > 0$, no matter how small, there is a whole number N, dependent on ε, such that for all $n > N$, a_n differs from a by less than ε:

$$|a_n - a| < \varepsilon.$$

The number a is called the *limit* of the sequence $\{a_n\}$, and we write

$$\lim_{n\to\infty} a_n = a.$$

A sequence that has no limit *diverges* (is *divergent*).

A note on terminology and history: When the distinguished Polish mathematician Antoni Zygmund, author of the text *Trigonometric Series*, came as a refugee to America, he was eager to learn about his adopted country. Among other things, he asked an American friend to explain baseball to him, a game totally unknown in Europe. He received a lengthy lecture. His only comment was that the World Series should be called the World Sequence. As we will see later, the word "series" in mathematics refers to the *sum* of the terms of a sequence.

Because many numbers are known only through a sequence of approximations, a question that arises immediately, and will be with us throughout this calculus course, is this: How can we decide whether a given sequence converges, and if it does converge, what is its limit? Each case has to be analyzed individually, but there are some rules of arithmetic for convergent sequences.

Theorem 1.6. *Suppose that $\{a_n\}$ and $\{b_n\}$ are convergent sequences,*

$$\lim_{n\to\infty} a_n = a, \qquad \lim_{n\to\infty} b_n = b.$$

Then

(a) $\lim_{n\to\infty}(a_n + b_n) = a + b.$
(b) $\lim_{n\to\infty}(a_n b_n) = ab.$
(c) If a is not zero, then for n large enough, $a_n \neq 0$, and $\lim_{n\to\infty} \frac{1}{a_n} = \frac{1}{a}$.

These rules certainly agree with our experience of computation with decimals. They assert that if the numbers a_n and b_n are close to the numbers a and b, then their sum, product, and reciprocals are close to the sum, product, and reciprocals of a and b themselves. In Problem 1.33 we show you how to prove these properties of convergent sequences.

Next we give some examples of convergent sequences.

Example 1.8. $a_n = \frac{1}{n}$. For any tolerance ε no matter how small, $\frac{1}{n}$ is within ε of 0 once n is greater than $\frac{1}{\varepsilon}$. So

$$\left|\frac{1}{n} - 0\right| < \varepsilon \quad \text{for } n > \frac{1}{\varepsilon}, \text{ and } \lim_{n\to\infty}\frac{1}{n} = 0.$$

Example 1.9. $a_n = \frac{1}{2^n}$. Since $2^n > n$ when $n > 2$, we see that $\frac{1}{2^n} < \frac{1}{n} < \varepsilon$ if n is large enough. So

$$\left|\frac{1}{2^n} - 0\right| < \varepsilon \quad \text{for } n \text{ sufficiently large, and } \lim_{n\to\infty}\frac{1}{2^n} = 0.$$

In these two examples the limit is zero, a rather simple number. Let us look at a very simple sequence whose limit is not zero.

Example 1.10. The limit of the constant sequence $a_n = 5$ for all $n = 1, 2, 3, \ldots$ is 5. The terms of the sequence do not differ from 5, so no matter how small ε is, $|a_n - 5| < \varepsilon$.

Here is a slightly more complicated example.

Example 1.11. Using algebra to rewrite the terms of the sequence, we obtain

$$\lim_{n\to\infty}\frac{5n+7}{n+1}=\lim_{n\to\infty}\left(\frac{5n+5}{n+1}+\frac{2}{n+1}\right)=\lim_{n\to\infty}\left(5+\frac{2}{n+1}\right).$$

Now by Theorem 1.6,

$$\lim_{n\to\infty}\left(5+\frac{2}{n+1}\right)=\lim_{n\to\infty}5+2\lim_{n\to\infty}\frac{1}{n+1}=5+2(0)=5.$$

As we just saw, the arithmetic rules for convergent sequences can help us evaluate limits of sequences by reducing them to known ones. The next theorem gives us a different way to use the behavior of known sequences to show convergence.

Theorem 1.7. The squeeze theorem. *Suppose that for all $n>N$,*

$$a_n\le b_n\le c_n,$$

and that $\lim_{n\to\infty}a_n=\lim_{n\to\infty}c_n=a$. Then $\lim_{n\to\infty}b_n=a$.

Proof. Subtracting a from the inequalities, we get

$$a_n-a\le b_n-a\le c_n-a.$$

Let $\varepsilon>0$ be any tolerance. Since $\{a_n\}$ and $\{c_n\}$ have limit a, there is a number N_1 such that when $n>N_1$, a_n is within ε of a, and there is a number N_2 such that when $n>N_2$, c_n is within ε of a. Let M be the largest of N, N_1, and N_2. Then when $n>M$, we get

$$-\varepsilon<a_n-a\le b_n-a\le c_n-a<\varepsilon.$$

So for the middle term, we see that $|b_n-a|<\varepsilon$. This shows that b_n converges to a. □

Example 1.12. Suppose $\frac{1}{2^n}\le a_n\le\frac{1}{n}$ for $n>2$. Since $\lim_{n\to\infty}\frac{1}{2^n}=\lim_{n\to\infty}\frac{1}{n}=0$, by the squeeze theorem, $\lim_{n\to\infty}a_n=0$ as well.

Example 1.13. Suppose $|a_n|\le|b_n|$ and $\lim_{n\to\infty}b_n=0$. By the squeeze theorem applied to

$$0\le|a_n|\le|b_n|,$$

we see that $\lim_{n\to\infty}|a_n|=0$. It is also true that $\lim_{n\to\infty}a_n=0$, since the distance between a_n and 0 is equal to the distance between $|a_n|$ and 0, which can be made arbitrarily small by taking n large enough.

1.3a Approximation of $\sqrt{2}$

Now let us apply what we have learned to construct a sequence of numbers that converges to the square root of 2. Let us start with an approximation s. How can we find a better one? The product of the numbers s and $\frac{2}{s}$ is 2. It follows that $\sqrt{2}$ lies between these two numbers, for if both were greater than $\sqrt{2}$, their product would be greater than 2, and if both of them were less than $\sqrt{2}$, their product would be less than 2. So a good guess for a better approximation is the arithmetic mean of the two numbers,

$$\text{new approximation} = \frac{s+\frac{2}{s}}{2}.$$

By the A-G inequality, this is greater than the geometric mean of the two numbers,

$$\sqrt{s\left(\frac{2}{s}\right)} < \frac{s+\frac{2}{s}}{2}.$$

This shows that our new approximation is greater than the square root of 2.

We generate a sequence of approximations $s_1, s_2, \ldots$ as follows:

$$s_{n+1} = \frac{1}{2}\left(s_n + \frac{2}{s_n}\right). \tag{1.7}$$

Starting with, say, $s_1 = 2$, we get

$$\begin{aligned}
s_1 &= 2 \\
s_2 &= 1.5 \\
s_3 &= 1.41666666666666\ldots \\
s_4 &= 1.41421568627451\ldots \\
s_5 &= 1.41421356237469\ldots \\
s_6 &= 1.41421356237309\ldots
\end{aligned}$$

The first twelve digits of s_5 and s_6 are the same. We surmise that they are the first twelve digits of $\sqrt{2}$. Squaring s_5, we get

$$s_5^2 \approx 2.00000000000451,$$

gratifyingly close to 2. So the numerical evidence suggests that the sequence $\{s_n\}$ defined above converges to $\sqrt{2}$. We are going to prove that this is so.

How much does s_{n+1} differ from $\sqrt{2}$?

$$s_{n+1} - \sqrt{2} = \frac{1}{2}\left(s_n + \frac{2}{s_n}\right) - \sqrt{2}.$$

Let us bring the fractions on the right side to a common denominator:

$$s_{n+1} - \sqrt{2} = \frac{1}{2s_n}\left(s_n^2 + 2 - 2s_n\sqrt{2}\right).$$

We recognize the expression in parentheses as a perfect square, $(s_n - \sqrt{2})^2$. So we can rewrite the above equation as

$$s_{n+1} - \sqrt{2} = \frac{1}{2s_n}\left(s_n - \sqrt{2}\right)^2.$$

Next we rewrite the right side giving

$$s_{n+1} - \sqrt{2} = \frac{1}{2}\left(s_n - \sqrt{2}\right)\left(\frac{s_n - \sqrt{2}}{s_n}\right).$$

Since s_n is greater than $\sqrt{2}$, the factor $(\frac{s_n - \sqrt{2}}{s_n})$ is less than one. Therefore, dropping it gives the inequality

$$0 < s_{n+1} - \sqrt{2} < \frac{1}{2}\left(s_n - \sqrt{2}\right).$$

Applying this repeatedly gives

$$0 < s_{n+1} - \sqrt{2} < \frac{1}{2^n}\left(s_1 - \sqrt{2}\right).$$

We have shown in the previous example that the sequence $\frac{1}{2^n}$ tends to the limit zero. It follows from Theorem 1.7, the squeeze theorem, that $s_{n+1} - \sqrt{2}$ tends to zero. This concludes the proof that $\lim_{n\to\infty} s_n = \sqrt{2}$.

1.3b Sequences and Series

One of the most useful tools for proving that a sequence converges to a limit is the monotone convergence theorem, which we discuss next.

Definition 1.3. A sequence $\{a_n\}$ is called *increasing* if $a_n \le a_{n+1}$. It is *decreasing* if $a_n \ge a_{n+1}$. The sequence is *monotonic* if it is either increasing or decreasing.

Definition 1.4. A sequence $\{a_n\}$ is called *bounded* if all numbers in the sequence are contained in some interval $[-B,B]$, so that $|a_n| \le B$. Every such number B is a *bound.*

If $a_n < K$ for all n, we say that $\{a_n\}$ is *bounded above* by K. If $K < a_n$ for all n, we say that $\{a_n\}$ is *bounded below* by K.

Example 1.14. The sequence $a_n = (-1)^n$,

$$a_1 = -1, \; a_2 = 1, \; -1, \; 1, \; -1, \; \ldots,$$

is bounded, since $|a_n| = |(-1)^n| = 1$. The sequence is also bounded above by 2 and bounded below by -3.

When showing that a sequence is bounded it is not necessary to find the smallest bound. A larger bound is often easier to verify.

Example 1.15. The sequence $\{5 + \frac{2}{n+1}\}$ is bounded. Since

$$0 \le \frac{2}{n+1} \le 1 \qquad (n = 1,2,3,\ldots),$$

we can see that $\left|5 + \frac{2}{n+1}\right| \le 6$. It is also true that $\left|5 + \frac{2}{n+1}\right| \le 100$.

The next theorem, which we help you prove in Problem 1.35, shows that being bounded is necessary for sequence convergence.

Theorem 1.8. *Every convergent sequence is bounded.*

The next theorem gives a very powerful and fundamental tool for proving sequence convergence.

Theorem 1.9. *An increasing sequence that is bounded converges to a limit.*

Proof. The proof is very similar to the proof of the existence of the least upper bound of a bounded set. We take the case that the sequence consists of positive numbers. For if not, $|a_n| < b$ for some b and the augmented sequence $\{a_n + b\}$ is an increasing sequence that consists of positive numbers. By Theorem 1.6, if the augmented sequence converges to the limit c, the original sequence converges to $c - b$.

Denote by w_n the integer part of a_n. Since the original sequence is increasing, so is the sequence of their integer parts. Since the original sequence is bounded, so are their integer parts. Therefore, w_{n+1} is greater than w_n for only a finite number of n. It follows that all integer parts w_n are equal for all n greater than some number N.

Denote by w the value of w_n for n greater than N. Next we look at the first decimal digit of a_n for n greater than N:

$$a_n = w.d_n \ldots .$$

Since the a_n form an increasing sequence, so do the digits d_n. It follows that the digits d_n are all equal for n greater than some number $N(1)$.

Denote this common value of d_n by c_1. Proceeding in this manner, we see that there is a number $N(k)$ such that for n greater than $N(k)$, the integer part and the first k digits of a_n are equal. Let us denote these common digits by $c_1, c_2, \ldots, c_k$, and denote by a the number whose integer part is w and whose kth digit is c_k for all k. Then for n greater than $N(k)$, a_n differs from a by less than 10^{-k}; this proves that the sequence $\{a_n\}$ converges to a. □

We claim that a decreasing, bounded sequence $\{b_n\}$ converges to a limit. To see this, define its negative, the sequence $a_n = -b_n$. This is a bounded *increasing* sequence, and therefore converges to a limit a. The sequence b_n then converges to $-a$. Theorem 1.9 and the analogous theorem for decreasing bounded sequences are often expressed as a single theorem:

Theorem 1.10. The monotone convergence theorem. *A bounded monotone sequence converges to a limit.*

Existence of Square Roots. The monotone convergence theorem is another of the workhorses of calculus. To illustrate its power, we show now how to use it to give a proof, different from the one in Sect. 1.2b, that every positive number has a square root. To keep notation to a minimum, we shall construct the square root of the number 2.

Denote as before by s_n the members of the sequence defined by

$$s_{n+1} = \frac{1}{2}\left(s_n + \frac{2}{s_n}\right). \tag{1.8}$$

We have pointed out earlier that for $n > 1$, s_n is greater than $\sqrt{2}$. Therefore, $\dfrac{2}{s_n}$ is less than $\sqrt{2}$, and hence less than s_n. It follows from Eq. (1.8) that

$$s_{n+1} < \frac{s_n + s_n}{2} = s_n.$$

This shows that $\{s_n\}$ is a decreasing sequence of positive numbers. We appeal to the monotone convergence theorem to conclude that the sequence $\{s_n\}$ converges to a limit. Denote this limit by s. We shall show that s is $\sqrt{2}$.

According to Theorem 1.6, the limit of the sequence on the right side of Eq. (1.8) is $\frac{1}{2}\left(s + \frac{2}{s}\right)$. This is equal to s, the limit of the left side of Eq. (1.8): $s = \frac{1}{2}\left(s + \frac{2}{s}\right)$. Multiply this equation by $2s$ to obtain $2s^2 = s^2 + 2$. Therefore $s^2 = 2$.

Geometric Sequences and Series. We define geometric sequences as follows.

Definition 1.5. Sequences of numbers that follow the pattern of multiplying by a fixed number to get the next term are called *geometric* sequences, or geometric progressions.

Example 1.16. The geometric sequences $1, 2, 4, 8, \dots, 2^n, \dots,$

$$\frac{1}{3}, -\frac{1}{6}, \frac{1}{12}, -\frac{1}{24}, \dots, \frac{1}{3}\left(-\frac{1}{2}\right)^n, \dots, \text{ and } 0.1, 0.01, 0.001, 0.0001, \dots, (0.1)^n, \dots$$

may be abbreviated $\{2^n\}$, $\{\frac{1}{3}\left(-\frac{1}{2}\right)^n\}$, $\{(0.1)^n\}$, $n = 0, 1, 2, \dots$.

Theorem 1.11. Geometric sequence. *The sequence* $\{r^n\}$

(a) converges if $|r| < 1$*, and in this case,* $\lim_{n\to\infty} r^n = 0$,
(b) converges if $r = 1$*, and in this case,* $\lim_{n\to\infty} 1^n = 1$,
(c) diverges for $r > 1$ *and for* $r \le -1$.

Proof. (a) If $0 \le r < 1$, then $\{r^n\}$ is a decreasing sequence that is bounded, $|r^n| \le 1$. Therefore, by the monotone convergence theorem it converges to a limit a. The sequence $r, r^2, r^3, \dots$ has the same limit as $1, r, r^2, r^3, \dots$, and so by Theorem 1.6,

$$a = \lim_{n\to\infty} r^{n+1} = \lim_{n\to\infty} rr^n = r \lim_{n\to\infty} r^n = ra,$$

and $a(r-1) = 0$. Now since $r \ne 1$, $a = 0$.
If $-1 < r < 0$, then each power r^n has the same distance to 0 as $|r|^n$, so again the limit is 0.

(b) For any tolerance ε, $|1^n - 1| = 0 < \varepsilon$, so the limit is clearly 1.

(c) To show that $\{r^n\}$ diverges for $|r| > 1$ or $r = -1$, suppose the limit exists: $\lim_{n\to\infty} r^n = a$. By the argument in part (a), the limit must be 0. But this is not possible. We know that $|r^n - 0| \ge 1$ for all n, i.e., the distance between r^n and 0 is always at least 1, so r^n does not tend to 0. □

Example 1.17. Recall that $n!$ denotes the product of the first n positive integers, $(1)(2)\cdots(n)$. We shall show that for every number b,

$$\lim_{n\to\infty} \frac{b^n}{n!} = 0.$$

Take an integer $N > |b|$, and decompose every integer n greater than N as $n = N + k$. Then $\frac{b^n}{n!} = \frac{b^N}{N!}\frac{b}{N+1}\cdots\frac{b}{N+k}$. The first factor $\frac{b^N}{N!}$ is a fixed number, and the other k factors each have absolute value less than $r = \dfrac{|b|}{N+1} < 1$. Since r^k tends to 0, the limit is 0 by the squeeze argument used in Example 1.12.

Definition 1.6. The numbers of a sequence $\{a_n\}$ can be added to make a new sequence $\{s_n\}$:

$$\begin{aligned} s_1 &= a_1 = \sum_{j=1}^{1} a_j \\ s_2 &= a_1 + a_2 = \sum_{j=1}^{2} a_j \\ &\dots \\ s_n &= a_1 + a_2 + \cdots + a_n = \sum_{j=1}^{n} a_j \\ &\dots \end{aligned}$$

called the sequence of *partial sums* of the *series* $\sum_{n=1}^{\infty} a_n$. If the limit

$$\sum_{j=1}^{\infty} a_j = a_1 + a_2 + a_3 + \cdots = \lim_{n\to\infty} s_n$$

exists, the series *converges*. Otherwise, it *diverges*. The numbers a_n are called the *terms* of the series.

Example 1.18. Take all $a_j = 1$, which gives the series $\sum_{j=1}^{\infty} 1$. The nth partial sum is $s_n = 1 + \cdots + 1 = n$. Since the s_n are not bounded, the series diverges by Theorem 1.8.

Example 1.18 suggests the following necessary condition for convergence.

Theorem 1.12. *If* $\sum_{n=1}^{\infty} a_n$ *converges, then* $\lim_{n\to\infty} a_n = 0$.

Proof. Let $s_n = a_1 + \cdots + a_n$. Since $\sum_{n=1}^{\infty} a_n$ converges, the limit $\lim_{n\to\infty} s_n = L$ exists, and for the shifted sequence $\lim_{n\to\infty} s_{n-1} = L$ as well. According to Theorem 1.6,

$$\lim_{n\to\infty} a_n = \lim_{n\to\infty} (s_n - s_{n-1}) = L - L = 0.$$

□

Example 1.19. The series $\sum_{n=1}^{\infty}\left(\frac{n}{2n+1}\right)$ diverges, because $\lim_{n\to\infty}\frac{n}{2n+1}=\frac{1}{2}$.

The following is one of the best-known and most beloved series.

Theorem 1.13. Geometric series. *If* $|x|<1$, *the sequence of partial sums*

$$s_n = 1+x+x^2+x^3+\cdots+x^n.$$

converges, and

$$\lim_{n\to\infty} s_n = \sum_{n=0}^{\infty} x^n = \frac{1}{1-x}, \qquad (|x|<1). \tag{1.9}$$

If $|x|\geq 1$, *the series diverges.*

Proof. By algebra, we see that $s_n(1-x)=(1+x+x^2+x^3+\cdots+x^n)(1-x)=1-x^{n+1}$. Therefore,

$$s_n = \frac{1-x^{n+1}}{1-x}, \qquad (x\neq 1). \tag{1.10}$$

According to Theorem 1.11, for $|x|<1$ we have $\lim_{n\to\infty} x^n = 0$ and

$$\lim_{n\to\infty} s_n = \lim_{n\to\infty}(1+x+x^2+x^3+\cdots+x^n) = \lim_{n\to\infty}\frac{1-x^{n+1}}{1-x} = \frac{1}{1-x}.$$

If $|x|\geq 1$, then x^n does not approach zero, so according to Theorem 1.12, the series diverges. □

Comparing Series. Next we show how to use monotone convergence and the arithmetic properties of sequences to determine convergence of some series. Consider the series

$$(a)\sum_{n=0}^{\infty}\frac{1}{2^n+1}, \qquad (b)\sum_{n=1}^{\infty}\frac{1}{2^n-1}.$$

For series (a), the numbers $\frac{1}{2^n+1}$ are positive, so the partial sums $\sum_{n=0}^{m}\frac{1}{2^n+1}$ form an increasing sequence. Since

$$\frac{1}{2^n+1} < \left(\frac{1}{2}\right)^n,$$

the partial sums satisfy $\sum_{n=0}^{m} \frac{1}{2^n+1} < \sum_{n=0}^{m} \left(\frac{1}{2}\right)^n$. Since $\sum_{n=0}^{\infty} \left(\frac{1}{2}\right)^n$ converges, its sequence of partial sums is bounded,

$$0 \le \sum_{n=0}^{m} \frac{1}{2^n+1} < \sum_{n=0}^{m} \left(\frac{1}{2}\right)^n \le 2.$$

By the monotone convergence theorem, the sequence of partial sums of $\sum_{n=0}^{\infty} \frac{1}{2^n+1}$ converges, so the series converges.

For series (b), the numbers $\frac{1}{2^n-1}$ are positive, so the partial sums $\sum_{n=1}^{m} \frac{1}{2^n-1}$ form an increasing sequence. Note that $\frac{1}{2^n-1}$ is not less than $\frac{1}{2^n}$ (it is slightly greater), so we cannot set up a comparison as we did for series (a). We look instead at the limit of the ratio of the terms,

$$\lim_{n\to\infty} \frac{\frac{1}{2^n-1}}{\frac{1}{2^n}} = \lim_{n\to\infty} \frac{1}{1-\frac{1}{2^n}} = 1.$$

Since the limit of the ratio is 1, the ratios eventually all get close to 1. So for every $R > 1$, there is a sufficiently large N such that

$$\frac{\frac{1}{2^n-1}}{\frac{1}{2^n}} < R$$

for all $n > N$. Therefore, $\frac{1}{2^n-1} < R\left(\frac{1}{2}\right)^n$ for all $n > N$. Since the partial sums of the series $R\sum_{n=N+1}^{\infty} \left(\frac{1}{2}\right)^n$ are bounded, so are the partial sums of $\sum_{n=N+1}^{\infty} \frac{1}{2^n-1}$. So this series converges. But then so does $\sum_{n=1}^{\infty} \frac{1}{2^n-1}$.

The arguments used for series (a) and (b) can be used to obtain the next two comparison theorems, which we ask you to prove in Problem 1.45.

Theorem 1.14. Comparison theorem. *Suppose that for all n,*

$$0 \le a_n \le b_n.$$

If $\sum_{n=1}^{\infty} b_n$ *converges, then* $\sum_{n=1}^{\infty} a_n$ *converges.*

Theorem 1.15. Limit comparison theorem. *Let $\sum_{n=1}^{\infty} a_n$ and $\sum_{n=1}^{\infty} b_n$ be series of positive terms. If $\lim_{n\to\infty} \frac{a_n}{b_n}$ exists and is a positive number, then $\sum_{n=1}^{\infty} a_n$ converges if and only if $\sum_{n=1}^{\infty} b_n$ converges.*

The comparison theorems are stated for terms that are positive or not negative. The next theorem is a handy result that sometimes allows us to use these theorems to deduce convergence of series with negative terms.

Theorem 1.16. *If $\sum_{j=1}^{\infty} |a_j|$ converges, then $\sum_{j=1}^{\infty} a_j$ also converges.*

Proof. Since $0 \le a_n + |a_n|$, the partial sums $s_m = \sum_{j=1}^{m} (a_n + |a_n|)$ are increasing. Since $a_n + |a_n| \le 2|a_n|$, s_m is less than the mth partial sum of $\sum_{j=1}^{\infty} 2|a_j|$, which converges. The sequence of partial sums s_m is increasing and bounded. Therefore, $\sum_{j=1}^{\infty} (a_n + |a_n|)$ converges. Since $a_j = (a_j + |a_j|) - |a_j|$, $\sum_{j=1}^{\infty} a_j$ converges by Theorem 1.6. □

Example 1.20. The series $\sum_{n=1}^{\infty} \frac{1}{(-2)^n n}$ does not have positive terms, however $\sum_{n=1}^{\infty} \left| \frac{1}{(-2)^n n} \right| = \sum_{n=1}^{\infty} \frac{1}{2^n n}$ does. Since $\frac{1}{2^n n} < \left(\frac{1}{2}\right)^n$, the series $\sum_{n=1}^{\infty} \left| \frac{1}{(-2)^n n} \right|$ converges by comparison with the geometric series $\sum_{n=1}^{\infty} \left(\frac{1}{2}\right)^n$. According to Theorem 1.16, $\sum_{n=1}^{\infty} \frac{1}{(-2)^n n}$ converges.

The next two examples show that the converse of Theorem 1.16 is not true. That is, a series $\sum_{n=1}^{\infty} a_n$ may converge while the series $\sum_{n=1}^{\infty} |a_n|$ diverges.

Example 1.21. Consider the series $\sum_{n=1}^{\infty} \frac{1}{n}$. This is known as the "harmonic series." Its sequence of partial sums is

$$s_1 = 1, \quad s_2 = s_1 + \frac{1}{2} = 1 + \frac{1}{2}, \quad s_3 = s_2 + \frac{1}{3} = 1 + \frac{1}{2} + \frac{1}{3}, \ldots.$$

By grouping the terms, we can see that

$$\begin{aligned} s_4 &= \quad s_2 + \tfrac{1}{3} + \tfrac{1}{4} \quad > 1.5 + \tfrac{2}{4} = 2 \\ s_8 &= s_4 + \tfrac{1}{5} + \tfrac{1}{6} + \tfrac{1}{7} + \tfrac{1}{8} > 2 + \tfrac{4}{8} \quad = 2.5 \\ s_{16} &= \quad s_8 + \tfrac{1}{9} + \cdots + \tfrac{1}{16} \quad > 2.5 + \tfrac{8}{16} = 3 \end{aligned}$$

and so forth. So s_n is an increasing sequence that is not bounded.

The harmonic series diverges. It is easy to see that the difference between its successive partial sums, $s_{n+1} - s_n = \frac{1}{n+1}$, can be made as small as we like by taking n large. However, this is *not* enough to ensure convergence of the series. We will revisit the harmonic series when we study improper integrals. The harmonic series is a good example of a series where the terms of the series $a_1, a_2, \ldots, a_n, \ldots$ decrease to zero and the series $\sum_{n=1}^{\infty} a_n$ diverges.

Now let us alternate the signs of the terms, to see that the resulting series does converge. Consider the series

$$\sum_{n=1}^{\infty} (-1)^{(n-1)} \frac{1}{n}. \tag{1.11}$$

The sequence of even partial sums $s_2, s_4, s_6, \ldots, s_{2k}, \ldots$ is an increasing sequence, since $s_{2k+2} - s_{2k} = \dfrac{1}{2k+1} - \dfrac{1}{2k+2} > 0$. It is bounded above by

$$1 > s_{2k} = 1 + \left(-\frac{1}{2} + \frac{1}{3}\right) + \cdots + \left(-\frac{1}{2k-2} + \frac{1}{2k-1}\right) - \frac{1}{2k}.$$

The sequence of odd partial sums $s_1, s_3, s_5, \ldots, s_{2k+1}, \ldots$ is a decreasing sequence, since $s_{2(k+1)+1} - s_{2k+1} = -\dfrac{1}{2k+2} + \dfrac{1}{2k+2+1} < 0$. It is bounded below by

$$\frac{1}{2} < s_{2k+1} = \left(1 - \frac{1}{2}\right) + \left(\frac{1}{3} - \frac{1}{4}\right) + \cdots + \frac{1}{2k+1}.$$

By the monotone convergence theorem, both $\{s_{2k}\}$ and $\{s_{2k+1}\}$ converge. Define $\lim_{k\to\infty} s_{2k} = L_1$ and $\lim_{k\to\infty} s_{2k+1} = L_2$. Then

$$L_2 - L_1 = \lim_{k\to\infty} (s_{2k+1} - s_{2k}) = \lim_{k\to\infty} \frac{1}{2k+1} = 0.$$

Therefore, s_k converges to $L_1 = L_2$, and $\sum_{n=1}^{\infty} (-1)^{(n-1)} \frac{1}{n}$ converges.

The same argument can be used to obtain the following more general result. We guide you through the steps in Problem 1.46.

Theorem 1.17. Alternating series theorem. *If*

$$a_1 \ge a_2 \ge a_3 \ge \cdots \ge a_n \ge \cdots \ge 0$$

and $\lim_{n\to\infty} a_n = 0$, *then* $\sum_{n=1}^{\infty} (-1)^n a_n$ *converges.*

Definition 1.7. If $\sum_{i=1}^{\infty} a_i$ converges and $\sum_{i=1}^{\infty} |a_i|$ diverges, we say that $\sum_{i=1}^{\infty} a_i$ converges *conditionally*. If $\sum_{i=1}^{\infty} |a_i|$ converges, we say that $\sum_{i=1}^{\infty} a_i$ converges *absolutely*.

Example 1.22. In the convergent geometric series

$$1 + \frac{1}{3} + \frac{1}{9} + \frac{1}{27} + \cdots$$

replace any number of plus signs by minus signs. According to Definition 1.7, the resulting series converges absolutely.

Example 1.23. We showed that $\sum_{n=1}^{\infty} (-1)^{(n-1)} \frac{1}{n}$ converges, and in Example 1.21, we showed that $\sum_{n=1}^{\infty} \frac{1}{n}$ diverges. Therefore, $\sum_{n=1}^{\infty} (-1)^{(n-1)} \frac{1}{n}$ converges conditionally.

Example 1.24. Because $\frac{1}{\sqrt{n}}$ decreases to 0, $\sum_{n=1}^{\infty} (-1)^{n-1} \frac{1}{\sqrt{n}}$ converges by the alternating series theorem, Theorem 1.17. Since $\sqrt{n} \le n$, $\frac{1}{\sqrt{n}} \ge \frac{1}{n}$, and the partial sums are given by

$$s_m = \sum_{n=1}^{m} \frac{1}{\sqrt{n}} \ge \sum_{n=1}^{m} \frac{1}{n}.$$

By Example 1.21, the partial sums of the harmonic series are not bounded. Therefore, the s_m are not bounded either; this shows that $\sum_{n=1}^{\infty} \frac{1}{\sqrt{n}}$ diverges. Thus $\sum_{n=1}^{\infty} (-1)^{n-1} \frac{1}{\sqrt{n}}$ converges conditionally.

Further Comparisons. Consider the series

$$\sum_{n=1}^{\infty} \frac{n}{2^n}. \tag{1.12}$$

Because the terms $a_n = \frac{n}{2^n}$ are positive, the sequence of partial sums is increasing. If the sequence of partial sums is bounded, then the series converges. Let us look at a few partial sums:

$$\begin{aligned} s_1 &= 0.5 \\ s_2 &= 1 \\ s_3 &= 1.375 \\ s_4 &= 1.625. \end{aligned}$$

We clearly need better information than this. Trying a limit comparison with $\sum_{n=1}^{\infty} \frac{1}{2^n}$ yields

$$\lim_{n\to\infty} \frac{\frac{n}{2^n}}{\frac{1}{2^n}} = \lim_{n\to\infty} n,$$

which does not exist, and so such a comparison is not helpful. We notice that when n is large, the terms $\frac{n}{2^n}$ grow by roughly a factor of $\frac{1}{2}$, i.e.,

$$\lim_{n\to\infty} \frac{\frac{n+1}{2^{n+1}}}{\frac{n}{2^n}} = \lim_{n\to\infty} \frac{n+1}{n}\frac{1}{2} = \frac{1}{2}.$$

This suggests comparing the series with a geometric series. Let r be any number greater than $\frac{1}{2}$ and less than 1. Since $\frac{a_{n+1}}{a_n}$ tends to $\frac{1}{2}$, there is some N such that for $n > N$,

$$\frac{a_{n+1}}{a_n} < r.$$

Multiply by a_n to get $a_{n+1} < ra_n$. Repeating the process gives

$$a_{N+k} < ra_{N+k-1} < \cdots < r^k a_N.$$

Since the geometric series $\sum_{k=0}^{\infty} r^k$ converges, the partial sums of the series $\sum_{k=0}^{\infty} a_{N+k}$ are bounded,

$$\sum_{k=0}^{m} a_{N+k} \le a_N \sum_{k=0}^{m} r^k.$$

Therefore, the partial sums of our series $\sum_{n=1}^{\infty} a_n$ are bounded, and it converges. The idea behind this example leads to the next theorem.

Theorem 1.18. Ratio test. *Suppose that* $\lim_{n\to\infty}\left|\frac{a_{n+1}}{a_n}\right| = L$. *Then*

(a) If $L < 1$, *the series* $\sum_{n=1}^{\infty} a_n$ *converges absolutely.*

(b) If $L > 1$, *the series* $\sum_{n=1}^{\infty} a_n$ *diverges.*

The case $L = 1$ *gives no information.*

In Problem 1.48, we ask you to prove the theorem by extending the argument we used for $\sum_{n=1}^{\infty} \frac{n}{2^n}$.

Example 1.25. Let us determine whether these series converge:

$$\sum_{n=1}^{\infty} \frac{n^5}{2^n}, \qquad \sum_{n=1}^{\infty} \frac{(-2)^n}{n!}, \qquad \sum_{n=1}^{\infty} \frac{2^n}{n^2}.$$

(a) $\lim_{n\to\infty} \frac{\frac{(n+1)^5}{2^{n+1}}}{\frac{n^5}{2^n}} = \lim_{n\to\infty} \frac{1}{2}\left(\frac{n+1}{n}\right)^5 = \frac{1}{2}$, so $\sum_{n=1}^{\infty} \frac{n^5}{2^n}$ converges by the ratio test.

(b) $\lim_{n\to\infty} \left|\frac{\frac{(-2)^{n+1}}{(n+1)!}}{\frac{(-2)^n}{n!}}\right| = \lim_{n\to\infty} \frac{2}{n+1} = 0$, so $\sum_{n=1}^{\infty} \frac{(-2)^n}{n!}$ converges absolutely by the ratio test.

(c) $\lim_{n\to\infty} \frac{\frac{2^{n+1}}{(n+1)^2}}{\frac{2^n}{n^2}} = \lim_{n\to\infty} 2\left(\frac{n}{n+1}\right)^2 = 2$, so $\sum_{n=1}^{\infty} \frac{2^n}{n^2}$ diverges by the ratio test.

Example 1.26. The harmonic series $\sum_{n=1}^{\infty} \frac{1}{n}$ diverges, as we know from Example 1.21. This is a case in which the ratio test would have offered no information, since

$$\frac{\frac{1}{n+1}}{\frac{1}{n}} = \frac{n}{n+1}$$

tends to 1.

We end this section with two very fundamental properties, the nested interval property and Cauchy's criterion for sequence convergence.

1.3c Nested Intervals

We can use the monotone convergence theorem to prove the *nested interval property* of numbers:

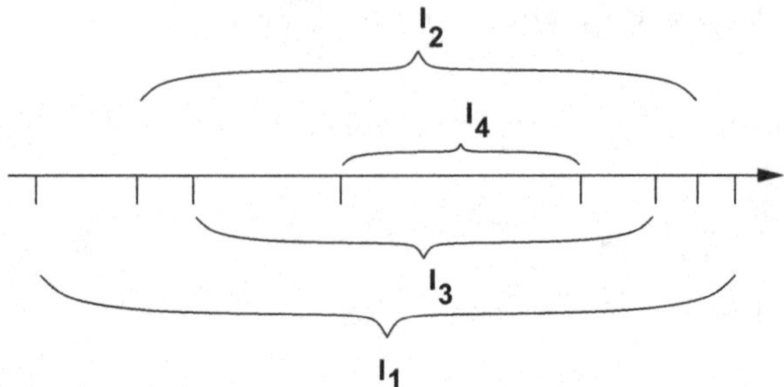

Fig. 1.17 Nested intervals

Theorem 1.19. Nested interval theorem. *If $I_1, I_2, I_3, \ldots$ is a sequence of closed intervals that are "nested," that is, each I_n contains I_{n+1} (Fig. 1.17), then the intervals I_n have at least one point in common. If the lengths of the intervals tend to 0, there is exactly one point in common.*

Proof. Denote by a_n and b_n the left and right endpoints of I_n. The assumption of nesting means that the sequence $\{a_n\}$ increases, the sequence $\{b_n\}$ decreases, and that each b_n is greater than every a_n. So by the monotone convergence theorem, $\{a_n\}$ converges to some limit a, and $\{b_n\}$ to some limit b. By the way we constructed the limits, we know that $a_n \leq a$ and $b \leq b_n$. Now, a cannot be greater than b, for otherwise, some a_n would be larger than some b_m, violating the nesting assumption. So a must be less than or equal to b, and $a_n \leq a \leq b \leq b_n$. If the lengths of the intervals I_n tend to zero as n tends to infinity, then the distance between a and b must be zero, and the intervals I_n have exactly one point in common, namely the point $a = b$. If the lengths of the I_n do not tend to zero, then $[a, b]$ belongs to all the intervals I_n, and the intervals have many points in common. □

The AGM of Two Numbers. In Sect. 1.1c, we saw that given two numbers $0 < g < a$, their arithmetic mean is $\dfrac{a+g}{2} = a_1$, their geometric mean is $\sqrt{ag} = g_1$, and $g \leq g_1 \leq a_1 \leq a$. Repeat the process of taking means, letting $\dfrac{a_1+g_1}{2} = a_2$ and $\sqrt{a_1 g_1} = g_2$. Then

$$g \leq g_1 \leq g_2 \leq a_2 \leq a_1 \leq a.$$

Continuing in this manner, we obtain a sequence of nested closed intervals, $[g_n, a_n]$.

Let us look at what happens to the width of the interval between g and a and how that compares to the width of the interval between $\sqrt{ag} = g_1$ and $\frac{a+g}{2} = a_1$. Using a little algebraic manipulation, we see that

$$\frac{a+g}{2} - \sqrt{ag} = \frac{a-g}{2} + g - \sqrt{ag} \le \frac{a-g}{2},$$

since $g \le \sqrt{ag}$. Thus

$$a_1 - g_1 \le \frac{1}{2}(a-g).$$

This means that the interval between the arithmetic and geometric means of a and g is less than or equal to half the length of the interval between a and g. This reduces the width of the intervals $[g_n, a_n]$ by at least half at each step, and so the lengths of the intervals tend to 0.

The nested interval theorem says that this process squeezes in on exactly one number. This number is called the *arithmetic–geometric mean* of a and g and is denoted by AGM(a,g):

$$\text{AGM}(a,g) = \lim_{n\to\infty} a_n = \lim_{n\to\infty} g_n.$$

The AGM may seem like a mathematical curiosity. Gauss invented the AGM and used it to give a very fast algorithm for computing π.

1.3d Cauchy Sequences

Sometimes the terms of a sequence appear to cluster tightly about a point on the number line, but we do not know the specific number they seem to be approaching. We present now a very general and very useful criterion for such a sequence to have a limit.

Definition 1.8. Cauchy's criterion. A sequence of numbers $\{a_n\}$ is called a *Cauchy sequence* if given any tolerance ε, no matter how small, there is an integer N such that for all integers n and m greater than N, a_n and a_m differ from each other by less than ε.

Examples of Cauchy sequences abound; you will see in Problem 1.52 that every convergent sequence is a Cauchy sequence. The next example shows how to verify directly that the sequence $\{\frac{1}{n}\}$ is a Cauchy sequence.

Example 1.27. Let ε be any tolerance and let N be a whole number greater than $\frac{1}{\varepsilon}$. Let $m > N$ and $n > N$. By the triangle inequality,

$$\left|\frac{1}{m}-\frac{1}{n}\right| \le \frac{1}{m}+\frac{1}{n} \le \frac{1}{N}+\frac{1}{N} < 2\varepsilon.$$

This can be made as small as desired. In fact, we can achieve the tolerance

$$\left|\frac{1}{m}-\frac{1}{n}\right| < \varepsilon$$

if we take n and m greater than $\frac{2}{\varepsilon}$. So $\{\frac{1}{n}\}$ is a Cauchy sequence.

Theorem 1.20. *Every Cauchy sequence converges.*

The proof has four steps. First, we show that *every* sequence has a monotone subsequence. This first step is worth recognizing as a "lemma," a key stepping-stone in the argument. Second, we show that every Cauchy sequence is bounded (and hence every subsequence as well). Third, we recognize that a Cauchy sequence has a monotone subsequence that is bounded and therefore converges. Fourth, we show that a Cauchy sequence converges to the same limit as its monotone subsequence.

Lemma 1.1. *Every infinite sequence of numbers has an infinite monotonic subsequence.*

Proof (of the Lemma). Let $a_1, a_2, \ldots$ be any sequence. We shall show that it contains either an increasing or a decreasing subsequence. We start by trying to construct an increasing subsequence. Start with a_1 and take any subsequent term in the original sequence that is greater than or equal to a_1 as the next term of the subsequence. Continue in this fashion. If we can continue indefinitely, we have the desired increasing subsequence. Suppose, on the other hand, that we get stuck after a finite number of steps at a_j because $a_n < a_j$ for all $n > j$. Then we try again to construct an increasing sequence, starting this time with a_{j+1}. If we can continue ad infinitum, we have an increasing subsequence. If, on the other hand, we get stuck at a_k because $a_n < a_k$ for all $n > k$, then we can again try to construct an increasing subsequence starting at a_{k+1}. Proceeding in this fashion, we either have success at some point, or an infinite sequence of failures. In the second case, the sequence of points $a_j, a_k, \ldots$ where the failures occurred constitutes a decreasing subsequence. This completes the proof of the lemma. □

Proof (of the Theorem). The lemma we just proved guarantees that a Cauchy sequence has a monotone subsequence. Next, we show that a Cauchy sequence is bounded. Being Cauchy ensures that there exists an N such that the terms from a_N onward are all within 1 of each other. This means that the largest of the numbers

$$1+a_1,\ 1+a_2,\ \ldots,\ 1+a_N$$

is an upper bound for $\{a_n\}$, and the smallest of the numbers

$$-1+a_1,\ -1+a_2,\ \dots,\ -1+a_N$$

is a lower bound of $\{a_n\}$. Now by the monotone convergence theorem, the monotone subsequence of $\{a_n\}$ converges to a limit a.

Next we show that not only a subsequence but the whole sequence converges to a. Let us write a_m for an element of the subsequence and a_n for any element of the sequence, and $a-a_n = a-a_m+a_m-a_n$. The triangle inequality gives

$$|a-a_n| \le |a-a_m| + |a_m-a_n|.$$

The first term on the right is less than any prescribed ε for m large, because the subsequence converges to a. The second is less than any prescribed ε for m and n both large, because this is a Cauchy sequence. This proves that the whole sequence also converges to a. □

Cauchy's criterion for convergence, that given any tolerance there is a place in the sequence beyond which *all* the terms are within that tolerance of each other, is a stronger requirement than just requiring that the difference between one term and the next tend to 0. For example, we saw in Example 1.21 that $s_{n+1}-s_n$ tends to 0 but s_n does not converge.

Problems

1.25. Find all the definitions in this section and copy them onto your paper. Illustrate each one with an example.

1.26. Find all the theorems in this section and copy them onto your paper. Illustrate each one with an example.

1.27. Find the first four approximations s_1,s_2,s_3,s_4 to $\sqrt{3}$ using $s_1=1$ as a first approximation and iterating

$$s_{n+1} = \frac{1}{2}\left(s_n + \frac{3}{s_n}\right).$$

What happens if you use $s_1=2$ instead to start?

1.28. In approximating $\sqrt{2}$, we used the fact that if $w_{n+1} < \frac{1}{2}w_n$ holds for each n, then

$$w_{n+1} < \frac{1}{2^n}w_1.$$

Explain why this is true.

1.29. We have said that if s is larger than $\sqrt{2}$, then $\frac{2}{s}$ is smaller than $\sqrt{2}$. Show that this is true.

1.30. If a number s is larger than the cube root $\sqrt[3]{2}$, is it true that $\frac{2}{s^2}$ is smaller?

1.31. Show that if $2 < s^2 < 2+p$, then $\sqrt{2} < s < \sqrt{2}+q$, where $q = \frac{p}{2^{1.5}}$.

1.32. Consider the sequences $a_n = -3n+1$ and $b_n = 3n+\frac{2}{n}$. If we carelessly try to write $\lim_{n\to\infty}(a_n+b_n) = \lim_{n\to\infty} a_n + \lim_{n\to\infty} b_n$, what does it seem to say? What goes wrong in this example?

1.33. Justify the following steps in the proof of parts of Theorem 1.6. Suppose that $\{a_n\}$ and $\{b_n\}$ are convergent sequences,

$$\lim_{n\to\infty} a_n = a, \qquad \lim_{n\to\infty} b_n = b.$$

(a) We want to prove that the sequence of sums $\{a_n+b_n\}$ converges and that $\lim_{n\to\infty}(a_n+b_n) = a+b$. Let $\varepsilon > 0$ be any tolerance. Show that:

(i) There is a number N_1 such that for all $n > N_1$, a_n is within ε of a, and there is a number N_2 such that for all $n > N_2$, b_n is within ε of b. Set N to be the larger of the two numbers N_1 and N_2. Then for $n > N$, $|a_n - a| < \varepsilon$ and $|b_n - b| < \varepsilon$.

(ii) For every n,

$$|(a_n+b_n)-(a+b)| \le |a_n - a| + |b_n - b|.$$

(iii) For all $n > N$, $a_n + b_n$ is within 2ε of $a+b$.

(iv) We have demonstrated that for $n > N$, $|(a_n+b_n)-(a+b)| \le 2\varepsilon$. Explain why this completes the proof.

(b) We want to prove that if a is not 0, then all but a finite number of the a_n differ from 0 and

$$\lim_{n\to\infty} \frac{1}{a_n} = \frac{1}{a}.$$

Let $\varepsilon > 0$ be any tolerance. Show that:

(i) There is a number N such that when $n > N$, a_n is within ε of a.

(ii) There is a number M such that for $n > M$, $a_n \ne 0$ and $\left|\frac{1}{a_n}\right|$ is bounded by some α.

(iii) For n larger than both M and N,

$$\left|\frac{1}{a_n} - \frac{1}{a}\right| = \left|\frac{a-a_n}{a_n a}\right| = |a-a_n|\left|\frac{1}{a_n}\right|\frac{1}{|a|} < \varepsilon\frac{\alpha}{|a|}.$$

Hence $\frac{1}{a_n}$ converges to $\frac{1}{a}$.

1.34. Solve $x^2 - x - 1 = 0$ as follows. Restate the equation as $x = 1 + \frac{1}{x}$, which suggests the sequence of approximations

$$x_0 = 1, \quad x_1 = 1 + \frac{1}{x_0}, \quad x_2 = 1 + \frac{1}{x_1}, \quad \ldots.$$

Explain the following items to prove that the sequence converges to a solution.

(a) $x_0 < x_2 < x_1$
(b) $x_0 < x_2 < x_4 < \cdots < x_5 < x_3 < x_1$
(c) The even sequence x_{2k} increases to a limit L, and the odd sequence x_{2k+1} decreases to a limit $R \geq L$.
(d) The distances $(x_{2k+3} - x_{2k+2})$ satisfy $(x_{2k+3} - x_{2k+2}) < \frac{1}{x_2^4}(x_{2k+1} - x_{2k})$.
(e) $R = L = \lim_{k\to\infty} x_k$ is a solution to $x^2 - x - 1 = 0$.

1.35. Suppose a sequence $\{a_n\}$ converges to a. Explain each of the following items, which prove Theorem 1.8.

(a) There is a number N such that for all $n > N$, $|a_n - a| < 1$.
(b) For all $n > N$, $|a_n| \leq |a| + 1$. *Hint*: $a_n = a + (a_n - a)$.
(c) Let α be the largest of the numbers

$$|a_1|, \ |a_2|, \ \ldots |a_N|, \ |a| + 1.$$

Then $|a_k| \leq \alpha$ for $k = 1, 2, 3, \ldots$.
(d) $\{a_n\}$ is bounded.

1.36. This problem explores the sum notation. Write out each finite sum.

(a) $\sum_{n=1}^{5} a_n$
(b) $\sum_{k=2}^{4} \frac{3}{k}$
(c) $\sum_{j=2}^{6} b_{j-1}$
(d) Rewrite the expression $t^2 + 2t^3 + 3t^4$ in the sum notation.
(e) Explain why $\sum_{n=1}^{10} n^2 = 105 + \sum_{n=3}^{9} n^2$, and why $\sum_{n=2}^{20} a_n = \sum_{k=0}^{18} a_{k+2}$.

1.37. Partial sums $s_1 = a_1$, $s_2 = a_1 + a_2$, and so forth are known to be given by $s_n = \frac{n}{n+2}$. Find a_1, a_2, and $\sum_{n=1}^{\infty} a_n$.

1.38. Use relation (1.10) to evaluate the sum $\sum_{k=0}^{n} \frac{1}{7^k}$.

1.39. Find the limit as n tends to infinity of $\frac{5}{7}+\frac{25}{49}+\frac{125}{343}+\cdots+\frac{5^n}{7^n}$.

1.40. Find the limit as n tends to infinity of $\frac{5}{7}+\frac{5}{49}+\frac{5}{343}+\cdots+\frac{5}{7^n}$.

1.41. Suppose the ratio test indicates that $\sum_{n=0}^{\infty} a_n$ converges. Use the ratio test to show that $\sum_{n=0}^{\infty} na_n$ also converges. What can you say about $\sum_{n=0}^{\infty}(-1)^n n^5 a_n$?

1.42. Why does the series $\sum_{n=1}^{\infty}\frac{n^2}{n^2+1}$ diverge?

1.43. Show that the infinite series $\sum_{n=1}^{\infty}\frac{1}{n^2}$ converges by verifying the following steps:

(a) $\frac{1}{n^2}<\frac{1}{n(n-1)}$ (b) $\frac{1}{n(n-1)}=\frac{1}{n-1}-\frac{1}{n}$ (c) $\sum_{n=2}^{k}\frac{1}{n(n-1)}=1-\frac{1}{k}$

1.44. For what numbers t does the sequence

$$s_n = 1-2t+2^2t^2-2^3t^3+\cdots+(-2)^n t^n$$

converge? What is the limit for those t?

1.45. Carry out the following steps to prove the comparison theorems, Theorems 1.14 and 1.15.

(a) Let $\{a_n\}$ and $\{b_n\}$ be sequences for which $0\le b_n\le a_n$. Use the monotone convergence theorem to show that if $\sum_{n=0}^{\infty} a_n$ converges, then $\sum_{n=0}^{\infty} b_n$ also converges.

(b) Let $\{a_n\}$ and $\{b_n\}$ be sequences of positive numbers for which $\lim_{n\to\infty}\frac{a_n}{b_n}$ exists and is a positive number, say L. First show that for n sufficiently large, $a_n\le(L+1)b_n$. Then explain why the convergence of $\sum_{n=N}^{\infty}(L+1)b_n$ implies that of $\sum_{n=0}^{\infty} a_n$.

1.46. Let $a_1\ge a_2\ge a_3\ge\cdots\ge a_n\ge\cdots\ge 0$ be a sequence with $\lim_{n\to\infty} a_n=0$. Let

$$s_n = a_1-a_2+a_3-\cdots+(-1)^{n+1}a_n.$$

(a) Explain why $a_{2k+1}-a_{2k+2}\ge 0$, and why $-a_{2k}+a_{2k+1}\le 0$.
(b) Explain why $s_2,s_4,s_6,\ldots,s_{2k},\ldots$ converges, and why $s_1,s_3,s_5,\ldots,s_{2k+1},\ldots$ converges.

(c) Show that $\lim_{k\to\infty}(s_{2k+1} - s_{2k}) = 0$.

(d) Explain why $\sum_{n=1}^{\infty}(-1)^{n-1}a_n$ converges.

1.47. Determine which of the following series converge absolutely, which converge, and which diverge.

(a) $\sum_{n=0}^{\infty}\frac{(-2)^n+1}{3^n}$

(b) $\sum_{n=1}^{\infty}\frac{1}{\sqrt[4]{n}}$

(c) $\sum_{n=1}^{\infty}\frac{(-1)^n}{\sqrt[4]{n}}$

(d) $\sum_{n=0}^{\infty}\frac{n}{\sqrt{n^2+1}}$

(e) $\sum_{n=0}^{\infty}\frac{n}{\sqrt{n^4+1}}$

(f) $\sum_{n=1}^{\infty}\frac{n^2}{(1.5)^n}$

1.48. We used series (1.12) to motivate the ratio test, Theorem 1.18. Extend the argument used in the example to create a proof of the theorem.

1.49. Determine which of the following series converge absolutely, which converge, and which diverge.

(a) $1+\frac{1}{2}-\frac{1}{4}+\frac{1}{8}+\frac{1}{16}-\frac{1}{32}+\frac{1}{64}+\frac{1}{128}-\frac{1}{256}+\cdots$

(b) $\sum_{n=1}^{\infty}10^{-n^2}$

(c) $\sum_{n=1}^{\infty}\frac{b^n}{n!}$ Are there any restrictions on b?

(d) $\sum_{n=0}^{\infty}\frac{n^{1/n}}{3^n}$ *Hint*: See Problem 1.13.

(e) $\sum_{n=0}^{\infty}\left(\frac{(-1)^n}{\sqrt{n}}+\frac{1}{2^n}\right)$

(f) $\sum_{n=0}^{\infty}\left(\frac{(-1)^n}{\sqrt{n}}+\frac{1}{2}\right)$

1.50. Suppose $\sum_{n=0}^{\infty}a_n^2$ and $\sum_{n=0}^{\infty}b_n^2$ both converge. Use the Cauchy–Schwarz inequality (see Problem 1.18) to explain the following.

(a) The partial sums $\sum_{n=0}^{k}a_nb_n$ satisfy $\sum_{n=0}^{k}a_nb_n \le \sqrt{\sum_{n=0}^{\infty}a_n^2}\sqrt{\sum_{n=0}^{\infty}b_n^2}$.

(b) If the numbers a_n and b_n are nonnegative, then $\sum_{n=0}^{\infty} a_n b_n$ converges.

(c) $\sum_{n=0}^{\infty} a_n b_n$ converges absolutely.

1.51. Let us set a_n equal to the n-place decimal expansion of some number x. For example, if $x = \sqrt{2}$, then we have $a_1 = 1.4$, $a_2 = 1.41$, $a_3 = 1.414$, etc. Is a_n a Cauchy sequence?

1.52. Prove that every convergent sequence is a Cauchy sequence.

1.4 The Number e

In your study of geometry you have come across the curious number denoted by the Greek letter π. The first six digits of π are

$$\pi = 3.14159\ldots.$$

In this section, we shall define another number of great importance, denoted by the letter e. This number is called Euler's constant in honor of the great mathematician who introduced it; its first six digits are

$$\mathrm{e} = 2.71828\ldots.$$

The number e is defined as the limit of the sequence

$$e_n = \left(1 + \frac{1}{n}\right)^n$$

as n tends to infinity. To make this a legitimate definition, we have to prove that this sequence has a limit.

Financial Motivation. Before we turn to the proof, let us first give a financial motivation for considering this limit. Suppose you invest one dollar at the interest rate of 100% per year. If the interest is compounded annually, a year later you will receive two dollars, that is, the original dollar invested plus another dollar for interest. If interest is compounded semiannually, you will receive at the end of the year $(1.5)^2 = 2.25$ dollars. That is 50% interest after six months, giving you a value of \$1.50, followed by another six months during which you earn 50% interest on your \$1.50. If interest is compounded n times a year, you will receive at the end of the year $\left(1 + \frac{1}{n}\right)^n$ dollars. The more frequently interest is compounded, the higher your return. This suggests the importance of the number e and indicates that the sequence $\left(1 + \frac{1}{n}\right)^n$ is increasing. Later, we shall show how e can be used to study arbitrary interest rates.

Monotonicity of e_n. Let us do a few numerical experiments before trying to prove anything about the sequence $\{e_n\}$. Using a calculator, we calculate the first ten terms of the sequence rounded down to three digits:

$$\begin{aligned} e_1 &= 2.000 \\ e_2 &= 2.250 \\ e_3 &= 2.370 \\ e_4 &= 2.441 \\ e_5 &= 2.488 \\ e_6 &= 2.521 \\ e_7 &= 2.546 \\ e_8 &= 2.565 \\ e_9 &= 2.581 \\ e_{10} &= 2.593 \end{aligned}$$

We notice immediately that this sequence of ten numbers is in increasing order. Just to check, let us do a few further calculations:

$$\begin{aligned} e_{100} &= 2.704 \\ e_{1000} &= 2.716 \\ e_{10000} &= 2.718 \end{aligned}$$

These numbers confirm our financial intuition that more frequent compounding results in a larger annual return, and that $\left(1+\frac{1}{n}\right)^n$ is an increasing sequence.

We shall now give a nonfinancial argument that the sequence e_n increases. We use the A-G inequality for $n+1$ numbers:

$$\left(a_1 a_2 \cdots a_{n+1}\right)^{1/(n+1)} \leq \frac{1}{n+1}\left(a_1 + a_2 + \cdots + a_{n+1}\right).$$

Take the $n+1$ numbers

$$\underbrace{\left(1+\frac{1}{n}\right), \ldots, \left(1+\frac{1}{n}\right)}_{n \text{ times}}, 1.$$

Their product is $\left(1+\dfrac{1}{n}\right)^n$, and their sum is $n\left(1+\dfrac{1}{n}\right)+1 = n+2$. So their geometric mean is $\left(1+\dfrac{1}{n}\right)^{n/(n+1)}$, and their arithmetic mean is $\dfrac{n+2}{n+1} = 1+\dfrac{1}{n+1}$. According to the A-G inequality,

$$\left(1+\frac{1}{n}\right)^{n/(n+1)} < 1+\frac{1}{n+1};$$

raising both sides to the power $n+1$ gives

$$\left(1+\frac{1}{n}\right)^n < \left(1+\frac{1}{n+1}\right)^{n+1},$$

proving that e_n is less than e_{n+1}. Therefore, the sequence $\{e_n\}$ is increasing.

Boundedness of e_n. To conclude that the sequence converges, we have to show that it is bounded. To accomplish this, we look at another sequence, $\{f_n\}$, defined as

$$f_n = \left(1+\frac{1}{n}\right)^{n+1}.$$

Using a calculator, we calculate the first ten terms of this sequence, rounded up to three digits:

$$\begin{aligned}
f_1 &= 4.000\\
f_2 &= 3.375\\
f_3 &= 3.161\\
f_4 &= 3.052\\
f_5 &= 2.986\\
f_6 &= 2.942\\
f_7 &= 2.911\\
f_8 &= 2.887\\
f_9 &= 2.868\\
f_{10} &= 2.854
\end{aligned}$$

These ten numbers are decreasing, suggesting that the whole infinite sequence f_n is decreasing. Further test calculations offer more evidence:

$$\begin{aligned}
f_{100} &= 2.732\\
f_{1000} &= 2.720\\
f_{10000} &= 2.719
\end{aligned}$$

Here is an intuitive demonstration that f_n is a decreasing sequence: Suppose you borrow \$1 from your family at no interest. If you return all that you owe a year later, you have nothing left. But if you return half of what you owe twice a year, you have left $(0.5)^2 = 0.25$. If you return a third of what you owe three times a year, at the end of the year you have left $(2/3)^3$. If you return $(1/n)$–th of what you owe n times a year, you end up with $\left(1-\frac{1}{n}\right)^n$ at the end of the year. This is an intuitive demonstration that the sequence $\{\left(1-\frac{1}{n}\right)^n\}$ is increasing. It follows that the sequence of reciprocals is decreasing:

$$\frac{1}{\left(1-\frac{1}{n}\right)^n} = \left(\frac{n}{n-1}\right)^n = \left(1+\frac{1}{m}\right)^{m+1}, \quad \text{where } m = n-1.$$

In Problem 1.54, we guide you through a nonfinancial argument for the inequality $f_n > f_{n+1}$.

The number f_n is $1+\frac{1}{n}$ raised to the power $n+1$; it is larger than e_n, which is $1+\frac{1}{n}$ raised to the lower power n:

$$e_n < f_n.$$

Since $\{f_n\}$ is a decreasing sequence, it follows that

$$e_n < f_n < f_1 = 4.$$

This proves that the sequence $\{e_n\}$ is monotonically increasing and bounded. It follows from the monotone convergence theorem, Theorem 1.10, that $\{e_n\}$ converges to a limit; this limit is called e.

The sequence $\{f_n\}$ is monotone decreasing and is bounded below by zero. Therefore it, too, tends to a limit; call it f. Next we show that f equals e. Each f_n is greater than e_n, so it follows that f is not less than e. To see that they are equal, we estimate the difference of f_n and e_n:

$$f_n - e_n = \left(1+\frac{1}{n}\right)^{n+1} - \left(1+\frac{1}{n}\right)^n = \left(1+\frac{1}{n}\right)^n\left(1+\frac{1}{n}-1\right) = \frac{e_n}{n}.$$

As we have seen, e_n is less than 4. Therefore, it follows that

$$f_n - e_n < \frac{4}{n}.$$

Since e is greater than e_n and f is less than f_n, it follows that also $f - \mathrm{e}$ is less than $\frac{4}{n}$. Since this is true for all n, $f - \mathrm{e}$ must be zero.

Even though the sequences e_n and f_n both converge to e, our calculations show that e_{1000} and f_{1000} are accurate only to two decimal places. The sequences $\{e_n\}$ and $\{f_n\}$ converge very slowly to e. Calculus can be used to develop sequences that converge more rapidly to e. In Sect. 4.3a, we show how to use knowledge of calculus to develop a sequence

$$g_n = 1 + 1 + \frac{1}{2!} + \frac{1}{3!} + \cdots + \frac{1}{n!}$$

that converges to e much more rapidly. In fact, g_9 gives e correct to six decimal places. In Problem 1.55, we lead you through an argument, which does not use calculus, to show that g_n converges to e. In Sect. 10.4, we shed some light on the sequences e_n and f_n and offer another way to use calculus to improve on them.

Problems

1.53. Explain the following items, which prove that $\lim_{n\to\infty} n^{1/n} = 1$.

(a) Use the fact that the sequence $e_n = \left(1+\frac{1}{n}\right)^n$ increases to e to show that $\left(1+\frac{1}{n-1}\right)^n < 6$.

(b) Deduce that the sequence $n^{1/n}$ is decreasing when $n \geq 6$.

(c) $1 \leq n^{1/n}$. Therefore, $r = \lim_{n\to\infty} n^{1/n}$ exists.

(d) Consider $(2n)^{1/(2n)}$ to show that $r > 1$ is not possible.

1.54. Apply the A-G inequality to the $n+1$ numbers $\left(1-\frac{1}{n}\right), \ldots, \left(1-\frac{1}{n}\right), 1$ to conclude that

$$\left(1-\frac{1}{n}\right)^{n/(n+1)} < \frac{n}{n+1}.$$

Take this inequality to the power $n+1$ and take its reciprocal to conclude that

$$\left(1+\frac{1}{n-1}\right)^n = f_{n-1} > f_n = \left(1+\frac{1}{n}\right)^{n+1}.$$

1.55. Set $g_n = 1+1+\frac{1}{2!}+\frac{1}{3!}+\cdots+\frac{1}{n!}$. Here are the first ten values:

$$\begin{aligned}
g_0 &= 1\\
g_1 &= 2\\
g_2 &= 2.5\\
g_3 &= 2.6666666666666\\
g_4 &= 2.7083333333333\\
g_5 &= 2.7166666666666\\
g_6 &= 2.7180555555555\\
g_7 &= 2.7182539682539\\
g_8 &= 2.7182787698412\\
g_9 &= 2.7182815255731
\end{aligned}$$

We know that $e_n = \left(1+\frac{1}{n}\right)^n$ converges to the limit e $= 2.718\ldots$ as n tends to infinity. Explain the following steps, which show that g_n also converges to e, that is,

$$\mathrm{e} = \sum_{n=0}^{\infty} \frac{1}{n!}.$$

(a) $n!$ is greater than 2^{n-1}. Explain why $g_n < 1+1+\frac{1}{2}+\frac{1}{4}+\cdots+\frac{1}{2^{n-1}}$. Explain why $g_n < 3$, and why g_n tends to a limit.

(b) Recall the binomial theorem $(a+b)^n = \sum_{k=0}^{n} \binom{n}{k} a^k b^{n-k}$. Let $a = \frac{1}{n}$ and $b = 1$ and show that

$$e_n = 1 + \sum_{k=1}^{n} \frac{n(n-1)\cdots(n-(k-1))}{k!} \frac{1}{n^k}.$$

Show that the kth term in e_n is less than the kth term in g_n. Use this to conclude that $e_n < g_n$.

In parts (c) and (d), we show that for large n, e_n is not much less than g_n.

(c) Write the difference $g_n - e_n$ as

$$g_n - e_n = \sum_{k=2}^{\infty} \frac{n^k - n(n-1)\cdots(n-(k-1))}{n^k k!},$$

and explain why this is less than $\sum_{k=2}^{\infty} \frac{n^k - (n-k)^k}{n^k k!} = \sum_{k=2}^{\infty} \frac{1-\left(1-\frac{k}{n}\right)^k}{k!}$.

(d) In this last part you need to explain through a sequence of steps, outlined below, why $g_n - e_n$ is less than $\frac{4}{n}$ and hence tends to 0. First recall that for $0 < x < 1$, we have the inequality

$$1 - x^k = (1-x)(1+x+\cdots+x^{k-1}) < (1-x)k.$$

Let $x = 1 - \frac{k}{n}$. Explain why $g_n - e_n < \frac{1}{n} \sum_{k=2}^{\infty} \frac{k^2}{k!}$. Using the convention that $0! = 1$, we have $\frac{k^2}{k!} = \frac{k}{k-1} \frac{1}{(k-2)!}$. Recall that $\frac{1}{(k-2)!}$ is less than $\frac{1}{2^{k-2}}$. Explain why

$$g_n - e_n < \frac{1}{n} \sum_{k=2}^{\infty} \frac{2}{2^{k-2}}.$$

Explain why $g_n - e_n < \frac{4}{n}$, and why this completes the proof that the sequences g_n and e_n have the same limit.

1.56. Let us find a way to calculate e to a tolerance of 10^{-20}. Let g_n denote the numbers in Problem 1.55. Explain why

$$\mathrm{e} - g_n < \frac{1}{n!} \left(\frac{1}{n+1} + \frac{1}{(n+1)^2} + \frac{1}{(n+1)^3} + \cdots \right).$$

Then explain how that gives $\mathrm{e} - g_n < \frac{1}{n!} \frac{1}{n}$. Finally, what would you take n to be so that g_n approximates e within 10^{-20}?

Chapter 2
Functions and Continuity

Abstract Calculus is the study of the rate of change and the total accumulation of processes described by functions. In this chapter we review some familiar notions of function and explore functions that are defined by sequences of functions.

2.1 The Notion of a Function

The idea of a function is the most important concept in mathematics. There are many sources of functions, and they carry information of a special kind. Some are based on observations, like the maximum daily temperature T in your town every day last year:

$$\text{temperature} = T(\text{day}).$$

Some express a causal relation between two quantities, such as the force f exerted by a spring as a function of the displacement:

$$\text{force} = f(\text{displacement}).$$

Research in science is motivated by finding functions to express such causal relations. A function might also express a purely arbitrary relationship like

$$F = \frac{9}{5}C + 32,$$

relating the Fahrenheit temperature scale to the Celsius scale. Or it could express a mathematical theorem:

$$r = -\frac{b}{2} + \frac{\sqrt{b^2-4}}{2},$$

where r is the larger root of the quadratic equation $x^2 + bx + 1 = 0$.

Functions can be represented in different ways. Some of these ways are familiar to you: graphs, tables, and equations. Other methods, such as representing a function

P.D. Lax and M.S. Terrell, *Calculus With Applications*, Undergraduate Texts in Mathematics, DOI 10.1007/978-1-4614-7946-8_2,

through a sequence of functions, or as a solution to a differential equation, are made possible by calculus.

Rather than starting with a definition of function, we shall first give a number of examples and then fit the definition to these.

Example 2.1. The vertical distance h (measured in kilometers) traveled by a rocket depends on the time t (measured in seconds) that has elapsed since the rocket was launched. Figure 2.1 graphically describes the relation between t and h.

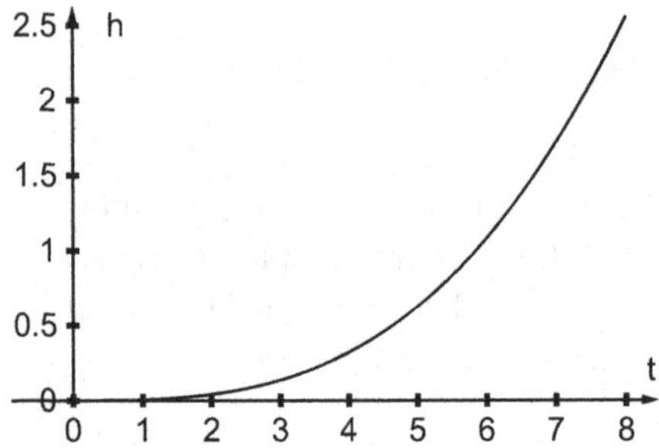

Fig. 2.1 Vertical distance traveled by a rocket. The horizontal axis gives the time elapsed since launch, in seconds. The vertical gives distance traveled, in kilometers

Example 2.2. The graph in Fig. 2.2 shows three related functions: U.S. consumption of oil, the price of oil unadjusted for inflation (the composite price), and the price of oil adjusted for inflation (in 2008 dollars).

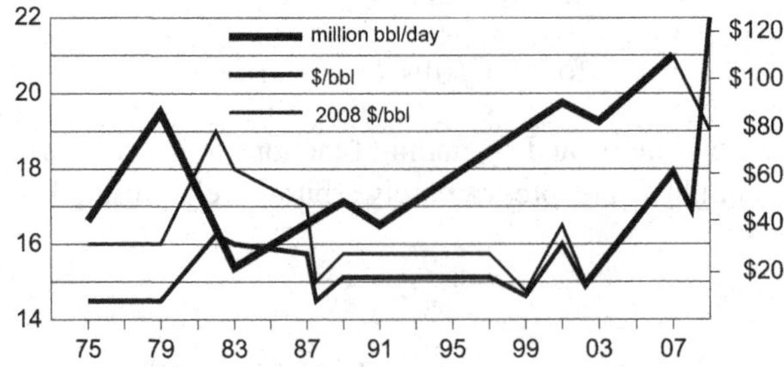

Fig. 2.2 Oil consumption, price, and inflation adjusted price

Example 2.3. The distance d traveled by a body falling freely from rest near the surface of the Earth, measured in meters, and t the time of fall measured in seconds.

$$d = 4.9t^2.$$

Example 2.4. The national debt D in billions of dollars in year y.

y	2004	2005	2006	2007	2008	2009	2010
D	7,354	7,905	8,451	8,951	9,654	10,413	13,954

In contrast, this is the table that appeared in the first edition of this book.

y	1955	1956	1957	1958	1959	1960	1961
D	76	82	85	90	98	103	105

Example 2.5. The volume V of a cube with edge length s is $V = s^3$.

Adjusted gross income	Tax rate (%)
0–8,375	10
8,375–34,000	15
34,000–82,400	25
82,400–171,850	28
171,850–373,650	33
373,650 and above	35

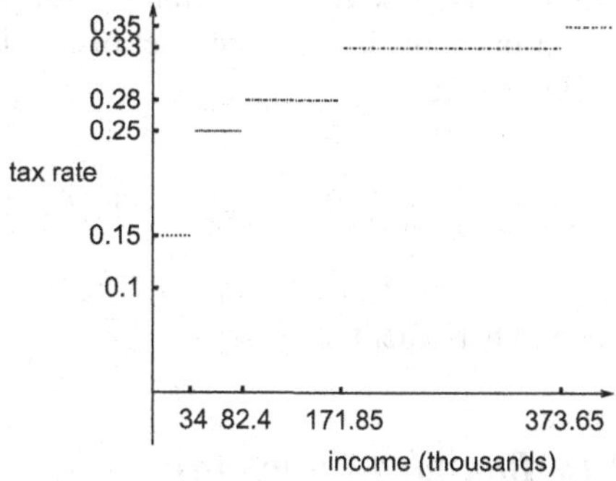

Fig. 2.3 *Left*: a table of tax rates in Example 2.6. *Right*: a graph of the tax rate by income level

Example 2.6. The Internal Revenue Service's 2010 tax rates for single-filing status are given in the table in Fig. 2.3. The tax rates can be described as a function. Let x be adjusted gross income in dollars, and $f(x)$ the rate at which each dollar within that income level is taxed. The function f can be described by the following rule

$$f(x) = \begin{cases} 0.10 & \text{for } 0 \le x \le 8{,}375 \\ 0.15 & \text{for } 8{,}375 < x \le 34{,}000 \\ 0.25 & \text{for } 34{,}000 < x \le 82{,}400 \\ 0.28 & \text{for } 82{,}400 < x \le 171{,}850 \\ 0.33 & \text{for } 171{,}850 < x \le 373{,}650 \\ 0.35 & \text{for } 373{,}650 < x. \end{cases}$$

The graph in Fig. 2.3 makes the jumps in tax rate and the levels of income subject to those rates easier to see. (To compute the tax on 10,000 dollars, for example, the first 8,375 is taxed at 10 %, and the next $10{,}000 - 8{,}375 = 1{,}625$ is taxed at 15 %. So the tax is $(0.15)(1{,}625) + (0.10)(8{,}375) = 1{,}121.25$.)

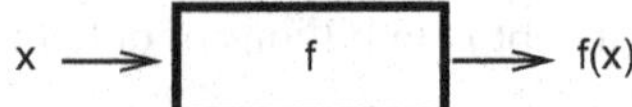

Fig. 2.4 A function can be thought of as a device in a box, with input and output

We can also think of a function as a box, as in Fig. 2.4. You drop in an input x, and out comes $f(x)$ as the output.

Definition 2.1. A *function* f is a rule that assigns to every number x in a collection D, a number $f(x)$. The set D is called the *domain* of the function, and $f(x)$ is called the *value* of the function at x. The set of all values of a function is called its *range*. The set of ordered pairs $(x, f(x))$ is called the *graph* of f.

When we describe a function by a rule, we assume, unless told otherwise, that the set of inputs is the largest set of numbers for which the rule makes sense. For example, take

$$f(x) = x^2 + 3, \qquad g(x) = \sqrt{x-1}, \qquad h(x) = \frac{1}{x^2 - 1}.$$

The domain of f is all numbers. The domain of g is $x \geq 1$, and the domain of h is any number other than 1 or -1.

2.1a Bounded Functions

Definition 2.2. We say that a function f is *bounded* if there is a positive number m such that for all values of f, $-m \leq f(x) \leq m$. We say that a function g is *bounded away from* 0 if there is a positive number p such that no value of g falls in the interval from $-p$ to p.

In Fig. 2.5, f is bounded because $-m \leq f(x) \leq m$ for all x, and g is bounded away from 0 because $0 < p \leq g(x)$ for all x.

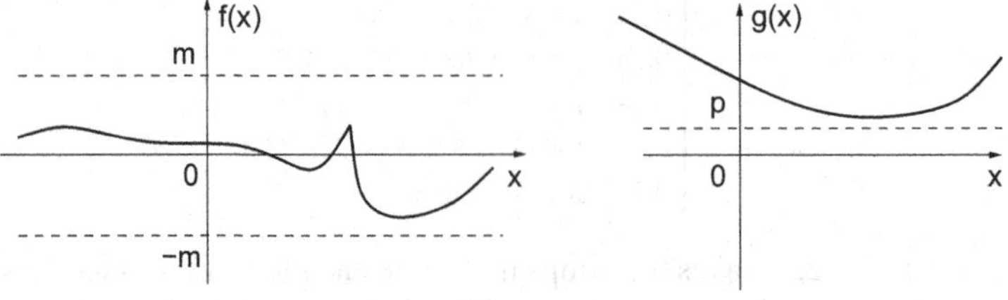

Fig. 2.5 *Left*: f is a bounded function. *Right*: g is bounded away from 0

A function that is not bounded, or is not bounded away from 0, may have one or both of those properties on a subset of its domain.

Example 2.7. Let $h(x) = \dfrac{1}{x^2 - 1}$. Then h is not bounded. It has arbitrarily large values (both positive and negative) as x tends to 1 or -1. Furthermore, h is not bounded away from 0, because $h(x)$ tends to 0 as x becomes arbitrarily large (positive or negative). However, if we restrict the domain of h to, say, the interval $[-0.8, 0.8]$, then h is both bounded and bounded away from 0 on $[-0.8, 0.8]$. See Fig. 2.6 for the graph of h.

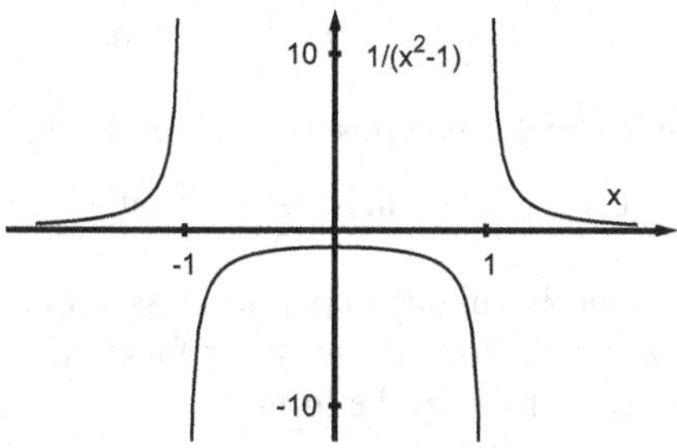

Fig. 2.6 The function $h(x) = \frac{1}{x^2-1}$ is neither bounded nor bounded away from 0

2.1b Arithmetic of Functions

Once you have functions, you can use them to make new functions. The sum of functions f and g is denoted by $f + g$, and the difference by $f - g$:

$$(f+g)(x) = f(x) + g(x), \qquad (f-g)(x) = f(x) - g(x).$$

The product and quotient of functions f and g are denoted by fg and $\dfrac{f}{g}$:

$$(fg)(x) = f(x)g(x), \qquad \frac{f}{g}(x) = \frac{f(x)}{g(x)} \quad \text{when} \quad g(x) \neq 0.$$

In applications, it makes sense to add or subtract two functions only if their values are measured in the same units. In our example about oil consumption and price, it makes sense to find the difference between the inflation-adjusted price and the nonadjusted price of oil. However, it does not make sense to subtract the price of oil from the number of barrels consumed.

Polynomials. Starting with the simplest functions, the constant functions

$$c(x) = c$$

and the identity function

$$i(x) = x,$$

we can build more complicated functions by forming sums and products. All functions that are obtained from constant and identity functions through repeated additions and multiplications are of the form

$$p(x) = a_n x^n + a_{n-1} x^{n-1} + \cdots + a_0,$$

where the a's are constants and n is a positive integer. Such a function is called a *polynomial*. A quotient of two polynomials, $\frac{p(x)}{q(x)}$, is called a *rational* function.

Linear Functions. A simple but highly important class of functions is that of *linear* functions. In Chap. 3, we show how to use linear functions to approximate other functions. Every linear function ℓ is of the form

$$\ell(x) = mx + b,$$

where m and b are some given numbers. A linear function is certainly simple from a computational point of view: to evaluate it, we need to perform one multiplication and one addition. Linear functions have the property that

$$\ell(x+h) = m(x+h) + b = \ell(x) + mh,$$

which means that when the input x in increased by h, the output changes by an amount that does not depend on x. The change in the output of a linear function is m times the change in the input (Fig. 2.7),

$$\ell(x+h) - \ell(x) = mh.$$

Example 2.8. In changing temperature from Celsius to Fahrenheit, we use the formula

$$F = \frac{9}{5}C + 32.$$

A change in temperature in degrees Celsius produces a change in temperature in degrees Fahrenheit that is always $\frac{9}{5}$ as large, independent of the temperature.

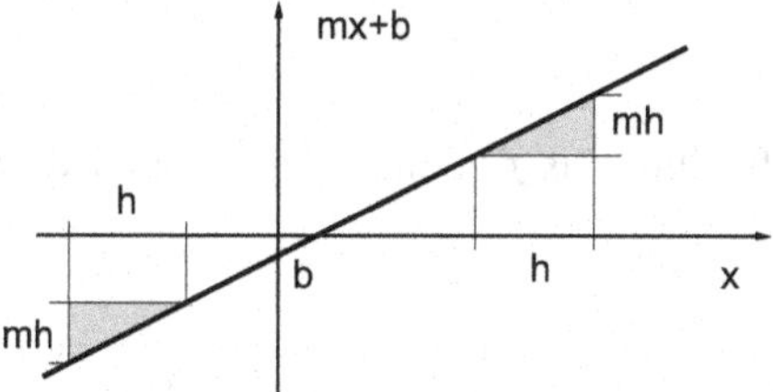

Fig. 2.7 The graph of a linear function $\ell(x) = mx + b$. The change in the output is m times the change in the input

You can completely determine a linear function if you know the function values at two different points. Suppose

$$y_1 = \ell(x_1) = mx_1 + b \quad \text{and } y_2 = \ell(x_2) = mx_2 + b.$$

By subtracting, we see that $y_2 - y_1 = m(x_2 - x_1)$. Solving for m, we get

$$m = \frac{y_2 - y_1}{x_2 - x_1}.$$

The number m is called the *slope* of the line through the points (x_1, y_1) and (x_2, y_2). Then b is also determined by the two points, because

$$b = y_1 - \frac{y_2 - y_1}{x_2 - x_1} x_1 \qquad (x_1 \neq x_2).$$

In addition to visualizing a linear function graphically, we can look at how numbers in the domain are mapped to numbers in the range. In this representation, m can be interpreted as a stretching factor.

Example 2.9. Figure 2.8 shows how the linear function $\ell(x) = 3x - 1$ maps the interval $[0, 2]$ onto the interval $[-1, 5]$, which is three times as long.

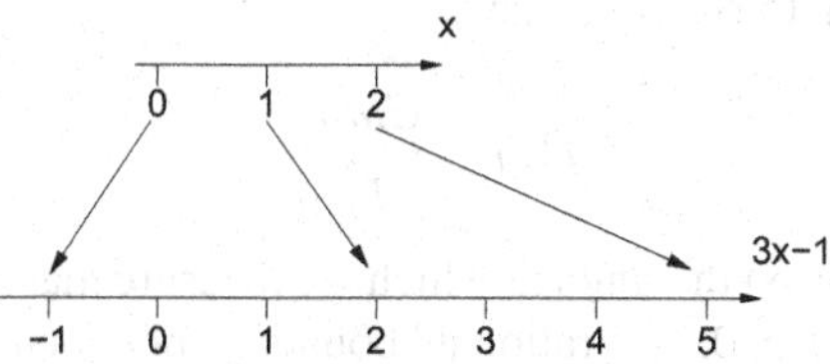

Fig. 2.8 The linear function $\ell(x) = 3x - 1$ as a mapping from $[0, 2]$ to $[-1, 5]$

There are many more important examples of functions to explore. We invite you to work on some in the Problems.

Problems

2.1. For each of these functions, is f bounded? is f bounded away from zero?

(a) $f(x) = x - \frac{1}{x} + 25$
(b) $f(x) = x^2 + 1$
(c) $f(x) = \frac{1}{x^2 + 1}$
(d) $f(x) = x^2 - 1$

2.2. Plot the national debt as given in Example 2.4 for the years 1955–1961. Is the national debt a linear function of time? Explain.

2.3. Let

$$f(x) = \frac{x^3 - 9x}{x^2 + 3x}, \quad g(x) = \frac{x^2 - 9}{x + 3}, \quad \text{and } h(x) = x - 3.$$

(a) Show that

$$f(x) = g(x) = h(x) \quad \text{when } x \neq 0, -3.$$

(b) Find the domains of f, g, and h.
(c) Sketch the graphs of f, g, and h.

2.4. Let $h(x) = \frac{1}{x^2 - 1}$ with domain $[-0.8, 0.8]$. Find bounds p and q on the range of h:

$$p \leq \frac{1}{x^2 - 1} \leq q.$$

2.5. Use the tax table or graph in Example 2.6 to find the total tax on an adjusted gross income of \$200,000.

2.6. The gravitational force between masses M and m with centers separated by distance r is, according to Newton's law,

$$f(r) = \frac{GMm}{r^2}.$$

The value of G depends on the units in which we measure mass, distance, and force. Take the domain to be $r > 0$. Is f rational? bounded? bounded away from 0?

2.7. Here is a less obvious example of a linear function. Imagine putting a rope around the Earth. Make it nice and snug. Now add 20 m to the length of the rope and arrange it concentrically around the Earth. Could you walk under it without hitting your head?

2.2 Continuity

In this section, we scrutinize the definition of function given in the previous section. According to that definition, a function f assigns a value $f(x)$ to each number x in the domain of f. Clearly, in order to find the value of $f(x)$, we have to know x. But what does knowing x mean? According to Chap. 1, we know x if we are able to produce as close an approximation to x as requested. This means that we never (or hardly ever) know x *exactly*. How then can we hope to determine $f(x)$? A way out of this dilemma is to remember that knowing $f(x)$ means being able to give as close an approximation to $f(x)$ as requested. So we can determine $f(x)$ *if approximate knowledge of x is sufficient for approximate determination of $f(x)$*. The notion of continuity captures this property of a function.

Definition 2.3. We say that a function f is *continuous* at c when:
for any tolerance $\varepsilon > 0$, there is a precision $\delta > 0$ such that $f(x)$ differs from $f(c)$ by less than ε whenever x differs from c by less than δ (Fig. 2.9).

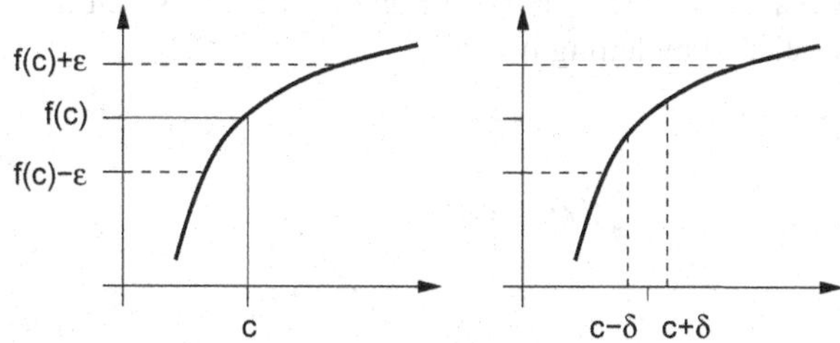

Fig. 2.9 *Left*: for any $\varepsilon > 0$, *Right*: we can find a $\delta > 0$

As a practical matter, f is continuous at c if all the values of f at points near c are very nearly $f(c)$. This leads to a useful observation about continuity: If f is continuous at c and $f(c) < m$, then it is also true that $f(x) < m$ for every x in some sufficiently small interval around c. To see this, take ε to be the distance between $f(c)$ and m, as in Fig. 2.10. Similarly, if $f(c) > m$, there is an entire interval of numbers x around c where $f(x) > m$.

Driver: "But officer, I only hit 90 mph for one instant!"
Officer: "Then you went more than 89 for an entire interval of time!"

Example 2.10. A constant function $f(x) = k$ is continuous at every point c in its domain. Approximate knowledge of c is sufficient for approximate knowledge of $f(c)$ because *all* inputs have the same output, k. As you can see in Fig. 2.11, for every x in the domain, $f(x)$ falls within ε of $f(c)$. No function can be more continuous than that!

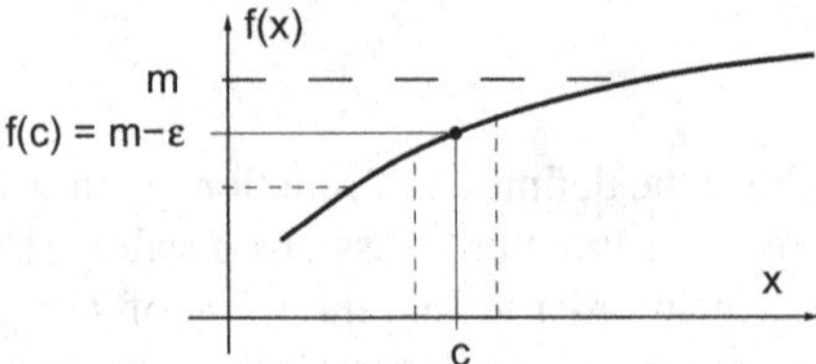

Fig. 2.10 If f is continuous, there will be an entire interval around c in which f is less than m

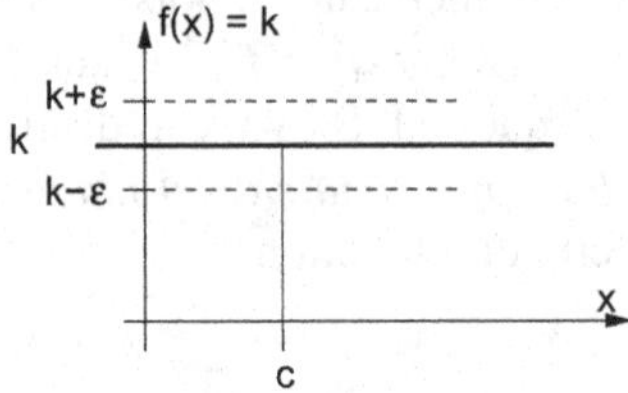

Fig. 2.11 A constant function

Example 2.11. The identity function $f(x) = x$ is continuous at every point c. Because $f(c) = c$, it is clear that approximate knowledge of c is sufficient to determine approximate knowledge of $f(c)$! Figure 2.12 shows that the definition for continuity is satisfied by letting $\delta = \varepsilon$.

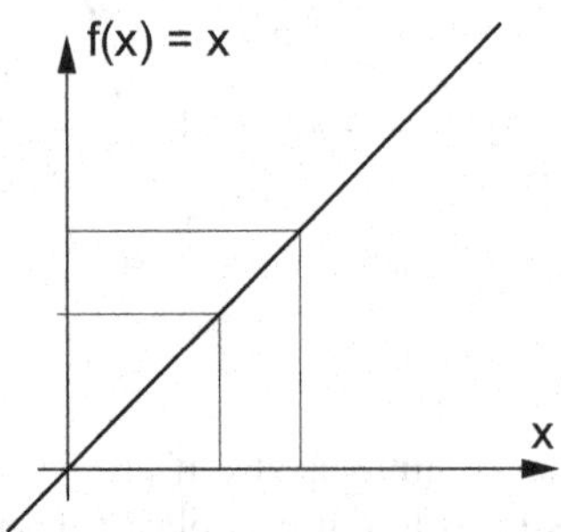

Fig. 2.12 For $f(x) = x$, $\delta = \varepsilon$ will do

A function can be continuous at some points in its domain but not at others.

Example 2.12. The graph of f in Fig. 2.3 shows the IRS 2010 tax rates for single filers. The rate is constant near 82,000. Small changes in income do not change the tax rate near 82,000. Thus f is continuous at 82,000. However, at 82,400, the situation is very different. Knowing that one's income is approximately 82,400 is not sufficient knowledge to determine the tax rate. Near 82,400, small changes in income result in very different tax rates. This is exactly the kind of outcome that continuity prohibits.

Inequalities and absolute values can be used to rewrite the definition of continuity at a point:

Restated definition. We say that a function f is *continuous at* c when:
for any tolerance $\varepsilon > 0$, there is a precision $\delta > 0$ such that

$$|f(x) - f(c)| < \varepsilon$$

whenever

$$|x - c| < \delta.$$

The precision δ depends on the tolerance ε.

2.2a Continuity at a Point Using Limits

The concept of the limit of a function gives another way to define continuity at a point.

Definition 2.4. The *limit* of a function $f(x)$ as x tends to c is L,

$$\lim_{x \to c} f(x) = L,$$

when:
for any tolerance $\varepsilon > 0$, there is a precision $\delta > 0$ such that $f(x)$ differs from L by less than ε whenever x differs from c by less than δ, $x \neq c$.

By comparing the definitions of limit as x tends to c and continuity at c, we find a new way to define continuity of f at c.

Alternative definition. We say that a function f is *continuous at* c when:

$$\lim_{x \to c} f(x) = f(c).$$

If f is not continuous at c, we say that f is discontinuous at c.

The limit of $f(x)$ as x tends to c can be completely described in terms of the limits of sequences of numbers. In fact, in evaluating $\lim_{x \to c} f(x)$, we often take a sequence of numbers $x_1, x_2, \ldots, x_n, \ldots$ that tend to c and we see whether the sequence

$f(x_1), f(x_2), \ldots, f(x_n), \ldots$ tends to some number L. In order for $\lim_{x \to c} f(x)$ to exist, we need to know that all sequences $\{x_i\}$ that tend to c result in sequences $\{f(x_i)\}$ that tend to L. In Problem 2.11, we ask you to explore the connection between the limit of a function at a point and limits of sequences of numbers. It will help you see why the next two theorems follow from the laws of arithmetic and the squeeze theorem, Theorems 1.6 and 1.7, for convergent sequences.

Theorem 2.1. *If* $\lim_{x \to c} f(x) = L_1$, $\lim_{x \to c} g(x) = L_2$, *and* $\lim_{x \to c} h(x) = L_3 \neq 0$, *then*

(a) $\lim_{x \to c} \left(f(x) + g(x)\right) = L_1 + L_2$,
(b) $\lim_{x \to c} \left(f(x)g(x)\right) = L_1 L_2$, *and*
(c) $\lim_{x \to c} \dfrac{f(x)}{h(x)} = \dfrac{L_1}{L_3}$.

Theorem 2.2. Squeeze theorem. *If*

$$f(x) \leq g(x) \leq h(x)$$

for all x in an open interval containing c, except possibly at $x = c$, and if $\lim_{x \to c} f(x) = \lim_{x \to c} h(x) = L$, *then* $\lim_{x \to c} g(x) = L$.

Combining Theorem 2.1 and the limit definition of continuity, one can prove the next theorem, as we ask you to do in Problem 2.12.

Theorem 2.3. *Suppose f, g, and h are continuous at c, and $h(c) \neq 0$. Then $f + g$, fg, and $\dfrac{f}{h}$ are continuous at c.*

We have noted before that any constant function, and the identity function, are continuous at each point c. According to Theorem 2.3, products and sums built from these functions are continuous at each c. Every polynomial

$$p(x) = a_n x^n + a_{n-1} x^{n-1} + \cdots + a_0$$

can be constructed by taking sums and products of functions that are continuous at c. This shows that polynomials are continuous at each c. It also follows from the theorem that a rational function $\dfrac{p(x)}{q(x)}$ is continuous at each number c for which $q(c) \neq 0$.

Example 2.13. Examples 2.10 and 2.11 explain why the constant function 3 and the function x are continuous. So according to Theorem 2.3, the rational function $f(x) = x^2 - \frac{1}{x} - 3 = \frac{x^3 - 1 - 3x}{x}$ is continuous at every point except 0.

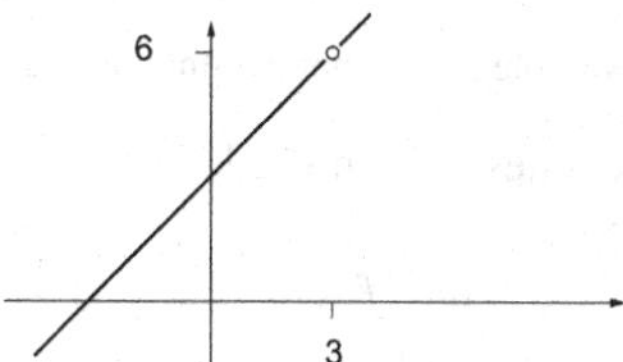

Fig. 2.13 The graph of $f(x) = \frac{x^2-9}{x-3}$

Sometimes a function is undefined at a point c, but the limit of $f(x)$ as x tends to c exists. For example, let

$$f(x) = \frac{x^2 - 9}{x - 3}.$$

Then f is not defined at 3. Notice, however, that

$$\text{for } x \neq 3, \qquad f(x) = \frac{x^2 - 9}{x - 3} = x + 3.$$

The graph of f looks like a straight line with a small hole at the point $x = 3$. (See Fig. 2.13.) The functions $\frac{x^2 - 9}{x - 3}$ and $x + 3$ are quite different at $x = 3$, but they are equal when $x \neq 3$. This means that their limits are the same as x tends to 3:

$$\lim_{x \to 3} \frac{x^2 - 9}{x - 3} = \lim_{x \to 3} (x + 3) = 6.$$

Example 2.14. Let $d(x)$ be defined as follows:

$$d(x) = \begin{cases} x & \text{for } x \leq 1, \\ x - 2 & \text{for } 1 < x. \end{cases}$$

Then d is not continuous at $x = 1$, because $d(1)$ equals 1, yet for x greater than 1 and no matter how close to 1, $d(x)$ is negative. A negative number is not close to 1. See Fig. 2.14.

It is useful to have a way to describe the behavior of $f(x)$ as x approaches c from one side or the other. If $f(x)$ tends to L as x approaches c from the right, $c < x$, we say that the *right-hand limit* of f at c is L, and write

$$\lim_{x \to c+} f(x) = L.$$

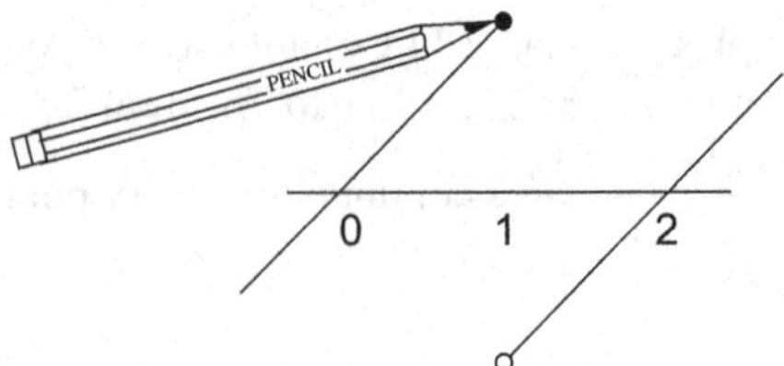

Fig. 2.14 The function $d(x)$ in Example 2.14 is not continuous at $x = 1$

If $f(x)$ tends to L as x approaches c from the left, $x < c$, we say that the *left-hand limit* of f at c is L, and write

$$\lim_{x \to c-} f(x) = L.$$

If $f(x)$ becomes arbitrarily large and positive as x tends to c, we write

$$\lim_{x \to c} f(x) = \infty,$$

and say that $f(x)$ tends to infinity as x tends to c. If $f(x)$ becomes arbitrarily large and negative as x tends to c, we write

$$\lim_{x \to c} f(x) = -\infty,$$

and say that $f(x)$ tends to minus infinity as x tends to c. Neither of these limits exists, but we use the notation to describe the behavior of the function near c. We also use the one-sided versions of these notations, as in Example 2.15.

Example 2.15. Let $f(x) = \dfrac{1}{x}$ for $x \neq 0$. Then $\lim_{x \to 0-} f(x) = -\infty$, $\lim_{x \to 0+} f(x) = \infty$.

It is also useful to have a way to describe one-sided continuity. If $\lim\limits_{x \to c-} f(x) = f(c)$, we say that f is *left continuous* at c. If $\lim\limits_{x \to c+} f(x) = f(c)$, we say that f is *right continuous* at c.

Example 2.16. The function d in Example 2.14 (see Fig. 2.14) is left continuous at 1, and not right continuous at 1:

$$\lim_{x \to 1-} d(x) = 1 = d(1), \qquad \lim_{x \to 1+} d(x) = -1 \neq d(1).$$

Left and right continuity give us a way to describe continuity on an interval that includes endpoints. For example, we say that f is *continuous on* $[a,b]$ if f is continuous at each c in (a,b) as well as right continuous at a, and left continuous at b.

2.2b Continuity on an Interval

Now we return to the question we considered at the start of this section: Is approximate knowledge of x sufficient for approximate knowledge of $f(x)$? We have seen that functions can be continuous at some points and not at others. The most interesting functions are the ones that are continuous at every point on *an interval* where they are defined.

Example 2.17. Let us analyze the continuity of the function $f(x) = x^2$ on the interval $[2,4]$. Let c be any point of this interval; how close must x be to c in order for $f(x)$ to differ from $f(c)$ by less than ε? Recall the identity

$$x^2 - c^2 = (x+c)(x-c).$$

On the left, we have the difference $f(x) - f(c)$ of two values of f. Since both x and c are between 2 and 4, we have $(x+c) \le 8$. It follows that

$$|f(x) - f(c)| = |x+c||x-c| \le 8|x-c|.$$

If we want x^2 to be within ε of c^2, it suffices to take x within $\frac{\varepsilon}{8}$ of c. That is, take $\delta = \frac{\varepsilon}{8}$ or less. This proves the continuity of f on $[2,4]$.

Example 2.18. In Chap. 1, we defined the number e through a sequence of approximations. Our intuition and experience tell us that we should get as good an approximation to e^2 as we desire by squaring a number that is close enough to e. But we do not need to rely on our intuition. Since e is between 2 and 4, the previous example shows that $f(x) = x^2$ is continuous at e. This means that if we want x^2 to be within $\varepsilon = \frac{1}{10^4}$ of e^2, it should suffice to take x within $\delta = \frac{\varepsilon}{8} = \frac{1}{8(10^4)}$ of e; in particular, $\delta < \frac{1}{10^5}$ should suffice. The list below shows squares of successively better decimal approximations to e. It confirms computationally what we proved theoretically.

$$\begin{aligned}
(2.7)^2 &= 7.29\\
(2.71)^2 &= 7.3441\\
(2.718)^2 &= 7.387524\\
(2.7182)^2 &= 7.38861124\\
(2.71828)^2 &= 7.3890461584\\
(2.718281)^2 &= 7.389051594961
\end{aligned}$$

Uniform Continuity. In Example 2.17, we showed that the difference between the squares of two numbers in $[2,4]$ will be within ε as long as the two numbers are within $\frac{\varepsilon}{8}$ of each other, no matter which two numbers in $[2,4]$ we are dealing with. Here is the general notion.

Definition 2.5. A function f is called *uniformly continuous* on an interval I if given any tolerance $\varepsilon > 0$, there is a precision $\delta > 0$ such that if x and z are in I and differ by less than δ, then $f(x)$ and $f(z)$ differ by less than ε.

Clearly, a function that is uniformly continuous on an interval is continuous at every point of that interval. It is a surprising mathematical fact that conversely, a function that is continuous at every point of a *closed* interval is uniformly continuous on that interval. We outline the proof of this theorem, Theorem 2.4, in Problem 2.21. Uniform continuity is a basic notion of calculus.

Theorem 2.4. *If a function f is continuous on $[a,b]$, then f is uniformly continuous on $[a,b]$.*

On a practical level, uniform continuity is a very helpful property for a function to have. When we evaluate a function with a calculator or computer, we round off the inputs, and we obtain outputs that are approximate. If f is uniformly continuous on $[a,b]$, then once we set a tolerance for the output, we can find a single level of precision for *all* the inputs in $[a,b]$, and the approximate outputs will be within the tolerance we have set.

2.2c Extreme and Intermediate Value Theorems

Next, we state and prove two key theorems about continuous functions on a closed interval.

Theorem 2.5. The intermediate value theorem. *If f is a continuous function on a closed interval $[a,b]$, then f takes on all values between $f(a)$ and $f(b)$.*

The theorem says in a careful way that the graph of f does not skip values.

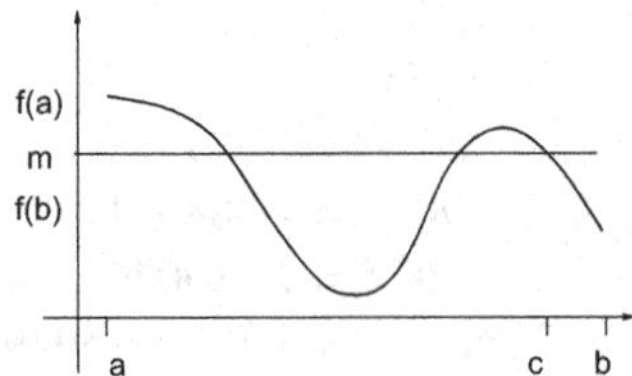

Fig. 2.15 The proof of the intermediate value theorem shows that there exists at least one number c between a and b at which $f(c) = m$

Proof. Let us take the case $f(a) > f(b)$; the opposite case can be treated analogously. Let m be any number between $f(a)$ and $f(b)$, and denote by V the set of those points x in the interval $a < x < b$ where $f(x)$ is greater than m. This set contains the point a, so it is not empty, and it is contained in $[a,b]$, so it is bounded. Denote by c the least upper bound of the set V. We claim that $f(c) = m$ (Fig. 2.15).

Suppose $f(c) < m$. Since f is continuous at c, there is a short interval to the left of c where $f(x) < m$ as well. These points x do not belong to V. And since c is an upper bound for V, no point to the right of c belongs to V. Therefore, every point of this short interval is an upper bound for V, a contradiction to c being the least upper bound.

On the other hand, suppose $f(c) > m$. Since $f(b)$ is less than m, c cannot be equal to b, and is strictly less than b. Since f is continuous at c, there is a short interval to the right of c where $f(x) > m$. But such points belong to V, so c could not be an upper bound for V.

Since according to the two arguments, $f(c)$ can be neither less nor greater than m, $f(c)$ must be equal to m. This proves the intermediate value theorem. □

Example 2.19. One use of the intermediate value theorem is in root-finding. Suppose we want to locate a solution to the equation

$$x^2 - \frac{1}{x} - 3 = 0.$$

Denote the left side by $f(x)$. With some experimentation we find that $f(1)$ is negative and $f(2)$ is positive. The function f is continuous on the interval $[1,2]$. By the intermediate value theorem, there is some number c between 1 and 2 such that $f(c) = 0$. In other words, f has a root in $[1,2]$.

Now let us bisect the interval into two subintervals, $[1,1.5]$ and $[1.5,2]$. We see that $f(1.5) = -1.416\ldots$ is negative, so f has a root in $[1.5,2]$. Bisecting again, we obtain $f(1.75) = -0.508\ldots$, which is again negative, so f has a root in $[1.75,2]$. Continuing in this manner, we can trap the root in an arbitrarily small interval.

Theorem 2.6. The extreme value theorem. *If f is a continuous function on a closed interval $[a,b]$, then f takes on both a maximum value and a minimum value at some points in $[a,b]$.*

One consequence of the extreme value theorem is that every function that is continuous on a closed interval is bounded. Although the extreme value theorem does not tell us how or where to find the bounds, it is still very useful.

Let us look at the graph of f and imagine a line parallel to the x-axis slid vertically upward until it just touches the graph of f at some last point of intersection, which is the maximum. Similarly, slide a line parallel to the x-axis vertically downward. The last point of intersection with the graph of f is the minimum value of f (Fig. 2.16).

We supplant now this intuitive argument by a mathematical proof of the existence of a maximum. The argument for a minimum is analogous.

Proof. Divide the interval $[a,b]$ into two closed subintervals of equal length. We compare the values of f on these two subintervals. It could be the case that there is a point on the first subinterval where the value of f is greater than at any point on the second subinterval. If there is no such point, then it must be the case that for every point x in the first subinterval there is a point z in the second subinterval where the value of f is at least as large as the value of f at x.

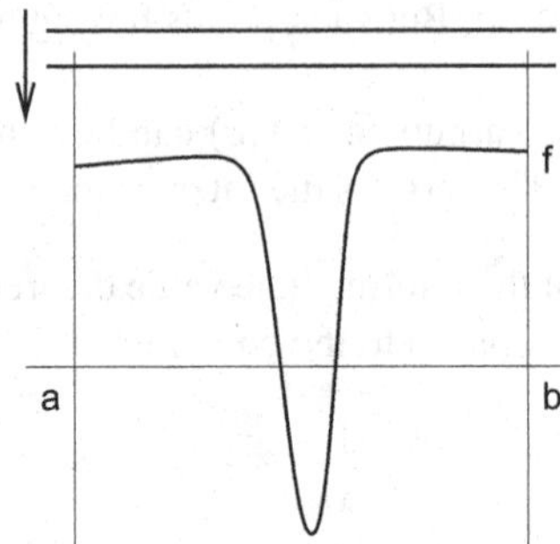

Fig. 2.16 A *horizontal line* moves down, seeking the minimum of a continuous function. The extreme value theorem guarantees that there is a last value where you can stop the moving line, keeping it in contact with the graph

In the first case, we choose the first subinterval, and in the second case, the second subinterval, and denote the chosen subinterval by I_1.

Key property of I_1: for every point x in $[a,b]$ but not in I_1, there is a point in I_1 where f is at least as large as $f(x)$.

Then we repeat the process of subdividing I_1 into two halves and choosing one of the halves according to the principle described above. Call the choice I_2. In this way, we construct a sequence of closed intervals $I_1, I_2, \ldots$, and so on. These intervals are *nested*; that is, the nth interval I_n is contained in the interval I_{n-1}, and its length is one-half of the length of I_{n-1}. Because of the way these intervals were chosen, for every point x in $[a,b]$ and every n, if x is not in I_n, then there is a point z in I_n where the value of f is at least as large as $f(x)$.

We appeal now to the nested interval Theorem 1.19, according to which the subintervals I_n have exactly one point in common; call this point c. We claim that the maximum value of the function f is $f(c)$. For suppose, to the contrary, that there is a point x in $[a,b]$ where the value of f is greater than $f(c)$. Since f is continuous at c, there would be an entire interval $[c-\delta, c+\delta]$ of numbers around c where f is less than $f(x)$. Since the lengths of the intervals I_n tend to zero, it follows that for n large enough, I_n would be contained in the interval $[c-\delta, c+\delta]$, so the value of f at every point of I_n would be smaller than $f(x)$. We can also take n sufficiently large that x is not in I_n. But this contradicts the key property of the intervals I_n established above. □

The extreme value theorem can be extended to open intervals in two special cases.

Corollary 2.1. *If f is continuous on an open interval (a,b) and $f(x)$ tends to infinity as x tends to each of the endpoints, then f has a minimum value at some point in (a,b).*

Similarly, if $f(x)$ tends to minus infinity as x tends to each of the endpoints, then f has a maximum value at some point in (a,b).

We invite you to prove this result in Problem 2.18.

Problems

2.8. Evaluate the following limits.

(a) $\lim_{x\to 4}\left(2x^3+3x+5\right)$
(b) $\lim_{x\to 0}\dfrac{x^2+2}{x^3-7}$
(c) $\lim_{x\to 5}\dfrac{x^2-25}{x-5}$

2.9. Evaluate the following limits.

(a) $\lim_{x\to 0}\dfrac{x^3-9x}{x^2+3x}$
(b) $\lim_{x\to -3}\dfrac{x^3-9x}{x^2+3x}$
(c) $\lim_{x\to 1}\dfrac{x^3-9x}{x^2+3x}$

2.10. Let $f(x)=\dfrac{|x|}{x}$ when $x\neq 0$, and $f(0)=1$.

(a) Sketch the graph of f.
(b) Is f continuous on $[0,1]$?
(c) Is f continuous on $[-1,0]$?
(d) Is f continuous on $[-1,1]$?

2.11. The limit of a function can be completely described in terms of the limits of sequences. To do this, show that these two statements are true:

(a) If $\lim_{x\to c} f(x)=L$ and x_n is any sequence tending to c, then $\lim_{n\to\infty} f(x_n)=L$.
(b) If $\lim_{n\to\infty} f(x_n)=L$ for every sequence x_n tending to c, then $\lim_{x\to c} f(x)=L$.

Conclude that in the discussion of continuity, $\lim_{x\to c} f(x)=f(c)$ is equivalent to $\lim_{n\to\infty} f(x_n)=f(c)$ for every sequence x_n tending to c.

2.12. Suppose that functions f, g, and h are each defined on an interval containing c, that they are continuous at c, and that $h(c) \neq 0$. Show that $f+g$, fg, and $\dfrac{f}{h}$ are continuous at c.

2.13. Let $f(x) = \dfrac{x^{32}+x^{10}-7}{x^2+2}$ on the interval $[-20,120]$. Is f bounded? Explain.

2.14. Show that the equation

$$\frac{x^6+x^4-1}{x^2+1} = 2$$

has a solution on the interval $[-2,2]$.

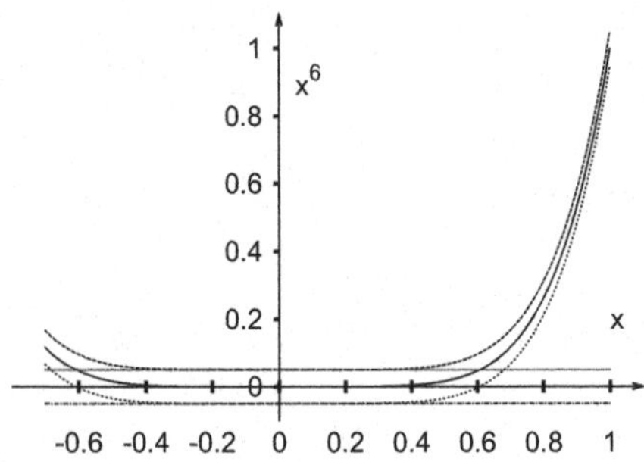

Fig. 2.17 Graphs are shown for x^6, $x^6+\frac{1}{20}$, $x^6-\frac{1}{20}$, and for the constant functions $\pm\frac{1}{20}$ on $[-0.7,1]$. See Problem 2.15

2.15. In Fig. 2.17, estimate the largest interval $[a,b]$ such that x^6 and 0 differ by less than $\frac{1}{20}$ on $[a,b]$.

2.16. Let $f(x) = \dfrac{1}{x}$. Show that on the interval $[3,5]$, $f(x)$ and $f(c)$ do not differ by more than $\frac{1}{9}|x-c|$. Copy the definition of uniform continuity onto your paper, and then explain why f is uniformly continuous on $[3,5]$.

2.17. You plan to compute the squares of numbers between 9 and 10 by squaring truncations of their decimal expansions. If you truncate after the eighth place, will this ensure that the outputs are within 10^{-7} of the true value?

2.18. Prove the first statement in Corollary 2.1, that if f is continuous on (a,b) and $f(x)$ tends to infinity as x tends to each of a and b, then f has a minimum value at some point in (a,b).

2.19. Explain why the function $f(x) = x^2 - \dfrac{1}{x} - 3 = \dfrac{x^3-1-3x}{x}$ is uniformly continuous on every interval $[a,b]$ not containing 0.

2.20. Let $f(x) = 3x+5$.

(a) Suppose each domain value x is rounded to x_{approx} and $|x - x_{\text{approx}}| < \frac{1}{10^m}$. How close is $f(x_{\text{approx}})$ to $f(x)$?
(b) If we want $|f(x) - f(x_{\text{approx}})| < \frac{1}{10^7}$, how close should x_{approx} be to x?
(c) On what interval can you use the level of precision you found in part (b)?

2.21. Explain the following steps to show that a function that is continuous at every point of a closed interval is uniformly continuous on that interval. It will be a proof by contradiction, so we assume that f is continuous, but not uniformly continuous, on $[a,b]$.

(a) There must be some $\varepsilon > 0$ and for each $n = 1,2,3,\ldots$, two numbers x_n, y_n in $[a,b]$ for which $|x_n - y_n| < \frac{1}{n}$ and $|f(x_n) - f(y_n)| \geq \varepsilon$.
(b) Use Lemma 1.1 and monotone convergence to show that a subsequence of the x_n (that is, a sequence consisting of some of the x_n) converges to some number c in $[a,b]$.
(c) To simplify notation, we can now take the symbols x_n to mean the subsequence, and y_n corresponding. Use the fact that $|x_n - y_n| < \frac{1}{n}$ to conclude that the y_n also converge to c.
(d) Use continuity of f and Problem 2.11 to show that $\lim_{n\to\infty} f(x_n) = f(c)$.
(e) Show that $\lim_{n\to\infty} f(x_n) = \lim_{n\to\infty} f(y_n)$, and that this contradicts our assumption that $|f(x_n) - f(y_n)| \geq \varepsilon$.

2.3 Composition and Inverses of Functions

In Sect. 2.1, we showed how to build new functions out of two others by adding, multiplying, and dividing them. In this section, we describe another way.

2.3a Composition

We start with a simple example:

A rocket is launched vertically from point L. The distance (in kilometers) of the rocket from the launch point at time t is $h(t)$. An observation post O is located 1 km from the launch site (Fig. 2.18). To determine the distance d of the rocket from the observation post as a function of time, we can use the Pythagorean theorem to express d as a function of h,

$$d(h) = \sqrt{1+h^2}.$$

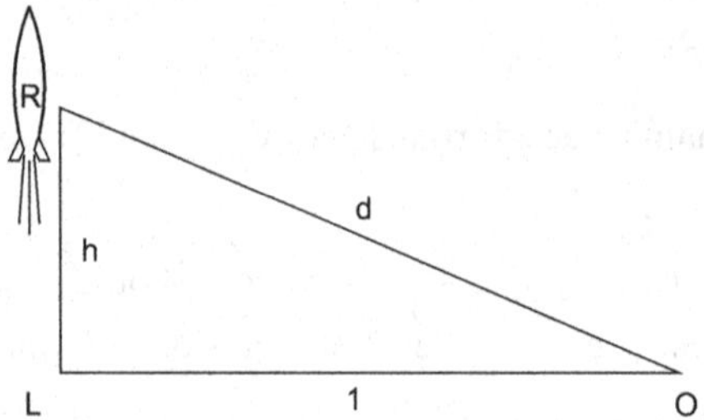

Fig. 2.18 Tracking the rocket from the observation post

Therefore, the distance from R to O at time t is

$$d(h(t)) = \sqrt{1 + (h(t))^2}.$$

The process that builds a new function in this way is called *composition*; the resulting function is called the *composition* of the two functions.

Definition 2.6. Let f and g be two functions, and suppose that the range of g is included in the domain of f. Then the composition of f with g, denoted by $f \circ g$, is defined by

$$(f \circ g)(x) = f(g(x)).$$

We also say that we have *composed* the functions.

The construction is well described by Fig. 2.19.

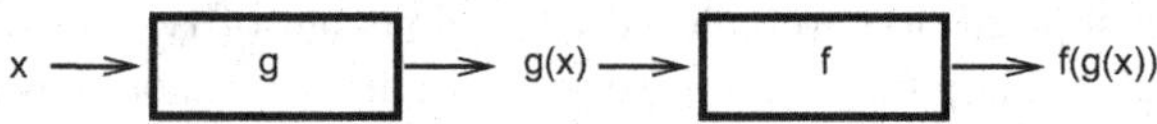

Fig. 2.19 Composition of functions, using the box picture of Fig. 2.4

Example 2.20. Let g and f be the linear functions $y = g(x) = 2x + 3$, and $z = f(y) = 3y + 1$. The composition $z = f(g(x)) = 3(2x + 3) + 1 = 6x + 10$ is illustrated in Fig. 2.20.

We saw in Fig. 2.8 that the linear function $mx + b$ stretches every interval by a factor of $|m|$. In Fig. 2.20, we see that when the linear functions are composed, these stretching factors are multiplied.

Example 2.21. The effect of composing a function f with $g(x) = x + 1$ depends on the order of composition. For example $f(g(x)) = f(x + 1)$ shifts the graph of

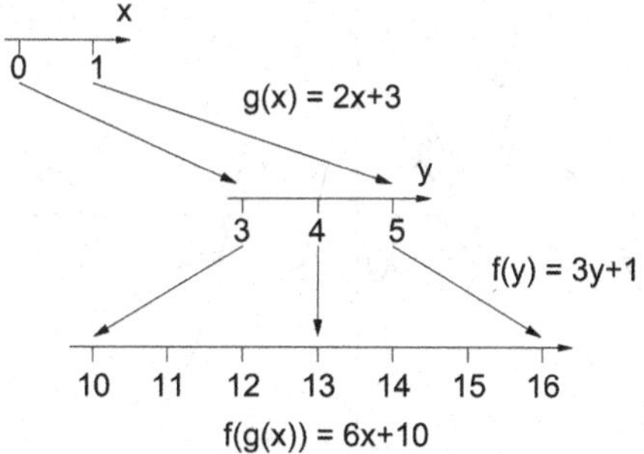

Fig. 2.20 A composition of two linear functions

f one unit to the left, since the output of f at x is the same as the output of $f \circ g$ at $x-1$. On the other hand, $g(f(x)) = f(x)+1$ shifts the graph of f up one unit. See Fig. 2.21.

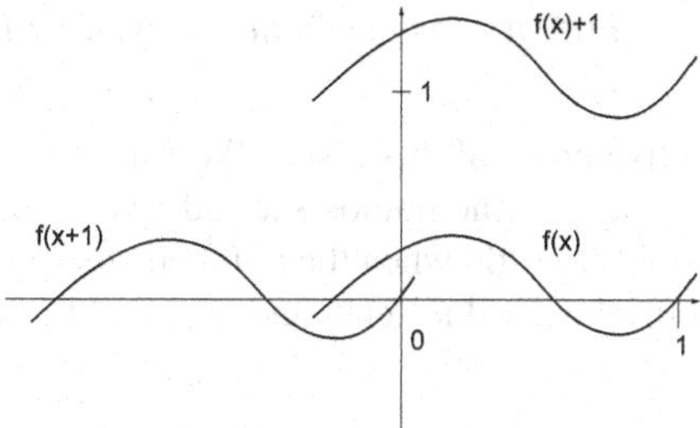

Fig. 2.21 Composition with the translation $x+1$, in Example 2.21. It makes a difference which function is applied first

Example 2.22. Let $h(x) = 3x$. The graph of $f(h(x))$ looks as though the domain of f has been compressed by a factor of 3. This is because the output of f at x is the same as the output of $f \circ h$ at $\frac{x}{3}$. If we compose f and h in the opposite order, the graph of $h(f(x)) = 3f(x)$ is the graph of f stretched by a factor of three in the vertical direction. See Fig. 2.22.

Example 2.23. Let $h(x) = -x$. The graph of $h(f(x)) = -f(x)$ is the reflection of the graph of f across the x-axis, while the graph of $f(h(x)) = f(-x)$ is the reflection of the graph of f across the y-axis.

Example 2.24. If $f(x) = \dfrac{1}{x+1}$ and $g(x) = x^2$, then

$$(f \circ g)(x) = \frac{1}{x^2+1} \quad \text{and} \quad (g \circ f)(x) = \left(\frac{1}{x+1}\right)^2 = \frac{1}{x^2+2x+1}.$$

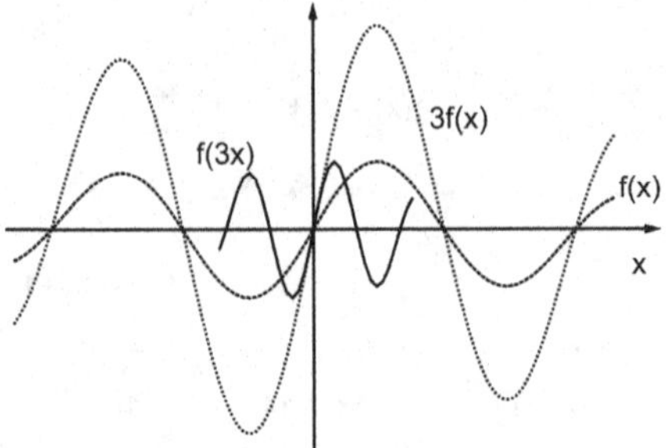

Fig. 2.22 Composition with multiplication $3x$ results in stretching or compressing the graph. See Example 2.22

Notice that $f \circ g$ and $g \circ f$ are quite different functions. Thus *composition is not a commutative operation*. This is not surprising: using the output of g as input for f is quite different from using the output of f as input for g.

Theorem 2.7. *The composition of two continuous functions is continuous.*

Proof. We give an intuitive proof of this result. We want to compare the values of $f(g(x))$ with those of $f(g(z))$ as the numbers x and z vary. Since f is continuous, these values will differ by very little when the numbers $g(x)$ and $g(z)$ are close. But since g is also continuous, those values $g(x)$ and $g(z)$ will be close whenever x and z are sufficiently close. □

Here is a related theorem about limits, which we show you how to prove in Problem 2.33.

Theorem 2.8. *Suppose $f \circ g$ is defined on an interval containing c, that $\lim_{x \to c} g(x) = L$, and that f is continuous at L. Then $\lim_{x \to c}(f \circ g)(x) = f(L)$, that is,*

$$\lim_{x \to c} f\big(g(x)\big) = f\big(\lim_{x \to c} g(x)\big).$$

2.3b Inverse Functions

We look at some examples of compositions of functions that undo each other.

Example 2.25. For $f(x) = 2x + 3$ and $g(x) = \frac{1}{2}x - \frac{3}{2}$, we see that

$$f(g(x)) = 2\left(\frac{1}{2}x - \frac{3}{2}\right) + 3 = x \quad \text{and } g(f(x)) = \frac{1}{2}(2x+3) - \frac{3}{2} = x.$$

Example 2.26. Let $f(x) = \dfrac{1}{x+1}$ when $x \neq -1$, and $g(x) = \dfrac{1-x}{x}$ when $x \neq 0$. Then if $x \neq 0$, we have

$$f(g(x)) = \frac{1}{\left(\frac{1-x}{x}\right)+1} = \frac{1}{\frac{1-x}{x}+\frac{x}{x}} = x.$$

You may also check that when $x \neq -1$, we have $g(f(x)) = x$.

In both of the examples above, we see that f applied to the output of g returns the input of g, and similarly, g applied to the output of f returns the input of f. We may ask the following question about a function: if we know the output, can we determine the input?

Definition 2.7. If a function g has the property that different inputs always lead to different outputs, i.e., if $x_1 \neq x_2$ implies $g(x_1) \neq g(x_2)$, then we can determine its input from the output. Such a function g is called *invertible*; its *inverse* f is defined in words: the domain of f is the range of g, and $f(y)$ is defined as the number x for which $g(x) = y$. We denote the inverse of g by g^{-1}.

By the way in which it is defined, we see that g^{-1} undoes, or reverses, g: it works backward from the output of g to the input. If g is invertible, then g^{-1} is also invertible, and its inverse is g. Furthermore, the composition of a function and its inverse, in either order, is the identity function:

$$(g \circ g^{-1})(y) = y \quad \text{and} \quad (g^{-1} \circ g)(x) = x.$$

Here is another example:

Example 2.27. Let $g(x) = x^2$, and restrict the domain of g to be $x \geq 0$. Since the squares of two different nonnegative numbers are different, g is invertible. Its inverse is $g^{-1}(x) = \sqrt{x}$. Note that if we had defined $g(x) = x^2$ and taken its domain to be all numbers, not just the nonnegative ones, then g would not have been invertible, since $(-x)^2 = x^2$. Thus, invertibility depends crucially on what we take to be the domain of the function (Fig. 2.23).

Monotonicity. The graph of a function can be very helpful in determining whether the function is invertible. If lines parallel to the x-axis intersect the graph in at most one point, then different domain values are assigned different range values, and the function is invertible. Two kinds of functions that pass this "horizontal line test" are the *increasing* functions and the *decreasing* functions.

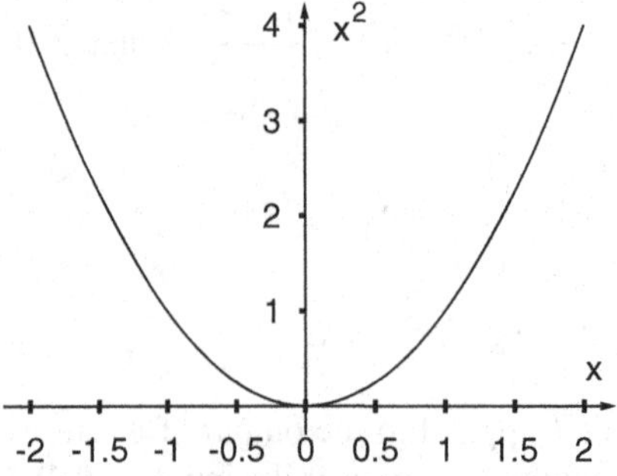

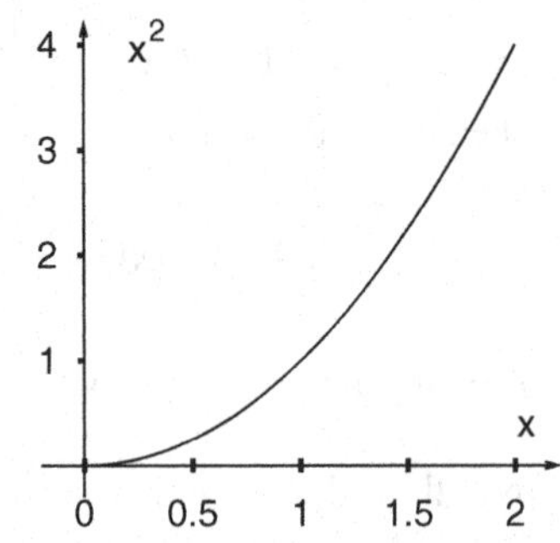

Fig. 2.23 *Left*: x^2 is plotted with the domain all numbers. *Right*: the domain is the positive numbers. Only one of these functions is invertible

Definition 2.8. An *increasing* function is one for which $f(a) < f(b)$ whenever $a < b$. A *decreasing* function is one for which $f(a) > f(b)$ whenever $a < b$. A *nondecreasing* function is one for which $f(a) \leq f(b)$ whenever $a < b$. A *nonincreasing* function is one for which $f(a) \geq f(b)$ whenever $a < b$.

Example 2.28. Suppose f is increasing and $f(x_1) > f(x_2)$. Which of the following is true?

(a) $x_1 = x_2$
(b) $x_1 > x_2$
(c) $x_1 < x_2$

Item (a) is certainly not true, because then we would have $f(x_1) = f(x_2)$. Item (b) is consistent with f increasing, but this does not resolve the question. If item (c) were true, then $f(x_1) < f(x_2)$, which is not possible. So it is (b) after all.

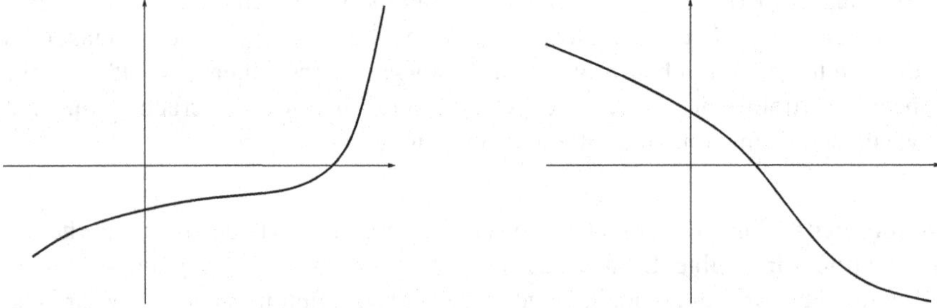

Fig. 2.24 Two graphs of monotonic functions. *Left*: increasing, *Right*: decreasing

Figure 2.24 shows the graphs of an increasing function and a decreasing function. Both pass the horizontal line test and both are invertible.

Definition 2.9. Functions that are either increasing or decreasing are called *strictly monotonic*. Function that are either nonincreasing or nondecreasing are called *monotonic*.

If f is strictly monotonic, then the graph of its inverse is simply the reflection of the graph of f across the line $y = x$ (Fig. 2.25).

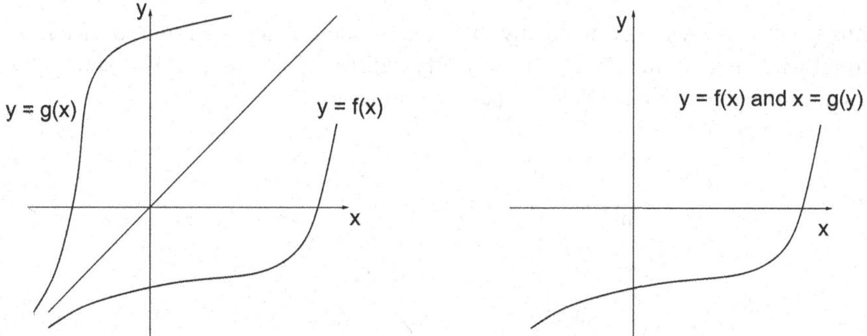

Fig. 2.25 *Left*: graphs of an increasing function f and its inverse g. *Right*: if you write $f(x) = y$ and $x = g(y)$, then the graph of $f(x) = y$ is also the graph of $g(y) = x$

The Inversion Theorem. The graphs suggest the following theorem:

Theorem 2.9. Inversion theorem. *Suppose that f is a continuous and strictly monotonic function defined on an interval $[a,b]$. Then its inverse g is a continuous strictly monotonic function defined on the closed interval between $f(a)$ and $f(b)$.*

Proof. A strictly monotonic function is invertible, because different inputs always result in different outputs. The inverse is strictly monotonic, as we ask you to show in Problem 2.30.

What remains to be shown is that the domain of the inverse function is precisely the closed interval between $f(a)$ and $f(b)$, no more no less, and that f^{-1} is continuous. According to the intermediate value theorem, for every m between $f(a)$ and $f(b)$, there is a number c such that $m = f(c)$. Thus every number between $f(a)$ and $f(b)$ is in the domain of the inverse function. On the other hand, the value $f(c)$ of a strictly monotonic function at the point c between a and b must lie between $f(a)$ and $f(b)$. This shows that the domain of f^{-1} is precisely the closed interval between $f(a)$ and $f(b)$.

Next, we show that f^{-1} is continuous. Let ε be any tolerance. Divide the interval $[a,b]$ into n subintervals of length less than $\frac{\varepsilon}{2}$, with endpoints $a = a_0, a_1, \ldots, a_n = b$.

The values $f(a_i)$ divide the range of f into an equal number of subintervals. Denote by δ the length of the smallest of these. See Fig. 2.26. Let y_1 and y_2 be numbers in the range that are within δ of each other. Then y_1 and y_2 are in either the same or adjacent subintervals of the range. Correspondingly, $f^{-1}(y_1)$ and $f^{-1}(y_2)$ lie in the same or adjacent subintervals of $[a,b]$. Since the lengths of the subintervals of $[a,b]$ were made less than $\frac{\varepsilon}{2}$, we have

$$|f^{-1}(y_1) - f^{-1}(y_2)| < \varepsilon.$$

Thus we have shown that given any tolerance ε, there is a δ such that if y_1 and y_2 differ by less than δ, then $f^{-1}(y_1)$ and $f^{-1}(y_2)$ differ by less than ε. This shows that f is uniformly continuous on $[a,b]$, hence continuous. □

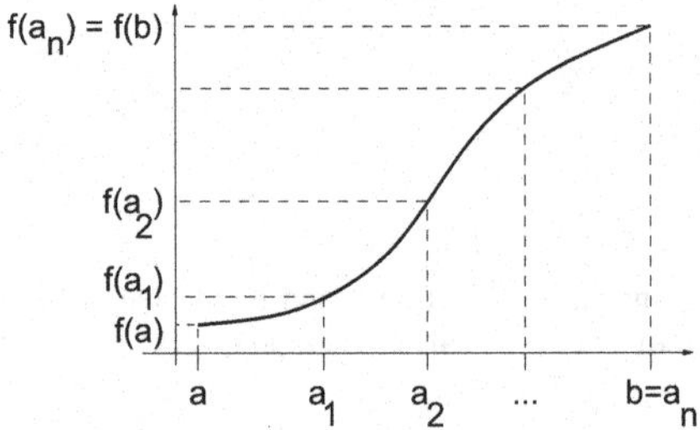

Fig. 2.26 The inverse of a continuous strictly monotonic function is continuous

As an application of the inversion theorem, take $f(x) = x^n$, n any positive integer. Then f is continuous and increasing on every interval $[0,b]$, so it has an inverse g. The value of g at a is the nth root of a and is written with a fractional exponent:

$$g(a) = a^{1/n}.$$

By the inversion theorem, the nth-root function is continuous and strictly monotonic. Then powers of such functions, such as $x^{2/3} = (x^{1/3})^2$, are continuous and strictly monotonic on $[0,b]$. Figure 2.27 shows some of these functions and their inverses.

We shall see later that many important functions can be defined as the inverse of a strictly monotonic continuous function and that we can make important deductions about a function f from properties of its inverse f^{-1}.

Problems

2.22. Find the inverse function of $f(x) = x^5$. Sketch the graphs of f and f^{-1}.

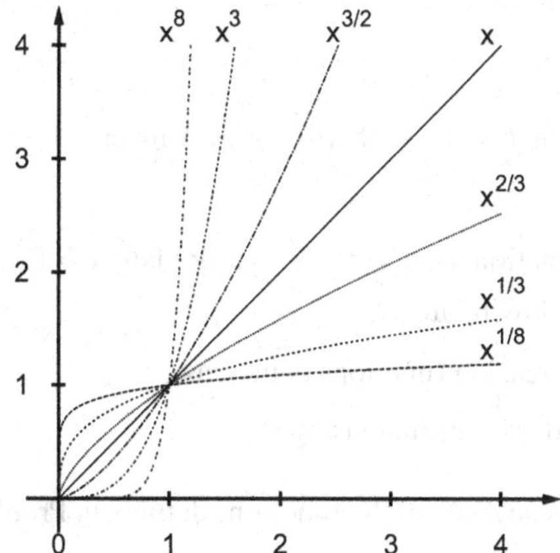

Fig. 2.27 The power functions

2.23. The volume of water V in a bottle is a function of the height H of the water, say $V = f(H)$. See Fig. 2.28. Similarly, the height of the water is a function of the volume of water in the bottle, say $H = g(V)$. Show that f and g are inverse functions.

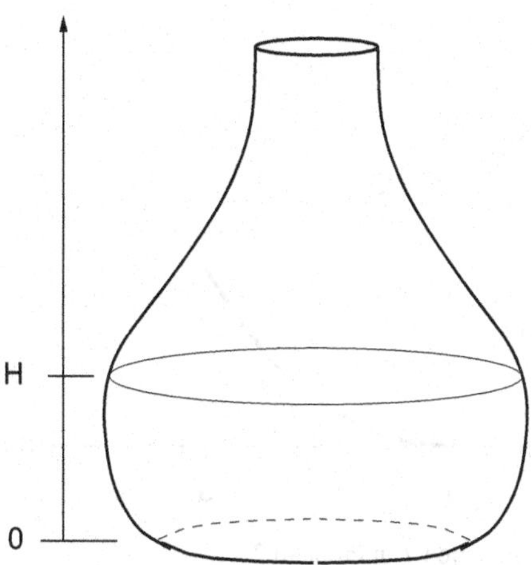

Fig. 2.28 A bottle of water for Problem 2.23

2.24. Let $f(x) = x$, $g(x) = x^2$, $h(x) = x^{1/5}$, and $k(x) = x^2 + 5$. Find formulas for the compositions

(a) $(h \circ g)(x)$
(b) $(g \circ h)(x)$
(c) $(f \circ g)(x)$
(d) $(k \circ h)(x)$

(e) $(h \circ k)(x)$
(f) $(k \circ g \circ h)(x)$

2.25. Is there a function $f(x) = x^a$ that is its own inverse function? Is there more than one such function?

2.26. Show that the function $f(x) = x - \frac{1}{x}$, on domain $x > 0$, is increasing by explaining each of the following items.

(a) The sum of two increasing functions is increasing.

(b) The functions x and $-\frac{1}{x}$ are increasing.

2.27. Tell how to compose some of the functions defined in Problem 2.24 to produce the functions

(a) $(x^2+5)^2+5$
(b) $(x^2+5)^2$
(c) x^4+5

2.28. The graph of a function f on $[0,a]$ is given in Fig. 2.29. Use the graph of f to sketch the graphs of the following functions.

(a) $f(x-a)$
(b) $f(x+a)$
(c) $f(-x)$
(d) $-f(x)$
(e) $f(-(x-a))$

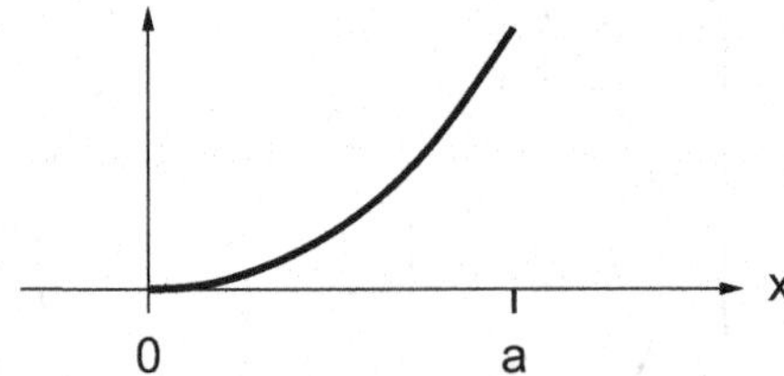

Fig. 2.29 The graph of the function f in Problem 2.28

2.29. Use the intermediate value theorem to show that the equation

$$\sqrt{x^2+1} = \sqrt[3]{x^5+2}$$

has a solution in $[-1,0]$.

2.30. (a) Show that the inverse of an increasing function is increasing. (b) Then state the analogous result for decreasing functions.

2.31. (a) Suppose f is increasing. Is $f \circ f$ increasing? Give a proof or a counterexample. (b) Suppose f is decreasing. Is $f \circ f$ decreasing? Give a proof or a counterexample.

2.32. Assume that functions f and g are increasing. Is fg increasing? If so give a proof, and if not, explain why not.

2.33. Prove Theorem 2.8 by explaining the following.

(a) Given any $\varepsilon > 0$, there is a $\delta > 0$ such that if $|z - L| < \delta$, then $|f(z) - f(L)| < \varepsilon$.
(b) For the δ in part (a), there is an $\eta > 0$ such that if $|x - c| < \eta$, then $|g(x) - L| < \delta$.
(c) Given any $\varepsilon > 0$, there is an $\eta > 0$ such that if $|x - c| < \eta$, then $|f(g(x)) - f(L)| < \varepsilon$.
(d) $\lim_{x \to c} f(g(x)) = f(L)$.

2.4 Sine and Cosine

It is often asserted that the importance of trigonometry lies in its usefulness for surveying and navigation. Since the proportion of our population engaged in these pursuits is rather small, one wonders what kind of stranglehold surveyors and navigators have over professional education to be able to enforce the universal teaching of this abstruse subject. Or is it merely inertia? The answer, of course, is that the importance of trigonometry lies elsewhere: in the description of *rotation* and *vibration*. It is an astonishing fact of mathematical physics that the vibration of as diverse a collection of objects as:

springs
strings
airplane wings
steel beams
light beams
and water streams
building sways
ocean waves
and sound waves ...

and many others are described in terms of trigonometric functions. That such diverse phenomena can be treated with a common tool is one of the most striking successes of calculus. Some simple and some not so simple examples will be discussed in the next chapters.

One also learns from older texts that there are six trigonometric functions:

sine, cosine, tangent, cotangent
secant, and cosecant.

This turns out to be a slight exaggeration. There are only two basic functions, sine and cosine; all the others can be defined in terms of them, when necessary.

Furthermore, sine and cosine are so closely related that each can be expressed in terms of the other; so one can say that there is really only one trigonometric function.

Geometric Definition. We shall describe the functions sine and cosine geometrically, using the circle of radius 1 in the Cartesian (x,y)-plane centered at the origin, which is called the *unit circle* (Fig. 2.30).

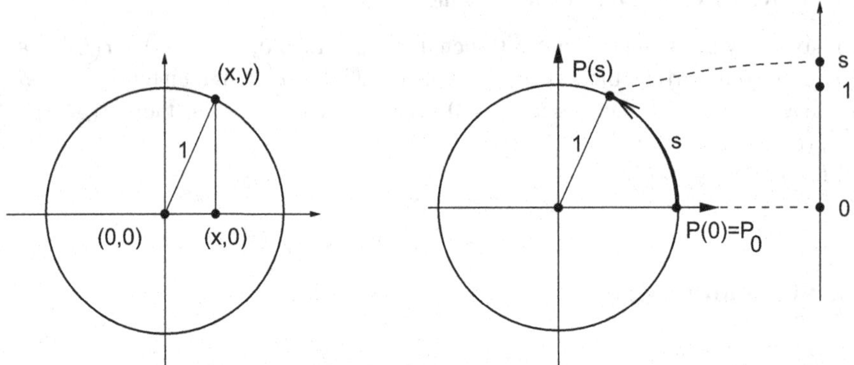

Fig. 2.30 *Left*: the unit circle. *Right*: measuring along the circumference using the same scale as on the axes is called radian measure

Let (x,y) be any point on the unit circle. The triangle with vertices $(0,0)$, $(x,0)$, and (x,y) is a right triangle. By the Pythagorean theorem,

$$x^2+y^2=1.$$

Let P_0 be the point $(1,0)$ on the unit circle. Let $P(s)$ be that point on the unit circle whose distance measured from P_0 counterclockwise along the arc of the unit circle is s.

You can imagine this distance along the arc with the aid of a very thin string of length s. Fasten one of its ends to the point P_0, and wrap the string counterclockwise around the circle. The other end of the string is at the point $P(s)$.

The two rays from the origin through the points P_0 and $P(s)$ form an angle. We define the size of this angle to be s, the length of the arc connecting P_0 and $P(s)$. Measuring along the circumference of the unit circle using the same scale as on the axes is called radian measure (Fig. 2.30). An angle of length 1 therefore has measure equal to one radian, and the radian measure of a right angle is $\dfrac{\pi}{2}$.

Definition 2.10. Denote the x- and y-coordinates of $P(s)$ by $x(s)$ and $y(s)$. We define

$$\cos s = x(s), \quad \sin s = y(s).$$

One immediate consequence of the definition is that $\cos s$ and $\sin s$ are continuous functions: The length of the chord between $P(s)$ and $P(s+\varepsilon)$ is less than ε. The differences Δx and Δy in the coordinates between $P(s)$ and $P(s+\varepsilon)$ are each less than the length of the chord. But these differences are also the changes in the cosine and sine:

$$|\Delta x| = |\cos(s+\varepsilon) - \cos s| < \varepsilon, \qquad |\Delta y| = |\sin(s+\varepsilon) - \sin s| < \varepsilon.$$

See Fig. 2.31.

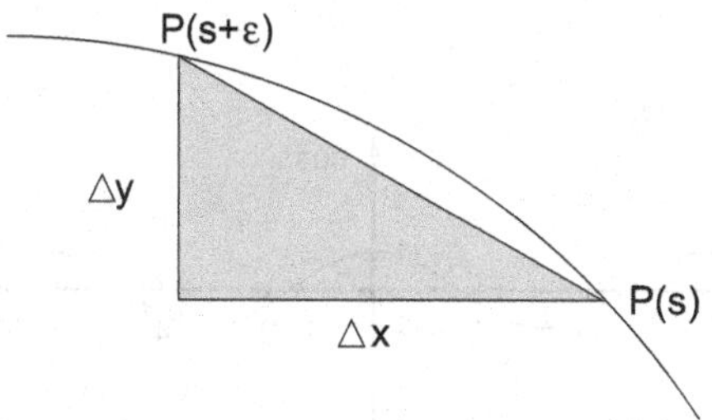

Fig. 2.31 A small arc of the unit circle with corresponding x and y increments. Observe that the x and y increments are smaller than the arc increment

We list this fact of continuity together with some other properties:

(a) The cosine and sine functions are continuous.
(b) $\cos^2 s + \sin^2 s = 1$. This is because the cosine and sine are the coordinates of a point of the unit circle, where $x^2 + y^2 = 1$.
(c) Since the circumference of the whole unit circle is 2π, when a string of length $s + 2\pi$ is wrapped around the unit circle in the manner described before, the endpoint $P(s+2\pi)$ coincides with the point $P(s)$. Therefore,

$$\cos(s+2\pi) = \cos s, \quad \sin(s+2\pi) = \sin s.$$

This property of the functions sine and cosine is called "periodicity," with period 2π. See Fig. 2.32.
(d) $\cos 0 = 1$, and the value $\cos s$ decreases to -1 as s increases to π. Then $\cos s$ increases again to 1 at $s = 2\pi$. $\sin 0 = 0$ and $\sin s$ also varies from -1 to 1.
(e) $P\left(\frac{\pi}{2}\right)$ lies one quarter of the circle from P_0. Therefore, $P\left(\frac{\pi}{2}\right) = (0,1)$, and

$$\cos\left(\frac{\pi}{2}\right) = 0, \qquad \sin\left(\frac{\pi}{2}\right) = 1.$$

(f) The point $P\left(\frac{\pi}{4}\right)$ is halfway along the arc between P_0 and $P\left(\frac{\pi}{2}\right)$. By symmetry, we see that $x\left(\frac{\pi}{4}\right) = y\left(\frac{\pi}{4}\right)$. By the Pythagorean theorem, $\left(x\left(\frac{\pi}{4}\right)\right)^2 + \left(y\left(\frac{\pi}{4}\right)\right)^2 = 1$. It follows that $(\cos\frac{\pi}{4})^2 = (\sin\frac{\pi}{4})^2 = \frac{1}{2}$, and so

$$\cos\left(\frac{\pi}{4}\right) = \sqrt{\frac{1}{2}}, \qquad \sin\left(\frac{\pi}{4}\right) = \sqrt{\frac{1}{2}}.$$

(g) For angles s and t, there are addition formulas

$$\cos(s+t) = \cos s \cos t - \sin s \sin t,$$
$$\sin(s+t) = \sin s \cos t + \cos s \sin t,$$

which will be discussed later.

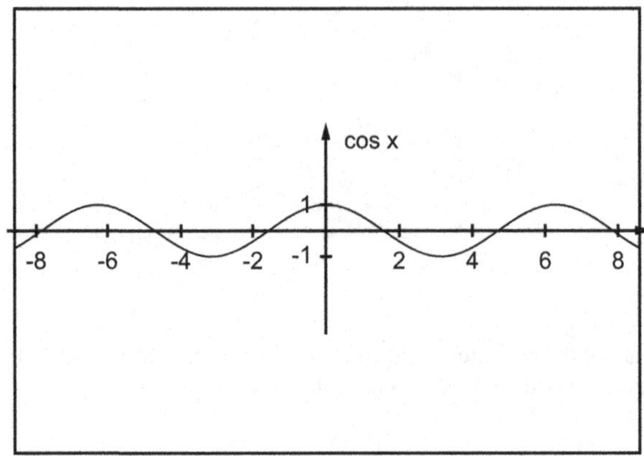

Fig. 2.32 Part of the graph of the cosine

Problems

2.34. On a sketch of the unit circle, mark the circumference at six equally spaced points. Are these subdivisions more, or less, than one radian each?

2.35. Which of the following pairs of numbers could be the cosine and sine of some angle?

(a) $(0.9, 0.1)$
(b) $(\sqrt{0.9}, \sqrt{0.1})$

2.36. Sketch the unit circle, and on it, mark the approximate location of points having angles of 1, 2, 6, 2π, and -0.6 from the horizontal axis.

2.37. The ancient Babylonians measured angles in degrees. They divided the full circle into 360 angles of equal size, each called one degree. So the size of a right angle in Babylonian units is 90 degrees. Since its size in modern units is $\frac{\pi}{2}$ radians, it follows that one radian equals $\dfrac{90}{\frac{\pi}{2}} = 57.295\ldots$ degrees. Let $c(x) = \cos\left(\dfrac{x}{57.295\ldots}\right)$ which is the cosine of an angle of x degrees. Sketch the graph of c as nearly as you can to scale, and explain how it differs from the graph of the cosine.

2.38. Which of the following functions are bounded, and which are bounded away from 0?

(a) $f(x) = \sin x$
(b) $f(x) = 5\sin x$
(c) $f(x) = \dfrac{1}{\sin x}$ for $x \neq n\pi$, $n = 0, \pm 1, \pm 2, \ldots$

2.39. A weight attached to a Slinky (a weak spring toy) oscillates up and down. Its position at time t is $y = 1 + 0.2\sin(3t)$ meters from the floor. What is the maximum height reached, and how much time elapses between successive maxima?

2.40. Use the intermediate value theorem to prove that the equation

$$x = \cos x$$

has a solution on the interval $[0, \frac{\pi}{2}]$.

2.41. Show that $\sin s$ is an increasing function on $[-\frac{\pi}{2}, \frac{\pi}{2}]$, and therefore has an inverse. Its inverse is denoted by $\sin^{-1}$.

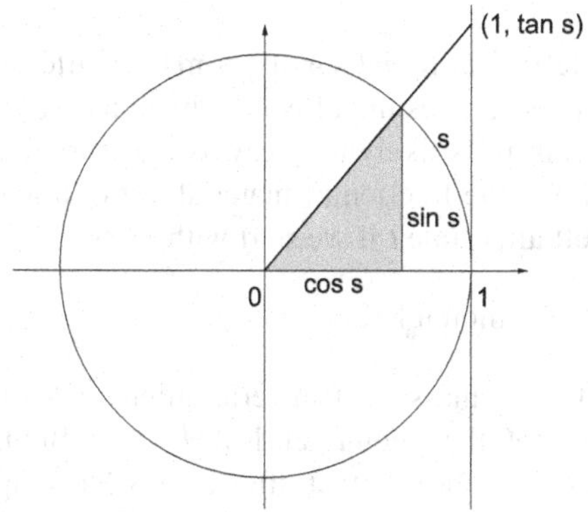

Fig. 2.33 The tangent of s. See Problems 2.42 and 2.43

2.42. Define the tangent function by $\tan s = \dfrac{\sin s}{\cos s}$ whenever the denominator is not 0. Refer to Fig. 2.33 to show that $\tan s$ is an increasing function on $(-\frac{\pi}{2}, \frac{\pi}{2})$. Show that $\tan x$ has a continuous inverse on $(-\infty, \infty)$. Its inverse is denoted by $\tan^{-1}$.

2.43. Set $z = \tan s$ and $y = \sin s$ in Fig. 2.33.

(a) Show that $\sin(\tan^{-1}(z)) = \dfrac{z}{\sqrt{1+z^2}}$.
(b) Express $\cos(\sin^{-1}(y))$ without using any trigonometric functions.

2.5 Exponential Function

We present two examples of naturally occurring functions f that arise in modeling growth and decay and satisfy the relation

$$f(t+s) = f(t)f(s). \tag{2.1}$$

We shall then show that all continuous functions satisfying this relation are exponential functions. Further natural examples of exponential functions are given in Chap. 10.

2.5a Radioactive Decay

Radioactive elements are not immutable. With the passage of time, they change into other elements. It is important to know how much of a given amount is left after time t has elapsed. To express this problem mathematically, we describe the decay by the following function:

Let $M(t)$ denote the fraction of material of a unit mass remaining after the elapse of time t. Assume that M is a continuous function of time, $M(0) = 1$, and that $0 < M(t) < 1$ for $t > 0$.

How much will be left of an initial supply of mass A after the elapse of time t? The number of atoms present does not affect the likelihood of any individual atom decaying. A solitary atom is as likely to decay as one buried among thousands of other atoms. Since $M(t)$ is the fraction of material left of a unit mass after time t, $AM(t)$ is the amount left after time t if we start with mass A:

$$(\text{amount left at time } t) = AM(t) \tag{2.2}$$

How much will be left of a mass A of material after time $s+t$ has elapsed? By definition of the function M, the amount left is $AM(s+t)$. But there is another way of answering this question. Observe that after time s has elapsed, the remaining mass is $AM(s)$, and then after an additional time t has elapsed, the amount left is $\big(AM(s)\big)M(t)$. These two answers must be the same, and therefore,

$$M(s+t) = M(s)M(t). \tag{2.3}$$

Since $M(s)$ and $M(t)$ are less than 1, M is decreasing, and $M(t)$ tends to zero as t tends to infinity. We assumed that M is a continuous function, and $M(0) = 1$. According to the intermediate value theorem, there is a number h for which $M(h) = 1/2$. Since M is decreasing, there is only one such number. Setting $s = h$ in relation (2.3), we get that

$$M(h+t) = \frac{1}{2}M(t).$$

In words: starting at any time t, let additional time h elapse, then the mass of the material is halved. The number h is called the *half-life* of the radioactive material.

For example, the half-life of radium-226 is about 1601 years, and the half-life of carbon-14 is about 5730 years.

2.5b Bacterial Growth

We turn to another example, the growth of a colony of bacteria. We describe the growth by the following function:

Let $P(t)$ be the size of the bacterial population of initial unit size after it has grown for time t. Assume that P is a continuous function of time, $P(0) = 1$, and that $P(t) > 1$ for $t > 0$.

If we supply ample nutrients so that the bacteria do not have to compete with each other, and if there is ample room for growth, then it is reasonable to conclude that the size of the colony at any time t is proportional to its initial size A, whatever that initial size is:

$$\text{(size at time } t) = AP(t), \tag{2.4}$$

What will be the size of a colony, of initial size A, after it has grown for time $s+t$? According to Eq. (2.4), the size will be $AP(s+t)$. But there is another way of calculating the size of the population. After time s has elapsed, the population size has grown to $AP(s)$. After an additional time t elapses, the size of the population will, according to Eq. (2.4), grow to $\big(AP(s)\big)P(t) = AP(s)P(t)$. The two answers must be the same, and therefore,

$$P(s+t) = P(s)P(t). \tag{2.5}$$

Since $P(t) > 1$, P is an increasing function, and $P(t)$ tends to infinity as t increases. We assumed that P is continuous and $P(0) = 1$, so by the intermediate value theorem, there is a value d for which $P(d) = 2$. Since P is an increasing function, there is only one such value. Setting $s = d$ in Eq. (2.5) gives

$$P(d+t) = 2P(t);$$

d is called the *doubling time* for the bacterial colony. Starting from any time t, the colony doubles after additional time d elapses.

2.5c Algebraic Definition

Next we show that every continuous function f that satisfies

$$f(x+y) = f(x)f(y) \quad \text{and} \quad a = f(1) > 0,$$

must be an exponential function $f(x) = a^x$. For example, $P(t)$ and $M(t)$ in the last section are such functions.

The relation $f(x+y) = f(x)f(y)$ is called the functional equation of the exponential function. If $y = x$, the equation gives

$$f(x+x) = f(2x) = f(x)f(x) = (f(x))^2 = f(x)^2,$$

where in the last form we have omitted unnecessary parentheses. When $y = 2x$, we get

$$f(x+2x) = f(x)f(2x) = f(x)f(x)^2 = f(x)^3.$$

Continuing in this fashion, we get

$$f(nx) = f(x)^n. \tag{2.6}$$

Take $x = 1$. Then

$$f(n) = f(n1) = f(1)^n = a^n.$$

This proves that $f(x) = a^x$ when x is any positive integer. Take $x = \frac{1}{n}$ in Eq. (2.6). We get $f(1) = a = f\left(\frac{1}{n}\right)^n$. Take the nth root of both sides. We get $f\left(\frac{1}{n}\right) = a^{1/n}$. This proves that $f(x) = a^x$ when x is any positive integer reciprocal. Next take $x = \frac{1}{p}$ in Eq. (2.6); we get

$$f\left(\frac{n}{p}\right) = f\left(\frac{1}{p}\right)^n = \left(a^{1/p}\right)^n = a^{n/p}.$$

So we have shown that for all positive rational numbers $r = \frac{n}{p}$,

$$f(r) = a^r.$$

In Problem 2.52, we ask you to show that $f(0) = 1$ and that $f(r) = a^r$ for all negative rational numbers r. Assume that f is continuous. Then it follows that $f(x) = a^x$ for irrational x as well, since x can be approximated by rational numbers.

The algebraic properties of the exponential functions a^x extend to all numbers x as well, where $a > 0$:

- $a^x a^y = a^{x+y}$
- $(a^x)^n = a^{nx}$
- $a^0 = 1$
- $a^{-x} = \frac{1}{a^x}$
- $a^x > 1$ for $x > 0$ and $a > 1$
- $a^x < 1$ for $x > 0$ and $0 < a < 1$

We can use these properties to show that for $a > 1$, $f(x) = a^x$ is an increasing function. Suppose $y > x$. Then $y - x > 0$, and $a^{y-x} > 1$. Since $a^{y-x} = \frac{a^y}{a^x}$, it follows that $a^y > a^x$. By a similar argument when $0 < a < 1$, we can show that a^x is decreasing.

2.5d Exponential Growth

Though it has a precise mathematical meaning, the phrase "exponential growth" is often used as a metaphor for any extremely rapid increase. Here is the mathematical basis of this phrase:

Theorem 2.10. Exponential growth. *For $a > 1$, the function a^x grows faster than x^k as x tends to infinity, no matter how large the exponent $k = 0, 1, 2, 3 \ldots$.*

In other words, the quotient $\frac{a^x}{x^k}$ tends to infinity as x tends to infinity (Fig. 2.34).

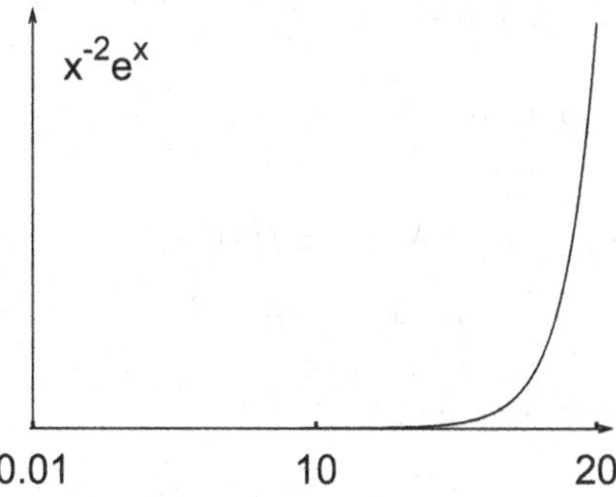

Fig. 2.34 The function $\frac{e^x}{x^2}$ plotted on $[0.01, 20]$. The *vertical scale* is compressed by a factor of 100,000

Proof. We first consider the case $k = 0$: that a^x tends to infinity for all a greater than 1. This is certainly true for $a = 10$, because $10^2 = 100$, $10^3 = 1000$, etc., clearly tend to infinity. It follows that a^x tends to infinity for all a greater than 10.

Consider the set of all numbers a for which a^x is bounded for all positive x. The set is not empty, because, for example, $a = 1$, and $a = \frac{1}{2}$ have this property. The set has an upper bound, because every number larger than 10 is *not* in the set. So the set of such a has a least upper bound. Denote the least upper bound by c. Since $a = 1$ lies in the set, c is not less than 1. We claim that c is 1. For suppose that c were greater than 1. Then b, the square root of c, and d, the square of c, would satisfy the inequalities

$$b < c < d.$$

Since d is greater than the least upper bound c, d^x tends to infinity with x. Since by definition, d is b^4, $b^{4x} = d^x$ tends to infinity with x. But since b is less than the least upper bound c, its powers remain bounded. This is a contradiction, so c must be 1. Therefore, a^x tends to infinity for all a greater than 1.

Next we consider the case $k = 1$: $\dfrac{a^x}{x}$ tends to infinity as x tends to infinity. Denote the function $\dfrac{a^x}{x}$ by $f(x)$. Then

$$f(x+1) = \frac{a^{x+1}}{x+1} = \frac{a^x}{x}\frac{a}{1+\frac{1}{x}} = f(x)\frac{a}{1+\frac{1}{x}}. \tag{2.7}$$

We claim that for large x, the factor $\dfrac{a}{1+\frac{1}{x}}$ is larger than 1: we know that $a > 1$, so in fact, $a > 1 + \dfrac{1}{m}$ for some integer m. Write $b = \dfrac{a}{1+\frac{1}{m}}$. Then for all $x \geq m$,

$$\frac{a}{1+\frac{1}{x}} \geq \frac{a}{1+\frac{1}{m}} = b > 1,$$

as claimed. Then by Eq. (2.7),

$$f(x+1) \geq f(x)b,$$
$$f(x+2) \geq f(x)b^2,$$

and continuing in this way, we see that

$$f(x+n) \geq f(x)b^n$$

for each positive integer n. Every large number X can be represented as some number x in $[m, m+1]$ plus a large positive integer n. Denote by M the minimum value of f in $[m, m+1]$. Then

$$f(X) = f(x+n) \geq f(x)b^n \geq Mb^n.$$

Since $b > 1$, this shows that $f(X)$ tends to infinity as X does.

In the cases $k > 1$, we argue as follows. Using the rules for the exponential function, we see that

$$\frac{a^x}{x^k} = \left(\frac{s^x}{x}\right)^k, \qquad \text{where} \quad s^k = a. \tag{2.8}$$

Since a is greater than 1, so is s. As we have already shown, $\dfrac{s^x}{x}$ tends to infinity as x does. Then so does its kth power. □

Later, in Sect. 4.1b, we shall give a much simpler proof of the theorem on exponential growth using calculus.

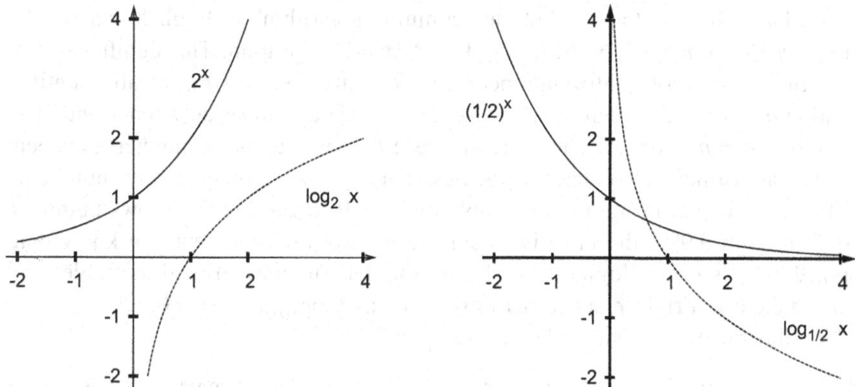

Fig. 2.35 *Left*: graphs of 2^x and $\log_2 x$. *Right*: graphs of $\left(\frac{1}{2}\right)^x$ and $\log_{1/2} x$

2.5e Logarithm

For a greater than 1, a^x is an increasing continuous function, and for $0 < a < 1$, a^x is decreasing. Hence for $a \neq 1$, a^x has a continuous inverse function, called the logarithm to the base a, which is defined by

$$\log_a y = x \quad \text{when} \quad y = a^x.$$

If $a > 1$, $\log_a$ is an increasing function. If $0 < a < 1$, $\log_a$ is a decreasing function. In either case, the domain of $\log_a$ is the range of a^x, all positive numbers (Fig. 2.35).

The exponential function is characterized by

$$a^x a^y = a^{x+y}.$$

Applying the function $\log_a$, we get

$$\log_a(a^x a^y) = x + y.$$

Take any two positive numbers u and v and denote their logarithms by x and y:

$$x = \log_a u, \qquad a^x = u, \qquad y = \log_a v, \qquad a^y = v. \tag{2.9}$$

We get

$$\log_a(uv) = \log_a u + \log_a v. \tag{2.10}$$

Calculations. The logarithm was invented by the Scottish scientist John Napier and expounded in a work published in 1614. Napier's logarithm was to the base e. In English, this is called called the "natural logarithm," a phrase that will be explained in the next section.

The base-10 logarithm, called the "common logarithm" in English, was introduced by Henry Briggs in 1617, based on Napier's logarithm. The significance of base ten is this: every positive number a can be written as $a = 10^n x$ (recall scientific notation $a = x \times 10^n$, where n is an integer and x is a number between 1 and 10). Then $\log_{10} a = n + \log_{10} x$. Therefore, the base-ten logarithms for numbers between 1 and 10 are sufficient to determine the base-ten logarithms for all positive numbers.

Table 2.1 is part of a traditional table of base-10 logarithms. It shows numbers 1.000 through 9.999, the last digit being read across the top row. We know that $\log_{10}(9.999)$ is nearly $\log_{10}(10) = 1$, and this tells us how to read the table: the entry in the lower right-hand corner must mean that $\log_{10}(9.999) = 0.99996$.

We illustrate multiplication by an example:

Example 2.29. What is the product of $a = 4279$ and $b = 78{,}520$? Write $a = 4.279 \times 10^3$. According to Table 2.1,

$$\log_{10}(4.279) = 0.63134.$$

Therefore, $\log_{10} a = 3.63134$. Similarly, $b = 7.852 \times 10^4$. According to the table, then,

$$\log_{10}(7.852) = 0.89498,$$

and therefore, $\log_{10} b = 4.89498$. To multiply a and b we use the fundamental property (2.10) of logarithms to write

$$\log_{10} ab = \log_{10} a + \log_{10} b = 3.63164 + 4.89498 = 8.52632.$$

By the table, the number whose base-10 logarithm is 0.52632 is, within a tolerance of 2×10^{-4}, equal to 3.360. This shows that the product ab is approximately 336,000,000, within a tolerance of 2×10^4.

Using a calculator, we get $ab = 335{,}987{,}080$, which is quite close to our approximate value calculated using base-10 logarithms.

No.	0	1	2	3	4	5	6	7	8	9
100	00000	00043	00087	00130	00173	00217	00260	00303	00346	00389
...	–	–	–	–	–	–	–	–	–	–
335	52504	52517	52530	52543	52556	52569	52582	52595	52608	52621
336	52634	52647	52660	52673	52686	52699	52711	52724	52737	52750
...	–	–	–	–	–	–	–	–	–	–
427	63043	63053	63063	63073	63083	63094	63104	63114	63124	63134
428	63144	63155	63165	63175	63185	63195	63205	63215	63225	63236
...	–	–	–	–	–	–	–	–	–	–
526	72099	72107	72115	72123	72132	72140	72148	72156	72165	72173
...	–	–	–	–	–	–	–	–	–	–
785	89487	89492	89498	89504	89509	89515	89520	89526	89531	89537
...	–	–	–	–	–	–	–	–	–	–
999	99957	99961	99965	99970	99974	99978	99981	99987	99991	99996
No.	0	1	2	3	4	5	6	7	8	9

Table 2.1 Excerpt from the $\log_{10}$ tables in Bowditch's practical navigator, 1868. We read, for example, $\log_{10}(3.358) = 0.52608$ from row no. 335, column 8

Division is carried out the same way, except we subtract the logarithms instead of adding them.

One cannot exaggerate the historical importance of being able to do arithmetic with base-ten logarithms. Multiplication and division by hand is a time-consuming, frustrating activity, prone to error.[1] For 350 years, no scientist, no engineer, no office, no laboratory was without a table of base-ten logarithms. Because of the force of habit, most scientific calculators have the base-ten logarithm available, although the main use of those logarithms is to perform multiplication and division. Of course, these arithmetic operations are performed by a calculator by pressing a button. The button labeled "log" often means $\log_{10}$. In the past, the symbol $\log x$, without any subscript, denoted the logarithm to base ten; the natural log of x was denoted by $\ln x$. Since in our time, multiplication and division are done by calculators, the base-ten logarithm is essentially dead, and rather naturally, $\log x$ has come to denote the natural logarithm of x.

Why Is the Natural Logarithm Natural? The explanation you will find in the usual calculus texts is that the inverse of the base-e logarithm, the base-e exponential function, is the most natural of all exponential functions because it has special properties related to calculus. Since Napier did not know what the inverse of the natural logarithm was, nor did he know calculus (he died about 25 years before Newton was born), his motivation must have been different. Here it is:

Suppose f and g are functions inverse to each other. That is, if $f(x) = y$, then $g(y) = x$. Then if we have a list of values $f(x_j) = y_j$ for the function f, it is also a list of values $x_j = g(y_j)$ for the function g. As an example, take the exponential function $f(x) = (10)^x = y$. Here is a list of its values for $x = 0, 1, 2, \ldots, 10$:

x	0	1	2	...	9	10
y	1	10	100	...	1,000,000,000	10,000,000,000

The inverse of the function $(10)^x = y$ is the base-10 logarithm, $\log_{10} y = x$. We have listed above its values for $y = 1, 10, 100, \ldots, 10{,}000{,}000{,}000$. The trouble with this list is that the values y for which $\log_{10} y$ is listed are very far apart, so we can get very little information about $\log_{10} y$ for values of y in between the listed values.

Next we take the base-2 exponential function $f(x) = 2^x = y$. Here is a list of its values for $x = 0, 1, 2, \ldots, 10$:

x	0	1	2	...	9	10
y	1	2	4	...	512	1024

The inverse of the function $2^x = y$ is the base-2 logarithm, $\log_2 y = x$. Here the values y for which the values of $\log_2 y$ are listed are not so far apart, but they are still quite far apart.

[1] There is a record of an educational conference in the Middle Ages on the topic, "Can one teach long division without flogging?"

Clearly, to make the listed values of the exponential function lie close together, we should choose the base small, but still greater than 1. So let us try the base $a = 1.01$. Here is a list of the values of $y = (1.01)^x$ for $x = 0, 1 \ldots 100$. Note that the evaluation of this exponential function for integer values of x requires just one multiplication for each value of x:

x	0	1	2	...	99	100
y	1	1.01	1.0201	...	2.6780	2.7048

The inverse of the function $(1.01)^x = y$ is the base-1.01 logarithm, $\log_{1.01} y = x$. The listed values y of the base-1.01 logarithm are close to each other, but the values of the logarithms are rather large: $\log_{1.01} 2.7048 = 100$. There is an easy trick to fix this. Instead of using 1.01 as the base, use $a = (1.01)^{100}$. Then

$$\left((1.01)^{100}\right)^x = (1.01)^{100x}.$$

We list values of a^x now for $x = 0, 0.01, 0.02, \ldots, 1.00$, which gives a table almost identical to the previous table:

x	0	0.01	0.02	...	0.99	1.00
y	1	1.01	1.0201	...	2.6780	2.7048

To further improve matters, we can take powers of numbers even closer to 1 as a base: Take as base $1+\frac{1}{n}$ raised to the power n, where n is a large number. As n tends to infinity, $\left(1+\dfrac{1}{n}\right)^n$ tends to e, the base of the natural logarithm.

Problems

2.44. Use the property $e^{x+y} = e^x e^y$ to find the relation between e^z and e^{-z}.

2.45. Suppose f is a function that satisfies the functional equation $f(x+y) = f(x)f(y)$, and suppose c is any number. Define a function $g(x) = f(cx)$. Explain why $g(x+y) = g(x)g(y)$.

2.46. A bacteria population is given by $p(t) = p(0)a^t$, where t is in days since the initial time. If the population was 1000 on day 3, and 200 on day 0, what was it on day 1?

2.47. A population of bacteria is given by $p(t) = 800(1.023)^t$, where t is in hours. What is the initial population? What is the doubling time for this population? How long will it take to quadruple?

2.48. Let P_0 be the initial principal deposited in an account. Write an expression for the account balance after 1 year in each of the following cases.

(a) 4 % simple interest,

(b) 4 % compounded quarterly (4 periods per year),
(c) 4 % compounded daily (365 periods per year),
(d) 4 % compounded continuously (number of periods tends to infinity),
(e) x % compounded continuously.

2.49. Calculate the product ab by hand, where a and b are as in Example 2.29.

2.50. Solve $e^{-x^2} = \frac{1}{2}$ for x.

2.51. Suppose $f(x) = ma^x$, and we know that

$$f\left(x+\frac{1}{2}\right) = 3f(x).$$

Find a.

2.52. Use the functional equation $f(x+y) = f(x)f(y)$ and $f(1) = a \neq 0$ to show that

(a) $f(0) = 1$,
(b) $f(r) = a^r$ for negative rational numbers r.

2.53. Suppose P satisfies the functional equation $P(x+y) = P(x)P(y)$, and that N is any positive integer. Prove that

$$P(0) + P(1) + P(2) + \cdots + P(N)$$

is a finite geometric series.

2.54. If b is the arithmetic mean of a and c, prove that e^b is the geometric mean of e^a and e^c.

2.55. Knowing that $e > 2$, explain why

(a) $e^{10} > 1000$,
(b) $\log 1000 < 10$,
(c) $\log 1{,}000{,}000 < 20$.

2.56. Let a denote a number greater than 1, $a = 1 + p$, where p is positive. Show that for all positive integers n, $a^n > 1 + pn$.

2.57. We know that $\frac{e^x}{x^2}$ tends to infinity as x does. In particular, it is eventually more than 1. Substitute $y = x^2$ and derive that

$$\log y < \sqrt{y}$$

for large y.

2.58. Use the relation $\log(uv) = \log u + \log v$ to show that $\log\left(\frac{x}{y}\right) = \log x - \log y$.

2.6 Sequences of Functions and Their Limits

We saw in Chap. 1 that we can only rarely present numbers exactly. In general, we describe them as limits of infinite sequences of numbers. What is true of numbers is also true of functions; we can rarely describe them exactly. We often describe them as limits of sequences of functions. It is not an exaggeration to say that almost all interesting functions are defined as limits of sequences of simpler functions. Therein lies the importance of the concept of a convergent sequence of functions.

Since most of the functions we shall study are continuous, we investigate next what convergence means for sequences of continuous functions. It turns out that with the right definition of convergence, continuity is preserved under the operation of taking limits.

First we look at some simple examples.

2.6a Sequences of Functions

Example 2.30. Consider the functions

$$f_0(x) = 1,\ f_1(x) = x,\ f_2(x) = x^2,\ f_3(x) = x^3,\ \dots, f_n(x) = x^n,\ \dots$$

on $[0,1]$. For each x in $[0,1]$, we get the following limits as n tends to infinity:

$$\lim_{n\to\infty} f_n(x) = \begin{cases} 0 & 0 \le x < 1, \\ 1 & x = 1. \end{cases}$$

Define f to be the function on $[0,1]$ given by $f(x) = \lim_{n\to\infty} f_n(x)$. The sequence of functions f_n converges to f, a discontinuous function. See Fig. 2.36.

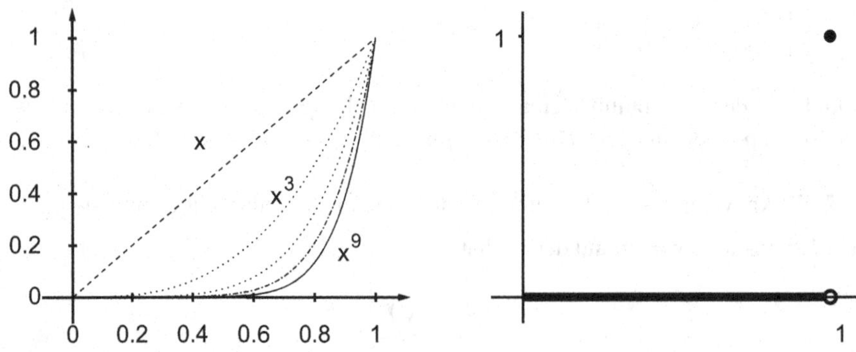

Fig. 2.36 *Left*: the functions $f_n(x) = x^n$ for $n = 1, 3, 5, 7$, and 9 are graphed on the interval $[0,1]$. *Right*: the discontinuous limit f. See Example 2.30

Example 2.30 shows that a sequence of continuous functions can converge to a discontinuous function. This is an undesirable outcome that we would like to avoid.

Example 2.31. Consider the functions $g_n(x) = x^n$ on $[0, \frac{1}{2}]$. The functions g_n are continuous on $[0, \frac{1}{2}]$ and converge to the constant function $g(x) = 0$, a continuous function.

These examples prompt us to make two definitions for sequence convergence. For a sequence of continuous functions $f_1, f_2, f_3, \ldots$ to converge to f, we certainly should require that for each x in their common domain, $\lim_{n\to\infty} f_n(x) = f(x)$.

Definition 2.11. A *sequence* of functions simply means a list $f_1, f_2, f_3, \ldots$ of functions with a common domain D. The sequence is said to *converge pointwise* to a function f on D if

$$\lim_{n\to\infty} f_n(x) = f(x) \qquad \text{for each } x \text{ in } D.$$

Uniform Convergence. We saw in Example 2.30 that a sequence of continuous functions may converge pointwise to a limit that is not continuous. We define a stronger form of convergence that avoids this trouble.

Definition 2.12. A sequence of functions $f_1, f_2, f_3, \ldots$ defined on a common domain D is said to *converge uniformly* on D to a limit function f if given any tolerance $\varepsilon > 0$, no matter how small, there is a whole number N depending on ε such that for all $n > N$, $f_n(x)$ differs from $f(x)$ by less than ε for all x in D.

To illustrate some benefits of uniform convergence, consider the problem of evaluating $f(x) = \cos x$. For instance, how would you compute $\cos(0.5)$ without using a calculator? We will see in Chap. 4 that one of the important applications of calculus is a method to generate a sequence of polynomial functions

$$p_n(x) = 1 - \frac{x^2}{2!} + \frac{x^4}{4!} - \cdots + k_n \frac{x^n}{n!} \qquad (k_n = 0,\ n \text{ odd, and } k_n = (-1)^{n/2},\ n \text{ even})$$

that converges uniformly to $\cos x$ on every closed interval $[-c, c]$. This means that once you set c and the tolerance ε, there is a polynomial p_n such that

$$|\cos x - p_n(x)| < \varepsilon \qquad \text{for all } x \text{ in } [-c, c].$$

In Chap. 4 we will see that we can get $|\cos x - p_n(x)| < \varepsilon$ for all x in $[-1, 1]$ by taking n such that $n! > \frac{1}{\varepsilon}$. For example, $\cos(0.3)$, $\cos(0.5)$, and $\cos(0.8)$ can each be approximated using $p_4(x) = 1 - \frac{x^2}{2!} + \frac{x^4}{4!}$, and since the convergence is uniform, the error in doing so will be less than $\dfrac{1}{24}$ in all cases. Evaluating $\cos x$ at an irrational number in $[-1,1]$ introduces an interesting complication. For example, $\cos\left(\frac{e}{3}\right)$ is approximated by

$$p_4\left(\frac{e}{3}\right) = 1 - \frac{1}{2}\left(\frac{e}{3}\right)^2 + \frac{1}{24}\left(\frac{e}{3}\right)^4.$$

Now we need to approximate $p_4\left(\dfrac{e}{3}\right)$ using some approximation to $\dfrac{e}{3}$, such as 0.9060939, which will introduce some error. Thinking ahead, there are many irrational numbers in $[-1,1]$ at which we would like to evaluate the cosine. Happily, p_4 is uniformly continuous on $[-1,1]$. We can find a single level of precision δ for the inputs, so that if z is within δ of x, then $p_4(z)$ is within ε of $p_4(x)$.

Looking at the big picture, we conclude that for a given tolerance ε, we can find n so large that $|\cos x - p_n(x)| < \dfrac{\varepsilon}{2}$ for all x in $[-1,1]$. Then we can find a precision δ such that if x and z are in $[-1,1]$ and differ by less than δ, then $p_n(x)$ and $p_n(z)$ will differ by less than $\dfrac{\varepsilon}{2}$. Using the triangle inequality, we get

$$|\cos x - p_n(z)| \leq |\cos x - p_n(x)| + |p_n(x) - p_n(z)| < \frac{\varepsilon}{2} + \frac{\varepsilon}{2} = \varepsilon.$$

Finding the right n and δ to meet a particular tolerance can be complicated, but we know in theory that it can be done. In short, approximate knowledge of the inputs and approximate knowledge of the function can be used to determine the function values within any given tolerance.[2] This is good news for computing.

Knowing that a sequence of continuous functions converges uniformly on $[a,b]$ guarantees that its limit function is continuous on $[a,b]$.

Theorem 2.11. *Let $\{f_n\}$ be a sequence of functions, each continuous on the closed interval $[a,b]$. If the sequence converges uniformly to f, then f is continuous on $[a,b]$.*

Proof. If f_n converges uniformly, then for n large enough,

$$|f_n(x) - f(x)| < \varepsilon$$

for all x in $[a,b]$. Since f_n is continuous on $[a,b]$, f_n is uniformly continuous on $[a,b]$ by Theorem 2.4. So for x_1 and x_2 close enough, say

$$|x_1 - x_2| < \delta,$$

2 There once was a function named g,
approximated closely by p.
When we put in x nearly,
we thought we'd pay dearly,
but $g(x)$ was as close as can be. –Anon.

This limerick expresses that $|g(x) - p(x_{\text{approx}})| \leq |g(x) - p(x)| + |p(x) - p(x_{\text{approx}})|$.

$f_n(x_1)$ and $f_n(x_2)$ will differ by less than ε. Next we see a nice use of the triangle inequality (Sect. 1.1b). The argument is that you can control the difference of function values at two points x_1 and x_2 by writing

$$f(x_1)-f(x_2)=f(x_1)-f_n(x_1)+f_n(x_1)-f_n(x_2)+f_n(x_2)-f(x_2)$$

and grouping these terms cleverly. We have, then, by the triangle inequality that

$$|f(x_1)-f(x_2)| \le |f(x_1)-f_n(x_1)|+|f_n(x_1)-f_n(x_2)|+|f_n(x_2)-f(x_2)|$$

Each of these terms is less than ε if $|x_1-x_2|<\delta$. This proves the uniform continuity of f on $[a,b]$. □

We now present examples of uniformly convergent sequences of continuous functions.

Example 2.32. The sequence of functions $f_n(x)=x^n$ on $[-c,c]$, where c is a positive number less than 1, converges pointwise to the function $f(x)=0$, because for each x in $[-c,c]$, x^n tends to 0 as n tends to infinity. To see why the sequence converges uniformly to f, look at the difference between $f_n(x)=x^n$ and 0 on $[-c,c]$. For any tolerance ε, we can find a whole number N such that $c^N<\varepsilon$, and hence $c^n<\varepsilon$ for every $n>N$ as well. Let x be any number between $-c$ and c. Then

$$|f_n(x)-0|=|x^n|\le c^n<\varepsilon.$$

Therefore, the difference between x^n and 0 is less than ε **for all** x in $[-c,c]$. That is, the sequence of functions converges uniformly. Note that the limit function, $f(x)=0$, is continuous, as guaranteed by the theorem (Fig. 2.37).

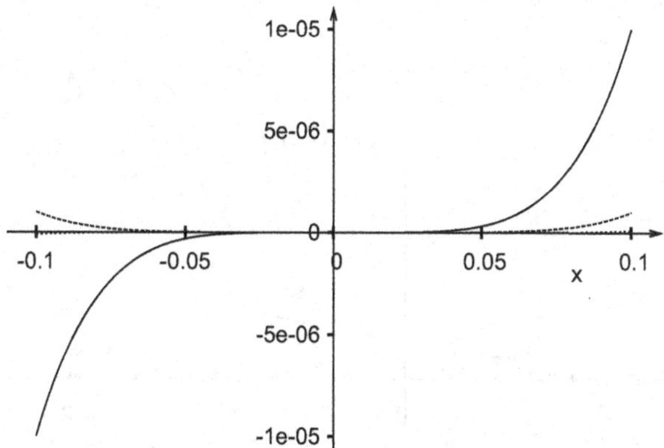

Fig. 2.37 The functions $f_n(x)=x^n$ for $n=5$, 6, and 7 are graphed on the interval $[-0.1,0.1]$. Note that the graph of f_7 is indistinguishable from the x-axis

Geometric Series. Consider the sequence of functions $\{f_n\}$ given by

$$f_n(x) = 1 + x + x^2 + \cdots + x^{n-1},$$

where x is in the interval $[-c, c]$ and $0 < c < 1$. The sum defining $f_n(x)$ is also given by the formula

$$f_n(x) = \frac{1 - x^n}{1 - x}.$$

For each x in $(-1, 1)$, $f_n(x)$ tends to $f(x) = \dfrac{1}{1-x}$, so the sequence f_n converges pointwise to f. To see why the f_n converge uniformly in $[-c, c]$, form the difference of $f_n(x)$ and $f(x)$. We get

$$f(x) - f_n(x) = \frac{x^n}{1 - x}.$$

For x in the interval $[-c, c]$, $|x|$ is not greater than c, and $|x^n|$ is not greater than c^n. It follows that

$$|f(x) - f_n(x)| = \frac{|x|^n}{1 - x} \leq \frac{c^n}{1 - c} \qquad \text{for all } x \text{ in } [-c, c].$$

Since c^n tends to zero, we can choose N so large that for n greater than N, $\dfrac{c^n}{1-c}$ is less than ε, and hence $f(x)$ differs from $f_n(x)$ by less than ε for all x in $[-c, c]$. This proves that f_n tends to f uniformly on the interval $[-c, c]$, $c < 1$. Note that $\dfrac{1}{1-x}$ is continuous on $[-c, c]$, as guaranteed by the theorem (Fig. 2.38).

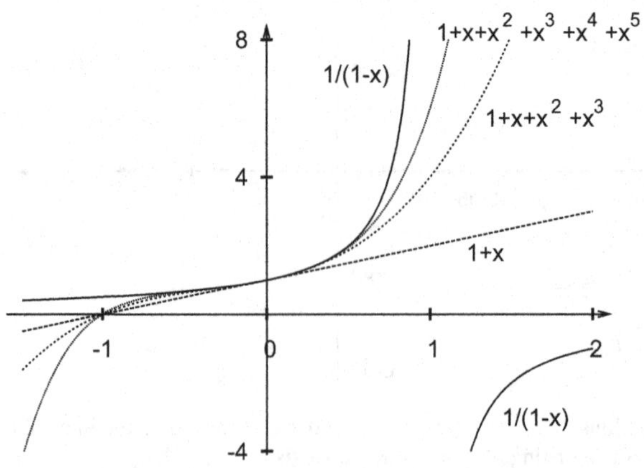

Fig. 2.38 The sequence of functions $f_n(x) = 1 + x + \cdots + x^n$ converges uniformly to $\frac{1}{1-x}$ on $[-c, c]$ when $c < 1$

Operations on Convergent Sequences of Functions. We can combine uniformly convergent sequences of continuous functions.

Theorem 2.12. *Suppose f_n and g_n are uniformly convergent sequences of continuous functions on $[a,b]$, converging to f and g. Then*

(a) $f_n + g_n$ converges uniformly to $f + g$.
(b) $f_n g_n$ converges uniformly to fg.
(c) If $f \neq 0$ on $[a,b]$, then for n large enough, $f_n \neq 0$ and $\frac{1}{f_n}$ tends to $\frac{1}{f}$ uniformly.
(d) If h is a continuous function with range contained in $[a,b]$, then $g_n \circ h$ converges uniformly to $g \circ h$.
(e) If k is a continuous function on a closed interval that contains the range of each g_n and g, then $k \circ g_n$ converges uniformly to $k \circ g$.

Proof. We give an outline of the proof of this theorem. For (a), use the triangle inequality:

$$|(f(x)+g(x)) - (f_n(x)+g_n(x))| \leq |f(x) - f_n(x)| + |g(x) - g_n(x)|.$$

For all x in $[a,b]$, the terms on the right are smaller than any given tolerance, provided that n is large enough. Figure 2.39 shows the idea. We guide you through the details of proving part (a) in Problem 2.61.

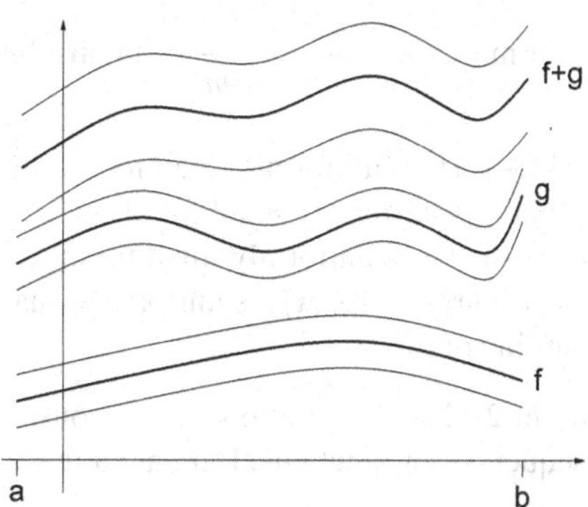

Fig. 2.39 Functions f_n are within ε of f for $n > N_1$, and the g_n are within ε of g for $n > N_2$. The sums $f_n + g_n$ are then within 2ε of $f+g$ for n larger than both N_1 and N_2

For (b), use

$$
\begin{aligned}
|f(x)g(x) - f_n(x)g_n(x)| &= |(f(x) - f_n(x))g(x) + f_n(x)(g(x) - g_n(x))| \\
&\le |f(x) - f_n(x)||g(x)| + |f_n(x)||g(x) - g_n(x)|.
\end{aligned}
$$

We can make the factors $|f(x) - f_n(x)|$ and $|g(x) - g_n(x)|$ small by taking n large. We check the factor $|f_n(x)|$. By the extreme value theorem (Theorem 2.6), $|f|$ has a maximum value M, so $-M \le f(x) \le M$. Since the f_n converge to f uniformly, they are within distance 1 of f for large n, and $-M-1 \le f_n(x) \le M+1$ for all x. Thus we have

$$
|f(x)g(x) - f_n(x)g_n(x)| \le |f(x) - f_n(x)||g(x)| + (M+1)|g(x) - g_n(x)|
$$

for large n, and this can be made arbitrarily small by taking n sufficiently large.

For (c): If f is not zero on an interval, then it is either positive at every point or negative at every point. For if it were positive at some point c and negative at another point d, then according to Theorem 2.5, the intermediate value theorem, $f(x)$ would be zero at some point x between c and d, contrary to our assumption about f. Take the case that f is positive. According to Theorem 2.6, the extreme value theorem, $f(x)$ takes on its minimum at some point of the closed interval $[a,b]$. This minimum is a positive number m, and $f(x) \ge m$ for all x in the interval. Since $f_n(x)$ tends uniformly to $f(x)$ on the interval, it follows that for n greater than some number N, $f_n(x)$ differs from $f(x)$ by less than $\frac{1}{2}m$. Since $f(x) \ge m$, $f_n(x) \ge \frac{1}{2}m$. We use

$$
\frac{1}{f_n(x)} - \frac{1}{f(x)} = \frac{f(x) - f_n(x)}{f_n(x)f(x)}.
$$

The right-hand side is not more than $\dfrac{|f(x) - f_n(x)|}{(\frac{1}{2}m)m}$ in absolute value, from which the result follows.

For (d) we use that $g(y) - g_n(y)$ is uniformly small for all y, and then take $y = h(x)$ to see that $g(h(x)) - g_n(h(x))$ is uniformly small for all x.

For (e) we use that $g(x) - g_n(x)$ is uniformly small for all x, and then use uniform continuity of k to see that $k(g(x)) - k(g_n(x))$ is uniformly small for all x.

This completes the outline of the proof. □

The beauty of Theorem 2.12 is that it allows us to construct a large variety of uniformly convergent sequences of functions. Here are a few examples.

Example 2.33. Let $g_n(x) = 1 + x + x^2 + \cdots + x^n$, and let $h(u) = -u^2$, where u is in $[-c,c]$, and $0 < c < 1$. Then

$$
g_n(h(u)) = 1 - u^2 + u^4 - u^6 + \cdots + (-u^2)^n
$$

converges uniformly in $[-c,c]$ to $\dfrac{1}{1+u^2}$.

Example 2.34. Let $r > 0$, a any number, and set

$$k_n(x) = 1 + \frac{x-a}{r} + \cdots + \left(\frac{x-a}{r}\right)^n.$$

Then $k_n(x) = g_n\left(\frac{x-a}{r}\right)$, where g_n is as in Example 2.33. The k_n converge uniformly to

$$\frac{1}{1 - \frac{x-a}{r}} = \frac{r}{r-x+a}$$

on every closed interval contained in $(a-r, a+r)$. This is true by part (d) of Theorem 2.12.

Example 2.35. Let $h(t) = \frac{1}{2}\cos t$, where $g_n(x)$ is as in Example 2.33. Then

$$g_n(h(t)) = 1 + \frac{1}{2}\cos t + \left(\frac{1}{2}\cos t\right)^2 + \cdots + \left(\frac{1}{2}\cos t\right)^n$$

converges uniformly to $\dfrac{2}{2-\cos t}$ for all t.

2.6b Series of Functions

Definition 2.13. The sequence of functions $\{f_n\}$ can be added to make a new sequence $\{s_n\}$, called the sequence of *partial sums* of $\{f_n\}$:

$$s_n = f_0 + f_1 + f_2 + \cdots + f_n = \sum_{j=0}^{n} f_j.$$

The sequence of functions $\{s_n\}$ is called a *series* and is denoted by

$$\sum_{j=0}^{\infty} f_j.$$

If $\lim_{n\to\infty} s_n(x)$ exists, denote it by $f(x)$, and we say that the series converges to $f(x)$ at x. We write

$$\sum_{j=0}^{\infty} f_j(x) = f(x).$$

If the sequence of partial sums converges uniformly on D, we say that the series converges uniformly on D.

We saw earlier that the sequence of partial sums of the geometric series

$$s_n(x) = 1 + x + x^2 + \cdots + x^n = \frac{1 - x^{n+1}}{1 - x}$$

converges uniformly to $\frac{1}{1-x}$ on every interval $[-c, c]$, if $0 < c < 1$. We often write

$$\sum_{k=0}^{\infty} x^k = 1 + x + x^2 + x^3 + \cdots = \frac{1}{1-x} \qquad (|x| < 1).$$

This series is of a special kind, a power series.

Definition 2.14. A *power series* is a series of the form

$$\sum_{k=0}^{\infty} a_k (x-a)^k.$$

The numbers a_n are called the coefficients. The number a is called the center of the power series.

Consider the power series

$$\sum_{n=1}^{\infty} \frac{x^n}{n} = x + \frac{x^2}{2} + \frac{x^3}{3} + \cdots.$$

For what values of x, if any, does the series converge? To find all values of x for which the series converges, we use the ratio test, Theorem 1.18. We compute the limit

$$\lim_{n\to\infty} \left| \frac{\frac{x^{n+1}}{n+1}}{\frac{x^{n+1}}{n+1}} \right| = \lim_{n\to\infty} |x| \frac{n+1}{n} = |x|.$$

According to the ratio test, if the limit is less than 1, then the series converges absolutely. Therefore, $\sum_{n=1}^{\infty} \frac{x^n}{n}$ converges for $|x| < 1$. Also, if the limit is greater than 1, then the series diverges, in this case for $|x| > 1$. The test gives no information when the limit is 1, in our case $|x| = 1$. So our next task is to investigate the convergence (or divergence) of $\sum_{n=1}^{\infty} \frac{x^n}{n}$ when $x = 1$ and when $x = -1$. At $x = 1$, we get $\sum_{n=1}^{\infty} \frac{1}{n}$, the well-known harmonic series. We saw in Example 1.21 that it diverges. At $x = -1$ we get the series $\sum_{n=1}^{\infty} \frac{(-1)^n}{n}$. It converges by the alternating series theorem, Theorem 1.17.

Therefore, $\sum_{n=1}^{\infty} \frac{x^n}{n}$ converges pointwise for all x in $[-1,1)$. We have not shown that the convergence is uniform, so we do not know whether the function $f(x) = \sum_{n=1}^{\infty} \frac{x^n}{n}$ is continuous.

Sometimes a sequence of functions converges to a function that we know by another rule. If so, we know a great deal about that limit function. But this is not always the case. Some sequences of functions, including power series, converge to functions that we know only through sequential approximation. The next two theorems give us important information about the limit function of a power series. The first tells us about its domain. The second tells us about its continuity.

Theorem 2.13. *For a power series* $\sum_{n=0}^{\infty} c_n(x-a)^n$*, one of the following must hold:*

(a) The series converges absolutely for every x.
(b) The series converges only at $x = a$.
(c) There is a positive number R*, called the radius of convergence, such that the series converges absolutely for* $|x-a| < R$ *and diverges for* $|x-a| > R$.

In case (c), the series might or might not converge at $x = a-R$ *and at* $x = a+R$.

Proof. Let us first point out that if the series converges at some $x_0 \neq a$, then it converges absolutely for every x that is closer to a, that is, $|x-a| < |x_0 - a|$. Here is why: The convergence of $\sum_{n=0}^{\infty} c_n(x_0-a)^n$ implies that the terms $c_n(x_0-a)^n$ tend to 0. In particular, there is an N such that $|c_n(x_0-a)^n| < 1$ for all $n > N$. If $0 < |x-a| < |x_0-a|$, set $r = \frac{|x_0-a|}{|x-a|}$. Then $r < 1$, and we get

$$\sum_{n=N+1}^{\infty} |c_n(x-a)^n| = \sum_{n=N+1}^{\infty} |c_n(x-a)^n| \left| \frac{(x_0-a)^n}{(x_0-a)^n} \right| \tag{2.11}$$

$$= \sum_{n=N+1}^{\infty} |c_n(x_0-a)^n| \left| \frac{(x-a)^n}{(x_0-a)^n} \right| \leq \sum_{n=N+1}^{\infty} r^n.$$

Therefore, $\sum_{n=0}^{\infty} c_n(x-a)^n$ converges absolutely by comparison with a geometric series.

Now consider the three possibilities we have listed in the theorem. It might happen that the series converges for every x. If so, it converges absolutely for every x by what we have just shown. This covers the first case.

The other possibility is that the series converges for some number x_0, but not for every number. If there is only one such x_0, then it must be a, since the series

$$c_0 + c_1(a-a) + c_2(a-a)^2 + \cdots = c_0$$

certainly converges. This covers the second case.

Finally, there may be an $x_0 \neq a$ for which the series converges, though the series does not converge for every number. We will use the least upper bound principle, Theorems 1.2 and 1.3, to describe R. Let S be the set of numbers x for which the series converges. Then S is not empty, because a and x_0 are in S, as well as every number closer to a than x_0. Also, S is bounded, because if there were arbitrarily large (positive or negative) numbers in S, then all numbers closer to a would be in S, i.e., S would be all the numbers. Therefore, S has a least upper bound M and a greatest lower bound m, which means that if

$$m < x < M,$$

then the series converges at x. We ask you in Problem 2.65 to show that m and M are the same distance from a:

$$m < a < M \qquad \text{and } a - m = M - a$$

and that the convergence is absolute in (m,M). Set $R = M - a$. This concludes the proof. □

Theorem 2.14. *A power series* $\sum_{n=0}^{\infty} a_n(x-a)^n$ *converges uniformly to its limit function on every closed interval* $|x-a| \le r$, *where* r *is less than the radius of convergence* R.

In particular, the limit function is continuous in $(a-R, a+R)$.

Proof. If the radius of convergence of $\sum_{n=0}^{\infty} a_n(x-a)^n = f(x)$ is $R = 0$, the series converges at only one point, $x = a$. The series is then just $f(a) = a_0 + 0 + \cdots$, which converges uniformly on that domain.

Suppose $R > 0$ or R is infinite, and take any positive $r < R$. Then the number $a + r$ is in the interval of convergence, so according to Theorem 2.13, $\sum_{n=0}^{\infty} a_n r^n$ converges absolutely. Then for every x with $|x-a| \le r$,

$$\left| f(x) - \sum_{n=0}^{k} |a_n(x-a)^n| \right| \le \sum_{n=k+1}^{\infty} |a_n(x-a)^n| \le \sum_{n=k+1}^{\infty} |a_n r^n|.$$

The last expression is independent of x and tends to 0 as k tends to infinity. Therefore, f is the uniform limit of its partial sums, which are continuous, on $|x-a| \le r$. According to Theorem 2.11, f is continuous on $[a-r, a+r]$.

Since every point of $(a-R, a+R)$ is contained in such a closed interval, f is continuous on $(a-R, a+R)$. □

The radius of convergence, R, of a power series can often be found by the ratio test. If that fails, there is another test, called the root test, which we describe in Problem 2.67.

Example 2.36. To find the interval of convergence of $\sum_{n=0}^{\infty} 2^n (x-3)^n$, we use the ratio test:

$$\lim_{n\to\infty} \left| \frac{2^{n+1}(x-3)^{n+1}}{2^n (x-3)^n} \right| = \lim_{n\to\infty} 2|x-3| = 2|x-3|.$$

When $2|x-3| < 1$, the series converges absolutely. When $2|x-3| > 1$, the series diverges. What happens when $2|x-3| = 1$?

(a) At $x = 2.5$, $2(x-3) = -1$, and $\sum_{n=0}^{\infty} 2^n (x-3)^n = \sum_{n=0}^{\infty} (-1)^n$ diverges.

(b) At $x = 3.5$, $2(x-3) = 1$, and $\sum_{n=0}^{\infty} 2^n (x-3)^n = \sum_{n=0}^{\infty} 1^n$ diverges.

Conclusion: $f(x) = \sum_{n=0}^{\infty} 2^n (x-3)^n$ converges for all x with $2|x-3| < 1$, i.e., in $(2.5, 3.5)$. Also, according to Theorem 2.14, the series converges uniformly to f on every closed interval $|x-3| \le r < \frac{1}{2}$, and f is continuous on $(2.5, 3.5)$.

Example 2.37. To find the interval of convergence of $\sum_{n=0}^{\infty} \frac{x^n}{n!}$, we use the ratio test:

$$\lim_{n\to\infty} \left| \frac{\frac{x^{n+1}}{(n+1)!}}{\frac{x^n}{n!}} \right| = \lim_{n\to\infty} \frac{|x|}{n+1} = 0 < 1.$$

Since $0 < 1$ for all x, the series converges for all x. It converges uniformly on every closed interval $|x-0| \le r$. So $f(x) = \sum_{n=0}^{\infty} \frac{x^n}{n!}$ is continuous on $(-\infty, \infty)$. In Chap. 4, we will see that this power series converges to a function that we know by another rule.

2.6c Approximating the Functions $\sqrt{x}$ *and* e^x

We close this section by looking at three examples of sequences of functions $\{f_n\}$ that are not power series that converge uniformly to the important functions $\sqrt{x}$, $|x|$, and e^x. In the case of e^x, we use the sequence of continuous functions $e_n(x) = \left(1 + \frac{x}{n}\right)^n$, and thus we prove that e^x is a continuous function.

Approximating $\sqrt{x}$. In Sect. 1.3a, we constructed a sequence of approximations $s_1, s_2, s_3, \ldots$ that converged to the square root of 2. There is nothing special about the number 2. The same construction can be used to generate a sequence of numbers that tends to the square root of any positive number x. Here is how:

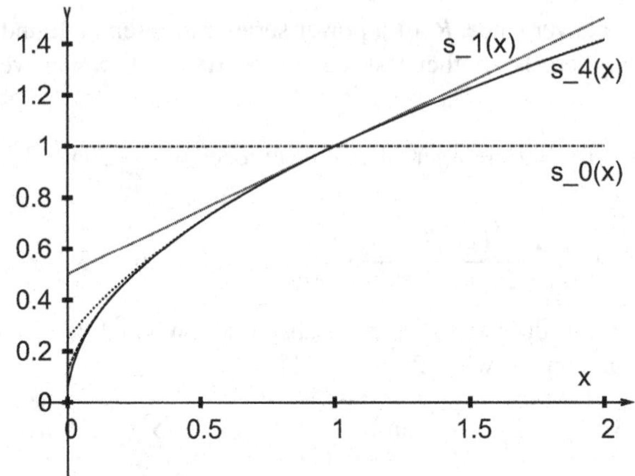

Fig. 2.40 The functions $s_n(x)$ converge to $\sqrt{x}$. The cases $0 \le n \le 4$ are shown. Note that $\sqrt{x}$ is not plotted

Suppose s is an approximation to the square root of x. To find a better approximation, we note that the product of s and $\frac{x}{s}$ is x. If s happens to be larger than $\frac{x}{s}$, then $s^2 > s\frac{x}{s} = x > \left(\frac{x}{s}\right)^2$, so $s > \sqrt{x} > \frac{x}{s}$, that is, the square root of x lies between these two. A similar argument shows that $\frac{x}{s} > \sqrt{x} > s$ if s happens to be less than $\frac{x}{s}$. So we take as the next approximation the arithmetic mean of the two:

$$\text{new approximation} = \frac{1}{2}\left(s + \frac{x}{s}\right).$$

Rather than start with an arbitrary first approximation, we start with $s_0 = 1$ and construct a sequence of approximations $s_1, s_2, \ldots$ as follows:

$$s_{n+1} = \frac{1}{2}\left(s_n + \frac{x}{s_n}\right).$$

The approximations s_n depend on the number x whose square root we seek; in other words, s_n is a function of x. How much does s_{n+1} differ from $\sqrt{x}$?

$$s_{n+1} - \sqrt{x} = \frac{1}{2}\left(s_n + \frac{x}{s_n}\right) - \sqrt{x}.$$

We bring the fractions on the right to a common denominator:

$$s_{n+1} - \sqrt{x} = \frac{1}{2s_n}(s_n^2 + x - 2s_n\sqrt{x}). \tag{2.12}$$

The expression in parentheses on the right is a perfect square, $(s_n - \sqrt{x})^2$. So we can rewrite Eq. (2.12) as

$$s_{n+1} - \sqrt{x} = \frac{1}{2s_n}(s_n - \sqrt{x})^2, \qquad (n \geq 0). \tag{2.13}$$

This formula implies that s_{n+1} is greater than $\sqrt{x}$ except when $s_n = \sqrt{x}$.

Since the denominator s_n on the right in Eq. (2.13) is greater than $s_n - \sqrt{x}$, we deduce that

$$s_{n+1} - \sqrt{x} < \frac{1}{2}(s_n - \sqrt{x}).$$

Applying this inequality n times, we get

$$s_{n+1} - \sqrt{x} < \frac{1}{2^n}(s_1 - \sqrt{x}) = \left(\frac{1}{2}\right)^n \left(\frac{1+x}{2} - \sqrt{x}\right). \tag{2.14}$$

Note that in Eq. (2.14), the factor $\frac{1+x}{2} - \sqrt{x}$ is less than $\frac{1+c}{2}$ whenever $x \leq c$. Therefore, inequality (2.14) implies

$$s_{n+1}(x) - \sqrt{x} \leq \frac{1+c}{2^n}.$$

It follows that the sequence of functions $s_n(x)$ converges uniformly to the function $\sqrt{x}$ over every finite interval $[0,c]$ of the positive axis (Fig. 2.40). The rate of convergence is even faster than what we have proved here, as we discuss in Sect. 5.3c.

Example 2.38. We show how to approximate $f(x) = |x|$ by a sequence of rational functions. Let $f_n(x) = s_n(x^2)$, where s_n is the sequence of functions derived in the preceding example that converge to $\sqrt{x}$. The $s_n(x)$ converge uniformly to $\sqrt{x}$, and x^2 is continuous on every closed interval. By Theorem 2.12, $s_n(x^2)$ converges uniformly to $\sqrt{x^2} = |x|$.

We indicate in Fig. 2.41 the graphs of $s_2(x^2)$, $s_3(x^2)$, and $s_5(x^2)$, which are rational approximations to $|x|$.

Approximating e^x**.** Take the functions $e_n(x) = \left(1 + \frac{x}{n}\right)^n$. We shall show, with your help, that they converge uniformly to the function e^x over every finite interval $[-c,c]$ (Fig. 2.42).

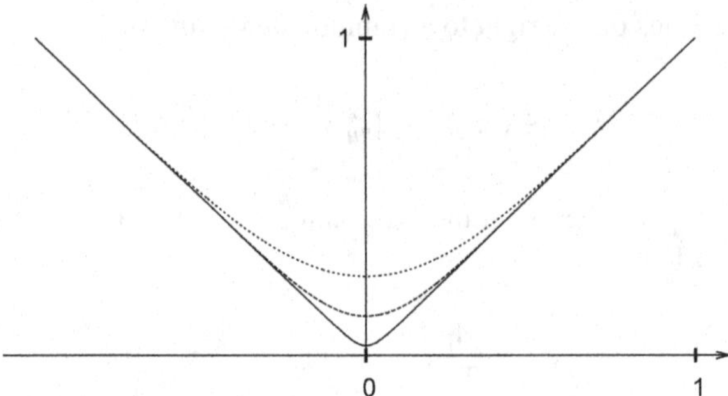

Fig. 2.41 Rational approximations of $|x|$ in Example 2.38

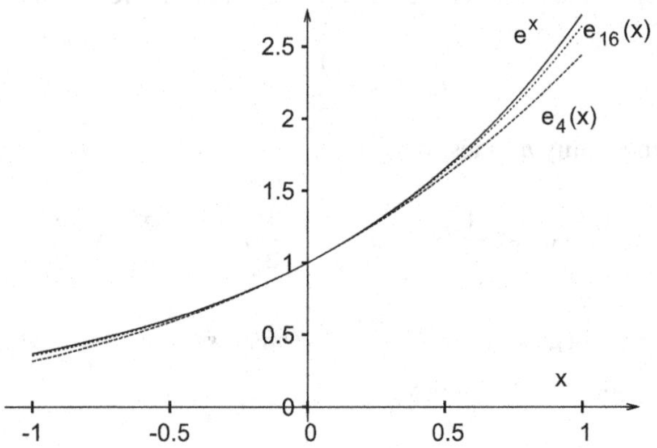

Fig. 2.42 The exponential function e^x and the functions $e_n(x) = \left(1+\frac{x}{n}\right)^n$ for $n=4$ and $n=16$ are graphed on the interval $[-1,1]$

Let us return to Sect. 1.4. There, we showed that the sequence of numbers $e_n = \left(1+\dfrac{1}{n}\right)^n$ is increasing and bounded, and therefore, by the monotone convergence theorem, it has a limit, a number that we have denoted by e.

We can show by similar arguments (see Problems 2.73 and 2.74) that for every positive x, the sequence of numbers $e_n(x)$ is increasing and bounded, whence by the monotone convergence theorem, it converges pointwise to a number $e(x)$ that depends on x. Note that $e_n(1) = e_n$, so $e(1) = \mathrm{e}$.

It remains to show that the limit function $e(x)$ is the exponential function e^x, and that convergence is uniform over every finite interval. To do this, we first show that $e(x) = \mathrm{e}^x$ when x is rational. We do this by showing that

$$e(r+s) = e(r)e(s)$$

for every pair of positive rational numbers r and s. We know from Sect. 2.5c that this relation implies that $e(x)$ is an exponential function for rational numbers.

Let r and s be any positive rational numbers. We can find a common denominator d such that

$$r = \frac{p}{d}, \qquad s = \frac{q}{d}$$

and p, q, and d are positive whole numbers. By manipulating $r+s$ algebraically, we obtain

$$e(r+s) = e\left(\frac{p}{d} + \frac{q}{d}\right) = e\left(\frac{1}{d}(p+q)\right).$$

We claim that

$$e(kx) = \big(e(x)\big)^k \tag{2.15}$$

for positive integers k. Here is a proof of the claim. Since $\left(1+\frac{x}{n}\right)^n$ converges to $e(x)$, for every positive integer k, $\left(1+\frac{kx}{n}\right)^n$ converges to $e(kx)$. Set $n = km$; we get that $\left(1+\frac{kx}{km}\right)^{km} = \left(1+\frac{x}{m}\right)^{mk}$ tends to $e(x)^k$. This proves Eq. (2.15).

Set $x = 1/d$ and $k = p+q$ in Eq. (2.15). We get

$$e\left(\frac{1}{d}(p+q)\right) = \left(e\left(\frac{1}{d}\right)\right)^{p+q} = \left(e\left(\frac{1}{d}\right)\right)^p \left(e\left(\frac{1}{d}\right)\right)^q = e\left(\frac{p}{d}\right)e\left(\frac{q}{d}\right) = e(r)e(s).$$

This concludes the proof that $e(x)$ is an exponential function a^x for x rational. Since $e(1) = \mathrm{e}$, it follows that $e(x) = \mathrm{e}^x$.

We turn now to showing that $e_n(x)$ converges *uniformly* to $e(x)$ on every finite interval $[-c, c]$. Our proof that the sequence $e_n(x)$ converges for every x as n tends to infinity used the monotone convergence theorem. Unfortunately, this gives no information as to how fast these sequences converge, and therefore it is useless in proving the uniformity of convergence. We will show that

$$\text{if} \quad -c \le x \le c, \quad \text{then} \quad e(x) - e_n(x) < \frac{k}{n},$$

for some constant k that depends on c. This is sufficient to prove the uniform convergence.

We make use of the following inequality:

$$a^n - b^n < (a-b)na^n \qquad \text{if} \quad 1 < b < a. \tag{2.16}$$

First we prove the inequality: We start from the observation that for all a and b,

$$a^n - b^n = (a-b)(a^{n-1} + a^{n-2}b + a^{n-3}b^2 + \cdots + b^{n-1}),$$

which we see by carrying out the multiplication on the right-hand side. Then in the case $0 < b < a$, we have for each power that $b^k < a^k$, so in the factor $(a^{n-1} + a^{n-2}b + a^{n-3}b^2 + \cdots + b^{n-1})$, there are n terms each less than a^{n-1}. This proves that $a^n - b^n < (a-b)na^{n-1}$. In the case $1 < a$, we may append one more factor of a, and this proves the inequality. We will use this inequality twice in two different ways to show uniform convergence.

Since $e_n(x)$ is an increasing sequence,

$$e(x) \geq \left(1+\frac{x}{n}\right)^n.$$

Take the nth root of this inequality and use Eq. (2.15) with $k=n$ to express the nth root of $e(x)$. We get

$$e\left(\frac{x}{n}\right) = (e(x))^{\frac{1}{n}} \geq 1+\frac{x}{n} \geq 1.$$

The first use of inequality (2.16) will be to show that for $n > x$,

$$e\left(\frac{x}{n}\right) < \frac{1}{1-\frac{x}{n}}. \tag{2.17}$$

Set $a = 1+\dfrac{x}{n}$ and $b=1$ in Eq. (2.16). We get

$$a^n - b^n = e_n(x) - 1 < (a-b)na^n = \frac{x}{n}n\left(1+\frac{x}{n}\right)^n = xe_n(x).$$

Letting n tend to infinity, we get in the limit $e(x)-1 < xe(x)$, or $(1-x)e(x) < 1$. Thus if $x<1$, then $1-x$ is positive, and we get $e(x) < \dfrac{1}{1-x}$. But if $n > x$, then $\dfrac{x}{n} < 1$, whence $e(\frac{x}{n}) < \dfrac{1}{1-\frac{x}{n}}$. This proves Eq. (2.17).

For the second use of inequality (2.16), set $a = e(\frac{x}{n})$ and $b = 1+\dfrac{x}{n}$. We get

$$e(x) - e_n(x) = \left(e(\tfrac{x}{n})\right)^n - \left(1+\frac{x}{n}\right)^n = a^n - b^n \leq (a-b)na^n$$

$$= \left(e\left(\frac{x}{n}\right) - \left(1+\frac{x}{n}\right)\right)n\left(e(\tfrac{x}{n})\right)^n = \left(e\left(\frac{x}{n}\right) - \left(1+\frac{x}{n}\right)\right)ne(x). \tag{2.18}$$

Combining the two results, set Eq. (2.17) into the right side of Eq. (2.18) to get

$$e(x) - e_n(x) < \left(\frac{1}{1-\frac{x}{n}} - \left(1+\frac{x}{n}\right)\right)ne(x) = \left(\frac{\frac{x^2}{n^2}}{1-\frac{x}{n}}\right)ne(x). \tag{2.19}$$

So for n greater than x,

$$e(x) - e_n(x) \leq \frac{1}{n}\frac{x^2 e(x)}{1-\frac{x}{n}}. \tag{2.20}$$

For $n > 2x$, the denominator on the right in Eq. (2.20) is greater than $\frac{1}{2}$, so

$$e(x) - e_n(x) < \frac{1}{n}2e(x)x^2 < \frac{2}{n}e(c)c^2$$

for every x in $[-c,c]$. This shows that as n tends to infinity, $e_n(x)$ tends to $e(x)$ uniformly on every finite x-interval. This concludes the proof. □

Example 2.39. We know that $g_n(x) = \left(1+\frac{x}{n}\right)^n$ converges uniformly to e^x for x in any interval $[a,b]$. By Theorem 2.12, then,

(a) $\left(1+\frac{x^2}{n}\right)^n = g_n(x^2)$ converges uniformly to e^{x^2};
(b) $\left(1-\frac{x}{n}\right)^n = g_n(-x)$ converges uniformly to e^{-x};
(c) $\log(g_n(x)) = n\log\left(1+\frac{x}{n}\right)$ converges uniformly to $\log(\mathrm{e}^x) = x$.

Problems

2.59. Use the identity $1+x+x^2+x^3+x^4 = \frac{1-x^5}{1-x}$ to estimate the accuracy of the approximation

$$1+x+x^2+x^3+x^4 \approx \frac{1}{1-x}$$

on $-\frac{1}{2} \le x \le \frac{1}{2}$.

2.60. In this problem, we explore another geometric meaning for geometric series. Refer to Fig. 2.43, where a line is drawn from the top point of the unit circle through the point (x,y) in the first quadrant of the circle. The point z where the line hits the axis is called the *stereographic projection* of the point (x,y). The shaded triangles are all similar. Justify the following statements.

(a) $z = \frac{x}{1-y}$.
(b) The height of the nth triangle is y times the height of the $(n-1)$st triangle.
(c) z is the sum of the series $z = x+xy+xy^2+xy^3+\cdots = \frac{x}{1-y}$.

2.61. We gave an outline of the proof of part (a) of Theorem 2.12. Let us fill in the details.

(a) Explain why

$$\left|f(x)+g(x)-(f_n(x)+g_n(x))\right| \le |f(x)-f_n(x)|+|g(x)-g_n(x)|$$

for all x.
(b) Explain why given any tolerance $\varepsilon > 0$, there is an N_1 such that $|f(x)-f_n(x)| < \frac{\varepsilon}{2}$ for all x when $n > N_1$, and why there is an N_2 such that $|g(x)-g_n(x)| < \frac{\varepsilon}{2}$ for all x when $n > N_2$.

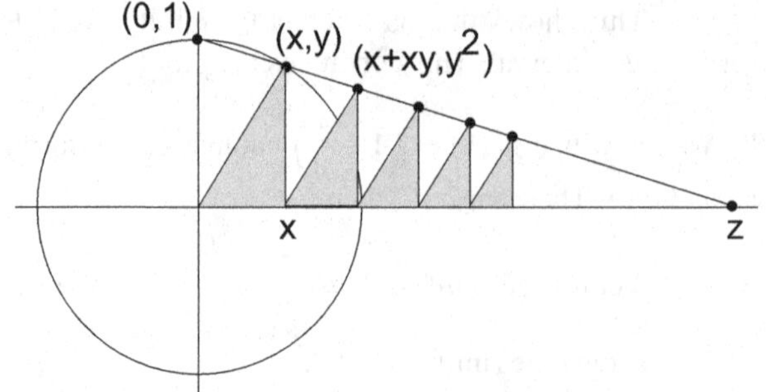

Fig. 2.43 Stereographic projection is the sum of a geometric series

(c) Explain why given any tolerance $\varepsilon > 0$, there is a number N such that

$$|f(x) - f_n(x)| + |g(x) - g_n(x)| < \varepsilon$$

for all x whenever $n > N$.
(d) Explain why given any tolerance $\varepsilon > 0$, there is a number N such that

$$\left| f(x) + g(x) - \left(f_n(x) + g_n(x)\right) \right| < \varepsilon$$

for all x whenever $n > N$.
(e) Explain why $f_n + g_n$ converges uniformly to $f + g$.

2.62. Use Theorem 2.12 to find an interval $a \le t \le b$ on which the convergence

$$1 + \mathrm{e}^{-t} + \mathrm{e}^{-2t} + \mathrm{e}^{-3t} + \cdots = \frac{1}{1 - \mathrm{e}^{-t}}$$

is uniform.

2.63. A power series $f(x) = \sum_{n=0}^{\infty} a_n(x-2)^n$ is known to converge at $x = 4$. At what other values of x must it converge? Find the largest open interval on which we can be sure that f is continuous.

2.64. For each pair of series, which one has the larger radius of convergence? In two cases, they have the same radius.

(a) $\sum_{n=0}^{\infty} x^n$ or $\sum_{n=0}^{\infty} 3^n x^n$
(b) $\sum_{n=0}^{\infty} x^n$ or $\sum_{n=0}^{\infty} \frac{x^n}{n!}$
(c) $\sum_{n=0}^{\infty} n(x-2)^n$ or $\sum_{n=0}^{\infty} (x-3)^n$
(d) $1 + x + \frac{x^2}{2!} + \frac{x^3}{3!} + \frac{x^4}{4!} + \cdots$ or $\frac{x^3}{3!} + \frac{x^4}{4!} + \frac{x^5}{5!} + \cdots$

2.65. Fill in the missing step that we have indicated in the proof of Theorem 2.13.

2.66. Which of these series represent a continuous function on (at least) $[-1,1]$?

(a) $\sum_{n=0}^{\infty} x^n$

(b) $\sum_{n=0}^{\infty} \left(\frac{1}{10}\right)^n x^n$

(c) $\sum_{n=0}^{\infty} \left(\frac{1}{10}\right)^n (x-2)^n$

(d) $1+x+\frac{x^2}{2!}+\frac{x^3}{3!}+\frac{x^4}{4!}+\cdots$

2.67. Consider a power series $\sum_{n=0}^{\infty} a_n x^n$. Suppose the limit $L = \lim_{n\to\infty} |a_n|^{1/n}$ exists and is positive. Justify the following steps, which prove that $1/L$ is the radius of convergence of the series. This is the *root test*.

(a) Let $\sum_{n=0}^{\infty} p_n$ be a series of positive numbers for which $\lim_{n\to\infty} p_n^{1/n} = \ell$ exists and $\ell < 1$. Show that there is a number r, $0 < \ell < r < 1$, such that for N large enough, $p_n < r^n$, $n > N$. Conclude that $\sum_{n=0}^{\infty} p_n$ converges.

(b) Let $\sum_{n=0}^{\infty} p_n$ be a series of positive numbers for which $\lim_{n\to\infty} p_n^{1/n} = \ell$ exists and $\ell > 1$. Show that there is a number r, $1 < r < \ell$, such that for N large enough, $p_n > r^n$, $n > N$. Conclude that $\sum_{n=0}^{\infty} p_n$ diverges.

(c) Taking $p_n = |a_n x^n|$ for different choices of x, show that $1/L$ is the radius of convergence of $\sum_{n=0}^{\infty} a_n x^n$.

2.68. Suppose $\{p_n\}$ is a positive sequence whose partial sums $p_1 + \cdots + p_n$ are less than nL for some number L. Use the root test (Problem 2.67) to show that the series $\sum_{n=1}^{\infty} (p_1 p_2 p_3 \cdots p_n) x^n$ converges in $|x| < 1/L$.

2.69. Suppose the root test (Problem 2.67) indicates that a series $\sum_{n=0}^{\infty} a_n x^n$ has radius of convergence R. Show that according to the root test, $\sum_{n=0}^{\infty} n a_n x^n$ also has radius of convergence R. (See Problem 1.53.)

2.70. For each of the following series, determine (i) the values of x for which the series converges; (ii) the largest open interval on which the sum is continuous.

(a) $\sum_{n=0}^{\infty} \frac{x^n}{2^n}$

(b) $\sum_{n=0}^{\infty} \frac{(x-3)^{2n}}{(2n)!}$

(c) $\sum_{n=0}^{\infty} \sqrt{n}x^n$

(d) $\sum_{n=0}^{\infty} \left(\frac{x^n}{2^n} + \sqrt{n}x^n\right)$

(e) $\sum_{n=1}^{\infty} \frac{2^n+7^n}{3^n+5^n}x^n$

2.71. For some of the following series it is possible to give an algebraic formula for the function to which the series converges. In those cases, give such a formula, and state the domain of the function where possible.

(a) $1 - t^2 + t^4 - t^6 + \cdots$

(b) $\sum_{n=3}^{\infty} x^n$ Note the 3.

(c) $\sum_{n=0}^{\infty} \sqrt{n}x^n$

(d) $\sum_{n=0}^{\infty} \left(\frac{t^n}{2^n} + 3^n t^{2n}\right)$

2.72. Our sequence of functions $s_n(x)$ approximating $\sqrt{x}$ was defined recursively. Write explicit expressions for $s_2(x)$ and $s_3(x)$, and verify that they are rational functions.

2.73. Use the method explained in Sect. 1.4 to show that for each $x > 0$, the sequence $e_n(x) = \left(1+\frac{x}{n}\right)^n$ is increasing.

2.74. Show that for each $x > 0$, the sequence $\{e_n(x)\}$ is bounded. *Hint*: For $x < 2$, $e_n(x) < \left(1+\frac{2}{n}\right)^n$. Set $n = 2m$ to conclude that $e_m(x) < e^2$.

2.75. Find a sequence of functions that converges to e^{-x} on every interval $[a,b]$ by composing the sequence $e_n(x) = \left(1+\frac{x}{n}\right)^n$ with a continuous function.

Chapter 3
The Derivative and Differentiation

Abstract Many interesting questions deal with the rate at which things change. Examples abound: What is the rate at which a population changes? How fast does radioactive material decay? At what rate is the national debt growing? At what rate does the temperature change as you move closer to a hot object? In this chapter, we define and discuss the concept of *rate of change*, which in mathematics, is called the *derivative*.

3.1 The Concept of Derivative

Among the instruments on the dashboard of a car there are two that indicate quantitative measurements: the odometer and the speedometer. We shall investigate the relation between these two (Fig. 3.1). To put the matter dramatically: Suppose your speedometer is broken; *is there any way of determining the speed of the car from the readings on the odometer* (so that, for example, you don't exceed the speed limit)?

Suppose the mileage reading at 2 o'clock was 5268, and 15 minutes later, it was 5280; then your *average* speed during that quarter-hour interval was

$$\frac{\text{distance covered}(\textit{miles})}{\text{time interval}(\textit{hours})} = \frac{5280-5268}{0.25} = \frac{12}{0.25} = 48\frac{\text{miles}}{\text{hour}}.$$

Denote by m the mileage reading as a function of time, i.e., the mileage reading at time t is $m(t)$. Then the *average speed* at time t, *averaged over a time interval of a quarter of an hour is*

$$\frac{m(t+0.25)-m(t)}{0.25}.$$

More generally, let h be any time interval. The *average speed over a time interval h* is

$$\frac{m(t+h)-m(t)}{h}.$$

P.D. Lax and M.S. Terrell, *Calculus With Applications*, Undergraduate Texts in Mathematics, 117
DOI 10.1007/978-1-4614-7946-8_3, © Springer Science+Business Media New York 2014

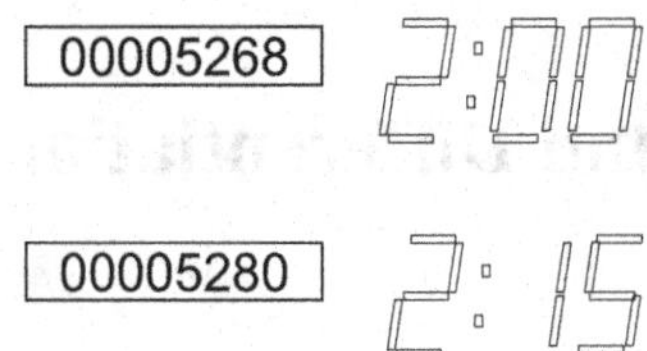

Fig. 3.1 Two readings from the instrument panel

The (unbroken) speedometer shows instantaneous speed, which is the limit of the average speed as h tends to 0.

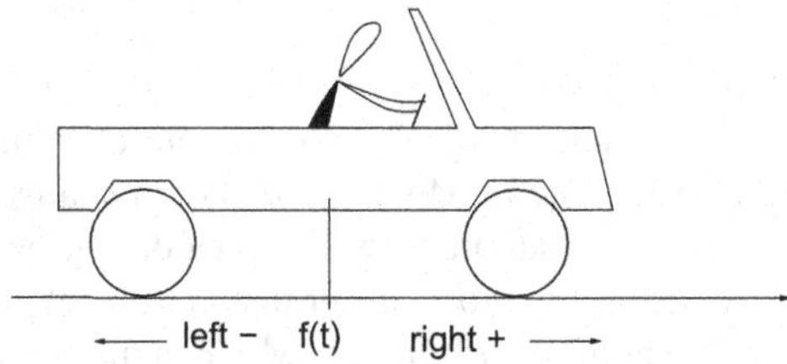

Fig. 3.2 The position $x = f(t)$ of the center of a car along a road marked as a number line

Velocity. Let $f(t)$ represent our *position* along a number line at time t (Fig. 3.2). The quotient $\dfrac{f(t+h)-f(t)}{h}$ is the average *velocity* during the time interval. This results in positive average velocity if the net change in position from an earlier to a later time is to the right, negative velocity if the net change in position is to the left, and zero when the positions are the same.

Example 3.1. Suppose the position function $f(t)$ is described by the formula

$$f(t) = 5000 + 35t + 2.5t^2.$$

Then the average velocity during the interval between t and $t+h$ is

$$\frac{f(t+h)-f(t)}{h} = \frac{5000+35(t+h)+2.5(t+h)^2-(5000+35t+2.5t^2)}{h}$$
$$= \frac{35h+5th+2.5h^2}{h} = 35+5t+2.5h.$$

Observe that as h tends to 0, the average velocity tends to $35+5t$. This quantity, the limit of average velocity over progressively shorter time intervals, is called the *instantaneous velocity*.

The process described above that derives velocity from position as a function of time is called *differentiation*. We now define this process without any reference to a physical model.

Definition 3.1. A function f is called *differentiable* at a if the *difference quotient*

$$\frac{f(a+h)-f(a)}{h}$$

tends to a limit as h tends to 0. This limit is called the *derivative* of f at a and is denoted by $f'(a)$:

$$f'(a)=\lim_{h\to 0}\frac{f(a+h)-f(a)}{h}. \tag{3.1}$$

Equation (3.1) can be interpreted as saying that the average rates of change in f computed over progressively smaller intervals containing a tend to a number $f'(a)$, the instantaneous rate of change in f at a.

The derivative of f, at points where it exists, yields another function f'. As we did for continuity, we extend the definition of derivative to allow for a *right* derivative at a, $\lim\limits_{h\to a+}\frac{f(a+h)-f(a)}{h}=f'_+(a)$, and a left derivative at a, $\lim\limits_{h\to a-}\frac{f(a+h)-f(a)}{h}=f'_-(a)$. We say that f is differentiable on $[a,b]$ if $f'(x)$ exists for each x in (a,b) and both $f'_+(a)$ and $f'_-(b)$ exist. If f is differentiable at every x in an interval I, we say that f is differentiable on I. If f is differentiable at every number x, we say that f is differentiable.

Next we look at examples where the derivative exists and can be easily found.

Example 3.2. Let $f(x)=c$ be any constant function. Using the definition to compute $f'(x)$, we see that

$$f'(x)=\lim_{h\to 0}\frac{f(x+h)-f(x)}{h}=\lim_{h\to 0}\frac{c-c}{h}=0.$$

Every constant function is differentiable, and its derivative is zero.

Example 3.3. Let $\ell(x)=mx+b$. Using the definition of the derivative, we obtain

$$\begin{aligned}\ell'(x)&=\lim_{h\to 0}\frac{\ell(x+h)-\ell(x)}{h}=\lim_{h\to 0}\frac{m(x+h)+b-(mx+b)}{h}\\&=\lim_{h\to 0}\frac{mh}{h}=\lim_{h\to 0}m=m,\end{aligned}$$

the slope of the graph.

Example 3.4. Let $f(x)=x^2$. Then

$$f'(x)=\lim_{h\to 0}\frac{(x+h)^2-x^2}{h}=\lim_{h\to 0}\frac{x^2+2xh+h^2-x^2}{h}=\lim_{h\to 0}(2x+h)=2x.$$

3.1a Graphical Interpretation

We saw in Example 3.3 that the derivative of a linear function ℓ is the slope of the graph of ℓ. Next we explore various graphical interpretations of the derivative for other functions.

Let f be differentiable at a. The points $(a, f(a))$ and $(a+h, f(a+h))$ determine a line called a *secant*. The slope of the secant is

$$\frac{f(a+h)-f(a)}{h}.$$

Since f is differentiable at a, these slopes tend to $f'(a)$ as h tends to 0, and the secants tend to the line through the point $(a, f(a))$ with slope $f'(a)$, called the line *tangent* to the graph of f at $(a, f(a))$ (Fig. 3.3).

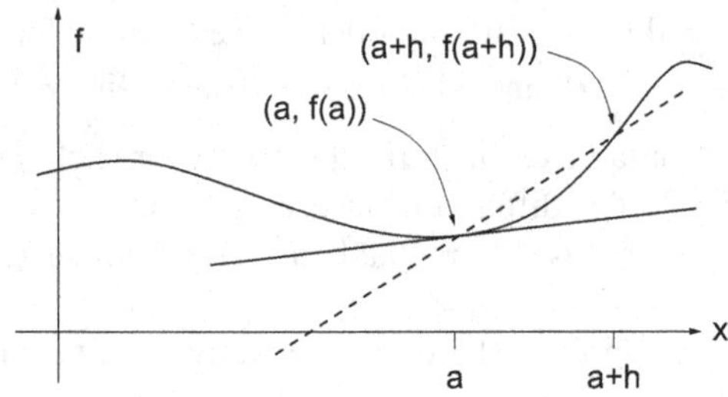

Fig. 3.3 A secant line and the tangent line through point $(a, f(a))$

If a function f is not differentiable at a, the limit of the slopes of secants does not exist. In some cases, the limit fails to exist because the secant slopes tend to ∞ or to $-\infty$. (See Fig. 3.4.) Then we say that the tangent line is vertical at $(a, f(a))$. If the limit fails to exist for some other reason, then there is no tangent line to the graph at $(a, f(a))$.

Example 3.5. $f(x) = x^{1/3}$ is not differentiable at $x = 0$, because $\lim_{h\to 0} \frac{(0+h)^{1/3}-0^{1/3}}{h} = \lim_{h\to 0} \frac{1}{h^{2/3}}$ does not exist. As h tends to zero, $\dfrac{1}{h^{2/3}}$ tends to positive infinity. Therefore, $f'(0)$ does not exist. Looking at the graph in Fig. 3.4, we see that the graph of f appears nearly vertical at $(0,0)$.

Example 3.6. The absolute value function

$$|x| = \begin{cases} x, & x \geq 0, \\ -x, & x < 0, \end{cases}$$

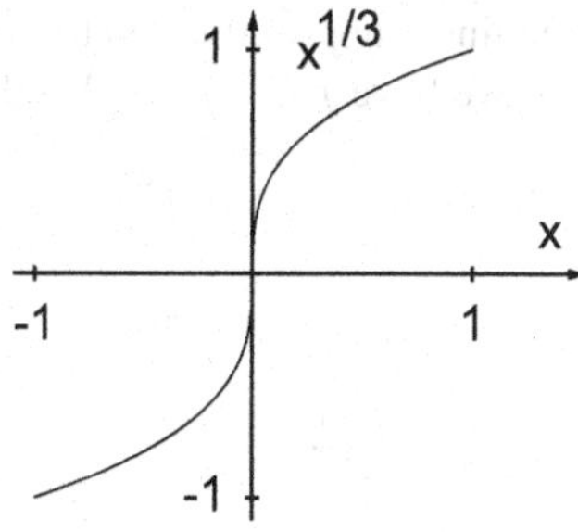

Fig. 3.4 The function $x^{1/3}$ is not differentiable at 0. See Example 3.5

is not differentiable at $x = 0$:

$$\frac{|0+h|-|0|}{h} = \begin{cases} \frac{h}{h} = 1, & h > 0, \\ \frac{-h}{h} = -1, & h < 0, \end{cases}$$

Therefore, $\lim_{h\to 0+} \frac{|0+h|-|0|}{h} = 1$ and $\lim_{h\to 0-} \frac{|0+h|-|0|}{h} = -1$. The left and right limits do not agree, and $\lim_{h\to 0} \frac{|0+h|-|0|}{h}$ does not exist. Looking at the graph of $|x|$ in Fig. 3.5, we see that to the left of $(0,0)$, the slope of the graph is -1, to the right the slope is 1, and at $(0,0)$ there is a sharp corner.

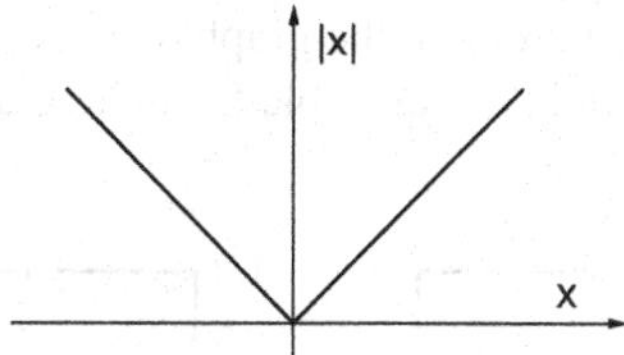

Fig. 3.5 The absolute value function is continuous, but it is not differentiable at 0. See Examples 3.6 and 3.8

Definition 3.2. The linear function

$$\ell(x) = f(a) + f'(a)(x-a)$$

is called the *linear approximation* to f at a.

One reason for calling ℓ the linear approximation to f at a is that it is the only linear function that has the following two properties:

(a) $\ell(a) = f(a)$,
(b) $\ell'(a) = f'(a)$.

Example 3.7. Let us find the linear approximation to $f(x) = x^2$ at -1 (Fig. 3.6). Since $f'(x) = (x^2)' = 2x$, we have $f'(-1) = -2$ and

$$\ell(x) = f(-1) + f'(-1)(x - (-1)) = (-1)^2 - 2(x+1) = -2x - 1.$$

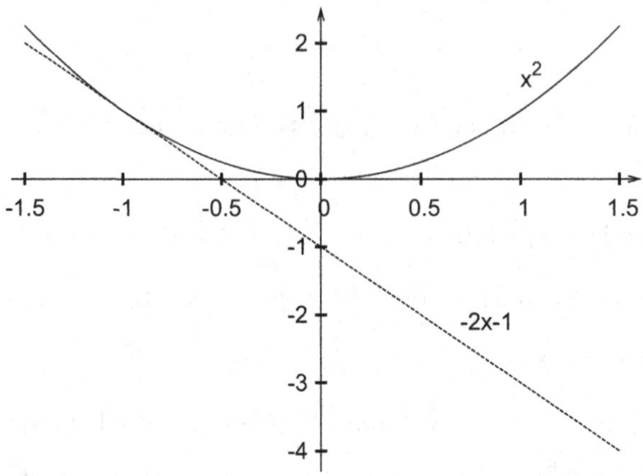

Fig. 3.6 The function $f(x) = x^2$ in Example 3.7 and its linear approximation $\ell(x) = -2x - 1$ at -1. The graph of ℓ is the tangent line at $(-1, 1)$

Figure 3.7 reveals another reason we call $\ell(x) = 2x + 1$ the linear approximation to x^2 at $x = -1$. The closer you look at the graph of x^2 near $x = -1$, the more linear it appears, becoming nearly indistinguishable from the graph of $\ell(x) = 2x + 1$.

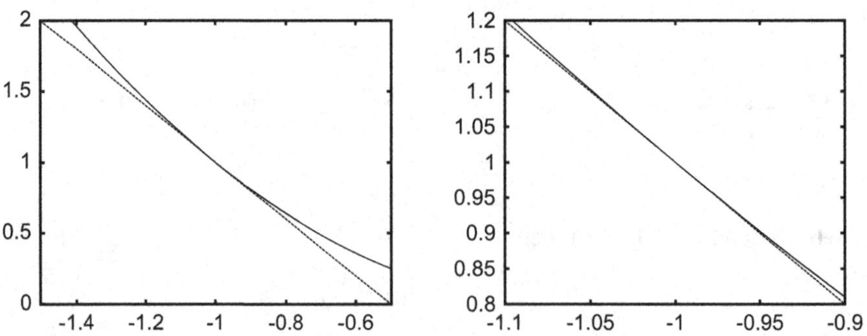

Fig. 3.7 A linear approximation to x^2 at -1 is seen at two different scales, as with a microscope. See Example 3.7

3.1b Differentiability and Continuity

It is not hard to see that a function that is differentiable at a is continuous at a. Differentiability requires that $f(a+h)$ tend to $f(a)$ at a rate proportional to h, whereas continuity merely requires that $f(a+h)$ tend to $f(a)$ as h tends to zero.

Theorem 3.1. Differentiability implies continuity. *A function that is differentiable at a is continuous at a.*

Proof. Since

$$f'(a) = \lim_{h\to 0} \frac{f(a+h)-f(a)}{h}$$

exists, the numerator must approach 0, and so $f(a+h)$ tends to $f(a)$ as h tends to 0. This is the definition of continuity of f at a. □

Next we show that continuity does not imply differentiability.

Example 3.8. The absolute value function is continuous at 0: as h tends to 0, $|h|$ does also. But as we saw in Example 3.6, $|x|$ is not differentiable at 0.

Example 3.9. The function $f(x) = x^{1/3}$ is continuous at $x = 0$, but $f'(0)$ does not exist, as we have seen in Example 3.5.

Example 3.10. Consider the function

$$f(x) = x\sin\left(\frac{\pi}{x}\right) \qquad \text{when } x \neq 0, \qquad f(0) = 0.$$

The graph of f is shown in Fig. 3.8. When $x \neq 0$,

$$-|x| \le x\sin\left(\frac{\pi}{x}\right) \le |x|.$$

As x tends to 0, both $-|x|$ and $|x|$ tend to 0. Therefore, by the squeeze theorem, Theorem 2.2,

$$\lim_{x\to 0} f(x) = \lim_{x\to 0} x\sin\left(\frac{\pi}{x}\right) = 0 = f(0),$$

and f is continuous at 0. On the other hand, the difference quotient

$$\frac{f(h)-f(0)}{h} = \frac{h\sin\left(\frac{\pi}{h}\right)-0}{h} = \sin\left(\frac{\pi}{h}\right)$$

takes on all values between 1 and -1 infinitely often as h tends to 0. Thus $\lim_{h\to 0} \sin\left(\frac{\pi}{h}\right)$ does not exist. Therefore, f is continuous at 0, but $f'(0)$ does not exist.

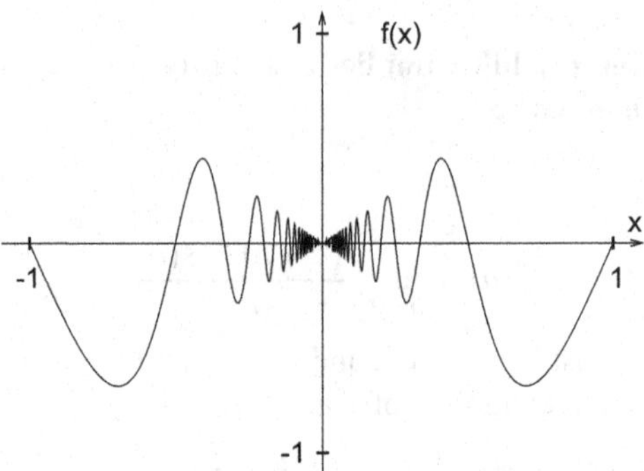

Fig. 3.8 The graph of $f(x) = x\sin\left(\frac{\pi}{x}\right)$, $f(0) = 0$. The function f is continuous but not differentiable at 0. See Example 3.10

In fact, a continuous function may fail to be differentiable at many points (See Fig. 3.9).

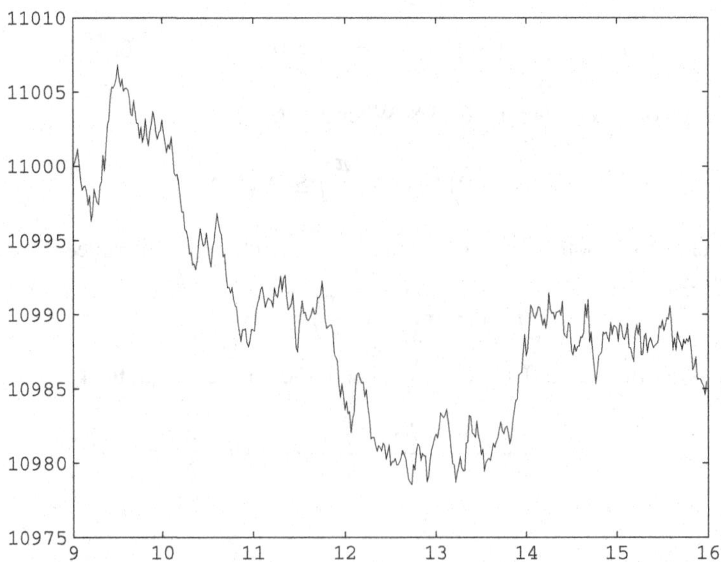

Fig. 3.9 A function tracking share prices in a stock market between 9:00 AM and 4:00 PM appears to have no derivative at any point

3.1c Some Uses for the Derivative

The Derivative as Stretching. Let $f(x) = x^2$. Figure 3.10 compares the length of the interval from x to $x+h$ to the length of the interval from x^2 to $(x+h)^2$:

$$\frac{(x+h)^2 - x^2}{h} = \frac{x^2 + 2xh + h^2 - x^2}{h} = \frac{2hx + h^2}{h} = 2x + h.$$

For small h, the interval $[x^2, (x+h)^2]$ is about $2x$ times as long as the interval from $[x, x+h]$, so $f'(x)$ is sometimes interpreted as a stretching factor.

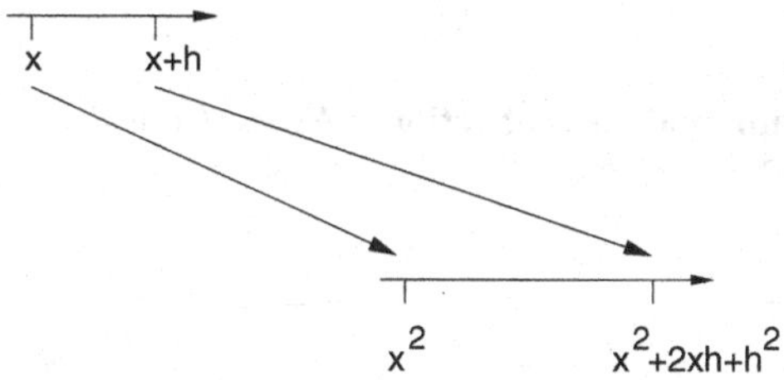

Fig. 3.10 The effect of squaring the numbers in a short interval

The Derivative as Sensitivity to Change. The derivative can be used to estimate how a small change in the input affects the output of f. When h is near zero, $\dfrac{f(a+h) - f(a)}{h}$ is nearly $f'(a)$, and so $f(a+h) - f(a)$ is approximately $hf'(a)$. The product $hf'(a)$ is called a *differential*, and it depends on both h and a. Let us look at an example.

Suppose we square 1.000 instead of 1.001, and 10,000 instead of 10,000.001. In each case, we have changed the inputs by 0.001. How sensitive is the output to a change in the input? Using our knowledge that $(x^2)' = 2x$ and the estimate $f(a+h) - f(a) \approx hf'(a)$, we see that

$$\begin{aligned} (1.001)^2 - (1)^2 &\approx 2(1)(0.001) = 0.002 \\ (10000.001)^2 - (10000)^2 &\approx 2(10000)(0.001) = 20. \end{aligned}$$

The function f is more sensitive to a change in the input at $x = 10{,}000$ than at $x = 1$. If we visualize the graph of x^2, we see why. The graph is much steeper at $x = 10{,}000$ than at $x = 1$.

The Derivative as a Density. Consider a rod of unit cross-sectional area, made from some material whose properties may vary along its length. Let $R(x)$ be the mass of that portion of the rod to the left of point x. Then the average density of the rod [mass/length] between x and $x+h$ is

$$\frac{R(x+h) - R(x)}{x+h-x} = \frac{R(x+h) - R(x)}{h} = \frac{\text{the mass within interval } [x, x+h]}{h}.$$

So the density *at* the point x is the limit of the average density taken over smaller thicknesses:

$$\lim_{h \to 0} \frac{R(x+h) - R(x)}{h} = R'(x).$$

The function $R'(x)$ is called the linear density of the rod at the point x (Fig. 3.11).

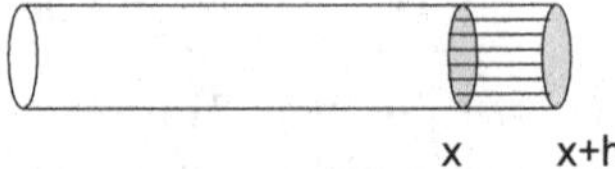

Fig. 3.11 The mass of the rod between x and $x+h$ is $R(x+h) - R(x)$

Volume of Revolution and Cross-Sectional Area. Consider a solid of revolution as shown in Fig. 3.12.

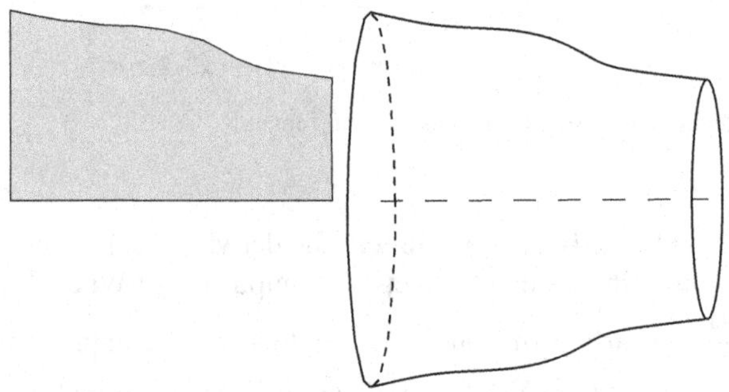

Fig. 3.12 *Left*: A region to be revolved around an axis to produce a solid of revolution. *Right*: The solid of revolution

Let $V(x)$ be the volume of the solid that lies to the left of the plane through x perpendicular to the axis of rotation, as in Fig. 3.13. Let $A(x)$ be the cross-sectional area of the solid at x and assume that A varies continuously with x. By the extreme value theorem, A has a maximum value A_M and a minimum A_m on the interval $[x, x+h]$. Consider the quotient

$$\frac{V(x+h) - V(x)}{x+h-x} = \frac{V(x+h) - V(x)}{h},$$

which is the volume of the solid between the planes divided by the distance between the planes.

Since this segment of the solid fits between larger and smaller cylinders, the numerator is bounded by the cylinder volumes

$$A_m h \le V(x+h) - V(x) \le A_M h$$

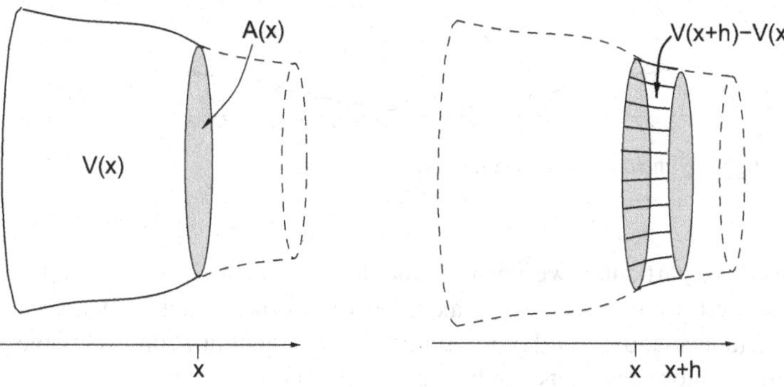

Fig. 3.13 *Left*: $V(x)$ is the volume to the left of x, and $A(x)$ is the area of the cross section at x. *Right*: $V(x+h)-V(x)$ is the volume between x and $x+h$

(taking $h > 0$), so

$$A_m \leq \frac{V(x+h)-V(x)}{h} \leq A_M.$$

Because $A(x)$ is continuous, A_M and A_m both tend to $A(x)$ as h tends to zero. Therefore,

$$V'(x) = A(x).$$

An Application of the Tangent to a Curve. Let us use our ability to find the tangent to a curve to investigate reflections from a mirror located in the (x,y)-plane whose parabolic shape is described by the equation

$$y = x^2.$$

First we state the laws of reflection.

- In a uniform medium, light travels in straight lines.
- When a ray of light impinges on a straight mirror, it is reflected; the angle i that the incident ray forms with the line *perpendicular* to the mirror equals the angle r that the reflected ray forms with the perpendicular; see Fig. 3.14. The angle i is called the *angle of incidence*; r, the *angle of reflection*.
- The same rule governs the reflection of light from a curved mirror:

$$\text{angle of incidence} = \text{angle of reflection};$$

in this case, the line perpendicular to the mirror is defined as the *line perpendicular to the tangent to the mirror at the point of incidence*.

In Sect. 5.4, we shall use calculus to deduce the laws of reflection from Fermat's principle that light takes the path that takes the least time.

We consider light rays that descend parallel to the y-axis, approaching a parabolic mirror as indicated in Fig. 3.15. We wish to calculate the path along which these rays

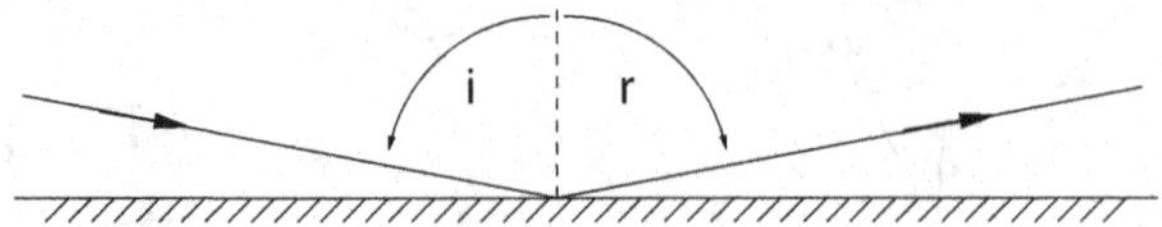

Fig. 3.14 A light ray incident on a mirror, $i = r$

are reflected. In particular, we wish to calculate the location of the point F where such a reflected ray intersects the y-axis. First a bit of geometry. Denote by P the point of incidence and by G the point where the tangent at P intersects the y-axis. The following geometric facts can be read off the figure:

- The angle FPG is complementary to the angle of reflection r.
- Since the incident ray is parallel to the y-axis, the angle FGP is equal to an angle that is complementary to the angle of incidence i.

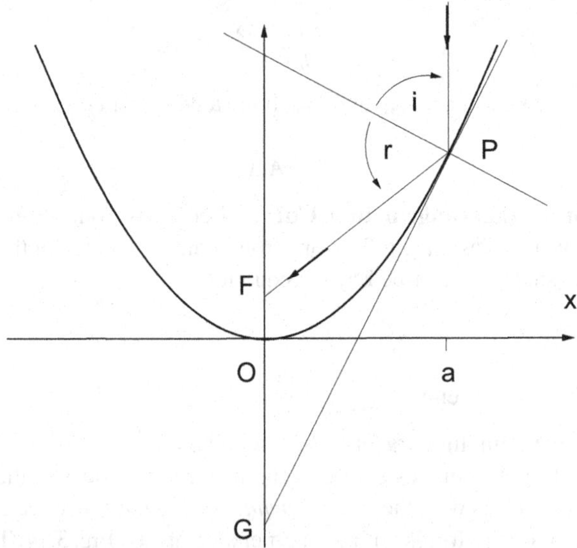

Fig. 3.15 The graph of $f(x) = x^2$, considered as a reflecting mirror. An incident light ray enters vertically from above and reflects to point F, also known as the focus. "Focus" is Latin for fireplace, and you can use a parabolic mirror to ignite an object at F

Since according to the law of reflection, r equals i, we conclude that the triangle FGP is isosceles, the angles at P and G being equal. From this, we conclude that the sides opposite are equal:

$$PF = FG.$$

Here PF denotes the distance from P to F. We shall calculate the length of these sides. We denote the x-coordinate of P by a. Since $y = x^2$, the y-coordinate of P is a^2. We denote the y-coordinate of the point F by k. By the Pythagorean theorem,

$$(PF)^2 = a^2 + (a^2 - k)^2.$$

Next we use calculus to calculate the y-coordinate of the point G. Since G is the intersection of the tangent with the y-axis, the y-coordinate of G is the value at $x = 0$ of the linear approximation ℓ to f at a,

$$\ell(x) = f(a) + f'(a)(x - a).$$

At $x = 0$,

$$\ell(0) = f(a) - f'(a)a.$$

In our case, $f(x) = x^2$, so $f'(x) = 2x$, and

$$\ell(0) = f(a) - f'(a)a = a^2 - 2a^2 = -a^2.$$

The length FG is the difference of the y-coordinates of F and G:

$$FG = k - (-a^2) = k + a^2.$$

Since $PF = FG$, we get

$$a^2 + (a^2 - k)^2 = (PF)^2 = (FG)^2 = (k + a^2)^2.$$

Now carry out the squarings, cancel common terms, and add $2a^2k$ to both sides. The result is $a^2 = 4a^2k$. So $k = \frac{1}{4}$. This gives the surprising result that the location of the point F is *the same* for all points P, i.e., *the reflections of all rays parallel to the y-axis pass through the point* $(0, \frac{1}{4})$. This point is called the *focus* of the parabola.

Rays coming from a very distant object such as one of the stars are very nearly parallel. Therefore, if a parabolic mirror is pointed so that its axis points in the direction of a star, all the rays will be reflected toward the focus; this principle is exploited in the construction of telescopes.

The rays from the sun are nearly parallel, and therefore they can be focused quite accurately by a parabolic mirror. This principle is exploited in the construction of solar furnaces.

Importance of the Derivative. One could argue persuasively that *changes* in magnitudes are often more important than the magnitudes themselves, and therefore, the rate at which the value of a function changes from point to point or moment to moment is more relevant than its actual value. For example, it is often more useful to know whether the outside temperature tomorrow is going to be higher, lower, or the same as today, than to know tomorrow's temperature without knowing today's.

Theoretical weather predictions are based on theories that relate the rate of change of meteorologically relevant quantities such as temperature, atmospheric pressure, and humidity to factors that cause the change. The mathematical formulation of these theories involves equations relating the derivatives of these meteorological variables to each other. These are called *differential equations*. A differential equation is one that relates an unknown function to one or more of its derivatives. Almost all physical theories and quantities arising in mechanics, optics, the theories of heat and of sound, etc., are expressed as differential equations. We shall explore examples from mechanics, population dynamics, and chemical kinetics in Chap. 10.

We conclude with a brief dictionary of familiar rates of change:

- *Speed* $\leftrightarrow$ rate of change of distance as a function of time.
- *Velocity* $\leftrightarrow$ rate of change of position as a function of time.
- *Acceleration* $\leftrightarrow$ rate of change of velocity as a function of time.
- *Angular velocity* $\leftrightarrow$ rate of change of angle as a function of time.
- *Density* $\leftrightarrow$ rate of change of mass as a function of volume.
- *Slope* $\leftrightarrow$ rate of change of height as a function of horizontal distance.
- *Current* $\leftrightarrow$ rate of change of the amount of electric charge as a function of time.
- *Marginal cost* $\leftrightarrow$ rate of change of production cost as a function of the number of items produced.

That so many words in common use denote rates of change of other quantities is an eloquent testimony to the importance of the notion of derivative.

Problems

3.1. Find the line tangent to the graph of $f(x) = 3x - 2$ at the point $a = 4$.

3.2. Find the line tangent to the graph of $f(x) = x^2$ at an arbitrary point a. Where does the tangent line intersect the x-axis? the y-axis?

3.3. The linear approximation to f at $a = 2$ is $\ell(x) = 5(x-2)+6$. Find $f(2)$ and $f'(2)$.

3.4. A metal rod lies along the x-axis. The mass of the part of the rod to the left of x is $R(x) = 25 + \frac{1}{5}x^3$. Find the average linear density of the part in $[2,5]$.

3.5. For each function f given below, find $f'(a)$ by forming the difference quotient $\dfrac{f(a+h)-f(a)}{h}$, and taking the limit as h tends to 0. Then find the line tangent to the graph of f at $x = a$.

(a) $f(x) = \sqrt{x},\ a = 4$.
(b) $f(x) = mx^2 + kx,\ a = 2$.
(c) $f(x) = x^3,\ a = -1$.

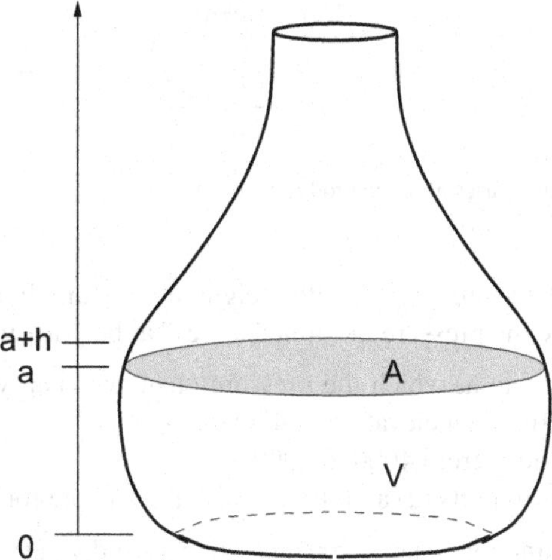

Fig. 3.16 A bottle of water for Problem 3.6

3.6. The volume of water in a bottle is a function of the depth of the water. Let $V(a)$ be the volume up to depth a. Let $A(a)$ be the cross-sectional area of the bottle at height a.

(a) What does $V(a+h)-V(a)$ represent? ($h>0$).
(b) Use Fig. 3.16 to put the following quantities in order:

$$A(a+h)h, \qquad A(a)h, \qquad V(a+h)-V(a).$$

Write your answer using an inequality.
(c) Using your inequality from part (b), explain why the quotient $\dfrac{V(a+h)-V(a)}{h}$ tends to $A(a)$ as h tends to 0.

3.7. Find a line that is tangent to the graphs of *both* functions $f(x)=x^2$ and $g(x)=x^2-2x$.

3.8. Find the point of intersection of the tangents to the graphs of $f_1(x)=x^2-2x$ and $f_2(x)=-x^2+1$ at the points (2,0) and (1,0), respectively.

3.9. The temperature of a rod at a given time varies along its length. See Fig. 3.17. Let $T(x)$ be the temperature at point x.

(a) Write an expression for the average rate of change in the temperature between points a and $a+h$ on the rod.
(b) Suppose $T'(a)$ is positive. Is it hotter to the left or to the right of a?
(c) If it is cooler just to the left of a, would you expect $T'(a)$ to be positive or negative?
(d) If the temperature is constant, what is T'?

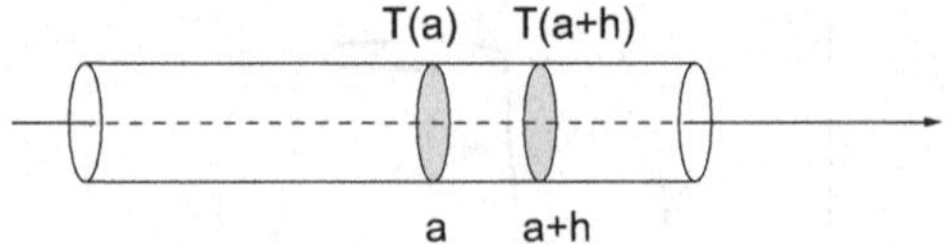

Fig. 3.17 The temperature varies along the rod in Problem 3.9

3.10. Atmospheric pressure varies with height above the Earth's surface. See Fig. 3.18. Let $p(x)$ be the pressure in "atmospheres" at height x in meters.

(a) Find the average rate at which the pressure changes when you move from an elevation of 2000 m to an elevation of 4000 m.
(b) Repeat (a) for a move from 4000 to 6000.
(c) Use your answers to parts (a) and (b) to make an estimate for $p'(4000)$.

3.11. Which of the functions $f(x) = 5x,\ g(x) = x^2$ would you say is more sensitive to change near $x = 3$?

3.12. Find all tangents to the graph of $f(x) = x^2 - x$ that go through the point $(2,1)$. Verify that no tangent goes through the point $(2,3)$. Can you find a geometric explanation for this?

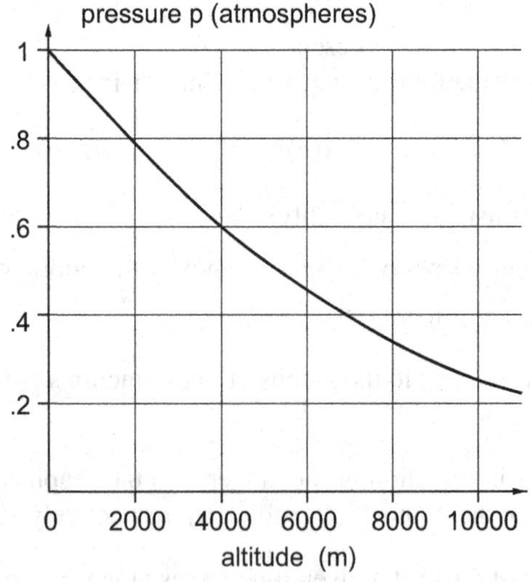

Fig. 3.18 Atmospheric pressure as a function of altitude, in Problem 3.10

3.13. Let $g(x) = x^2 \sin\left(\frac{\pi}{x}\right),\ g(0) = 0.$

(a) Sketch the graph of g on $-1 \le x \le 1$.
(b) Show that g is continuous at 0.
(c) Show that g is differentiable at 0 and find $g'(0)$.

3.14. Show that a function that is not continuous at a cannot be differentiable at a.

3.15. Let $f(x) = x^{2/3}$. Does the one-sided derivative $f'_+(0)$ exist? Is f differentiable on $[0,1]$?

3.2 Differentiation Rules

We often have occasion to form new functions out of given functions by addition, multiplication, division, composition, and inversion. In this section, we show that the sums, products, quotients, compositions, and inverses of differentiable functions are likewise differentiable, and we shall see how to express their derivatives in terms of the component functions and their derivatives.

3.2a Sums, Products, and Quotients

Theorem 3.2. Derivative of sums, differences, and constant multiples. *If f and g are differentiable at x, and c is any constant, then $f+g$, $f-g$, and cf are differentiable at x, and*

$$\begin{aligned}(f+g)'(x) &= f'(x)+g'(x)\\(f-g)'(x) &= f'(x)-g'(x)\\(cf)'(x) &= cf'(x)\end{aligned}$$

Proof. The proofs of all three assertions are straightforward and are based on limit rules (Theorem 2.1) and observations that relate the difference quotients of $f+g$, $f-g$, and cf to those of f and g. We also use the existence of the derivatives $f'(x)$ and $g'(x)$ (Fig. 3.19):

$$\begin{aligned}(f+g)'(x) &= \lim_{h\to 0}\left(\frac{f(x+h)+g(x+h)-(f(x)+g(x))}{h}\right)\\&= \lim_{h\to 0}\left(\frac{f(x+h)-f(x)}{h}+\frac{g(x+h)-g(x)}{h}\right)\\&= \lim_{h\to 0}\frac{f(x+h)-f(x)}{h}+\lim_{h\to 0}\frac{g(x+h)-g(x)}{h} = f'(x)+g'(x).\end{aligned}$$

The proof that $(f-g)'(x) = f'(x)-g'(x)$ is similar. Next

$$(cf)'(x) = \lim_{h\to 0}\frac{cf(x+h)-cf(x)}{h} = c\lim_{h\to 0}\frac{f(x+h)-f(x)}{h} = cf'(x).$$

□

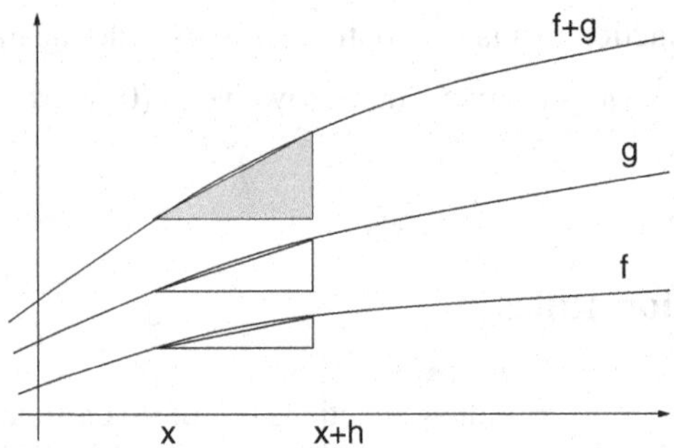

Fig. 3.19 The slope of the graph of $f+g$ at x is the sum of the slopes of the graphs of f and g

Quite analogously, the sum of finitely many functions that are differentiable at x is differentiable at x.

Theorem 3.3. Product rule. *If f and g are differentiable at x, then their product is differentiable at x, and*

$$(fg)'(x) = f(x)g'(x) + f'(x)g(x). \tag{3.2}$$

Proof. For the first two steps see Fig. 3.20.

$$(fg)'(x) = \lim_{h\to 0} \frac{f(x+h)g(x+h) - f(x)g(x)}{h}$$

$$= \lim_{h\to 0} \frac{f(x+h)g(x+h) - f(x+h)g(x) + f(x+h)g(x) - f(x)g(x)}{h}$$

$$= \lim_{h\to 0} \left(f(x+h)\frac{g(x+h) - g(x)}{h} + \frac{f(x+h) - f(x)}{h} g(x) \right)$$

$$= \lim_{h\to 0} f(x+h) \lim_{h\to 0} \frac{g(x+h) - g(x)}{h} + \left(\lim_{h\to 0} \frac{f(x+h) - f(x)}{h} \right) g(x)$$

$$= f(x)g'(x) + f'(x)g(x).$$

In the last two steps we used that f and g are differentiable at x, and that f is continuous at x. □

Example 3.11. Note that the function x^2 is a product. By the product rule,

$$(x^2)' = (xx)' = xx' + x'x = x + x = 2x.$$

Example 3.12. The function x^3 is also a product, so

$$(x^3)' = (x^2)'x + x^2x' = 2xx + x^2 = 3x^2.$$

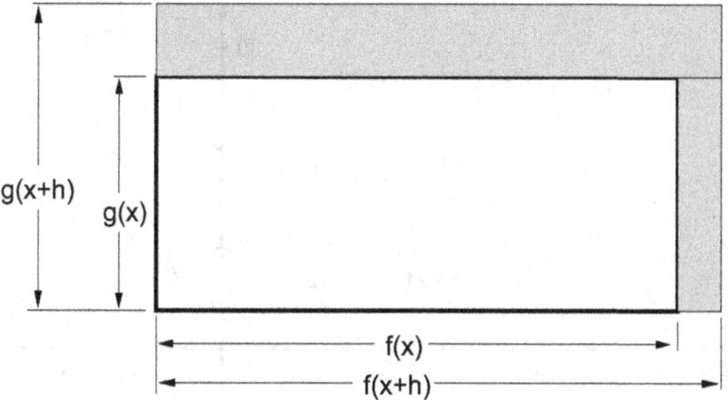

Fig. 3.20 The difference of the products $f(x+h)g(x+h) - f(x)g(x)$ was expressed as a sum of the two shaded areas in the proof of Theorem 3.3

Theorem 3.4. Power rule. *For every positive integer n,*

$$(x^n)' = nx^{n-1}. \tag{3.3}$$

Proof. We use mathematical induction. We prove the inductive step: if the result holds for $n-1$, then it holds for n. Since the result holds for $n = 1$, namely $(x)' = 1x^0$, its validity will then follow for $n = 2, 3, \ldots$. This is the type of step we made in Example 3.12, in going from knowledge of $(x^2)'$ to $(x^3)'$.

To prove the inductive step, we write $x^n = x^{n-1}x$ and apply the product rule:

$$(x^n)' = (x^{n-1}x)' = (x^{n-1})'x + x^{n-1}x'.$$

Using the assumed validity of $(x^{n-1})' = (n-1)x^{n-2}$, we get

$$(x^n)' = (x^{n-1})'x + x^{n-1}x' = (n-1)x^{n-2}x + x^{n-1} = (n-1)x^{n-1} + x^{n-1} = nx^{n-1}.$$

□

Let p be a polynomial function of the form

$$p(x) = a_n x^n + a_{n-1}x^{n-1} + \cdots + a_0.$$

Using the rules for differentiating a sum, a constant multiple, and the formula already verified for the derivative of x^n, we obtain

$$p'(x) = na_n x^{n-1} + (n-1)a_{n-1}x^{n-2} + \cdots + a_1.$$

Example 3.13. If $p(x) = 2x^4 + 7x^3 + 6x^2 + 2x + 15$, then

$$p'(x) = 4(2)x^3 + 3(7)x^2 + 2(6)x + 1(2)x^0 + 0 = 8x^3 + 21x^2 + 12x + 2.$$

The functions p and p' are graphed in Fig. 3.21.

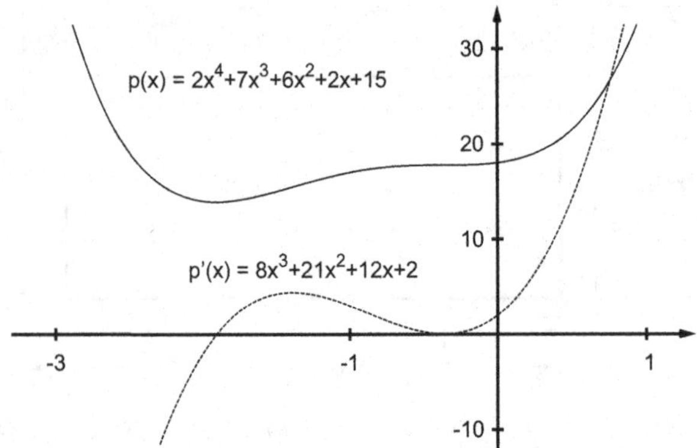

Fig. 3.21 Graphs of the polynomial $p(x)$ and its derivative $p'(x)$ in Example 3.13

We now turn to division.

Theorem 3.5. Reciprocal rule. *If f is differentiable at x and $f(x) \neq 0$, then $\frac{1}{f}$ is differentiable at x, and*

$$\left(\frac{1}{f}\right)'(x) = -\frac{f'(x)}{\left(f(x)\right)^2}. \tag{3.4}$$

Proof. $\left(\frac{1}{f}\right)'(x) = \lim_{h\to 0} \frac{\frac{1}{f(x+h)} - \frac{1}{f(x)}}{h}$. By adding the fractions in the numerator and regrouping terms, we get

$$\left(\frac{1}{f}\right)'(x) = \lim_{h\to 0} \frac{\frac{f(x)-f(x+h)}{f(x+h)f(x)}}{h} = \lim_{h\to 0} \frac{f(x)-f(x+h)}{hf(x+h)f(x)}$$

$$= \lim_{h\to 0} \left(\frac{f(x+h)-f(x)}{h} \frac{-1}{f(x)f(x+h)}\right) = \lim_{h\to 0} \frac{f(x+h)-f(x)}{h} \lim_{h\to 0} \frac{-1}{f(x)f(x+h)},$$

assuming that these last two limits exist. The left limit exists and is $f'(x)$. For the limit on the right, f is continuous at x, and $f(x) \neq 0$. Hence there is an interval around x where f is not 0, so $\lim_{h\to 0} \frac{-1}{f(x+h)} = \frac{-1}{f(x)}$. By the limit laws, the product is equal to $f'(x)\frac{-1}{f(x)f(x)} = -\frac{f'(x)}{(f(x))^2}$. □

Example 3.14. By the reciprocal rule, $\left(\frac{1}{x^2+1}\right)' = -\frac{(x^2+1)'}{(x^2+1)^2} = -\frac{2x}{(x^2+1)^2}$.

Example 3.15. We use the reciprocal rule to calculate $(x^{-3})'$ when $x \neq 0$:

$$(x^{-3})' = \left(\frac{1}{x^3}\right)' = -\frac{(x^3)'}{(x^3)^2} = -\frac{3x^2}{(x^3)^2} = -3x^{-4}.$$

Example 3.16. The power rule can be extended to negative integers by the reciprocal rule when $x \neq 0$:

$$(x^{-n})' = \left(\frac{1}{x^n}\right)' = -\frac{(x^n)'}{(x^n)^2} = -\frac{nx^{n-1}}{(x^n)^2} = -nx^{-n-1}.$$

This shows that the power rule, previously known for positive integer exponents, is also valid when the exponent is a negative integer.

More generally, any quotient may be viewed as a product: $\frac{f}{g} = f\frac{1}{g}$. So if $g(x) \neq 0$, by the product rule we have

$$\left(\frac{f}{g}\right)'(x) = f(x)\frac{-g'(x)}{(g(x)^2)} + f'(x)\frac{1}{g(x)} = \frac{g(x)f'(x) - f(x)g'(x)}{(g(x))^2}.$$

This proves the quotient rule:

Theorem 3.6. Quotient rule. *If f and g are differentiable at x and $g(x) \neq 0$, then their quotient is differentiable at x, and*

$$\left(\frac{f}{g}\right)'(x) = \frac{g(x)f'(x) - f(x)g'(x)}{(g(x))^2}.$$

Since we know how to differentiate polynomials, we can use the quotient rule to differentiate any rational function.

Example 3.17. For $f(x) = x$ and $g(x) = x^2 + 1$, the derivative of the quotient is

$$\left(\frac{x}{x^2+1}\right)' = \frac{(x^2+1)(x)' - x(x^2+1)'}{(x^2+1)^2} = \frac{(x^2+1)1 - x(2x)}{(x^2+1)^2} = \frac{1-x^2}{(x^2+1)^2}.$$

3.2b Derivative of Compositions of Functions

As we saw in Sect. 2.3a, when we compose linear functions, the rates of change, i.e., the derivatives, are multiplied. Roughly speaking, this applies for all functions (Fig. 3.22).

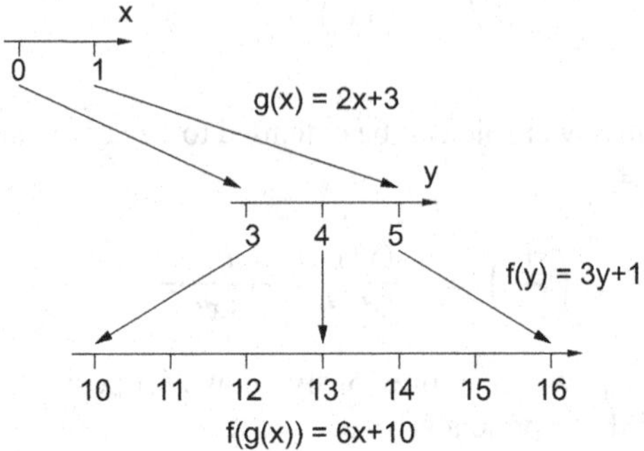

Fig. 3.22 A composite of two linear functions. The rate of change of $f \circ g$ is the product of the rates of change: $(f \circ g)' = f'g' = (3)(2) = 6$

For continuous f and g, recall the continuity of $f(g(x))$: a small change in x causes a small change in $g(x)$, and that causes a small change in $f(g(x))$. Now with differentiable f and g we can quantify these small changes.

Theorem 3.7. Chain rule. *If f is differentiable at $g(x)$ and g is differentiable at x, then $f \circ g$ is differentiable at x and*

$$(f \circ g)'(x) = f'(g(x))g'(x). \tag{3.5}$$

Proof. We distinguish two cases:

(a) $g'(x) \neq 0$
(b) $g'(x) = 0$

In case (a), $g(x+h) - g(x)$ is not zero for h small enough. Let

$$k = g(x+h) - g(x),$$

and write

$$\frac{f(g(x+h)) - f(g(x))}{h} = \frac{f(g(x)+k) - f(g(x))}{k} \frac{g(x+h) - g(x)}{h}. \tag{3.6}$$

The first factor on the right tends to $f'(g(x))$, and the second factor to $g'(x)$, giving formula (3.5).

In case (b), the chain rule will be confirmed if we show that $(f(g(x)))'$ is zero, that is, if we show that the left-hand side of Eq. (3.6) tends to zero as h tends to zero. The problem is the k in the denominator of the first factor on the right side of Eq. (3.6). When k is not zero, the factors on the right tend to the product of $f'(g(x))$ and $g'(x)$, and therefore tend to zero. Now if k is zero, the difference between $g(x+h)$ and $g(x)$ is zero, and so the difference quotient of the left is zero. So as h tends to zero, k tends to zero or is zero sometimes, and in either event the left side of Eq. (3.6) tends to zero. □

Example 3.18. $(x^2-x+5)^4$ is $f \circ g$, with $g(x) = x^2-x+5$ and $f(y) = y^4$. Using the power rule and the chain rule, we get

$$((x^2-x+5)^4)' = 4(x^2-x+5)^3(x^2-x+5)' = 4(x^2-x+5)^3(2x-1)$$

Derivative of the Inverse of a Function. Suppose f is a strictly monotonic function whose derivative is not zero on some open interval, and that f and g are inverse to each other,

$$f(g(x)) = x.$$

In Problem 3.35, we help you to verify that g is differentiable. Anticipating this result, differentiate both sides. By the chain rule, we obtain

$$f'(g(x))g'(x) = 1.$$

Therefore, the derivative of g at x is the reciprocal of the derivative of f at $g(x)$:

$$g'(x) = \frac{1}{f'(g(x))} \tag{3.7}$$

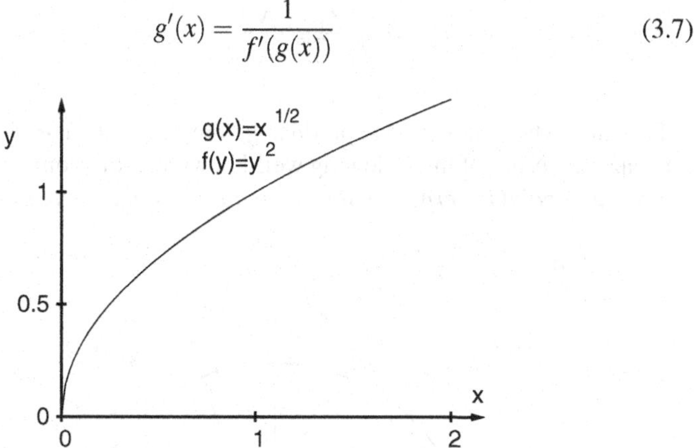

Fig. 3.23 The graph of a function g and its inverse f, viewed together. See Example 3.19

Example 3.19. Let $f(y) = y^2$. Then f is invertible on the interval $y > 0$. The inverse is (Fig. 3.23)

$$y = g(x) = x^{1/2} = \sqrt{x}.$$

Since $f'(y) = 2y$, Eq. (3.7) gives

$$g'(x) = \frac{1}{f'(g(x))} = \frac{1}{2g(x)} = \frac{1}{2\sqrt{x}}.$$

Therefore,

$$\left(x^{1/2}\right)' = \frac{1}{2}x^{-1/2}.$$

This shows that the power rule $(x^n)' = nx^{n-1}$ holds for $n = \dfrac{1}{2}$.

Previously, we proved the power rule for integer exponents only.

Example 3.20. The graph of the function $f(x) = \sqrt{1-x^2} = (1-x^2)^{1/2}$, defined on the interval $-1 \le x \le 1$, is a semicircle with radius 1, centered at the origin. See Fig. 3.24. By the chain rule and the power rule for the exponent $n = \frac{1}{2}$, we have

$$f'(x) = \frac{1}{2}(1-x^2)^{-1/2}(-2x) = \frac{-x}{(1-x^2)^{1/2}} \qquad (-1 < x < 1).$$

The slope of the tangent at the point $(a, \sqrt{1-a^2})$ is then $\dfrac{-a}{(1-a^2)^{1/2}}$. The slope of the line through the origin and the point $(a, \sqrt{1-a^2})$ is $\dfrac{\sqrt{1-a^2}}{a}$. The product of these two slopes is

$$-\frac{a}{\sqrt{1-a^2}} \cdot \frac{\sqrt{1-a^2}}{a} = -1.$$

Two lines whose slopes have product -1 are perpendicular. Thus we have given an analytic proof of the following well-known fact of geometry: *the tangent to a circle at a point is perpendicular to the radius of the circle through that point.*

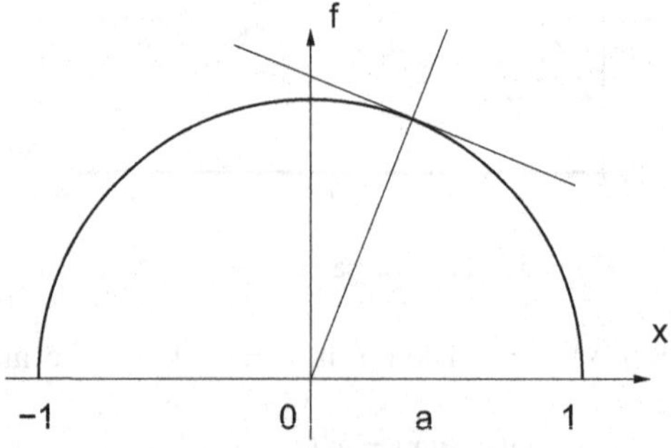

Fig. 3.24 The graph of $f(x) = \sqrt{1-x^2}$ on $(-1,1)$ is half of the unit circle. The tangent line at $(a, f(a))$ is perpendicular to the radial line through $(a, f(a))$, as in Example 3.20

Example 3.21. Let $f(y) = y^k$, k a positive integer. This function is invertible on $y > 0$. Its inverse is

$$g(x) = x^{1/k}.$$

We have $f'(y) = ky^{k-1}$, so by Eq. (3.7),

$$g'(x) = \frac{1}{f'(g(x))} = \frac{1}{kg(x)^{k-1}} = \frac{1}{k}\frac{1}{(x^{1/k})^{k-1}} = \frac{1}{k}\frac{1}{x^{1-1/k}} = \frac{1}{k}x^{(1/k)-1}.$$

This shows that the rule (3.3) for differentiating x^n holds for $n = \dfrac{1}{k}$.

Theorem 3.8. Power rule for rational exponents. *For every rational number $r \neq 0$ and for every $x > 0$, $(x^r)' = rx^{r-1}$.*

Proof. Write $r = \dfrac{p}{q}$, where p and q are integers and $q > 0$. By Example 3.21, the chain rule, and the power rule for integers, we have

$$(x^r)' = ((x^p)^{1/q})' = \frac{1}{q}(x^p)^{(1/q)-1}(px^{p-1}) = \frac{p}{q}x^{(p/q)-p+p-1} = rx^{r-1}.$$

□

Example 3.22. The function $(x^3+1)^{2/3}$ is a composition of f with g, where $g(x) = x^3+1$, $f(y) = y^{2/3}$. Using the power rule for rational exponents and the chain rule, we get

$$\left((x^3+1)^{2/3}\right)' = \frac{2}{3}(x^3+1)^{-1/3}(x^3+1)' = \frac{2}{3}(x^3+1)^{-1/3}3x^2 = \frac{2x^2}{(x^3+1)^{1/3}}.$$

3.2c Higher Derivatives and Notation

Functions may be written in very different ways, and it is helpful to have different ways to denote the derivative. If we use y to represent the function $y = f(x)$, then we can represent $f'(x)$ in any of the following ways:

$$f'(x) = y' = \frac{\mathrm{d}y}{\mathrm{d}x} = \frac{\mathrm{d}}{\mathrm{d}x}f(x). \tag{3.8}$$

When we want to indicate the variable with respect to which we are differentiating, we may use the $\dfrac{\mathrm{d}}{\mathrm{d}x}$ notation. Let u and v be differentiable functions of x. Here are differentiation rules rewritten in this notation:

(a) $\dfrac{\mathrm{d}}{\mathrm{d}x}(u+v) = \dfrac{\mathrm{d}u}{\mathrm{d}x} + \dfrac{\mathrm{d}v}{\mathrm{d}x}$.

(b) If c is constant, then $\dfrac{\mathrm{d}}{\mathrm{d}x}(cu) = c\dfrac{\mathrm{d}u}{\mathrm{d}x}$, and more generally, $\dfrac{\mathrm{d}}{\mathrm{d}x}(uv) = u\dfrac{\mathrm{d}v}{\mathrm{d}x} + v\dfrac{\mathrm{d}u}{\mathrm{d}x}$.

(c) $\dfrac{\mathrm{d}}{\mathrm{d}x}\left(\dfrac{u}{v}\right) = \dfrac{v\frac{\mathrm{d}u}{\mathrm{d}x} - u\frac{\mathrm{d}v}{\mathrm{d}x}}{v^2}$.

(d) If u is a function of y, and y is a function of x, then $\dfrac{\mathrm{d}u}{\mathrm{d}x} = \dfrac{\mathrm{d}u}{\mathrm{d}y}\dfrac{\mathrm{d}y}{\mathrm{d}x}$.

For example, the volume of a right circular cylinder with base radius r and height h is given by the formula $V = \pi r^2 h$.

(a) If the height remains constant, then the rate at which V changes with respect to r is $\dfrac{\mathrm{d}V}{\mathrm{d}r} = 2\pi rh$.

(b) If the radius remains constant, then the rate at which V changes with respect to changes in h is $\dfrac{\mathrm{d}V}{\mathrm{d}h} = \pi r^2$.

(c) If both the radius and height change as functions of time, then the rate at which V changes with respect to t is

$$\frac{\mathrm{d}V}{\mathrm{d}t} = \frac{\mathrm{d}}{\mathrm{d}t}(\pi r^2 h) = \pi\left(r^2\frac{\mathrm{d}h}{\mathrm{d}t} + h\frac{\mathrm{d}}{\mathrm{d}t}(r^2)\right) = \pi\left(r^2\frac{\mathrm{d}h}{\mathrm{d}t} + 2hr\frac{\mathrm{d}r}{\mathrm{d}t}\right).$$

Part (c) used first the product rule, then the chain rule.

When the derivative of the derivative of f exists at x, it is called the second derivative of f and is written in any of the following ways:

$$(f')'(x) = f''(x) = y'' = \frac{\mathrm{d}}{\mathrm{d}x}\left(\frac{\mathrm{d}y}{\mathrm{d}x}\right) = \frac{\mathrm{d}^2 y}{\mathrm{d}x^2} = \frac{\mathrm{d}^2}{\mathrm{d}x^2} f(x).$$

Similar notation is used with higher derivatives, sometimes with superscripts to indicate the order. If u is a function of x, the higher derivatives in the $\dfrac{\mathrm{d}}{\mathrm{d}x}$ notation are written

$$u''' = \frac{\mathrm{d}}{\mathrm{d}x}\left(\frac{\mathrm{d}}{\mathrm{d}x}\left(\frac{\mathrm{d}u}{\mathrm{d}x}\right)\right) = \frac{\mathrm{d}^3 u}{\mathrm{d}x^3}, \quad u'''' = u^{(4)} = \frac{\mathrm{d}^4 u}{\mathrm{d}x^4}, \quad \ldots, \quad u^{(k)} = \frac{\mathrm{d}^k u}{\mathrm{d}x^k}.$$

Example 3.23. Let $f(x) = x^4$. Then

$$f'(x) = 4x^3, \quad f''(x) = 12x^2, \quad f'''(x) = 24x, \quad f^{(4)}(x) = 24, \quad f^{(5)}(x) = 0.$$

It is sometimes useful to be able to give meaning to $\mathrm{d}y$ and $\mathrm{d}x$ so that their quotient $\dfrac{\mathrm{d}y}{\mathrm{d}x}$ equals $f'(x)$. We do this by making the following definition. Let $\mathrm{d}x$ be a new independent variable and define $\mathrm{d}y = f'(x)\,\mathrm{d}x$. We call $\mathrm{d}y$ the differential of y, and it is a function of both x and $\mathrm{d}x$.

Problems

3.16. Find the derivatives of the following functions:

(a) $x^5 - 3x^4 + 0.5x^2 - 17$
(b) $\dfrac{x+1}{x-1}$
(c) $\dfrac{x^2-1}{x^2-2x+1}$
(d) $\dfrac{\sqrt{x}}{x+1}$

3.17. Find the derivatives of the following functions:

(a) $\sqrt{x^3+1}$
(b) $\left(x+\frac{1}{x}\right)^3$
(c) $\sqrt{1+\sqrt{x}}$
(d) $(\sqrt{x}+1)(\sqrt{x}-1)$

3.18. Calculate $\dfrac{d}{dx}\left(\dfrac{1-x^2}{1+x^2}\right)$ by the quotient rule and $\dfrac{d}{dx}\left((1-x^2)(1+x^2)^{-1}\right)$ by the product rule and show that the results agree.

3.19. A particle is moving along a number line, and its position at time t is $x = f(t) = t - t^3$.

(a) Find the position of the particle at $t = 0$ and at $t = 2$.
(b) Find the velocity of the particle at $t = 0$ and at $t = 2$.
(c) In which direction is the particle moving at $t = 0$? at $t = 2$?

3.20. Find the acceleration $f''(t)$ of an object whose velocity at time t is given by $f'(t) = t^3 - \frac{1}{2}t^2$.

3.21. Find the first six derivatives $f', \dots, f^{(6)}$ for each of these functions. In which cases will the seventh and higher derivatives be identically 0?

(a) $f(x) = x^3$
(b) $f(t) = t^3 + 5t^2$
(c) $f(r) = r^6$
(d) $f(x) = x^{-1}$
(e) $f(t) = t^{-3} + t^3$
(f) $f(r) = 6 + r + r^8$

3.22. Derive a formula for the derivative of a triple product by applying the product rule twice: show that if f, g, and h are differentiable at x, then fgh is too, and

$$(fgh)'(x) = f(x)g(x)h'(x) + f(x)g'(x)h(x) + f'(x)g(x)h(x).$$

3.23. Use the chain rule to find the derivatives indicated.

(a) $f(t) = 1 + t + t^2$. Find $(f(t)^2)'$ without squaring f.
(b) If $g'(3) = 0$, find $\left(g(t)^6\right)'$ when $t = 3$.
(c) If $h'(3) = 4$ and $h(3) = 5$, find $\left(h(t)^6\right)'$ when $t = 3$.

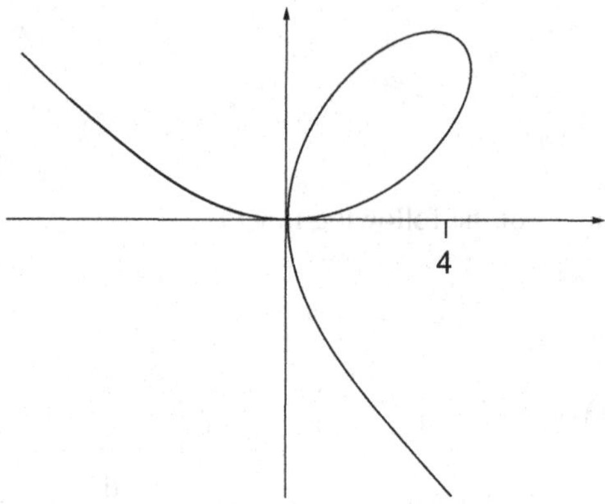

Fig. 3.25 The folium of Descartes, in Problem 3.24

3.24. The set of points (x, y) that satisfy $x^3 + y^3 - 9xy = 0$ lie on the curve shown in Fig. 3.25.

(a) Suppose a portion of the curve is the graph of a function $y(x)$. Use the chain rule to show that $y' = \dfrac{-x^2 + 9y}{y^2 - 9x}$.
(b) Verify that $(2, 4)$ is on the curve. If $y(x)$ is a function having $y(2) = 4$, find $y'(2)$ and use the linear approximation of y to estimate $y(1.97)$.
(c) Verify that each line $y = mx$ other than $y = -x$ intersects the curve at one point. Find the point.
(d) Find the other two points $(2, y)$ on the curve besides $(2, 4)$. For functions $y(x)$ passing through those points, does $y'(2)$ have the same value as you found for $(2, 4)$?

3.25. The gravitational force on a rocket at distance r from the center of a star is $F = -\dfrac{GmM}{r^2}$, where m and M are the rocket and star masses, and G is a constant.

(a) Find $\dfrac{dF}{dr}$.
(b) If the distance r depends on time t according to $r(t) = 2000000 + 1000t$, find the time rate of change of the force, $\dfrac{dF}{dt}$.
(c) If the distance r depends on time t in some manner yet to be specified, express the derivative $\dfrac{dF}{dt}$ in terms of $\dfrac{dr}{dt}$.

3.26. Verify that the gravitational force in Problem 3.25 can be written $F = -\phi'(r)$, where $\phi(r) = \dfrac{GmM}{r}$ is called the potential energy. Explain the equation

$$\frac{dF}{dt} = -\frac{d^2\phi}{dr^2}\frac{dr}{dt}.$$

3.27. Suppose that the volume V of a spherical raindrop grows at a rate, with respect to time t, that is proportional to the surface area of the raindrop. Show that the radius of the raindrop changes at a constant rate.

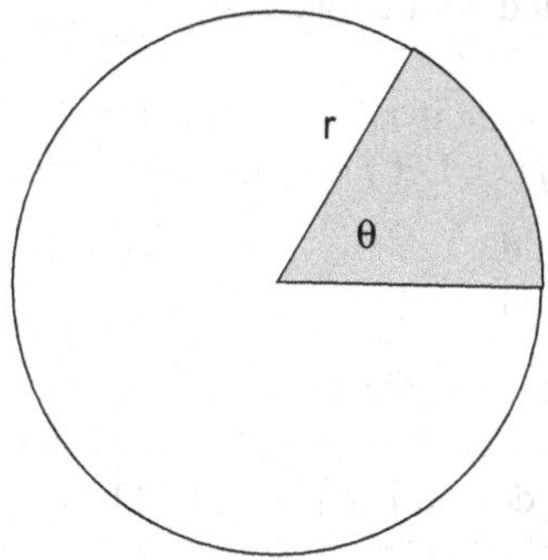

Fig. 3.26 The areas of a sector of a circle varies with θ and r. See Problem 3.28

3.28. The area of a sector of a circle is given by $A = \frac{1}{2}r^2\theta$, where r is the radius of the circle and θ is the central angle measured in radians. See Fig. 3.26.

(a) Find the rate of change of A with respect to r if θ remains constant.
(b) Find the rate of change of A with respect to θ if r remains constant.
(c) Find the rate of change of r with respect to θ if A remains constant.
(d) Find the rate of change of θ with respect to r if A remains constant.
(e) Suppose r and θ change with time. Find the rate of change of A with respect to time in terms of r, θ, $\dfrac{dr}{dt}$, and $\dfrac{d\theta}{dt}$.

3.29. The air–fuel mixture is compressed in an engine. The pressure of the mixture is given by $P = k\rho^{7/5}$, where ρ is the density of the mixture and k is a constant. Write a relation between the rates of change with respect to time, $\dfrac{dP}{dt}$ and $\dfrac{d\rho}{dt}$.

3.30. The line tangent to the graph of f at $(2,7)$ has slope $\frac{1}{3}$. Find the equation of the line tangent to the graph of f^{-1} at $(7,2)$.

3.31. Let $f(x) = x^3 + 2x^2 + 3x + 1$, and denote by g the inverse of f. Check that $f(1) = 7$ and calculate $g'(7)$.

3.32. Let $f(x) = \sqrt{x^2 - 1}$ for $x > 1$, and $g(y) = \sqrt{y^2 + 1}$ for $y > 0$.

(a) Show that f and g are inverses.
(b) Calculate f' and g'.
(c) Verify that $f'(g(y))g'(y) = 1$ and $g'(f(x))f'(x) = 1$.

3.33. In this problem, all the numbers are positive. What is the relation between a and b so that the functions $f(x) = x^a$ and $g(y) = y^b$ are inverses? Show that if

$$\frac{1}{p} + \frac{1}{q} = 1,$$

then the *derivatives* of $\frac{x^p}{p}$ and $\frac{y^q}{q}$ are inverses.

3.34. A function f is called *even* if $f(-x) = f(x)$ for all x in the domain of f, and it is called *odd* if $f(-x) = -f(x)$. Let f be a differentiable function. Show that

(a) If f is even, then f' is odd.
(b) If f is odd, then f' is even.

3.35. Suppose f is strictly monotonic and $f'(x) \neq 0$ on an open interval. In this exercise we ask you to prove that the inverse of f is differentiable. Denote the inverse of f by g, set $y = f(x)$, and denote $f(x+h) - f(x)$ by k.

(a) Explain why $g(y+k) = x+h$.
(b) Show that if $h \neq 0$, then $k \neq 0$.
(c) Explain why

$$\frac{g(y+k) - g(y)}{k} = \frac{h}{k} = \frac{h}{y+k-y} = \frac{h}{f(x+h) - f(x)}$$

tends to $\frac{1}{f'(x)}$ as k tends to 0.

3.3 Derivative of e^x and $\log x$

3.3a Derivative of e^x

We show now that e^x has a special property.

Theorem 3.9. $(\mathrm{e}^x)' = \mathrm{e}^x$.

Proof. Since $\mathrm{e}^x\mathrm{e}^y = \mathrm{e}^{x+y}$, we can write the difference quotient as

$$\frac{\mathrm{e}^{x+h} - \mathrm{e}^x}{h} = \frac{\mathrm{e}^x\mathrm{e}^h - \mathrm{e}^x}{h} = \mathrm{e}^x\frac{\mathrm{e}^h - 1}{h}.$$

Taking the limit as h tends to zero, we see that the derivative of e^x at x is e^x times the derivative of e^x at 0. That is,

$$e'(x) = e(x)e'(0), \tag{3.9}$$

where $e(x)$ denotes the function e^x.

We have therefore only to determine the value $e'(0)$, which is the limit of $\dfrac{e^h - 1}{h}$ as h tends to 0. Recall that in Sect. 1.4, the number e was defined using an increasing sequence $e_n = \left(1+\dfrac{1}{n}\right)^n$ and a decreasing sequence $f_n = \left(1+\dfrac{1}{n}\right)^{n+1}$. So we have inequalities

$$\left(1+\frac{1}{n}\right)^n < e < \left(1+\frac{1}{n}\right)^{n+1} < \left(1+\frac{1}{n-1}\right)^n \tag{3.10}$$

for all integers $n > 1$. Since these approximations contain integer powers, we first take $h = \dfrac{1}{n}$ as a special sequence of h tending to zero. Raise each term in Eq. (3.10) to the power h. We get

$$1+\frac{1}{n} \le e^h \le 1+\frac{1}{n-1}.$$

Subtract 1 from each term:

$$\frac{1}{n} \le e^h - 1 \le \frac{1}{n-1}.$$

Divide each term by $h = \dfrac{1}{n}$:

$$1 \le \frac{e^h - 1}{h} \le \frac{n}{n-1}.$$

As n tends to infinity, the right-hand term tends to 1, so by the squeeze theorem, Theorem 1.7, the center term tends to 1 also. In other words, the derivative of e^x at $x = 0$ is equal to 1, that is, $e'(0) = 1$. Setting this into relation (3.9), we deduce that

$$(e^x)' = e'(x) = e(x)e'(0) = e(x) = e^x.$$

The derivative of the function e^x is e^x itself!

This does not quite finish the proof that $e'(0) = 1$, since we have taken h to be of the special form $h = \dfrac{1}{n}$. In Problem 3.51, we guide you to fill this gap. □

By the chain rule,

$$(e^{kx})' = e^{kx}(kx)' = ke^{kx} \tag{3.11}$$

for every constant k.

Let a be any positive number. To find the derivative of a^x, we write a as $\mathrm{e}^{\log a}$. Then $a^x = (\mathrm{e}^{\log a})^x = \mathrm{e}^{(\log a)x}$. Using the chain rule and the fact that the derivative of e^x is e^x, we get that

$$(a^x)' = \left(\mathrm{e}^{(\log a)x}\right)' = (\log a)\mathrm{e}^{(\log a)x} = (\log a)a^x.$$

3.3b Derivative of $\log x$

Next we show how to use our knowledge of the derivative of e^x and the chain rule to compute the derivative of $\log x$. We know that the natural logarithm is the inverse function of e^x,

$$x = \mathrm{e}^{\log x} \qquad (x > 0).$$

The expression $\mathrm{e}^{\log x}$ is a composition of two functions, where the logarithm is applied first and then the exponential. Thus by the chain rule,

$$1 = (x)' = \left(\mathrm{e}^{\log x}\right)' = \mathrm{e}^{\log x}(\log x)' = x(\log x)'.$$

Solving for $(\log x)'$, we get

$$(\log x)' = \frac{1}{x}, \qquad x > 0.$$

The function $\log x$ is a rather complicated function, but its derivative is very simple.

When x is negative, we have $\big(\log(-x)\big)' = \dfrac{1}{-x}(-x)' = \dfrac{1}{x}$. So we see more generally that

$$(\log|x|)' = \frac{1}{x}, \qquad x \neq 0.$$

The derivative of $\log x$ can be used to compute the derivative of $\log_a x$: We have the identity $x = a^{\log_a x}$. Applying the function log to each side, we get $\log x = (\log a)(\log_a x)$. Hence

$$\log_a x = \frac{1}{\log a}\log x.$$

Differentiate to get

$$(\log_a x)' = \frac{1}{\log a}(\log x)' = \frac{1}{\log a}\frac{1}{x}.$$

Now that we know the derivatives of e^x and $\log x$, we may use the chain rule to compute the derivatives of many more functions.

Example 3.24. The function e^{x^2+1} is a composition $f(g(x))$, where $g(x) = x^2+1$ and $f(x) = \mathrm{e}^x$. By the chain rule,

$$(\mathrm{e}^{x^2+1})' = \mathrm{e}^{x^2+1}(x^2+1)' = \mathrm{e}^{x^2+1}2x.$$

In the $\frac{d}{dx}$ notation, if $y = e^{x^2+1}$ and $u = x^2+1$, then

$$\frac{dy}{dx} = \frac{dy}{du}\frac{du}{dx} = e^u(2x) = e^{x^2+1}2x.$$

More generally,

$$(e^{f(x)})' = e^{f(x)}f'(x)$$

for any differentiable function f. Similarly, by the chain rule,

$$(f(e^x))' = f'(e^x)e^x$$

for any differentiable function f.

Example 3.25. $(\sqrt{1-e^x})' = \frac{1}{2}(1-e^x)^{-\frac{1}{2}}(1-e^x)' = \frac{1}{2}(1-e^x)^{-\frac{1}{2}}(-e^x)$.

Using the chain rule, we also see that

$$\left(\log|f(x)|\right)' = \frac{1}{f(x)}f'(x) = \frac{f'(x)}{f(x)}, \qquad f(x) \neq 0,$$

and

$$(f(\log|x|))' = f'(\log|x|)\frac{1}{x}, \qquad x \neq 0.$$

Example 3.26. $(\log(x^2+1))' = \frac{1}{x^2+1}(x^2+1)' = \frac{1}{x^2+1}2x.$

In the $\frac{d}{dx}$ notation, if $y = \log u$ and $u = x^2+1$, then $\frac{dy}{dx} = \frac{dy}{du}\frac{du}{dx} = \frac{1}{u}2x = \frac{2x}{x^2+1}$.

Example 3.27. For $x > 1$, we have $\log x$ positive, and

$$(\log(\log x))' = \frac{1}{\log x}(\log x)' = \frac{1}{\log x}\frac{1}{x}.$$

3.3c Power Rule

Recall that we have proved the power rule for rational exponents. We now present a proof valid for arbitrary exponents.

Theorem 3.10. *If $r \neq 0$ and $x > 0$, then x^r is differentiable and*

$$(x^r)' = rx^{r-1}.$$

Proof. For $x>0$, $x^r = \mathrm{e}^{\log(x^r)} = \mathrm{e}^{r\log x}$, so

$$(x^r)' = (\mathrm{e}^{r\log x})' = \mathrm{e}^{r\log x}(r\log x)' = \mathrm{e}^{r\log x}\frac{r}{x} = x^r\frac{r}{x} = rx^{r-1}.$$

□

Example 3.28.

$$(x^\pi)' = \pi x^{\pi-1}, \qquad (x^\mathrm{e})' = \mathrm{e}x^{\mathrm{e}-1}.$$

3.3d The Differential Equation $y' = ky$

At the end of Sect. 3.3a, we showed that the function $y = \mathrm{e}^{kx}$ satisfies the differential equation

$$y' = ky. \tag{3.12}$$

For any constant c, $y = c\mathrm{e}^{kx}$ satisfies the same equation. We show now that these are the only functions that satisfy Eq. (3.12).

Theorem 3.11. *Suppose y is a function of x for which*

$$\frac{\mathrm{d}y}{\mathrm{d}x} = ky,$$

where k is a constant. Then there is a number c such that $y = c\mathrm{e}^{kx}$.

Proof. We need to show that the function $\frac{y}{\mathrm{e}^{kx}}$ is a constant. For this reason, we consider the derivative

$$\frac{\mathrm{d}}{\mathrm{d}x}\left(\frac{y}{\mathrm{e}^{kx}}\right) = \frac{\mathrm{d}}{\mathrm{d}x}\left(y\mathrm{e}^{-kx}\right) = \frac{\mathrm{d}y}{\mathrm{d}x}\mathrm{e}^{-kx} - yk\mathrm{e}^{-kx} = \left(\frac{\mathrm{d}y}{\mathrm{d}x} - ky\right)\mathrm{e}^{-kx} = 0.$$

In Sect. 4.1, we will show that the only functions having derivative equal to 0 on an interval are the constant functions. Now, since $\frac{\mathrm{d}}{\mathrm{d}x}\left(\frac{y}{\mathrm{e}^{kx}}\right)$ is 0 for all x, there is a constant c such that

$$\frac{y}{\mathrm{e}^{kx}} = c.$$

So $y = c\mathrm{e}^{kx}$, as claimed. □

Example 3.29. We find all solutions to the equation $y' = y$ having $y = 1$ when $x = 0$. By Theorem 3.11, there is a number c such that $y = c\mathrm{e}^x$ for all x. Take $x = 0$. Then $y = 1 = c\mathrm{e}^0 = c$. Since $c = 1$, there is only one such function, $y = \mathrm{e}^x$.

We close this section with an example of how to *deduce properties of functions from knowledge of the differential equations they satisfy*. Here we deduce from the differential equation $y' = ky$, the addition formula $a^{x+m} = a^x a^m$ for the exponential functions.

Let $y = a^{x+m}$, where m is any number and $a > 0$. By the chain rule,

$$y' = a^{x+m}(\log a) = (\log a)y.$$

So y satisfies the differential equation $y' = ky$, where k is the constant $k = \log a$. By Theorem 3.11, there is a number c such that

$$y = ce^{(\log a)x} = ca^x.$$

Since $y = a^{x+m}$, this shows that

$$a^{x+m} = ca^x. \tag{3.13}$$

To determine c, we set $x = 0$ in Eq. (3.13), giving

$$a^m = c.$$

Setting this into Eq. (3.13), we get

$$a^{x+m} = a^m a^x,$$

the addition formula for exponential functions.

This result shows that $y' = ky$ is the basic property of exponential functions.

Problems

3.36. Compute the first and second derivatives of

(a) e^x
(b) e^{3x}
(c) e^{-3x}
(d) e^{-x^2}
(e) $e^{-1/x}$

3.37. Write an expression for the nth derivative $\left(e^{-t/10}\right)^{(n)}$.

3.38. Find the length marked h in Fig. 3.27 in terms of the length k shown there. The length of the horizontal segment is 1.

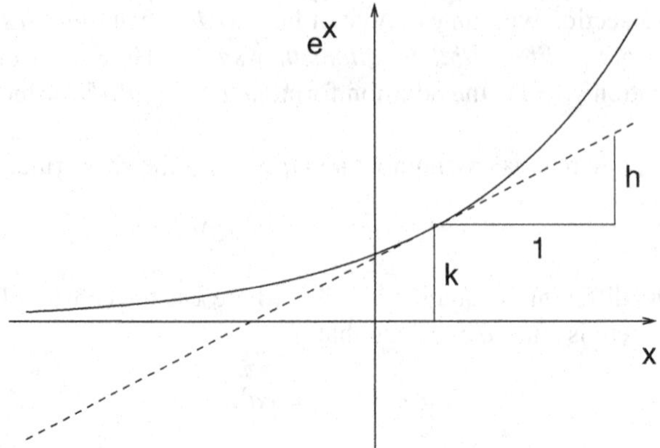

Fig. 3.27 A graph of e^x and one of the tangent lines, for Problem 3.38

3.39. Find the derivative $(x^2 + e^2 + 2^e + 2^x + e^x + x^e)'$.

3.40. Find the linear function $\ell(x) = mx + b$ whose graph is the tangent line to the graph of e^{-x} at $x = 0$.

3.41. Compute the derivatives of the following functions.

(a) $\log|3x|$
(b) $\log(x^2)$
(c) $\log(3x) - \log x$
(d) e^{-e^x}
(e) $\log(1 + e^{-x})$

3.42. Compute the derivatives of the following functions.

(a) $(\log x)'$
(b) $(\log(\log x))'$
(c) $(\log(\log(\log x)))'$
(d) $(\log(\log(\log(\log x))))'$

3.43. Find the equation of the line tangent to the graph of $\log x$ at $x = 1$.

3.44.

(a) Find all the solutions to the differential equation $f'(t) = -\frac{1}{10} f(t)$.
(b) Find the solution if $f(0) = 2$.
(c) Find the solution if $f(10) = 5$.

3.45. Suppose that at time $t = 1$, the population of a bacterial colony is 100, and that at any time, the rate at which the population grows per hour is 1.5 times the size of the population. Find the population at time $t = 3$ h.

3.46. Suppose that at time $t = 0$, the amount of carbon-14 in a sample of wood is 100, and that the half-life of carbon-14 is 5730 years. Find the amount of carbon-14 at time $t = 10{,}000$ years.

3.47. Suppose f and g are positive functions. Obtain the product rule for $(fg)'$ by differentiating $\log(fg) = \log f + \log g$.

3.48. The product rule for a product of many functions can be deduced inductively from the product formula for two functions. Here is an alternative approach. Justify each step of the following argument.

(a) Suppose $y = f(x)g(x)h(x)k(x)$. Then $|y| = |f||g||h||k|$, and when all are nonzero,

$$\log|y| = \log|f| + \log|g| + \log|h| + \log|k|.$$

(b) $\frac{1}{y}y' = \frac{1}{f}f' + \frac{1}{g}g' + \frac{1}{h}h' + \frac{1}{k}k'$.
(c) $y' = f'ghk + fg'hk + fgh'k + fghk'$.

3.49. Evaluate $\frac{\mathrm{d}}{\mathrm{d}x}\log\left(\frac{\sqrt{x^2+1}\sqrt[3]{x^4-1}}{\sqrt[5]{x^2-1}}\right)$.

3.50. Define two functions by $y(x) = \frac{1}{\mathrm{e}^x + \mathrm{e}^{-x}}$ and $z(x) = \mathrm{e}^x - \mathrm{e}^{-x}$. Verify the algebraic equation $z^2 = \frac{1}{y^2} - 4$ and the following differential equations.

(a) $y' = -zy^2$ and $z' = \frac{1}{y}$.
(b) Use part (a) and the chain rule to show that $y'' = y - 8y^3$.

3.51. This problem fills the gap in the proof that $e'(0) = 1$. Recall from Eq. (3.10) that e is between the increasing sequence e_n and the decreasing sequence f_n. Explain the following items.

(a) If $h > 0$ is not of the form $\frac{1}{n}$, then there is an integer n for which $\frac{1}{n} < h < \frac{1}{n-1}$.

(b) Using Eq. (3.10), one has $\left(1+\frac{1}{n}\right)^{nh} < \mathrm{e}^h < \left(1+\frac{1}{n-2}\right)^{(n-1)h}$.

(c) $(n-1)h < 1 < nh$.
(d) $1 + \frac{1}{n} < \mathrm{e}^h < 1 + \frac{1}{n-2}$.
(e) $n - 1 < \frac{1}{h} < n$, and $\frac{n-1}{n} < \frac{\mathrm{e}^h - 1}{h} < \frac{n}{n-2}$.

Conclude from this that $\frac{\mathrm{e}^h - 1}{h}$ tends to 1 as h tends to zero.

3.4 Derivatives of the Trigonometric Functions

We saw in Sect. 2.4 that the sine and cosine functions are continuous and periodic. These functions often arise when we study natural phenomena that fluctuate in a periodic way, for example vibrating strings, waves, and the orbits of planets. The derivatives of the sine and cosine have special properties, as was the case with the exponential function, and they solve important differential equations.

3.4a Sine and Cosine

Recall that we defined $\cos t$ and $\sin t$ to be the x- and y-coordinates of the point $P(t)$ on the unit circle, so that the arc from $(1,0)$ to $P(t)$ is t units along the circle. This definition with a bit of geometry will help us find the derivatives of sine and cosine.

Theorem 3.12. $\sin' t = \cos t \quad \text{and} \quad \cos' t = -\sin t.$

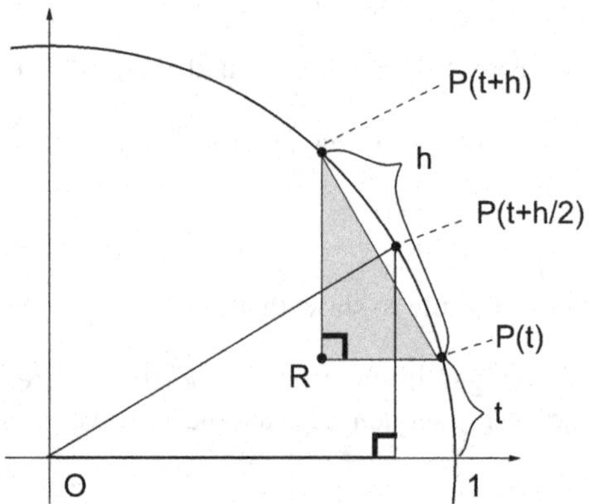

Fig. 3.28 The two right triangles are similar because their corresponding sides are perpendicular

Proof. In Fig. 3.28, $P(t)$ is on the unit circle at distance t along the circle from the point $(1,0)$, and $P(t+h)$ is another point on the unit circle at distance $t+h$ along the circle from $(1,0)$. Draw a vertical line from the point $P(t+h)$, and a horizontal line from the point $P(t)$. Denote their intersection by R. The three points form a right triangle, as indicated. The horizontal leg has length $\cos(t)-\cos(t+h)$. Denote the length of the hypotenuse by c, the distance between $P(t+h)$ and $P(t)$. As h tends to zero, the ratio $\dfrac{h}{c}$ tends to 1.

Consider another right triangle whose vertices are O, $P\left(t+\frac{h}{2}\right)$, and the intersection of the vertical line drawn from $P\left(t+\frac{h}{2}\right)$ with the x-axis. The sides of this triangle are perpendicular to the sides of the small triangle considered before, and the two triangles are similar. Therefore,

$$\frac{\sin\left(t+\frac{h}{2}\right)}{1} = \frac{\cos t - \cos(t+h)}{c} = \frac{\cos t - \cos(t+h)}{h}\frac{h}{c}.$$

As h tends to zero, $\sin\left(t+\dfrac{h}{2}\right)$ tends to $\sin t$. The first factor on the right tends to $-\cos' t$, and the second factor tends to 1. Therefore, $\cos' t = -\sin t$.

To find the derivative of $\sin t$, use a similar argument:

$$\frac{\cos\left(t+\frac{h}{2}\right)}{1} = \frac{\sin(t+h)-\sin t}{c} = \frac{\sin(t+h)-\sin t}{h}\frac{h}{c}.$$

As h tends to 0, we have $\cos t = \sin' t$. □

Once you have the derivatives of sine and cosine, the derivatives of the tangent, cotangent, secant, and cosecant functions are easily derived using the quotient rule.

Example 3.30.

$$(\tan t)' = \left(\frac{\sin t}{\cos t}\right)' = \frac{\cos t\cos t - \sin t(-\sin t)}{(\cos t)^2} = \frac{1}{(\cos t)^2} = \sec^2 t.$$

When trigonometric functions are composed with other functions, we use the chain rule to compute the derivative.

Example 3.31. $\left(\sin(e^x)\right)' = \cos(e^x)\left(e^x\right)' = e^x\cos(e^x)$.

Example 3.32. $\left(e^{\sin x}\right)' = e^{\sin x}\left(\sin x\right)' = e^{\sin x}\cos x$.

Example 3.33. $\left(\log(\tan x)\right)' = \dfrac{1}{\tan x}\dfrac{1}{\cos^2 x} = \dfrac{1}{\cos x\,\sin x}$.

We close this section with an alternative way of determining the derivatives of the sine and cosine functions, based on the interpretation of derivative as a velocity. Suppose a particle Q moves with velocity 1 along a ray that makes angle a with the x-axis, as shown in Fig. 3.29. Then the shadow of the particle on the x-axis travels with velocity $\cos a$. Similarly, the shadow of Q on the vertical axis travels with velocity $\sin a$.

Suppose that particle P travels with velocity 1 on the unit circle, starting at the point $(1,0)$ and moving in a counterclockwise direction. Then its position at time t is $(\cos t, \sin t)$, and its shadows on the x- and y-axes are $\cos t$ and $\sin t$ respectively. We now find the velocities of these shadows, i.e., $\cos' t$ and $\sin' t$.

Let s be a particular time, and denote by L the tangent line to the unit circle at the point $(\cos s, \sin s)$. Denote by $Q(t)$ the position of a particle Q that passes through the

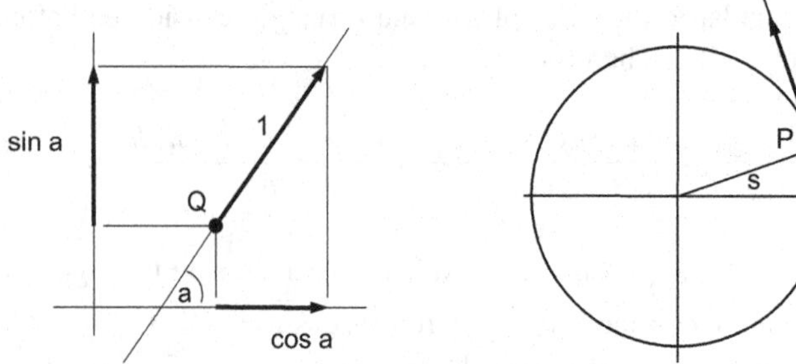

Fig. 3.29 *Left*: Q moves with velocity 1 along a line. *Right*: P moves with velocity 1 along a circle. At time s, the velocities of particles P and Q are equal

point of tangency at time s and travels along L with velocity 1. At time s, particles P and Q are at the same point $(\cos s, \sin s)$ and are moving in the same direction with unit velocity. Hence at time s, the shadows of P and Q on the x-axis travel at the same velocity, and the shadows of P and Q on the y-axis travel at the same velocity.

The tangent to the circle at $(\cos s, \sin s)$ is perpendicular to the line through the origin and $P(s)$. Therefore, the tangent line makes the angle $s + \frac{\pi}{2}$ with the horizontal ray. The shadows of Q on the horizontal and vertical axes travel with the constant velocities $\cos\left(s + \frac{\pi}{2}\right)$ and $\sin\left(s + \frac{\pi}{2}\right)$ respectively. These are the velocities at which the shadows of P travel at time s. Therefore, at time s,

$$\cos' s = \cos\left(s + \frac{\pi}{2}\right) = -\sin s, \qquad \sin' s = \sin\left(s + \frac{\pi}{2}\right) = \cos s.$$

3.4b The Differential Equation $y'' + y = 0$

We turn to some other interesting properties of sine and cosine. Recall that we saw earlier that $y = e^t$ is the only solution to $y' = y$ satisfying $y = 1$ when $t = 0$. We shall derive a similar relation for the sine and cosine. Differentiating the expressions for the derivatives of sine and cosine, we get

$$\sin'' t = \left(\sin' t\right)' = \cos' t = -\sin t, \text{ and } \cos'' t = \left(\cos' t\right)' = (-\sin t)' = -\cos t.$$

Here we have used the notation f'' to denote the derivative of the derivative of f, called the second derivative of f. Notice that the equation

$$f'' + f = 0 \tag{3.14}$$

is satisfied by both $f(t) = \sin t$ and $f(t) = \cos t$.

Consider the function $f(t) = 2\sin t - 3\cos t$. Then

$$f' = 2\cos t + 3\sin t, \qquad f'' = -2\sin t + 3\cos t = -f,$$

so we have $f'' + f = 0$.

More generally, every function of the form $f(t) = u\cos t + v\sin t$, with u and v constant, solves Eq. (3.14). Next we address the question whether there are any other solutions to the differential equation $f'' + f = 0$. We will see that there are no others.

First we shall derive some properties of solutions of Eq. (3.14). Multiply the equation by $2f'$. We obtain

$$2f'f'' + 2f'f = 0. \tag{3.15}$$

We recognize the first term on the left as the derivative of $(f')^2$, and the second term as the derivative of f^2. So the left side of Eq. (3.15) is the derivative of $(f')^2 + f^2$. According to Eq. (3.15), this derivative is zero. As we noted already using Corollary 4.1 of the mean value theorem, a function whose derivative is zero over the entire real line is a constant. So it follows that every function f that satisfies the differential equation (3.14) satisfies

$$(f')^2 + f^2 = \text{constant}. \tag{3.16}$$

Theorem 3.13. *Denote by f a solution of Eq. (3.14) for which $f(c)$ and $f'(c)$ are both zero at some point c. Then $f(t) = 0$ for all t.*

Proof. We have shown that all solutions f of Eq. (3.14) satisfy Eq. (3.16) for all t. At $t = c$, both $f(c)$ and $f'(c)$ are zero, so it follows that the constant in Eq. (3.16) is zero. But then it follows that $f(t)$ is zero for all t. □

Theorem 3.14. *Suppose f_1 and f_2 are two solutions of*

$$f'' + f = 0$$

and that there is a number c for which $f_1(c) = f_2(c)$ and $f_1'(c) = f_2'(c)$. Then $f_1(t) = f_2(t)$ for every t.

Proof. Since f_1 and f_2 are solutions of Eq. (3.14), so is their difference $f = f_1 - f_2$. Since we have assumed that f_1 and f_2 are equal at c and also that their first derivative are equal at c, it follows that $f(c)$ and $f'(c)$ are both zero. According to Theorem 3.13, $f(t) = 0$ for all t. It follows that $f_1(t) = f_2(t)$ for all t. □

Let f be any solution of Eq. (3.14). Set $f(0) = u$, $f'(0) = v$. We saw earlier that $g(t) = u\cos t + v\sin t$ is a solution of Eq. (3.14). The values of g and g' at $t = 0$ are

$$g(0) = u, \qquad g'(0) = v.$$

The functions $f(t)$ and $g(t)$ are both solutions of Eq. (3.14), and $f(0) = g(0)$, $f'(0) = g'(0)$. According to Theorem 3.14, $f(t) = g(t)$ for all t. This shows that all solutions of Eq. (3.14) are of the form $g(t) = u\cos t + v\sin t$.

Next we see that we can deduce properties of the sine and cosine functions from our knowledge of the differential equation that they solve, $f'' + f = 0$.

Theorem 3.15. *Addition law for the cosine and sine.*

$$\cos(t+s) = \cos s\cos t - \sin s\sin t \tag{3.17}$$

and

$$\sin(t+s) = \sin s\cos t + \cos s\sin t. \tag{3.18}$$

Proof. If you differentiate the addition law (3.17) for the cosine with respect to s, you obtain (the negative of) the addition law for the sine. So it suffices to prove the addition law for the cosine.

We have seen that every combination of the form $u\cos t + v\sin t$ is a solution of Eq. (3.14). Take any number s, and set $u = \cos s$, $v = \sin s$. Then the function

$$b(t) = \cos s\cos t - \sin s\sin t$$

is a solution of Eq. (3.14). We next show that

$$a(t) = \cos(s+t),$$

viewed as a function of t, is also a solution. The chain rule gives

$$\begin{aligned}
a'(t) &= \cos'(s+t)\frac{\mathrm{d}(s+t)}{\mathrm{d}t} = -\sin(s+t),\\
a''(t) &= -\sin'(s+t)\frac{\mathrm{d}(s+t)}{\mathrm{d}t} = -\cos(s+t) = -a(t).
\end{aligned}$$

Compare then $a(t)$ with $b(t)$. We have

$$a(0) = \cos s = b(0)$$

and

$$a'(0) = -\sin s = b'(0).$$

By Theorem 3.14, then, $a(t)$ is identically equal to $b(t)$. Since $a(t)$ is the function on the left side of Eq. (3.17) and $b(t)$ is the function on the right side, this concludes the proof. □

3.4c Derivatives of Inverse Trigonometric Functions

Inverse Sine. Over the interval $[-\frac{\pi}{2}, \frac{\pi}{2}]$, the sine function is increasing. Therefore, it has an inverse, called the arcsine function, denoted by $\sin^{-1}$.

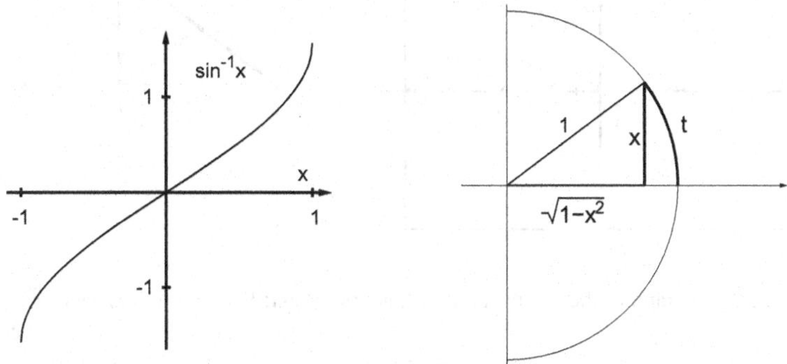

Fig. 3.30 *Left*: The graph of the inverse sine function. *Right*: For $t = \sin^{-1} x$, $\cos t = \sqrt{1-x^2}$

To find the derivative of $\sin^{-1}(x)$, we differentiate

$$\sin(\sin^{-1} x) = x$$

with respect to x, giving

$$\cos(\sin^{-1} x)\left(\sin^{-1} x\right)' = 1.$$

This shows that

$$\left(\sin^{-1} x\right)' = \frac{1}{\cos(\sin^{-1} x)}, \qquad -1 < x < 1.$$

The expression $\cos(\sin^{-1} x)$ can be simplified. Let $t = \sin^{-1} x$. Then $x = \sin t$. Since $\cos^2 t + \sin^2 t = 1$, we get (See Fig. 3.30)

$$\cos(\sin^{-1} x) = \cos t = \pm\sqrt{1-\sin^2 t} = \sqrt{1-x^2}.$$

For $-\frac{\pi}{2} \le t \le \frac{\pi}{2}$, $\cos t \ge 0$, so we have used the positive root,

$$\left(\sin^{-1} x\right)' = \frac{1}{\sqrt{1-x^2}}, \qquad -1 < x < 1. \tag{3.19}$$

Notice that $\sin^{-1}$ is differentiable on the open interval $(-1, 1)$. The slope of its graph approaches infinity at x tends to 1 from the left, and also as x tends to -1 from the right.

Inverse Tangent. Similarly, the tangent function restricted to the interval $-\frac{\pi}{2} < t < \frac{\pi}{2}$ has an inverse, called the arctangent function, or $\tan^{-1}$.

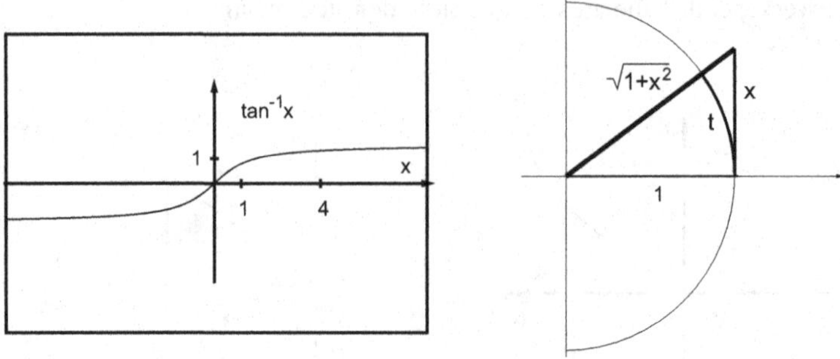

Fig. 3.31 *Left*: The graph of the inverse tangent function. *Right*: For $t = \tan^{-1} x$, $\cos t = \dfrac{1}{\sqrt{1+x^2}}$

To find the derivative of $\tan^{-1} x$ for $-\infty < x < \infty$, we differentiate

$$\tan(\tan^{-1} x) = x$$

with respect to x:

$$\sec^2(\tan^{-1} x)(\tan^{-1} x)' = 1,$$

so $(\tan^{-1} x)' = \cos^2(\tan^{-1} x)$. Let $t = \tan^{-1} x$. (See Fig. 3.31.) Then

$$x^2 = \tan^2 t = \frac{\sin^2 t}{\cos^2 t} = \frac{1-\cos^2 t}{\cos^2 t} = \frac{1}{\cos^2 t} - 1.$$

This shows that $\cos^2(\tan^{-1} x) = \dfrac{1}{1+x^2}$. Therefore,

$$(\tan^{-1} x)' = \frac{1}{1+x^2}.$$

Example 3.34. $\left(\tan^{-1} e^x\right)' = \dfrac{1}{1+(e^x)^2}(e^x)' = \dfrac{e^x}{1+e^{2x}}.$

Example 3.35. $\left(\sin^{-1} x^2\right)' = \dfrac{1}{\sqrt{1-(x^2)^2}}(x^2)' = \dfrac{2x}{\sqrt{1-x^4}}.$

Example 3.36. $\left(\log(\sin^{-1} x)\right)' = \dfrac{1}{\sin^{-1} x}\left(\sin^{-1} x\right)' = \dfrac{1}{\sin^{-1} x}\dfrac{1}{\sqrt{1-x^2}}.$

In Problem 3.59, we guide you through a similar derivation of

$$(\sec^{-1} x)' = \frac{1}{x\sqrt{x^2-1}}, \quad (0 < x < \frac{\pi}{2}).$$

3.4d The Differential Equation $y'' - y = 0$

We shall show that the differential equation $y'' - y = 0$ plays the same role for exponential functions as the differential equation $y'' + y = 0$ does for trigonometric functions. We introduce two new functions.

Definition 3.3. The *hyperbolic cosine* and *hyperbolic sine*, denoted by cosh and sinh, are defined as follows:

$$\cosh x = \frac{e^x + e^{-x}}{2}, \qquad \sinh x = \frac{e^x - e^{-x}}{2}. \tag{3.20}$$

Their derivatives are easily calculated, as we ask you to check in Problem 3.61:

$$\cosh' x = \sinh x, \quad \text{and} \quad \sinh' x = \cosh x. \tag{3.21}$$

Therefore,

$$\cosh'' x = \cosh x \quad \text{and} \quad \sinh'' x = \sinh x.$$

That means that cosh and sinh are solutions to the differential equation

$$f'' - f = 0. \tag{3.22}$$

For any constants u and v, the functions $u\cosh x$ and $v\sinh x$ also satisfy $f'' - f = 0$, as does their sum $f(x) = u\cosh x + v\sinh x$:

$$(u\cosh x + v\sinh x)'' = (u\sinh x + v\cosh x)' = u\cosh x + v\sinh x.$$

The value of $f(x) = u\cosh x + v\sinh x$ and of its derivative $f'(x)$ at $x = 0$ are

$$f(0) = u, \qquad f'(0) = v.$$

We show now that $u\cosh x + v\sinh x$ is the only solution of Eq. (3.22) with these values at $x = 0$. The proof is based on the following theorem.

Theorem 3.16. *Suppose f_1 and f_2 are two solutions of*

$$f'' - f = 0$$

having the same value and derivatives at $x = a$: $f_1(a) = f_2(a)$ and $f_1'(a) = f_2'(a)$. Then $f_1(x) = f_2(x)$ for every x.

Proof. Set $g = f_1 - f_2$. Then $g'' - g = 0$, and $g(a) = g'(a) = 0$. It follows that

$$(g' + g)' = g'' + g' = g + g'.$$

We appeal now to Theorem 3.11, which says that if a function y satisfies the equation $y'(x) = ky(x)$ for all x, then $y = ce^{kx}$. Since we have shown that $y = g + g'$ satisfies $y' = y$, it follows that $g + g' = ce^x$. But both g and g' are zero at $x = a$. It follows that the constant c is 0, and therefore $g + g' = 0$. We appeal once more to Theorem 3.11 to deduce that $g(x)$ is of the form be^{-x}. But since $g(a) = 0$, b must be zero, which makes $g(x) = 0$ for all x. Since g was defined as $f_1 - f_2$, it follows that f_1 and f_2 are the same function. □

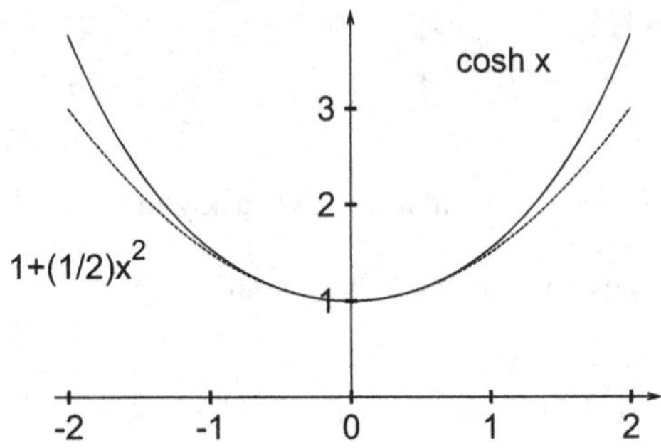

Fig. 3.32 The graphs of $\cosh x$ and $1 + \frac{1}{2}x^2$

A graph of cosh is shown in Fig. 3.32. Just as the cosine and sine satisfy the relation $\cos^2 x + \sin^2 x = 1$ and are sometimes called "circular" functions, the functions $\cosh x$ and $\sinh x$ satisfy the algebraic relation

$$\cosh^2 x - \sinh^2 x = 1, \tag{3.23}$$

suggesting the name "hyperbolic." We ask you to verify Eq. (3.23) in Problem 3.63 (Fig. 3.33).

The following addition laws can be deduced from the addition law for the exponential function, or from the differential equation $f'' - f = 0$, as we ask you to try in Problem 3.68:

$$\begin{aligned} \cosh(x+y) &= \cosh x \cosh y + \sinh x \sinh y, \\ \sinh(x+y) &= \sinh x \cosh y + \cosh x \sinh y. \end{aligned} \tag{3.24}$$

Formulas (3.21), (3.23), and (3.24) are very similar to corresponding formulas for sine and cosine; they differ only by a minus sign in Eqs. (3.21) and (3.23). This indicates a deep connection between the trigonometric functions and the exponential function. The nature of this deep connection will be explored in Chap. 9.

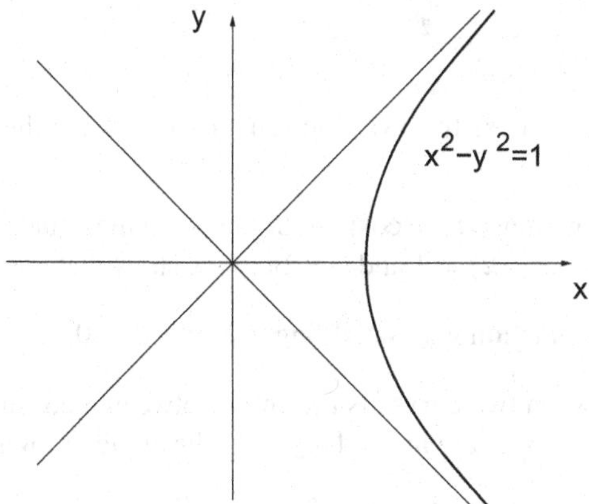

Fig. 3.33 For each t, the point $(\cosh t, \sinh t)$ lies on the hyperbola $x^2 - y^2 = 1$

Problems

3.52. Find the derivatives of the following functions.

(a) $(\cos x)'$
(b) $(\cos(2x))'$
(c) $(\cos^2 x)'$

3.53. Find the derivatives of the following functions.

(a) $\log|\sin x|$
(b) $e^{\tan^{-1} x}$
(c) $\tan^{-1}(5x^2)$
(d) $\log(e^{2x}+1)$
(e) $e^{(\log x)(\cos x)}$

3.54. Find the derivatives of the following functions.

(a) $\log|\tan x|$
(b) $\tan^{-1}(e^{2x})$
(c) $(e^{kx})^2$
(d) $1+\sin^2 x$
(e) $1-\cos^2 x$

3.55. Use the product and quotient rules and your knowledge of sin′ and cos′ to prove

(a) $(\sec x)' = \left(\dfrac{1}{\cos x}\right)' = \sec x \tan x$

(b) $(\csc x)' = \left(\dfrac{1}{\sin x}\right)' = -\csc x \cot x$

(c) $(\cot x)' = \left(\dfrac{\sin x}{\cos x}\right)' = -\csc^2 x$

3.56. The periodicity of $\cos t$ follows from Theorem 3.14 and the differential equation $f'' + f = 0$:

(a) Show that the function $f(t) = \cos(t + 2\pi)$ satisfies the equation.
(b) Using the values $\cos(2\pi) = 1$ and $\sin(2\pi) = 0$, show that $\cos(t + 2\pi) = \cos t$.

3.57. Determine the function $y(x)$ satisfying $y'' + y = 0$, $y(0) = -2$, and $y'(0) = 3$.

3.58. The angle between two curves is the angle between their tangents at the point of intersection. If the slopes are m_1 and m_2, then the angle of intersection θ is

$$\theta = \tan^{-1}\left(\frac{m_2 - m_1}{1 + m_1 m_2}\right).$$

(a) Use Theorem 3.15 to derive the formula for θ.
(b) Find the angles at which the graphs of $y = x^3 + 1$ and $y = x^3 + x^2$ intersect.

3.59. Derive the derivatives for additional inverse trigonometric functions.

(a) Give a description of the inverse secant function $\sec^{-1}$ similar to our presentation of $\tan^{-1}$. Explain why $\sec x$ is invertible for $0 \le x < \dfrac{\pi}{2}$ and for $\dfrac{\pi}{2} < x \le \pi$ and sketch the graph of $\sec^{-1} y$, $|y| > 1$. Then show that

$$(\sec^{-1} y)' = \frac{1}{|y|\sqrt{y^2 - 1}}, \qquad (|y| > 1).$$

(b) Use the equation $\sin^{-1} x + \cos^{-1} x = \dfrac{\pi}{2}$ to find $(\cos^{-1} x)'$.
(c) Use the equation $\sec^{-1} x + \csc^{-1} x = \dfrac{\pi}{2}$ to find $(\csc^{-1} x)'$.
(d) Use the equation $\tan^{-1} x + \tan^{-1}\left(\dfrac{1}{x}\right) = \dfrac{\pi}{2}$ to find $\left(\tan^{-1}\dfrac{1}{x}\right)'$. Compare your answer to the answer you would obtain using the chain rule.

3.60. Evaluate or simplify the following.

(a) $\cosh(\log 2)$
(b) $\dfrac{d}{dt}\big(\cosh(6t)\big)$
(c) $\cosh^2(5t) - \sinh^2(5t)$
(d) $\cosh t + \sinh t$
(e) $\sinh x + \sinh(-x)$

3.61. Derive the formulas $\sinh' x = \cosh x$ and $\cosh' x = \sinh x$ from the definitions of cosh and sinh.

3.62. Verify that at $x=0$, the functions $\cosh x$ and $1+\frac{1}{2}x^2$ and their first two derivatives are equal. (See Fig. 3.32.)

3.63. Verify $\cosh^2 x - \sinh^2 x = 1$ using the definitions of cosh and sinh.

3.64. Use the definitions

$$\tanh t = \frac{\sinh t}{\cosh t} = \frac{e^t - e^{-t}}{e^t + e^{-t}}, \qquad \operatorname{sech} t = \frac{1}{\cosh t} = \frac{2}{e^t + e^{-t}}$$

to verify that

$$\tanh' t = \operatorname{sech}^2 t = 1 - \tanh^2 t.$$

3.65. Suppose y is a differentiable function of x and that

$$y + y^5 + \sin y = 3x^2 - 3.$$

(a) Verify that the point $(x,y) = (1,0)$ satisfies this equation.
(b) Explain why $\frac{dy}{dx} + 5y^4\frac{dy}{dx} + \cos y\frac{dy}{dx} = 6x$.
(c) Use the result of part (b) to find $\frac{dy}{dx}$ when $x=1$, $y=0$.
(d) Sketch the line tangent to the curve at $(1,0)$.
(e) Estimate $y(1.01)$.

3.66. The function $\sinh x$ is increasing, so it has an inverse.

(a) Derive the formula

$$\sinh^{-1} y = \log\left(y + \sqrt{1+y^2}\right)$$

by writing first $\frac{e^x - e^{-x}}{2} = y$ as $(e^x)^2 - 1 = 2ye^x$, and then treating this expression as a quadratic equation in e^x.
(b) Deduce that $\left(\sinh^{-1}\right)' y = \frac{1}{\sqrt{1+y^2}}$.

3.67. Determine the function $y(x)$ having $y'' - y = 0$, $y(0) = -2$, and $y'(0) = 3$.

3.68. Use a method similar to the proof of Theorem 3.15 to prove the addition law

$$\cosh(x+y) = \cosh(x)\cosh(y) + \sinh(x)\sinh(y).$$

Differentiate the result with respect to x to deduce the addition law

$$\sinh(x+y) = \sinh(x)\cosh(y) + \cosh(x)\sinh(y).$$

Use the two addition laws to prove that

$$\tanh(x+y) = \frac{\tanh x + \tanh y}{1 + \tanh x \tanh y},$$

where the hyperbolic tangent function tanh is defined in Problem 3.64.

3.5 Derivatives of Power Series

We saw in Chap. 2 that uniform convergence of a sequence of functions interacts well with continuity, that is, if a sequence of continuous functions f_n converges uniformly to f, then f is continuous. Now we ask a similar question about differentiability: If the f_n are differentiable, do the f_n' converge to f'? Unfortunately, in general the answer to these questions is no. Here is an example.

Example 3.37. Let $f_n(x) = \dfrac{\sin nx}{n^{1/2}}$. Since $-n^{-1/2} \le f_n(x) \le n^{-1/2}$, the limit $\lim_{n\to\infty} f_n(x) = f(x)$ is equal to 0 for all x. In fact, $|f_n(x)| \le n^{-1/2}$ shows that f_n converges uniformly to f. Note that $f'(x) = 0$. Now $f_n'(x) = n^{1/2}\cos nx$, and for $x = 2\pi/n$, we see that $f_n'(x) = n^{1/2}$ does not tend to 0. So $f_n'(x)$ does not even converge pointwise to $f'(x)$. Thus

$$\left(\lim_{n\to\infty} f_n(x)\right)' \ne \lim_{n\to\infty} f_n'(x).$$

However, we are in luck if we restrict our attention to power series. For simplicity, we consider the case that the center of the power series is $a = 0$.

Theorem 3.17. Term-by-term differentiation. *If the power series*

$$f(x) = \sum_{n=0}^{\infty} a_n x^n$$

converges on $-R < x < R$, then f is differentiable on $(-R,R)$ and

$$f'(x) = \sum_{n=1}^{\infty} n a_n x^{n-1}.$$

Proof. Let x be in $(-R,R)$ and choose $\delta > 0$ small enough that $|x| + \delta$ is also in $(-R,R)$. According to Theorem 2.13, $\sum_{n=0}^{\infty} |a_n(|x|+\delta)^n|$ converges. We ask you in Problem 3.72 to verify that for n large enough,

$$|nx^{n-1}| \le (|x|+\delta)^n.$$

Therefore, for n large enough, $|na_nx^{n-1}| \le |a_n(|x|+\delta)^n|$, and so $\sum_{n=1}^{\infty} |na_nx^{n-1}|$ converges by comparison with $\sum_{n=0}^{\infty} |a_n(|x|+\delta)^n|$. We conclude that

$$g(x) = \sum_{n=1}^{\infty} na_nx^{n-1}, \qquad -R < x < R,$$

converges.

Next we show that f is differentiable at each x in $(-R,R)$ and that $f'(x) = g(x)$. We need to show that the difference quotient of f tends to g:

$$\begin{aligned} \left| \frac{f(x+h)-f(x)-g(x)h}{h} \right| &= \left| \sum_{n=1}^{\infty} \frac{a_n(x+h)^n - a_nx^n - na_nx^{n-1}h}{h} \right| \\ &\le \frac{1}{|h|} \sum_{n=1}^{\infty} |a_n||(x+h)^n - x^n - nx^{n-1}h|. \end{aligned}$$

The next step in the proof was written up by R. Výborny in the *American Mathematical Monthly*, 1987, and we guide you through the details in Problem 3.71. Let H be chosen such that $0 < |h| < H$ and $|x|+H$ is in $(-R,R)$. Then $K = \sum_{n=0}^{\infty} |a_n(|x|+H)^n|$ converges and is positive. By the binomial theorem and some algebra, you can show in Problem 3.71 that

$$|(x+h)^n - x^n - nx^{n-1}h| \le \frac{|h|^2}{|H|^2}(|x|+H)^n.$$

Using this result, we get

$$\left| \frac{f(x+h)-f(x)-g(x)h}{h} \right| \le \frac{1}{|h|}\frac{|h|^2}{|H|^2} \sum_{n=0}^{\infty} |a_n(|x|+H)^n| = |h|\frac{K}{|H|^2}.$$

As h tends to 0, the right-hand side tends to zero, so the left-hand side tends to 0 and $f'(x) = g(x)$. □

Example 3.38. To find the sum of the series $\sum_{n=0}^{\infty} \frac{n(n-1)}{2^n}$, notice that it is $f''\left(\frac{1}{2}\right)$, where

$$f(x) = \sum_{n=0}^{\infty} x^n = \frac{1}{1-x}, \quad f'(x) = \sum_{n=1}^{\infty} nx^{n-1}, \quad f''(x) = \sum_{n=2}^{\infty} n(n-1)x^{n-2} = \frac{2}{(1-x)^3}.$$

Therefore, $\sum_{n=0}^{\infty} \frac{n(n-1)}{2^n} = \frac{2}{(1-\frac{1}{2})^3} = 16.$

Example 3.39. $f(x) = \sum_{n=0}^{\infty} \frac{x^n}{n!}$ converges for all x. According to Theorem 3.17,

$$f'(x) = \sum_{n=1}^{\infty} n \frac{x^{n-1}}{n!} = \sum_{n=1}^{\infty} \frac{x^{n-1}}{(n-1)!} = \sum_{n=0}^{\infty} \frac{x^n}{n!},$$

so f is its own derivative. We saw in Theorem 3.11 that the only functions that satisfy $y' = y$ are of the form $y = ke^x$. Since $f(0) = 1$, $f(x) = e^x$. In Sect. 4.3a, we will show by a different argument that $\sum_{n=0}^{\infty} \frac{x^n}{n!} = e^x$.

Example 3.40. Let $f(x) = x - \frac{x^3}{3!} + \frac{x^5}{5!} + \cdots = \sum_{n=0}^{\infty} (-1)^n \frac{x^{2n+1}}{(2n+1)!}$, which converges for all x. According to Theorem 3.17,

$$f'(x) = 1 - \frac{x^2}{2!} + \frac{x^4}{4!} - \cdots \quad \text{and} \quad f''(x) = -x + \frac{x^3}{3!} - \cdots = -f(x).$$

In Theorem 3.14 of Sect. 3.4b, we have shown that a function that satisfies $f'' = -f$ is of the form $f(x) = u\cos x + v\sin x$ for some numbers u and v. In Problem 3.69, we ask you to determine u and v. In Sect. 4.3a, we will show by a different argument that $f(x) = \sin x$ and $f'(x) = \cos x$.

Problems

3.69. Find the numbers u and v for which $x - \frac{x^3}{3!} + \frac{x^5}{5!} - \cdots = u\cos x + v\sin x$.

3.70. Use the technique demonstrated in Example 3.38 to evaluate the following series.

(a) $\sum_{n=1}^{\infty} \frac{n}{2^n}$

(b) $\sum_{n=1}^{\infty} \frac{n}{2^{n-1}}$

(c) $\sum_{n=1}^{\infty} \frac{n^2}{2^{n-1}}$

3.71. Explain the following steps that are used in the differentiation of power series.

(a) $(x+h)^n - x^n - nx^{n-1}h = \binom{n}{2} x^{n-2}h^2 + \binom{n}{3} x^{n-3}h^3 + \cdots + h^n.$

(b) Let $0 < |h| < H$. Then

$$|(x+h)^n - x^n - nx^{n-1}h| \le \binom{n}{2}|x|^{n-2}\frac{|h|^2}{H^2}H^2 + \binom{n}{3}|x|^{n-3}\frac{|h|^3}{H^3}H^3 + \cdots + \frac{|h|^n}{H^n}H^n.$$

(c) Therefore

$$|(x+h)^n - x^n - nx^{n-1}h| \le \frac{|h|^2}{H^2}\left(\binom{n}{2}|x|^{n-2}H^2 + \binom{n}{3}|x|^{n-3}H^3 + \cdots + H^n\right).$$

(d) The last expression in part (c) is less than or equal to $\frac{|h|^2}{H^2}(|x|+H)^n$.

3.72. Explain why for n large enough, $|nx^{n-1}| \le (|x|+\delta)^n$.
Hint: You will need the limit of $n^{1/n}$ from Problem 1.53.

3.73. Differentiate the geometric series to show that $\sum_{n=1}^{\infty} nx^{n-1} = \left(\sum_{n=0}^{\infty} x^n\right)^2$. For what numbers x does this hold?

Chapter 4
The Theory of Differentiable Functions

Abstract In this chapter, we put the derivative to work in analyzing functions. We will see how to find optimal values of functions and how to construct polynomial approximations to the function itself.

4.1 The Mean Value Theorem

At the beginning of Chap. 3, we posed a question about traveling with a broken speedometer: "Is there any way of determining the speed of the car from readings on the odometer?" In answering the question, we were led to the concept of the derivative of f at a, $f'(a)$, the instantaneous rate of change in f at a. Next we examine implications of the derivative. The mean value theorem for derivatives provides an important link between the derivative of f on an interval and the behavior of f over the interval. The mean value theorem says that if the distance that we traveled between 2:00 p.m. and 4:00 p.m. is 90 miles, then there must be at least one point in time when we traveled at exactly 45 mph. The conclusion that the average velocity over an interval must be equal to the instantaneous velocity at some point appears to be just common sense, but some of the most commonsensible theorems require somewhat sophisticated proofs. So it is with the mean value theorem, which we now state precisely.

Theorem 4.1. Mean value theorem. *Suppose that a function f is continuous on the closed interval $[a,b]$ and differentiable on the open interval (a,b). Then there exists a number c in the interval (a,b) where*

$$f'(c) = \frac{f(b)-f(a)}{b-a}.$$

P.D. Lax and M.S. Terrell, *Calculus With Applications*, Undergraduate Texts in Mathematics, 171
DOI 10.1007/978-1-4614-7946-8_4, © Springer Science+Business Media New York 2014

The theorem has an interesting geometric interpretation: Given points $(a, f(a))$ and $(b, f(b))$ on the graph, there is a point between them, $(c, f(c))$, at which the tangent is parallel to the secant through $(a, f(a))$ and $(b, f(b))$ (Fig. 4.1).

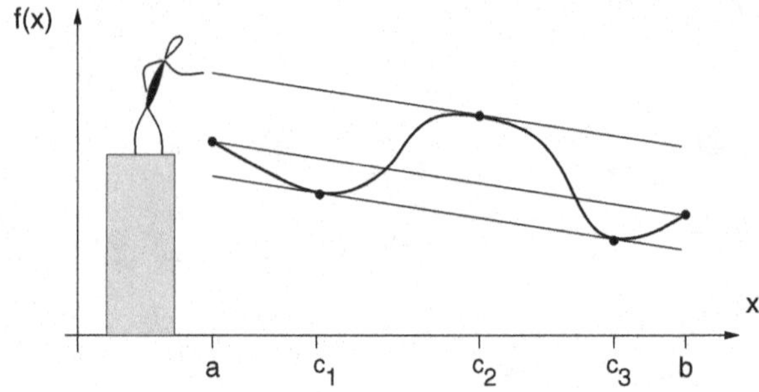

Fig. 4.1 Move the secant vertically as far as possible without breaking contact with the graph

To see how such a point c may be found, take a duplicate copy of the secant line and raise or lower it vertically to the point at which it loses contact with the graph. We notice two things about this point. First: that it occurs at a point $(c, f(c))$ on the graph of f that is farthest from the secant line. Second: that $f'(c) = \dfrac{f(b) - f(a)}{b - a}$. This suggests the key observation in the proof of the mean value theorem: that we find the desired point c by finding the point on the graph that is farthest from the secant line. The proof of the mean value theorem relies on the following result, which is important enough to present separately as a lemma.

Lemma 4.1. *Suppose a function f is defined on an open interval (a, b) and reaches its maximum or minimum at c. If $f'(c)$ exists, then $f'(c) = 0$.*

Proof. We show that $f'(c) = 0$ by eliminating the possibilities that it is positive or negative. Suppose that $f'(c) > 0$. Since the limit of $\dfrac{f(c+h) - f(c)}{h}$ as h tends to 0 exists and is positive, it follows that for h small enough, $\dfrac{f(c+h) - f(c)}{h}$ is also positive. This implies that for all h small enough and *positive*,

$$f(c+h) > f(c).$$

But for h small enough, $c+h$ belongs to (a, b), so that the above inequality violates the assumption that $f(c)$ is the maximum of f on (a, b). Therefore, it is not possible that $f'(c) > 0$.

Similarly, we show that $f'(c) < 0$ is not possible. Suppose that $f'(c)$ is negative. Then for all small h, $\dfrac{f(c+h) - f(c)}{h}$ is also negative. Taking small *negative* h, then the numerator $f(c+h) - f(c)$ is positive, $c+h$ is in (a, b), and

$$f(c+h) > f(c).$$

This contradicts the assumption that $f(c)$ is a maximum.

So only $f'(c) = 0$ is consistent with f achieving its maximum at c. A similar argument shows that if the minimum of f on (a,b) occurs at c, then $f'(c) = 0$. □

Now we are ready to prove the mean value theorem.

Proof. Let $\ell(x) = f(a) + \frac{f(b)-f(a)}{b-a}(x-a)$ be the linear function that is the secant through $(a, f(a))$ and $(b, f(b))$. Define d to be the difference between f and ℓ:

$$d(x) = f(x) - \ell(x) = f(x) - \left(f(a) + \frac{f(b)-f(a)}{b-a}(x-a)\right).$$

Since $f(x)$ and $\ell(x)$ have the same values at the endpoints, $d(x)$ is zero at the endpoints a and b. Since both f and ℓ are continuous on $[a,b]$ and differentiable on (a,b), so is d. By the extreme value theorem, Theorem 2.6, d has a maximum and a minimum on $[a,b]$. We consider two possibilities. First, it may happen that both the maximum and the minimum of d occur at the endpoints of the interval. In this case, every $d(x)$ is between $d(a)$ and $d(b)$, but both of these numbers are 0, so this would imply that $d(x) = 0$ and that $f(x) = \ell(x)$. In this case, $f'(x) = \frac{f(b)-f(a)}{b-a}$ for all x between a and b. Thus every number c in (a,b) satisfies the requirements of the theorem.

The other possibility is that either the maximum or the minimum of d occurs at some point c in (a,b). Since d is differentiable on (a,b), we have by Lemma 4.1 that $d'(c) = 0$:

$$0 = d'(c) = f'(c) - \ell'(c) = f'(c) - \frac{f(b)-f(a)}{b-a}.$$

it follows that

$$f'(c) = \frac{f(b)-f(a)}{b-a}.$$

□

Let us see how the mean value theorem provides a way to use information about the derivative on an interval to get information about the function.

Corollary 4.1. *A function whose derivative is zero at every point of an interval is constant on that interval.*

Proof. Let $a \neq b$ be any two points in the interval. The function is differentiable at a and b and every point in between. So by the mean value theorem, there exists at least one c between a and b at which $f(b) - f(a) = f'(c)(b-a)$. Since $f'(c) = 0$ for every c, it follows that $f(a) = f(b)$. Since f has the same value at any two points in the interval, it is constant. □

If two functions have the same derivatives throughout an interval, then $f' - g' = (f-g)' = 0$ and $f - g$ is a constant function. In Chap. 3, we used this result to find all the solutions to the differential equations $y' = y$, $y'' + y = 0$, and $y'' - y = 0$. Here are some additional ways to use this corollary.

Example 4.1. Suppose f is a function for which $f'(x) = 3x^2$. What can f be? One possibility is x^3. Therefore, $f(x) - x^3$ has derivative zero everywhere. According to Corollary 4.1, $f(x) - x^3$ is a constant c. Therefore, $f(x) = x^3 + c$.

Example 4.2. Suppose again that f is a function for which $f'(x) = 3x^2$, and that we now know in addition that $f(1) = 2$. By the previous example, we know that $f(x) = x^3 + c$ for some number c. Since

$$f(1) = 2 = 1^3 + c,$$

it follows that $c = 1$, and $f(x) = x^3 + 1$ is the only function satisfying both requirements.

Example 4.3. Suppose f is a function for which $f'(x) = -x^{-2}$. What can f be? The domain of f' does not include 0, so two intervals to which we can apply the corollary are x positive, and x negative.

Arguing as in Example 4.1, we conclude that $f(x) = \frac{1}{x} + a$ for x positive, and $f(x) = \frac{1}{x} + b$ for x negative, where a and b are arbitrary numbers.

The mean value theorem also enables us to determine intervals on which a function f is increasing, or on which it is decreasing, by considering the sign of f'.

Corollary 4.2. Criteria for Monotonicity. *If $f' > 0$ on an interval then f is increasing on that interval. If $f' < 0$ on an interval then f is decreasing on that interval. If $f' \geq 0$ on an interval then f is nondecreasing on that interval. If $f' \leq 0$ on an interval thin f is nonincreasing on that interval.*

Proof. Take any two points a and b in the interval such that $a < b$. By the mean value theorem, there exists a point c between a and b such that $f'(c) = \dfrac{f(b) - f(a)}{b - a}$. Since $b > a$, it follows that the sign of $f(b) - f(a)$ is the same as the sign of $f'(c)$. If $f' > 0$, then $f(b) - f(a) > 0$, and f is increasing. If $f' < 0$, then $f(b) - f(a) < 0$, and f is decreasing. The proof for the nonstrict inequalities is analogous. □

4.1a Using the First Derivative for Optimization

Finding Extreme Values on a Closed Interval. The next two examples show how to apply Lemma 4.1 to find extreme values on closed intervals.

Example 4.4. We find the largest and smallest values of

$$f(x) = 2x^3 + 3x^2 - 12x \qquad \text{on } [-4,3].$$

By the extreme value theorem, f must attain a maximum and a minimum on the interval. The extreme values can occur either at the endpoints, $x = -4$ and $x = 3$, or in $(-4,3)$. At the endpoints, we have $f(-4) = -32$ and $f(3) = 45$. If f attains a maximum or minimum at c in $(-4,3)$, then by Lemma 4.1, the derivative $f'(c)$ must be equal to 0. Next, we identify all points at which the derivative is 0: we have

$$0 = f'(x) = 6x^2 + 6x - 12 = 6(x+2)(x-1)$$

when $x = -2$ or $x = 1$. Both of these lie in $(-4,3)$. The possibilities for extreme points are then

$$f(1) = -7, \qquad f(-2) = 20, \qquad f(-4) = -32, \qquad f(3) = 45.$$

The left endpoint yields the smallest value of f on $-4,3$, and the right endpoint yields the largest value. See Fig. 4.2.

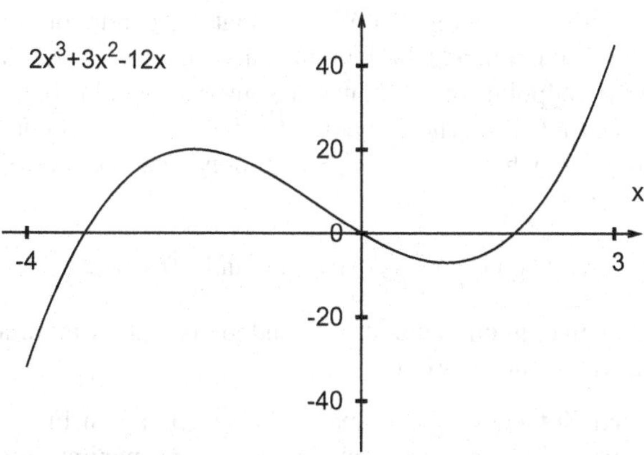

Fig. 4.2 The maximum and minimum of f on $[-4,3]$ occur at the endpoints in Example 4.4

Example 4.5. Suppose we change the domain of $f(x) = 2x^3 + 3x^2 - 12x$ to $[-3,3]$. Then the possibilities are

$$f(1) = -7, \qquad f(-2) = 20, \qquad f(-3) = 9, \qquad f(3) = 45.$$

The maximum is at the right endpoint, and the minimum is now at $x = 1$.

Example 4.6. Find the largest and smallest values of

$$f(x) = x^{2/3} \qquad \text{on } [-1,1].$$

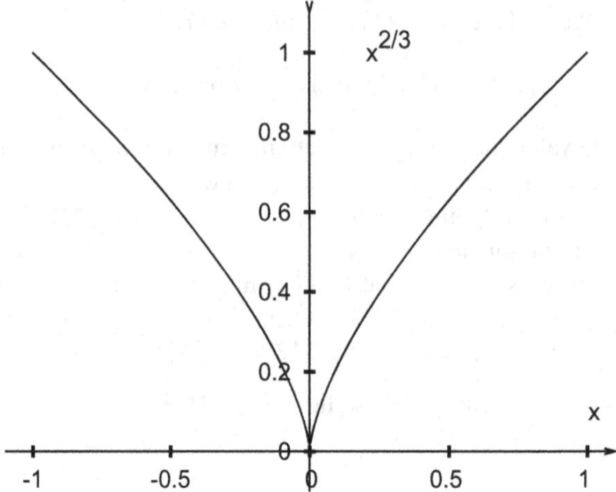

Fig. 4.3 The maximum occurs at both endpoints, and the minimum occurs at a point at which the derivative does not exist, in Example 4.6

The graph of f is shown in Fig. 4.3. We note that f is continuous on $[-1,1]$, and so it has both a maximum and a minimum value on $[-1,1]$. The extreme values can occur at the endpoints or at an interior point c in $(-1,1)$ where $f'(c)=0$ or where $f'(c)$ does not exist. The derivative $f'(x)=\frac{2}{3}x^{-1/3}$ does not exist at $x=0$. There are no points where $f'(x)=0$. So the only candidates for the maximum and minimum are

$$f(-1)=1, \quad f(1)=1, \quad \text{and} \quad f(0)=0.$$

So f attains its maximum value at both endpoints and its minimum at $x=0$, where the derivative fails to exist.

Local and Global Extrema. The graph of the function f in Fig. 4.2 shows that there may be points of interest on a graph that are not the maximum or minimum of f, but are relative or local extrema.

Definition 4.1. Local and global extrema. A function f has a *local maximum* $f(c)$ at c if there is a positive number h such that $f(x)\le f(c)$ whenever $c-h\le x\le c+h$. A function f has a *local minimum* $f(c)$ at c if there is a positive number h such that $f(x)\ge f(c)$ whenever $c-h\le x\le c+h$. A function f has an *absolute, or global, maximum* $f(c)$ at c if $f(x)\le f(c)$ for all x in the domain of f. A function f has an *absolute, or global minimum* $f(c)$ at c if $f(x)\ge f(c)$for all x in the domain of f.

Most functions we encounter in calculus have nonzero derivatives at most points in their domains. Points at which the derivative is zero or does not exist are called *critical points* of the function. Endpoints of the domain (if there are any) and critical points subdivide the domain into smaller intervals on which the derivative is either positive or negative. Next we show how the monotonicity criteria can be used to identify points in the domain of f at which f has extreme values. This result is often called the *first derivative test.*

Theorem 4.2. First derivative test. *Suppose that f is continuous on an interval containing c, and that $f'(x)$ is positive for x less than c, and negative for x greater than c. Then f reaches its maximum on the interval at c.*

A similar characterization holds for the minimum.

The proof follows from the criterion for monotonicity and the definitions of maximum and minimum. We ask you to write it out in Problem 4.6. Figure 4.4 shows some examples. Whether an extremum on an interval is local or absolute depends on whether the interval is the entire domain.

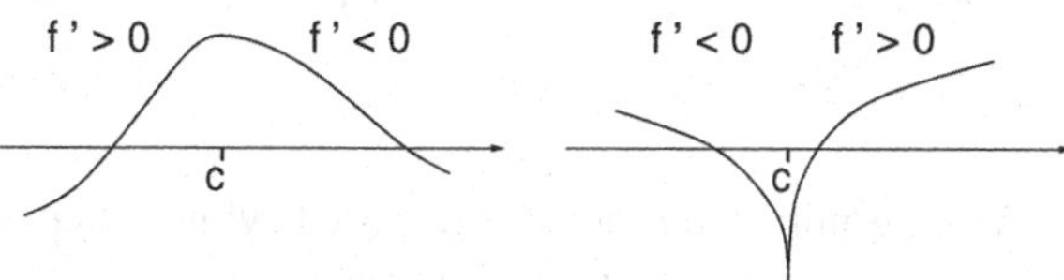

Fig. 4.4 An illustration of the first derivative test. *Left*: f has a maximum at c. *Right*: f has a minimum

Example 4.7. Consider the quadratic function

$$f(x) = x^2 + bx + c.$$

We can rewrite this by "completing the square":

$$f(x) = x^2 + bx + c = \left(x + \frac{1}{2}b\right)^2 - \frac{1}{4}b^2 + c. \tag{4.1}$$

We see at a glance that the minimum of $f(x)$ is achieved at $x = -\frac{1}{2}b$. We next show how to derive this result by calculus. First we look for critical points of f: $f'(x) = 2x + b$ is zero when $x = -\frac{1}{2}b$. Next, we check the sign of the first derivative:

$$f'(x) \begin{cases} < 0 & \text{for } x < -\frac{1}{2}b, \\ = 0 & \text{for } x = -\frac{1}{2}b, \\ > 0 & \text{for } x > -\frac{1}{2}b. \end{cases}$$

By the first derivative test, f achieves an absolute minimum at $x = -\dfrac{1}{2}b$.

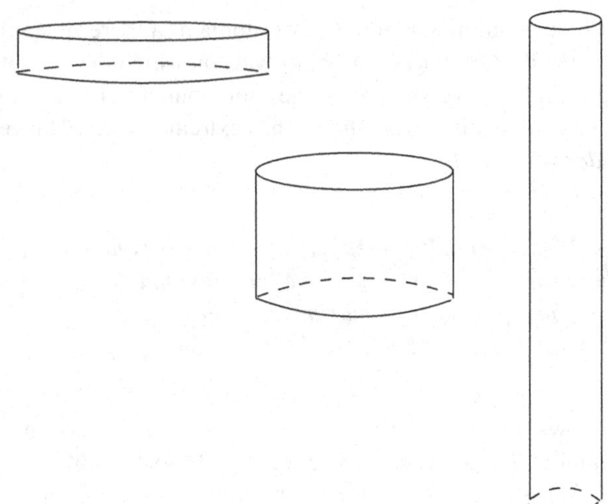

Fig. 4.5 Example 4.8 illustrated. Three cylinders have the same surface area $A = 2\pi$, but unequal volumes. *Top left* : $r = \frac{7}{8}$, $V = 0.6442\ldots$ *Center* : $r = \frac{2}{3}$, $V = 1.1635\ldots$ *Right* : $r = \frac{1}{4}$, $V = 0.7363\ldots$

Example 4.8. We determine the shape of the closed cylinder that has the largest volume among all cylinders of given surface area A.

We plan to accomplish this by expressing the volume as a function of one variable, the radius r. Let h be the height. Then the surface area is

$$A = 2\pi r^2 + 2\pi rh = 2\pi r(r+h), \tag{4.2}$$

and the volume is $V = \pi r^2 h$. We have expressed V as a function of two variables, r and h, but we can eliminate one by means of the constraint (4.2). Solving for h, we see that $h = \dfrac{A}{2\pi r} - r$, so that

$$V = f(r) = \pi r^2 \left(\frac{A}{2\pi r} - r\right) = \frac{Ar}{2} - \pi r^3, \qquad r > 0.$$

The derivative $f'(r) = \frac{1}{2}A - 3\pi r^2$ is zero when $r = r_0 = \sqrt{\dfrac{A}{6\pi}}$. Since $f' > 0$ for smaller values of r, and $f' < 0$ for larger r, f has an absolute maximum at r_0. To determine the shape of this cylinder, evaluate h in terms of r_0: the height of the cylinder of largest volume is

$$h = \frac{A}{2\pi r} - r = \frac{6\pi r_0^2}{2\pi r_0} - r_0 = 2r_0.$$

That is, for cylinders of a given surface area, the volume is greatest when the diameter of the cylinder equals the height. See Fig. 4.5.

The surface area of the cylinder is proportional to the amount of material needed to manufacture a cylindrical container. The above shape is *optimal* in the sense that it encloses the largest volume for a given amount of material. Examine the cans in the supermarket and determine which brands use the optimal shape.

4.1b Using Calculus to Prove Inequalities

Next, we show how calculus makes it easier to derive some inequalities that we obtained before we had calculus.

Exponential Growth. We showed in Theorem 2.10 that as x tends to infinity, the exponential function grows faster than any power of x, in the sense that for a fixed n, $\frac{e^x}{x^n}$ tends to infinity as x tends to infinity. Here is a simple calculus proof of this fact.

Differentiate $f(x) = \frac{e^x}{x^n}$. Using the product rule, we get

$$f'(x) = \frac{e^x}{x^n} - n\frac{e^x}{x^{n+1}} = f(x) - n\frac{f(x)}{x} = f(x)\frac{x-n}{x}. \tag{4.3}$$

This shows that the derivative of $f(x)$ is negative for $0 < x < n$ and is positive for x greater than n. So $f(x)$ is decreasing as x goes from 0 to n and increasing from then on. It follows that $f(x)$ reaches its minimum at $x = n$. This means that

$$f(x) = \frac{e^x}{x^n} \geq \frac{e^n}{n^n}, \qquad (x > 0).$$

Multiply this inequality by x. It follows that

$$\frac{e^x}{x^{n-1}} \geq x\frac{e^n}{n^n}.$$

The number $\frac{e^n}{n^n}$ is fixed by our original choice of n, so the function on the right tends to infinity as x tends to infinity. Therefore, so does the function on the left. Since n is arbitrary, this proves our contention.

The A-G Inequality. Recall from Sect. 1.1c that the arithmetic–geometric inequality says that for any two positive numbers a and b,

$$\frac{a+b}{2} - (ab)^{1/2} > 0, \tag{4.4}$$

unless $a = b$. Let us see how to use calculus to obtain Eq. (4.4). Let a be the smaller of the two numbers: $a < b$. Define the function $f(x)$ to be

$$f(x) = \frac{a+x}{2} - (ax)^{1/2}. \tag{4.5}$$

Then $f(b) = \dfrac{a+b}{2} - (ab)^{1/2}$, $f(a)$ is zero, and the A-G inequality can be formulated thus:

$$f(b) \text{ is greater than } f(a).$$

Since $a < b$, this will follow if we can show that $f(x)$ is an increasing function between a and b. The calculus criterion for a function to be increasing is for its derivative to be positive. The derivative is

$$f'(x) = \frac{1}{2} - \frac{1}{2}\frac{a^{1/2}}{x^{1/2}},$$

which is positive for x greater than a. This completes the proof of the A-G inequality for two numbers.

How about three numbers? The A-G inequality for three positive numbers a, b, and c states that

$$\frac{a+b+c}{3} - (abc)^{1/3} > 0, \tag{4.6}$$

unless all three numbers a, b, c are equal.

Rewrite Eq. (4.6) as

$$abc \le \left(\frac{a+b+c}{3}\right)^3$$

and divide by c:

$$ab \le \frac{1}{c}\left(\frac{a+b+c}{3}\right)^3. \tag{4.7}$$

Keep a and b fixed and define the function $f(x)$ as the right side of Eq. (4.7) with c replaced by x, where $x > 0$:

$$f(x) = \frac{1}{x}\left(\frac{a+b+x}{3}\right)^3. \tag{4.8}$$

As x tends to 0, $f(x)$ tends to infinity, and as x tends to infinity, $f(x)$ tends to infinity. Therefore, $f(x)$ attains its absolute minimum for some $x > 0$. We shall use the calculus criterion to find that minimum value. Differentiate f. Using the product rule and chain rule, we get

$$f'(x) = \frac{1}{x}\left(\frac{a+b+x}{3}\right)^2 - \frac{1}{x^2}\left(\frac{a+b+x}{3}\right)^3.$$

We see that $f'(x)$ is zero if

$$x = \frac{a+b+x}{3},$$

so the minimum occurs at $x = \frac{1}{2}(a+b)$. The value of f at this point is

$$f\left(\frac{1}{2}(a+b)\right) = \frac{2}{a+b}\left(\frac{a+b+\frac{1}{2}(a+b)}{3}\right)^3 = \left(\frac{a+b}{2}\right)^2.$$

According to the A-G inequality for $n = 2$, this is greater than or equal to ab. So at its minimum value, $f(x)$ is not less than ab. Therefore, it is not less than ab at any other point c. This completes the proof of the A-G inequality for three numbers.

The A-G inequality can be proved inductively for every n by a similar argument, and we ask you to do so in Problem 4.15.

4.1c A Generalized Mean Value Theorem

At the start of this section, we noted that the mean value theorem guarantees that if there is an interval over which your average velocity was 30 mph, then there was at least one moment in that interval when your velocity was exactly 30 mph. In this section we see that the mean value theorem can be used to prove a somewhat surprising variation: If during an interval of time you have traveled five times as far as your friend, then there has to be at least one moment when you were traveling exactly five times as fast as your friend.

Theorem 4.3. Generalized mean value. *Suppose f and g are differentiable on (a,b) and continuous on $[a,b]$. If $g'(x) \neq 0$ in (a,b), then there exists a point c in (a,b) such that*

$$\frac{f'(c)}{g'(c)} = \frac{f(b)-f(a)}{g(b)-g(a)}.$$

Proof. Let

$$H(x) = \big(f(x)-f(a)\big)\big(g(b)-g(a)\big) - \big(g(x)-g(a)\big)\big(f(b)-f(a)\big).$$

Then H is differentiable on (a,b) and continuous on $[a,b]$, and $H(a) = H(b) = 0$. By the mean value theorem, there exists a point c in (a,b) where

$$0 = \frac{H(b)-H(a)}{b-a} = H'(c) = f'(c)\big(g(b)-g(a)\big) - g'(c)\big(f(b)-f(a)\big),$$

and so $f'(c)\big(g(b)-g(a)\big) = g'(c)\big(f(b)-f(a)\big)$. Since $g' \neq 0$ in (a,b), neither of $g'(c)$, $\big(g(b)-g(a)\big)$ is 0. To complete the proof, divide both sides by $g'(c)\big(g(b)-g(a)\big)$. □

This variation of the mean value theorem can be used to prove the following technique (l'Hospitals' rule) for evaluating some limits.

Theorem 4.4. *Suppose* $\lim_{x\to a} f(x) = 0$, $\lim_{x\to a} g(x) = 0$, *and that* $\lim_{x\to a} \frac{f'(x)}{g'(x)}$ *exists. Then*

$$\lim_{x\to a} \frac{f(x)}{g(x)} = \lim_{x\to a} \frac{f'(x)}{g'(x)}.$$

Proof. Since $\lim_{x\to a} \dfrac{f'(x)}{g'(x)}$ exists, there is an interval around a (perhaps excluding a) where $f'(x)$ and $g'(x)$ exist and $g'(x) \neq 0$. Define two new functions F and G that agree with f and g for $x \neq a$, and set $F(a) = G(a) = 0$. By Theorem 4.3 applied to F and G, there is a point c between a and x such that

$$\frac{f(x)}{g(x)} = \frac{F(x)}{G(x)} = \frac{F(x)-F(a)}{G(x)-G(a)} = \frac{F'(c)}{G'(c)} = \frac{f'(c)}{g'(c)}.$$

Since c is between a and x and $\lim_{x\to a} \dfrac{f'(x)}{g'(x)}$ exists, it follows that $\lim_{x\to a} \dfrac{f'(x)}{g'(x)} = \lim_{x\to a} \dfrac{f'(c)}{g'(c)} = \lim_{x\to a} \dfrac{f(x)}{g(x)}$. □

Example 4.9. The limit $\lim_{x\to 1} \dfrac{\log x}{x^2-1}$ satisfies $\lim_{x\to 1} \log x = 0$ and $\lim_{x\to 1}(x^2-1) = 0$, and both $\log x$ and x^2-1 are differentiable near 1. Therefore,

$$\lim_{x\to 1} \frac{\log x}{x^2-1} = \lim_{x\to 1} \frac{(\log x)'}{(x^2-1)'},$$

provided that the last limit exists. But it does exist, because

$$\lim_{x\to 1} \frac{(\log x)'}{(x^2-1)'} = \lim_{x\to 1} \frac{1/x}{2x} = \frac{1}{2}.$$

Therefore, $\lim_{x\to 1} \dfrac{\log x}{x^2-1} = \dfrac{1}{2}$. See Fig. 4.6.

Another version of this theorem may be found in Problem 4.23.

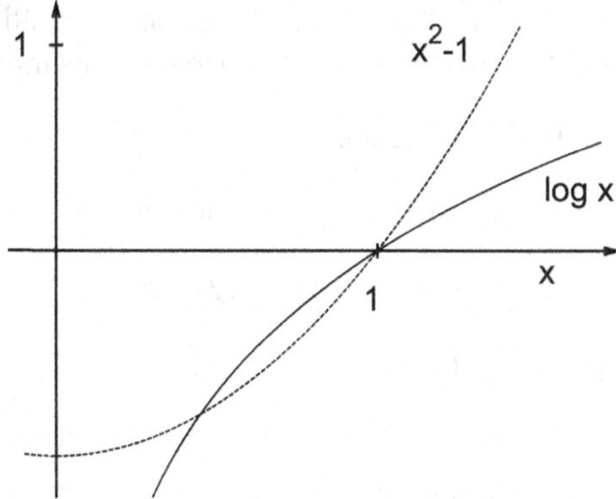

Fig. 4.6 Graphs of the functions in Example 4.9. The ratio of values is equal to the ratio of slopes as x tends to 1

Problems

4.1. Suppose $f(2) = 6$, and $0.4 \le f'(x) \le 0.5$ for x in $[2, 2.2]$. Use the mean value theorem to estimate $f(2.1)$.

4.2. Suppose $g(2) = 6$ and $-0.6 \le g'(x) \le -0.5$ for x in $[1.8, 2]$. Use the mean value theorem to estimate $g(1.8)$.

4.3. If $h'(x) = 2\cos(3x) - 3\sin(2x)$, what could $h(x)$ be? If, in addition, $h(0) = 0$, what could $h(x)$ be?

4.4. If $k'(t) = 2 - 2e^{-3t}$, what could $k(t)$ be? If, in addition, $k(0) = 0$, what could $k(t)$ be?

4.5. Consider $f(x) = \dfrac{x}{x^2+1}$.

(a) Find $f'(x)$.
(b) In which interval(s) does f increase?
(c) In which interval(s) does f decrease?
(d) Find the minimum value of f in $[-10, 10]$.
(e) Find the maximum value of f in $[-10, 10]$.

4.6. Justify the following steps to prove the first derivative test, Theorem 4.2. Suppose that $f'(x)$ is positive for $x < c$ and negative for $x > c$. We need to show that f reaches its maximum at c.

(a) Explain why f is increasing for all $x < c$, and why f is decreasing for all $x > c$.
(b) Use the continuity of f at c and the fact that f is increasing when $x < c$ to explain why $f(x)$ cannot be greater than $f(c)$ for any $x < c$. Similarly explain why $f(x)$ cannot be greater than $f(c)$ for any $x > c$.
(c) Explain why $f(c)$ is a maximum on S.

(d) Revise the argument to show that if $f'(x)$ is negative for all x less than c and $f'(x)$ is positive for all x greater than c, then f reaches its minimum at $f(c)$.

4.7. Rework Problem 1.10 using calculus.

4.8. Find the maximum and minimum values of the function

$$f(x) = 2x^3 - 3x^2 - 12x + 8$$

on each of the following intervals.

(a) $[-2.5, 4]$
(b) $[-2, 3]$
(c) $[-2.25, 3.75]$

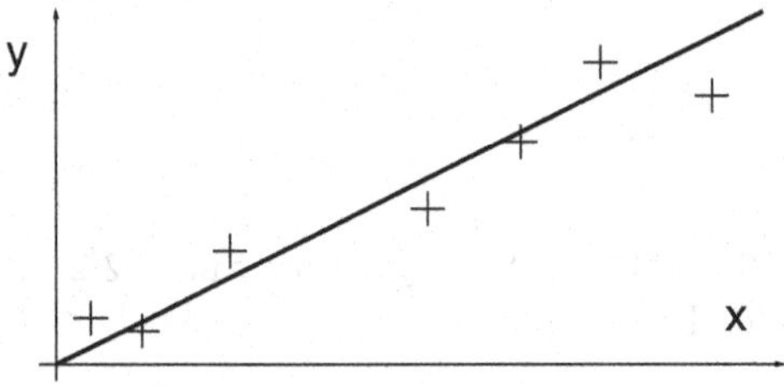

Fig. 4.7 In Problem 4.9 we find a slope to fit given data

4.9. Suppose an experiment is carried out to determine the value of the constant m in the equation

$$y = mx$$

relating two physical quantities. Let $(x_1, y_1),\ (x_2, y_2), \ldots, (x_n, y_n)$ be the measured values. Find the value of m, in terms of the x_i and y_i, that minimizes E, the sum of the squares of the errors between the observed measurements and the linear function $y = mx$:

$$E = (y_1 - mx_1)^2 + (y_2 - mx_2)^2 + \cdots + (y_n - mx_n)^2.$$

See Fig. 4.7.

4.10. Consider an open cardboard box whose bottom is a square of edge length x, and whose height is y. The volume V and surface area S of the box are given by

$$V = x^2 y, \qquad S = x^2 + 4xy.$$

Among all boxes with given volume, find the one with smallest surface area. Show that this box is squat, i.e., $y < x$.

4.11. Consider a particle of unit mass moving on a number line whose position at time t is given by $x(t) = 3t - t^2$. Find the time when the particle's position x is maximal.

4.12. Find the point on the graph of $y = \frac{1}{2}x^2$ closest to the point $(6,0)$.

4.13. What is the largest amount by which a positive number can exceed its cube?

4.14. Find the positive number x such that the sum of x and its reciprocal is as small as possible,

(a) Using calculus, and
(b) By the A-G inequality.

4.15. Use calculus to prove, by induction, the A-G inequality for n positive numbers.

4.16. Let $w_1, \dots, w_n$ be positive numbers whose sum is 1, and $a_1, \dots, a_n$ any positive numbers. Prove an extension of the A-G inequality:

$$a_1^{w_1} a_2^{w_2} \cdots a_n^{w_n} \le w_1 a_1 + w_2 a_2 + \cdots + w_n a_n,$$

with equality only in the case $a_1 = a_2 = \cdots = a_n$. Try an inductive proof with one of the a's as the variable.

4.17. Suppose $g'(x) \le h'(x)$ for $0 < x$ and $g(0) = h(0)$. Prove that $g(x) \le h(x)$ for $0 < x$.

4.18. Here we apply Problem 4.17 to find polynomial bounds for the cosine and sine.

(a) Show that $g'(x) \le h'(x)$ for $g(x) = \sin x$ and $h(x) = x$ and deduce that

$$\sin x \le x \qquad (x > 0). \tag{4.9}$$

(b) Rewrite Eq. (4.9) as $\left(-\cos x\right)' \le \left(\frac{x^2}{2} - 1\right)'$ and deduce that $1 - \frac{x^2}{2} \le \cos x$.
(c) Continue along these lines to derive

$$1 - \frac{x^2}{2} \le \cos x \le 1 - \frac{x^2}{2} + \frac{x^4}{4!},$$

and in particular, estimate $\cos(0.2)$ with a tolerance of 0.001.
(d) Extend the previous argument to derive

$$1 - \frac{x^2}{2} + \frac{x^4}{4!} - \frac{x^6}{6!} \le \cos x \le 1 - \frac{x^2}{2} + \frac{x^4}{4!} - \frac{x^6}{6!} + \frac{x^8}{8!}.$$

4.19. Denote the exponential function e^x by $e(x)$. Use $e' = e$ and Problem 4.17 to show the following.

(a) For $x > 0$, $1 < e(x)$.
(b) $1 < e'(x)$ for $0 < x$, and deduce that $1 + x < e(x)$.
(c) Rewrite this as $1 + x < e'(x)$, and deduce that $1 + x + \frac{1}{2}x^2 < e(x)$ for $0 < x$.
(d) For all n and all $x > 0$, $1 + x + \frac{x^2}{2} + \cdots + \frac{x^n}{n!} < e(x)$.

4.20. Evaluate $\lim_{x\to 0} \frac{\sin x}{x}$, first using the mean value theorem to write $\sin x = \sin x - \sin 0 = \cos(c)x$, and then using Theorem 4.4.

4.21. Evaluate $\lim_{x\to 0} \frac{x}{e^x - 1}$, first by recognizing the quotient as a reciprocal derivative, and then using Theorem 4.4.

4.22. Evaluate $\lim_{x\to 0} \frac{\sin x - x\cos x}{x^3}$ using Theorem 4.4 twice.

4.23. Substitute $f(x) = F(1/x)$ and $g(x) = G(1/x)$ into Theorem 4.4 to prove the following version of Theorem 4.4.

Suppose $\lim_{y\to\infty} F(y) = 0$, $\lim_{y\to\infty} G(y) = 0$, and that $\lim_{y\to\infty} \frac{F'(y)}{G'(y)}$ exists. Then

$$\lim_{y\to\infty} \frac{F(y)}{G(y)} = \lim_{y\to\infty} \frac{F'(y)}{G'(y)}.$$

You will need to take $a = 0$ in the theorem. Explain how to extend f and g as odd functions, so that the theorem can be applied.

4.24. Use the result of Problem 4.23 and the exponential growth theorem where needed to evaluate the following limits.

(a) $\lim_{y\to\infty} e^{-1/y}$

(b) $\lim_{y\to\infty} y^2 e^{-y}$

(c) $\lim_{y\to\infty} \frac{e^{-y}}{1 - e^{-1/y}}$

4.2 Higher Derivatives

Many of the functions f we have presented in examples so far have the property that their derivatives f' also turned out to be differentiable. Such functions are called *twice differentiable*. Similarly, we define a three-times differentiable function f as one whose second derivative is differentiable.

Definition 4.2. A function f is called *n times differentiable* at x if its $(n-1)$st derivative is differentiable at x. The resulting function is called the nth derivative of f and is denoted by

$$f^{(n)} \quad \text{or} \quad \frac{d^n f}{dx^n}.$$

Definition 4.3. A function f is called *continuously differentiable* on an interval if f' exists and is continuous on the interval. A function f is *n times continuously differentiable* if the nth derivative exists and is continuous on the interval.

As we saw in Chap. 3, if $x(t)$ denotes the position of a particle at time t, then the rate of change of position $\frac{dx}{dt}$ is the velocity v of the particle. The derivative of velocity is called *acceleration*. Thus

$$\text{acceleration} = \frac{dv}{dt} = \frac{d^2x}{dt^2};$$

in words, *acceleration is the second derivative of position.*

The geometric interpretation of the second derivative is no less interesting than the physical interpretation. We note that a linear function $f(x) = mx + b$ has second derivative zero. Therefore, a function with nonzero second derivative is not linear. Since a linear function can be characterized as one whose graph is a straight line, it follows that if $f'' \neq 0$, then the graph of f is not flat but curved. This fact suggests that the size of $f''(x)$ measures in some sense the deviation of the graph of f from a straight line at the point x.

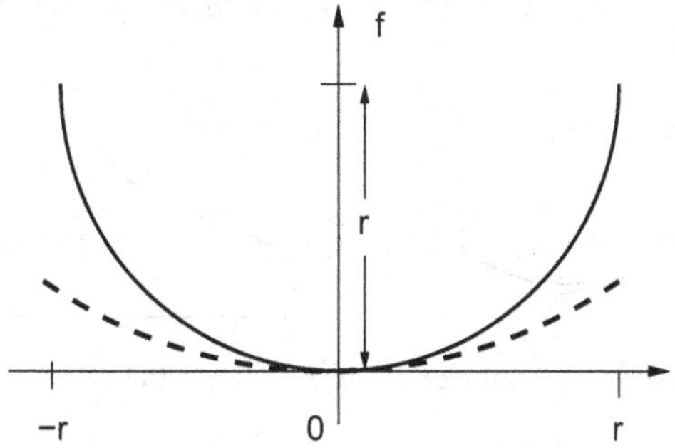

Fig. 4.8 The second derivative illustrated: large r corresponds to small curvature. The dotted arc has a larger value of r. See Example 4.10

Example 4.10. The graph of the function

$$f(x) = r - \sqrt{r^2 - x^2}, \qquad -r < x < r,$$

is a semicircle of radius r; see Fig. 4.8. The larger the value of r, the closer this semicircle lies to the x-axis, for values of x in a fixed interval about the origin, as illustrated in Fig. 4.8. We have

$$f'(x) = \frac{x}{\sqrt{r^2 - x^2}}, \qquad f''(x) = \frac{1}{\sqrt{r^2 - x^2}} + \frac{x^2}{(r^2 - x^2)^{3/2}} = \frac{r^2}{(r^2 - x^2)^{3/2}}.$$

The value of f'' at $x = 0$ is $f''(0) = \dfrac{r^2}{(r^2)^{3/2}} = \dfrac{1}{r}$. The larger r is, the smaller is the value of $f''(0)$, so in this case, the smallness of $f''(0)$ indeed indicates that the graph of f is close to a straight line.

What Does f'' Tell Us About f? The goal of this section is to explain the following result: *over a short interval, every twice continuously differentiable function can be exceedingly well approximated by a quadratic polynomial.* Note the simplification implied here, that a complicated function can sometimes be replaced by a simple one.

We shall use the monotonicity criterion to relate knowledge about f'' to knowledge about f. For example, if $f'' > 0$, then by monotonicity, f' is increasing. Graphically, this means that the slopes of the tangents to the graph of f are increasing as you move from left to right. Similarly, if $f'' < 0$, then f' is decreasing, and the slopes of the tangents to the graph of f are decreasing as you move from left to right. In Fig. 4.9, some quick sketches of tangent lines with increasing (and decreasing) slopes suggest that the graph of the underlying function opens upward if $f'' > 0$, and that it opens downward if $f'' < 0$.

Rather than trust a few sketches, we shall investigate this question: if we have information about f'', what can we say about f itself? Suppose we have an estimate for f'',

$$m \le f''(x) \le M \quad \text{on } [a,b]. \tag{4.10}$$

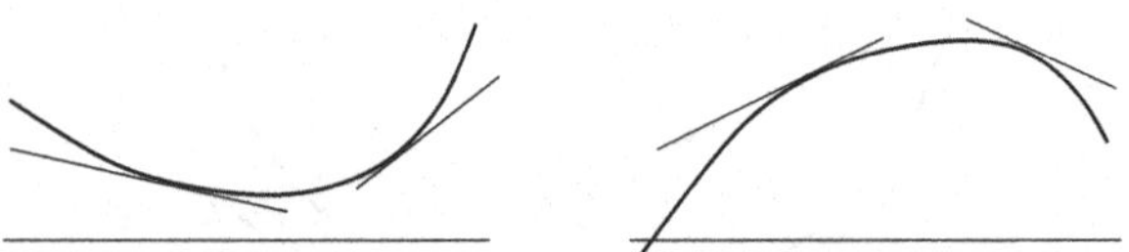

Fig. 4.9 *Left*: $f'' > 0$ and increasing slopes. *Right*: $f'' < 0$ and decreasing slopes

The inequality on the right is equivalent to $M - f''(x) \ge 0$. Note that $M - f''(x)$ is the derivative of $Mx - f'(x)$. So by the monotonicity criterion, $Mx - f'(x)$ is a nondecreasing function, and we conclude that

$$Ma - f'(a) \le Mx - f'(x) \text{ on } [a,b].$$

This inequality can be rewritten as follows:

$$f'(x) - f'(a) \le M(x-a) \text{ on } [a,b].$$

Note that the function on left-hand side is the derivative of $f(x) - xf'(a)$, and the function on the right-hand side is the derivative of $\frac{1}{2}M(x-a)^2$. Taking their difference, again by the monotonicity criterion it follows that $\frac{1}{2}M(x-a)^2 - (f(x) - xf'(a))$ is a nondecreasing function. Since a is less than or equal to x, we have

$$\frac{1}{2}M(a-a)^2 - (f(a) - af'(a)) \le \frac{M}{2}(x-a)^2 - (f(x) - xf'(a)).$$

Rewrite this last inequality by taking the term $f(x)$ to the left-hand side and all other terms to the right-hand side, giving

$$f(x) \le f(a) + f'(a)(x-a) + \frac{M}{2}(x-a)^2.$$

Remark. This is the first step in our stated goal, since the function on the right-hand side is a quadratic polynomial.

By an analogous argument we can deduce from $m \le f''(x)$ and repeated uses of monotonicity that

$$f(a) + f'(a)(x-a) + \frac{m}{2}(x-a)^2 \le f(x)$$

for all x in $[a,b]$. We can combine the two inequalities into one statement. If $f''(x)$ is bounded below by m and above by M on $[a,b]$, then f itself is bounded below and above by two quadratic polynomials:

$$f(a)+f'(a)(x-a)+\frac{m}{2}(x-a)^2 \le f(x) \le f(a)+f'(a)(x-a)+\frac{M}{2}(x-a)^2. \quad (4.11)$$

The upper and lower bounds differ inasmuch as one contains the constant m and the other M. It follows that there is a number H between m and M such that

$$f(x) = f(a) + f'(a)(x-a) + \frac{H}{2}(x-a)^2. \quad (4.12)$$

Suppose next that f'' is continuous on $[a,b]$ and that m and M are the minimum and maximum values of f'' on the interval. Take $x = b$ in Eq. (4.12). It follows again from the intermediate value theorem that there is a point c between a and b such that

$$f(b) = f(a) + f'(a)(b-a) + \frac{1}{2}f''(c)(b-a)^2.$$

This equation provides a rich source of observations about f, and we obtain the following generalization of the mean value theorem:

Theorem 4.5. Linear approximation. *Let f be twice continuously differentiable on an interval containing a and b. Then there is a point c between a and b such that*

$$f(b) = f(a) + f'(a)(b-a) + \frac{f''(c)}{2}(b-a)^2. \quad (4.13)$$

We proved the linear approximation theorem for $a < b$. It is also true for $a > b$. The proof is outlined in Problem 4.32.

Example 4.11. Let us use Theorem 4.5 to estimate $\log(1.1)$. Let $f(x) = \log(1+x)$. Then $f'(x) = \dfrac{1}{1+x}$ and $f''(x) = -\dfrac{1}{(1+x)^2}$. Taking $a = 0$ and $b = 0.1$, we get $f(0) = 0,\ f'(0) = 1$, and

$$\log(1.1) = 0 + 1(0.1-0) - \frac{1}{(1+c)^2}\frac{(0.1)^2}{2},$$

where c is a number between 0 and 0.1. Since $-\dfrac{1}{(1)^2} \le f''(c) \le -\dfrac{1}{(1.1)^2}$, we get

$$0.095 = 0.1 - \frac{0.01}{2} \le \log(1.1) \le 0.1 - \frac{1}{(1.1)^2}\frac{0.01}{2} = 0.0958\ldots < 0.096.$$

See what your calculator says about the natural logarithm of 1.1.

Example 4.12. Let $f(x) = \log(1+x)$. We approximate f by two quadratic polynomials on $[0, 0.5]$. From Example 4.11, we have $f(0) = 0,\ f'(0) = 1$, the minimum of $f''(x)$ on $[0, 0.5]$ is -1, and the maximum is $-\dfrac{1}{(1+0.5)^2} = -\dfrac{4}{9}$. Therefore (See Fig. 4.10),

$$x - \frac{x^2}{2} \le \log(1+x) \le x - \frac{4}{9}\frac{x^2}{2}, \quad (0 \le x \le 0.5).$$

In the linear approximation theorem, suppose b is close to a. Then c is even closer to a, and since f'' is continuous, $f''(c)$ is close to $f''(a)$. We express this by writing

$$f''(c) = f''(a) + s,$$

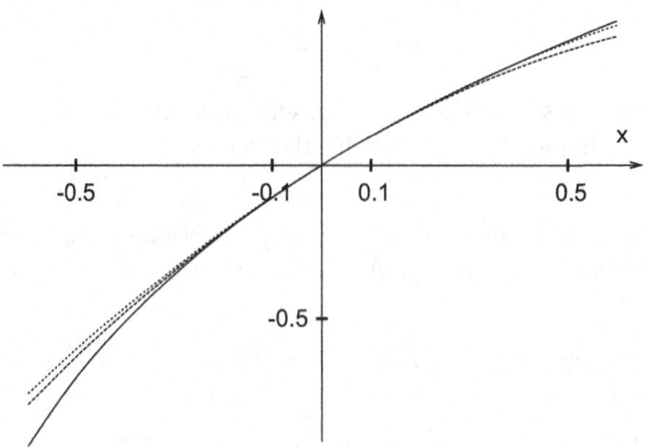

Fig. 4.10 The graphs of $\log(1+x)$ and the two quadratic polynomials of Example 4.12

where s denotes a quantity that is small when b is close to a. Substituting this into Eq. (4.13), we get

$$f(b) = f(a) + f'(a)(b-a) + \frac{1}{2}f''(a)(b-a)^2 + \frac{1}{2}s(b-a)^2.$$

This formula shows that for b close to a, the first three terms on the right are a very good approximation to $f(b)$. Therefore, it follows from the linear approximation theorem, as we have stated, that *over a short interval, every twice continuously differentiable function can be exceedingly well approximated by a quadratic polynomial.*

4.2a Second Derivative Test

The linear approximation theorem, Theorem 4.5, has applications to optimization. The next two theorems are sometimes referred to as the *second derivative test* for local extrema:

Theorem 4.6. Local minimum theorem. *Let f be a twice continuously differentiable function on an open interval containing a, and suppose that $f'(a) = 0$ and $f''(a) > 0$. Then f has a local minimum at a, i.e.,*

$$f(a) < f(b)$$

for all points $b \neq a$ sufficiently close to a.

Proof. We have $f''(a) > 0$, so by the continuity of f'', $f''(x) > 0$ for all x close enough to a. Choose b so close to a that $f''(c) > 0$ for all c between a and b. According to the linear approximation theorem, since $f'(a) = 0$ and $f''(c) > 0$, we get

$$f(b) = f(a) + \frac{f''(c)}{2}(b-a)^2 > f(a),$$

as asserted. □

We suggest a way for you to prove the analogous maximum theorem in Problem 4.38:

Theorem 4.7. Local maximum theorem. *Let f be a twice continuously differentiable function defined on an open interval containing a, and suppose that $f'(a) = 0$ and $f''(a) < 0$. Then f has a local maximum at a, i.e.,*

$$f(a) > f(b)$$

for all points $b \neq a$ sufficiently close to a.

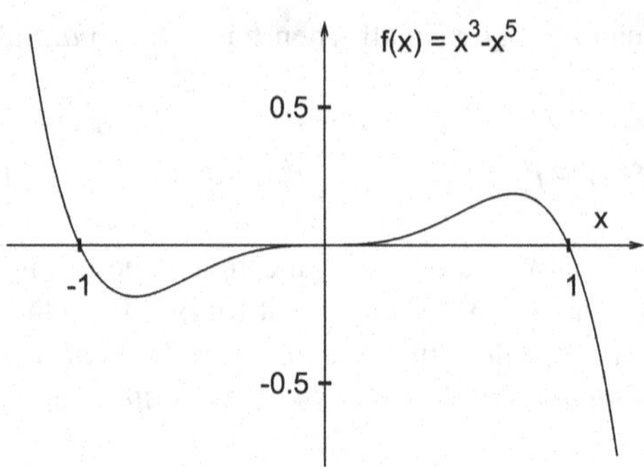

Fig. 4.11 A local maximum and a local minimum, in Example 4.13

Example 4.13. The polynomial $f(x) = x^3 - x^5$ has $f'(x) = x^2(3 - 5x^2)$, which is 0 at three numbers $x_1 = -\sqrt{\frac{3}{5}}$, $x_2 = 0$, and $x_3 = \sqrt{\frac{3}{5}}$. The second derivative is $f''(x) = 6x - 20x^3 = 2x(3 - 10x^2)$. So

$$f''(x_1) = -2\sqrt{\frac{3}{5}}(3-6) > 0, \quad f''(x_2) = 0, \quad f''(x_3) = 2\sqrt{\frac{3}{5}}(3-6) < 0.$$

We conclude that f has a local minimum at x_1 and a local maximum at x_3. However, $f''(x_2) = 0$ does not give any information about the possibility of local extrema at x_2. The graph of f is drawn in Fig. 4.11.

4.2b Convex Functions

We give further applications of the linear approximation theorem.

Suppose f'' is nonnegative in an interval containing a and b. Then the last term on the right in Eq. (4.13) is nonnegative, so omitting it yields the inequality

$$f(b) \geq f(a) + f'(a)(b-a).$$

This inequality has a striking geometric interpretation. We notice that the quantity on the right is the value at b of the linear function

$$l(x) = f(a) + f'(a)(x-a).$$

The graph of this linear function is the line tangent to the graph of f at $(a, f(a))$. So

$$f(b) \geq f(a) + f'(a)(b-a)$$

asserts that the graph of f lies above its tangent lines.

Definition 4.4. A function whose graph lies above its tangents is called *convex.*

In this language, the inequality says that *every function whose second derivative is positive is convex.* Convex functions have another interesting property:

Theorem 4.8. Convexity theorem. *Let f be a twice continuously differentiable function on $[a,b]$, and suppose that $f'' > 0$ there. Then for every x satisfying $a < x < b$,*

$$f(x) < f(b)\frac{x-a}{b-a} + f(a)\frac{b-x}{b-a}. \tag{4.14}$$

This theorem has an illuminating geometric interpretation. Denote by $\ell(x)$ the function on the right-hand side of inequality (4.14). Then ℓ is a linear function whose values at $x = a$ and at $x = b$ agree with the values of f at these points. Thus the graph of ℓ is the *secant line* of f on $[a,b]$. Therefore, inequality (4.14) says that *the graph of a convex function f on an interval $[a,b]$ lies below the secant line* (Fig. 4.12).

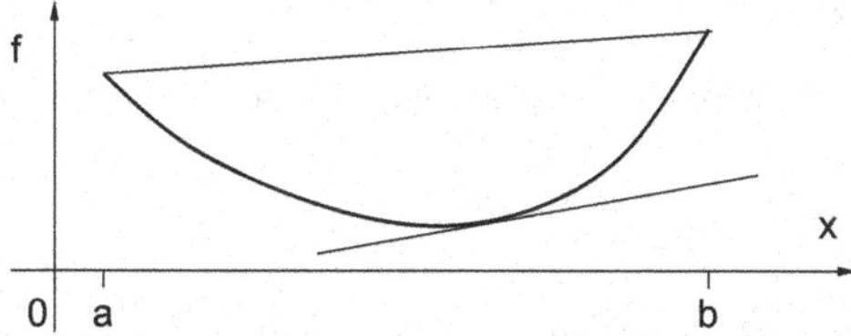

Fig. 4.12 The graph of a convex function lies above its tangent lines and below the secant on $[a,b]$

Proof. We wish to show that $f(x) - \ell(x) \le 0$ on $[a,b]$. According to the extreme value theorem, Theorem 2.6, $f - \ell$ reaches a maximum at some point c in $[a,b]$. The point c could be either at an endpoint or in (a,b). We show now that c is not in (a,b). For if it were, then the first derivative of $f - \ell$ would be zero at c. The second derivative of $f(x) - \ell(x)$ at c is given by

$$f''(c) - \ell''(c) = f''(c) - 0,$$

since ℓ is linear. We have assumed that f'' is positive. According to the local minimum theorem, Theorem 4.6, the function $f - \ell$ has a local minimum at c. This shows that the point c where the maximum occurs cannot be in the interior of $[a,b]$.

The only alternative remaining is that c is one of the endpoints. At an endpoint, f and ℓ have the same value. This shows that the maximum of $f - \ell$ is 0, and that at all points x of $[a,b]$ other than the endpoints,

$$f(x) - \ell(x) < 0.$$

This completes the proof of the convexity theorem. □

Definition 4.5. A function whose graph lies *below* its tangent is called *concave.*

The following analogues of results for convex functions hold: *every function whose second derivative is negative is concave, and the graph of a concave function lies above its secant* (Fig. 4.13).

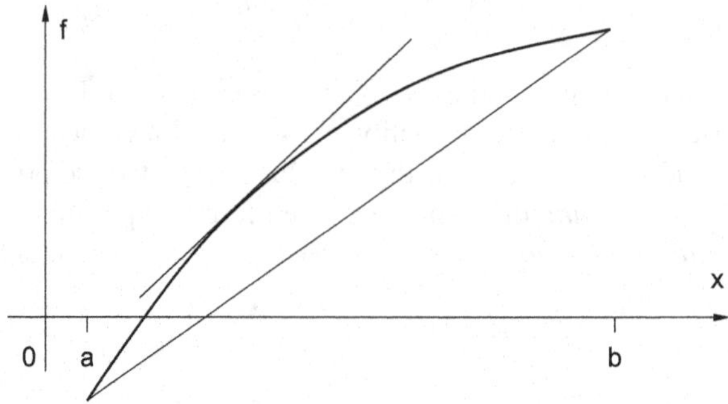

Fig. 4.13 The graph of a concave function f lies above the secant on $[a,b]$, and below each of the tangent lines

Example 4.14. We have seen in Example 4.11 that the second derivative of the function $\log(1+x)$ is negative. It follows that $\log(1+x)$ is a concave function.

Problems

4.25. A particle has position $x = f(t)$, and at time $t = 0$, the position and velocity are 0 and 3, respectively.. The acceleration is between 9.8 and 9.81 for all t. Give bounds on $f(t)$.

4.26. Recall from the chain rule that if f and g are differentiable inverse functions, $f(g(x)) = x$, then

$$f'(g(x)) = \frac{1}{g'(x)}.$$

Find a relation for the second derivatives.

4.27. Find all local extreme values of $f(x) = 2x^3 - 3x^2 + 12x$. On what intervals is f convex? concave? Sketch a graph of f based on this information.

4.28. Over which intervals are the following functions convex?

(a) $f(x) = 5x^4 - 3x^3 + x^2 - 1$
(b) $f(x) = \dfrac{x+1}{x-1}$
(c) $f(x) = \sqrt{x}$
(d) $f(x) = \dfrac{1}{\sqrt{x}}$
(e) $f(x) = \sqrt{1-x^2}$
(f) $f(x) = \mathrm{e}^{-x^2}$

4.29. Are the linear approximations to $f(x) = x^2 - 3x + 5$ above or below the graph?

4.30. Is the secant line for $f(x) = -x^2 - 3x + 5$ on $[0,7]$ above or below the graph?

4.31. Find an interval $(0,b)$ where $\mathrm{e}^{-1/x}$ is convex. Sketch the graph on $(0,\infty)$.

4.32. We proved the linear approximation theorem, Theorem 4.5, for a twice continuously differentiable function f on an interval containing a and b, where $a < b$,

$$f(b) = f(a) + f'(a)(b-a) + \frac{f''(c)}{2}(b-a)^2. \tag{4.15}$$

In this problem we show how the case $a > b$ follows from this. Given f'' continuous on $[a,b]$, define the function g as $g(x) = f(a+b-x)$.

(a) Show that g is defined in the interval $[a,b]$.
(b) Show that g'' is continuous in $[a,b]$.
(c) Show that

$$g(a) = f(b), \quad g'(a) = -f'(b), \quad g''(a) = f''(b),$$

$$g(b) = f(a), \quad g'(b) = -f'(a), \quad g''(b) = f''(a).$$

(d) Write equation (4.15) for the function g. Then use results from part (c) to conclude that Eq. (4.15) holds for $b < a$.

4.33. Is e^f convex when f is convex?

4.34. Give an example of convex functions f and g for which $f \circ g$ is not convex.

4.35. Show, using the linear approximation theorem, that for f'' continuous on an interval containing a and b,

$$\frac{f(a)+f(b)}{2} \text{ differs from } f\left(\frac{a+b}{2}\right) \text{ by less than } \frac{M}{8}(b-a)^2,$$

where M is an upper bound for $|f''|$ on $[a,b]$.

4.36. Suppose f'' is continuous on an interval that contains a and b. Use the linear approximation theorem to explain why

$$\frac{f(b)-f(a)}{b-a} = \frac{f'(b)+f'(a)}{2} + s(b-a),$$

where s is small when b is close to a.

4.37. Let f have continuous first and second derivatives in $a < x < b$. Prove that

(a) $f'(x) = \lim_{h\to 0} \frac{f(x+h)-f(x-h)}{2h}$

(b) $f''(x) = \lim_{h\to 0} \frac{f(x+h)-2f(x)+f(x-h)}{h^2}$

4.38. Prove Theorem 4.7 by applying Theorem 4.6 to the function $-f$.

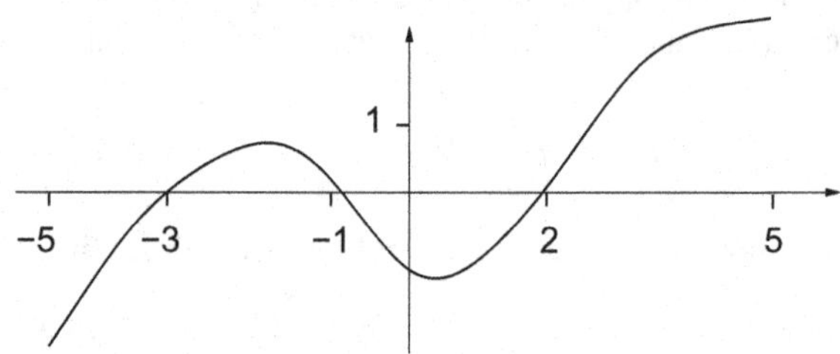

Fig. 4.14 The graph of f in Problems 4.39 and 4.40

4.39. The graph of a function f is given on $[-5,5]$ in Fig. 4.14. Use the graph to find, approximately, the intervals on which $f' > 0$, $f' < 0$, $f'' > 0$, $f'' < 0$.

4.40. Use approximations (see Problem 4.37)

$$f'(x) \approx \frac{f(x+h)-f(x-h)}{2h}, \qquad f''(x) \approx \frac{f(x+h)-2f(x)+f(x-h)}{h^2},$$

with $h = 1$, to estimate $f'(-1)$ and $f''(0.5)$ for the function graphed in Fig. 4.14.

4.41. Let $f(x) = e^{-\frac{x^2}{2}}$ for all x and $g(x) = e^{-1/x}$ for $x > 0$.

(a) Use your calculator or computer to graph f and g.
(b) Use calculus to find the intervals on which f is increasing, decreasing, convex, concave, and locate any extreme values or critical points.
(c) Use calculus to find the intervals on which g is increasing, decreasing, convex, concave.

4.3 Taylor's Theorem

We saw in Sect. 4.2 that bounds on the second derivative, $m \le f''(x) \le M$ in $[a,b]$, enabled us to find two quadratic polynomial functions that bound f:

$$f(a)+f'(a)(x-a)+\frac{m}{2}(x-a)^2 \le f(x) \le f(a)+f'(a)(x-a)+\frac{M}{2}(x-a)^2.$$

Now we are ready to tackle the general problem: if we are given upper and lower bounds for the nth derivative $f^{(n)}(x)$ on $[a,b]$, find nth-degree polynomial functions that are upper and lower bounds for $f(x)$. Generalizing the result we obtained for second derivatives, we surmise that the following result holds:

Theorem 4.9. Taylor's inequality. *Suppose that f is an n-times continuously differentiable function on $[a,b]$, and denote by m and M the minimum and maximum, respectively, of $f^{(n)}$ over $[a,b]$; that is,*

$$m \le f^{(n)}(x) \le M, \qquad x \text{ in } [a,b].$$

Then Taylor's inequality

$$f(a)+f'(a)(x-a)+\frac{f''(a)}{2!}(x-a)^2+\cdots+\frac{f^{(n-1)}(a)}{(n-1)!}(x-a)^{n-1}+\frac{m}{n!}(x-a)^n \tag{4.16}$$

$$\le f(x)$$

$$\le f(a)+f'(a)(x-a)+\frac{f''(a)}{2!}(x-a)^2+\cdots+\frac{f^{(n-1)}(a)}{(n-1)!}(x-a)^{n-1}+\frac{M}{n!}(x-a)^n$$

holds for all x in $[a,b]$.

The polynomials on the left and right sides of Taylor's inequality (4.16) are identical up through the next-to-last terms. We call the identical parts Taylor polynomials.

Definition 4.6. If f is n times differentiable at a, the *Taylor polynomials* at a are

$$\begin{aligned}
t_0(x) &= f(a)\\
t_1(x) &= f(a)+f'(a)(x-a)\\
t_2(x) &= f(a)+f'(a)(x-a)+f''(a)\frac{(x-a)^2}{2!}\\
&\cdots\\
t_n(x) &= f(a)+f'(a)(x-a)+f''(a)\frac{(x-a)^2}{2!}+\cdots+f^{(n)}(a)\frac{(x-a)^n}{n!}\\
&\cdots
\end{aligned}$$

Proof of Theorem 4.9. We prove Taylor's inequality for all n inductively, i.e., we first show that it is true for $n = 1$, and then we show that if the result is true for any particular number n, then it is true for $n+1$. By the mean value theorem, we know that for some c between a and x,

$$f(x) = f(a) + f'(c)(x-a).$$

Since f' is continuous on $[a,b]$, it attains a maximum M and minimum m on that interval. Then since $a \le x$, it follows that

$$f(a) + m(x-a) \le f(x) \le f(a) + M(x-a), \qquad (a \le x \le b).$$

Thus the theorem holds for $n = 1$. Next, we show that if the result holds for n, then it holds for $n+1$. Assume that Taylor's inequality holds for every function whose nth derivative is bounded on $[a,b]$. If f is $(n+1)$ times continuously differentiable, then there are bounds

$$m \le f^{(n+1)}(x) \le M, \qquad (a \le x \le b).$$

Since $f^{(n+1)}$ is the nth derivative of f', we can apply the inductive hypothesis to the function f' to obtain

$$f'(a) + f''(a)(x-a) + \cdots + \frac{m}{n!}(x-a)^n \le f'(x) \le f'(a) + f''(a)(x-a) + \cdots + \frac{M}{n!}(x-a)^n.$$

The sum on the right is the derivative of

$$t_n(x) + \frac{M}{(n+1)!}(x-a)^{n+1}.$$

Since

$$\left(t_n(x) + \frac{M}{(n+1)!}(x-a)^{n+1} \right)' - f'(x) \ge 0,$$

we see that

$$\left(t_n(x) + \frac{M}{(n+1)!}(x-a)^{n+1} \right) - f(x)$$

is a nondecreasing function on $[a,b]$. At $x = a$, the difference

$$\left(t_n(a) + \frac{M}{(n+1)!}(a-a)^{n+1} \right) - f(a)$$

is zero. So for $x > a$,

$$0 \le \left(t_n(x) + \frac{M}{(n+1)!}(x-a)^{n+1} \right) - f(x).$$

It follows that

$$f(x) \le t_n(x) + \frac{M}{(n+1)!}(x-a)^{n+1},$$

which is the right half of Taylor's inequality. The left half follows in a similar manner. Thus, we have shown that if Taylor's inequality holds for n, it holds for $n+1$. Since the inequality holds for $n=1$, by induction it must hold for all positive integers. □

Example 4.15. We write Taylor's inequality for $f(x) = \sin x$ on $[0,4]$, where $a=0$ and $n=5$. The first four derivatives are

$$f'(x) = \cos x,\quad f''(x) = -\sin x,\quad f'''(x) = -\cos x,\quad f''''(x) = \sin x.$$

Evaluate at $a=0$: $f(0)=0,\ f'(0)=1,\ f''(0)=0,\ f'''(0)=-1,\ f''''(0)=0.$ Then the fourth Taylor polynomial $t_4(x) = x - \frac{x^3}{3!}$. We have $f^{(5)}(x) = \cos x$, so $-1 \le f^{(5)}(x) \le 1$ on $[0,4]$. Therefore,

$$x - \frac{x^3}{3!} - \frac{x^5}{5!} \le \sin x \le x - \frac{x^3}{3!} + \frac{x^5}{5!}.$$

Figure 4.15 contains the graphs of the three functions.

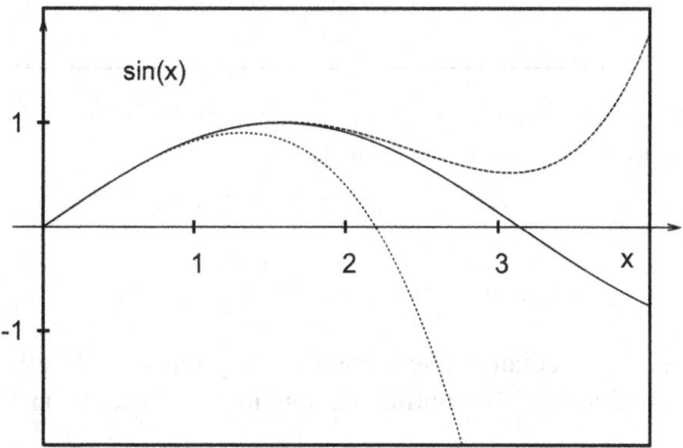

Fig. 4.15 Taylor's inequality for the case in Example 4.15: $x - \frac{x^3}{6} - \frac{x^5}{120} \le \sin x \le x - \frac{x^3}{6} + \frac{x^5}{120}$, where $0 \le x \le 4$

Example 4.16. Let us write Taylor's inequality for $f(x) = \log x$ on the interval $[1,3]$, where $a=1$ and $n=4$:

$$f'(x) = \frac{1}{x},\quad f''(x) = -\frac{1}{x^2},\quad f'''(x) = 2!\frac{1}{x^3},\quad f''''(x) = -3!\frac{1}{x^4},$$

$$f(1) = 0,\qquad f'(1) = 1,\qquad f''(1) = -1,\qquad f'''(1) = 2,$$

and since $f''''(x)$ is increasing, $-3! \le f''''(x) \le -3!\frac{1}{3^4}$. According to Taylor's inequality (see Fig. 4.16),

$$(x-1) - \frac{1}{2}(x-1)^2 + \frac{1}{3}(x-1)^3 - \frac{1}{4}(x-1)^4$$
$$\le \log x \le (x-1) - \frac{1}{2}(x-1)^2 + \frac{1}{3}(x-1)^3 - \frac{1}{3^4 4}(x-1)^4.$$

Just as we saw in Sect. 4.2, the upper and lower bounds in Taylor's inequality differ inasmuch as one contains the constant m and the other M. So, given $x > a$, there is a number H between m and M such that

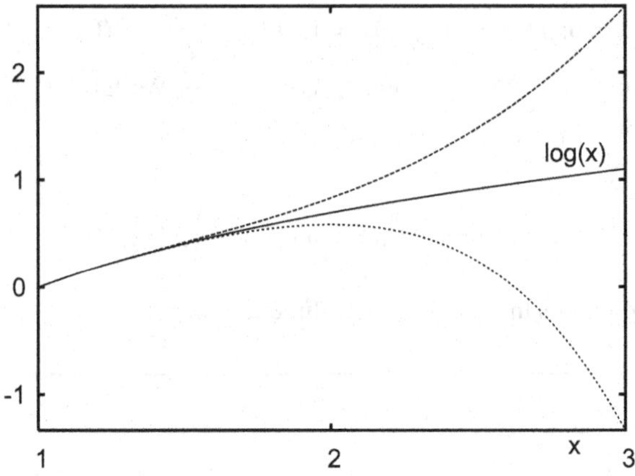

Fig. 4.16 Taylor's inequality for $\log x$ as in Example 4.16

$$f(x) = t_{n-1}(x) + H\frac{(x-a)^n}{n!}.$$

According to the intermediate value theorem, every number H between the minimum m and maximum M of the continuous function $f^{(n)}$ is taken on at some point c between a and b. Now when $x = b$, $f(b) = t_{n-1}(b) + f^{(n)}(c)\frac{(b-a)^n}{n!}$. The difference

$$f(b) - t_{n-1}(b) = f^{(n)}(c)\frac{(b-a)^n}{n!}$$

is called the *remainder*. We express our results with the following theorem.

Theorem 4.10. Taylor's formula with remainder. *Let f be an n-times continuously differentiable function on an interval containing a and b. Then*

$$f(b) = f(a) + f'(a)(b-a) + \cdots + f^{(n-1)}(a)\frac{(b-a)^{n-1}}{(n-1)!} + f^{(n)}(c)\frac{(b-a)^n}{n!}, \tag{4.17}$$

where c lies between a and b.

In the derivation of this theorem we have exploited the fact that $a < b$. It is not hard to show that the theorem remains true if $a > b$. We ask you to do so in Problem 4.50. Here are some applications of Taylor's formula.

Example 4.17. Let $f(x) = x^m$, m a positive integer. Then

$$f^{(k)}(x) = m(m-1)\cdots(m-k+1)x^{m-k}.$$

In particular, $f^{(m)}(x) = m!$, and higher derivatives are 0. Therefore, according to Taylor's formula with $b = 1+y$, $a = 1$, and any $n \geq m$,

$$(1+y)^m = 1 + my + \frac{m(m-1)}{2!}y^2 + \cdots + y^m = \sum_{k=0}^{m} \binom{m}{k} y^k.$$

Example 4.17 is nothing but the binomial expansion, revealed here as a special case of Taylor's formula.

Taylor's inequality

$$t_{n-1}(x) + m\frac{(x-a)^n}{n!} \leq f(x) \leq t_{n-1}(x) + M\frac{(x-a)^n}{n!}$$

is an approximation to f on $[a,b]$. The polynomials on the left- and right-hand sides of Taylor's inequality differ only in the last terms. That difference is due to the variation in the maximum and minimum value of $f^{(n)}$ on $[a,b]$, which leads to the next definition.

Definition 4.7. Denote by C_n the *oscillation* of $f^{(n)}$ on the interval $[a,b]$, i.e.,

$$C_n = M_n - m_n,$$

where M_n is the maximum, m_n the minimum of $f^{(n)}$ over $[a,b]$.

We derive now a useful variant of Taylor's inequality. Taylor's formula (4.17),

$$f(b) = f(a) + f'(a)(b-a) + \cdots + f^{(n-1)}(a)\frac{(b-a)^{n-1}}{(n-1)!} + f^{(n)}(c)\frac{(b-a)^n}{n!},$$

differs from Taylor's polynomial

$$t_n(b) = f(a) + f'(a)(b-a) + \cdots + f^{(n-1)}(a)\frac{(b-a)^{n-1}}{(n-1)!} + f^{(n)}(a)\frac{(b-a)^n}{n!}$$

in that the last term has $f^{(n)}$ evaluated at c rather than a. Since $f^{(n)}(c)$ and $f^{(n)}(a)$ differ by at most the oscillation C_n, we see that

$$|f(x)-t_n(x)| \leq \frac{C_n}{n!}(x-a)^n \leq \frac{C_n}{n!}(b-a)^n \qquad \text{for all } x \text{ in } [a,b]. \tag{4.18}$$

Suppose the function f is *infinitely differentiable*, i.e., has derivatives of all orders. Suppose further that

$$\lim_{n\to\infty} \frac{C_n}{n!}(b-a)^n = 0. \tag{4.19}$$

Then as n gets larger and larger, $t_n(x)$ tends to $f(x)$. We can state this result in the following spectacular form.

Theorem 4.11. Taylor's theorem. *Let f be an infinitely differentiable function on an interval $[a,b]$. Denote by C_n the oscillation of $f^{(n)}$, and suppose that $\lim_{n\to\infty} \frac{C_n}{n!}(b-a)^n = 0$. Then f can be represented at every point of $[a,b]$ by the Taylor series*

$$f(x) = \lim_{n\to\infty} t_n(x) = \sum_{k=0}^{\infty} \frac{1}{k!} f^{(k)}(a)(x-a)^k,$$

and the Taylor polynomials converge uniformly to f on $[a,b]$. There is an analogous theorem for the interval $[b,a]$ when $b<a$.

Proof. The meaning of the infinite sum on the right is this: Form the nth Taylor polynomials $t_n(x)$ of f at a and take the limit of this sequence of functions as n tends to infinity. Since $|f(x)-t_n(x)| \leq \frac{C_n}{n!}(b-a)^n$ and $\lim_{n\to\infty} \frac{C_n}{n!}(b-a)^n = 0$, the sequence $t_n(x)$ tends to $f(x)$ as n tends to infinity. The convergence is uniform on $[a,b]$, because the estimate that we derived for $|f(x)-t_n(x)|$ does not depend on x.

We ask you to verify the proof of the theorem in the case $b<a$ in Problem 4.50. □

4.3a Examples of Taylor Series

The Sine. Let $f(x) = \sin x$ and $a = 0$. The derivatives are

$$f(x) = \sin x,\quad f'(x) = \cos x,\quad f''(x) = -\sin x,\quad f'''(x) = -\cos x,\quad f''''(x) = \sin x,$$

and so forth. So at 0,

$$f(0) = 0,\quad f'(0) = 1,\quad f''(0) = 0,\quad f'''(0) = -1,\quad f''''(0) = 0,\quad f^{(5)}(0) = 1,$$

etc. The nth Taylor polynomial at $a = 0$ for $\sin x$ is (Fig. 4.17)

$$t_n(x) = x - \frac{x^3}{3!} + \frac{x^5}{5!} - \frac{x^7}{7!} + \cdots + \sin^{(n)}(0)\frac{x^n}{n!},$$

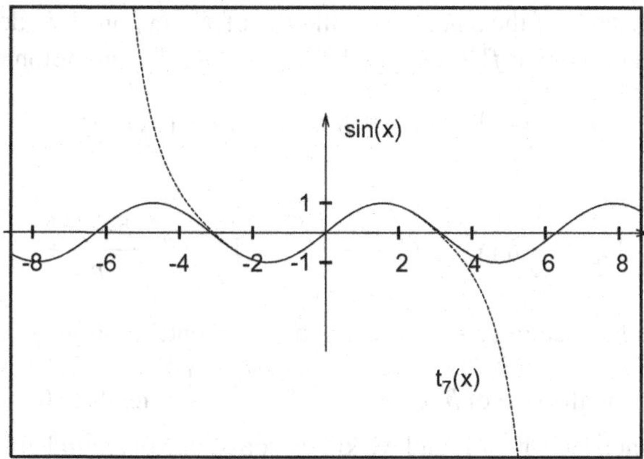

Fig. 4.17 Graphs of $\sin x$ and the Taylor polynomial $t_7(x) = x - \frac{x^3}{3!} + \frac{x^5}{5!} - \frac{x^7}{7!}$

where the last coefficient is 0, 1, or -1, depending on n. The oscillation C_n is equal to 2, because the sine and cosine have minimum -1 and maximum 1. Then on $[0,b]$,

$$C_n \frac{(b-a)^n}{n!} = 2\frac{b^n}{n!}.$$

For any number b, the terms on the right tend to 0 as n tends to infinity by Example 1.17. Therefore, the Taylor series for $\sin x$ at $a = 0$,

$$\sin x = x - \frac{x^3}{3!} + \frac{x^5}{5!} - \frac{x^7}{7!} + \cdots, \tag{4.20}$$

converges for all x in $[0,b]$ for every b, and therefore on $[0,\infty)$. By a similar argument, $\sin x$ converges for all x in $[b,0]$ when $b < 0$, and therefore on $(-\infty,\infty)$.

The Logarithm. Let $f(x) = \log x$ and $a = 1$. As we saw in Example 4.16, the derivatives follow a pattern,

$$f(x) = \log x, \ \ f'(x) = x^{-1}, \ \ f''(x) = -x^{-2}, \ \ f'''(x) = 2!x^{-3}, \ \ f''''(x) = -3!x^{-4}, \ldots.$$

At $a = 1$,

$$f(1) = 0, \ \ f'(1) = 1, \ \ f''(1) = -1, \ \ f'''(1) = 2!, \ \ f''''(1) = -3!, \ldots.$$

The nth Taylor polynomial of $\log x$ at $a = 1$ is

$$t_n(x) = (x-1) - \frac{1}{2}(x-1)^2 + \frac{1}{3}(x-1)^3 + \cdots + \frac{(-1)^{n-1}}{n}(x-1)^n.$$

In contrast to the case of the sine, the oscillation of $f^{(n)}(x)$ on $[1,b]$ depends on the value of b. Each derivative $f^{(n)}(x) = (-1)^{n-1}(n-1)!x^{-n}$ is monotonic, so

$$C_n = |f^{(n)}(1) - f^{(n)}(b)| = (n-1)!|1-b^{-n}|.$$

On $[1,b]$,

$$|\log(x) - t_n(x)| \le C_n \frac{(b-1)^n}{n!} = |1-b^{-n}| \frac{(b-1)^n}{n}. \tag{4.21}$$

Since $b > 1$, the first factor $|1-b^{-n}|$ tends to 1 as n tends to infinity. How $\dfrac{(b-1)^n}{n}$ behaves depends on the size of b: for $1 < b \le 2$, $\dfrac{(b-1)^n}{n}$ tends to 0 as n tends to infinity. If $b > 2$, then $(b-1) > 1$, and we know from the exponential growth theorem, Theorem 2.10, that $\dfrac{(b-1)^n}{n}$ tends to infinity. Hence the Taylor series converges to $\log x$ in $[1,2]$.

We show by another method in Example 7.35 that $|\log x - t_n(x)|$ also tends to 0 for $0 < x \le 1$. Given that future result, we have

$$\log x = \sum_{n=1}^{\infty} (-1)^{n-1} \frac{(x-1)^n}{n} \qquad (0 < x \le 2).$$

Figure 4.18 shows part of the graphs of $\log x$ and $t_4(x)$.

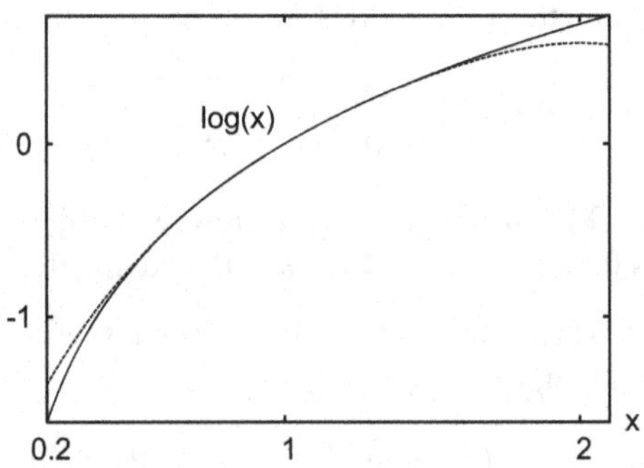

Fig. 4.18 Graphs of $\log x$ and the Taylor polynomial $t_4(x) = (x-1) - \frac{(x-1)^2}{2} + \frac{(x-1)^3}{3} - \frac{(x-1)^4}{4}$

Remark. According to the ratio test, using

$$\lim_{n\to\infty} \left| \frac{\frac{(x-1)^{n+1}}{n+1}}{\frac{(x-1)^n}{n}} \right| = |x-1| < 1,$$

we see that the Taylor series for the logarithm, $\lim_{n\to\infty} t_n(x) = \sum_{n=1}^{\infty} (-1)^{n-1} \dfrac{(x-1)^n}{n}$, converges uniformly on every closed interval in $(0,2)$. Checking the endpoints, we see that the power series converges at $x = 2$ (alternating series theorem) and diverges at $x = 0$ (harmonic series). But this does not tell us that $t_n(x)$ converges to $\log x$ for $0 < x \le 2$. To show that a Taylor series $\sum_{n=0}^{\infty} f^{(n)}(a)\frac{(x-a)^n}{n!}$ converges to the function f from which it is derived, it is necessary to show that $|f(x) - t_n(x)|$ tends to 0 as n tends to infinity. Examining the oscillation is one way to do this. Another way is to examine the behavior of the remainders

$$|f(b) - t_n(b)| = \left| f^{(n+1)}(c)\frac{(b-a)^{n+1}}{n+1!} \right|$$

from Taylor's formula, as we do in the next example. After we study integration in Chap. 7, we will have an integral formula for the remainder that gives another way to estimate the remainder.

The Exponential Function. Let $f(x) = \mathrm{e}^x$ and $a = 0$. Since $f^{(n)}(x) = \mathrm{e}^x$, it follows that

$$f(0) = 1,\ \ f'(0) = 1,\ \ f''(0) = 1,\ \ f'''(0) = 1, \dots,$$

and the nth Taylor polynomial for e^x at $a = 0$ is

$$t_n(x) = 1 + x + \frac{x^2}{2!} + \frac{x^3}{3!} + \cdots + \frac{x^n}{n!}.$$

We want to show that $\lim_{n\to\infty} |\mathrm{e}^x - t_n(x)| = 0$ for all x. According to Taylor's formula,

$$|\mathrm{e}^x - t_n(x)| = \left| \mathrm{e}^c \frac{x^{n+1}}{n+1!} \right|$$

for some c between 0 and x. Suppose x is in $[-b, b]$. Then

$$|\mathrm{e}^x - t_n(x)| \le \mathrm{e}^b \frac{b^{n+1}}{n+1!}.$$

We saw in Example 1.17 that $\lim_{n\to\infty} \frac{b^n}{n!} = 0$ for every number b. Hence $\mathrm{e}^b \frac{b^{n+1}}{n+1!}$ tends to 0. The Taylor series

$$\sum_{k=0}^{\infty} \frac{x^k}{k!} = 1 + x + \frac{x^2}{2!} + \frac{x^3}{3!} + \cdots$$

converges to e^x for all x in $[-b, b]$. Since b is arbitrary, the series converges to e^x for all x. Figure 4.19 shows graphs of e^x, $t_3(x)$, and $t_4(x)$.

The Binomial Series. Here we point out that the binomial expansions as in Example 4.17 have a generalization to every real exponent. We prove the following.

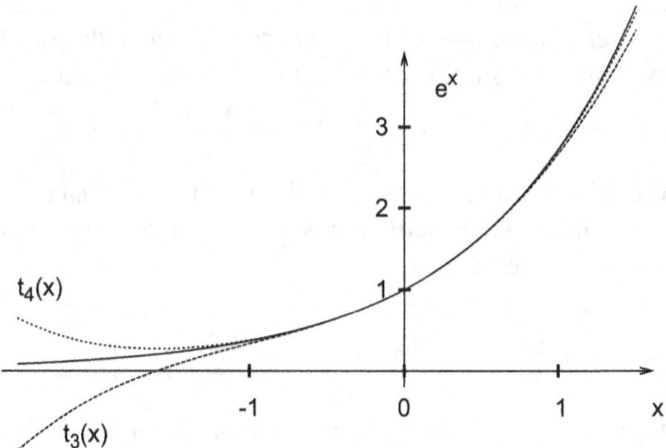

Fig. 4.19 Graphs of e^x with Taylor polynomials $t_3(x) = 1+x+\frac{x^2}{2!}+\frac{x^3}{3!}$ and $t_4(x) = t_3(x)+\frac{x^4}{4!}$ on $[-2.5, 1.5]$

Theorem 4.12. The binomial theorem. *If ℓ is any number and $|y| < 1$, then*

$$(1+y)^\ell = \sum_{k=0}^{\infty} \binom{\ell}{k} y^k,$$

where the binomial coefficients are defined by

$$\binom{\ell}{0} = 1, \qquad \binom{\ell}{k} = \frac{\ell(\ell-1)\cdots(\ell-k+1)}{k!} \quad (k>0).$$

Proof. Let $f(y) = (1+y)^\ell$. The nth derivative of f is

$$f^{(n)}(y) = \ell(\ell-1)\cdots(\ell-n+1)(1+y)^{\ell-n}. \tag{4.22}$$

If $|y| < 1$, the power series

$$g(y) = \sum_{k=0}^{\infty} \binom{\ell}{k} y^k = \sum_{k=0}^{\infty} \frac{\ell(\ell-1)\cdots(\ell-k+1)}{k!} y^k$$

converges by the ratio test, since

$$\lim_{k\to\infty} \left| \frac{\binom{\ell}{k+1} y^{k+1}}{\binom{\ell}{k} y^k} \right| = \lim_{k\to\infty} \frac{\frac{\ell(\ell-1)\cdots(\ell-k)}{(k+1)!} y^{k+1}}{\frac{\ell(\ell-1)\cdots(\ell-k+1)}{k!} y^k} = \lim_{k\to\infty} \left| \frac{\ell-k}{k+1} \right| |y| = |y|.$$

We want to show that $g(y) = (1+y)^\ell$ for $|y| < 1$. According to Theorem 3.17, we can differentiate $g(y)$ term by term to get

$$g'(y) = \sum_{k=0}^{\infty} \frac{\ell(\ell-1)\cdots(\ell-k+1)}{k!} L y^{k-1} = \sum_{k=0}^{\infty} \binom{\ell}{k-1} y^{k-1}.$$

Multiply $g(y)$ by y and add $g'(y)$ to get

$$(1+y)g'(y) = \sum_{k=0}^{\infty} \left((k+1)\binom{\ell}{k+1} + k\binom{\ell}{k} \right) y^k = \sum_{k=0}^{\infty} \ell \binom{\ell}{k} y^k = \ell g(y).$$

Now let us examine

$$\frac{\mathrm{d}}{\mathrm{d}y} \frac{g(y)}{(1+y)^{\ell}} = \frac{(1+y)^{\ell} g'(y) - g(y)\ell(1+y)^{\ell-1}}{((1+y)^{\ell})^2}$$

$$= \frac{\ell(1+y)^{\ell-1}}{(1+y)^{2\ell}} \left((1+y)g'(y) - \ell g(y) \right) = 0.$$

Therefore, $\dfrac{g(y)}{(1+y)^{\ell}}$ is constant. But at $y = 0$, we know that $\dfrac{g(0)}{(1)^{\ell}} = 1$. Therefore, the power series $g(y)$ equals $(1+y)^{\ell}$ for $|y| < 1$. □

This generalization of the binomial theorem to noninteger exponents was derived by Newton. This shows that Newton was familiar with Taylor's theorem, even though Taylor's book appeared 50 years after Newton's.

Problems

4.42. Find the Taylor polynomials $t_2(x)$ and $t_3(x)$ for $f(x) = 1 + x + x^2 + x^3 + x^4$ in powers of x.

4.43. Find the Taylor series for $\cos x$ in powers of x. For what values of x does the series converge to $\cos x$?

4.44. Find the Taylor series for $\cos(3x)$ in powers of x. For what values of x does the series converge to $\cos(3x)$?

4.45. Compare Taylor polynomials t_3 and t_4 for $\sin x$ in powers of x. Give the best estimate you can of $|\sin x - t_3(x)|$.

4.46. Find the Taylor polynomial of degree 4 and estimate the remainder in

$$\tan^{-1} x = x - \frac{x^3}{3} + (\text{remainder}) \quad x \text{ in } [-\frac{1}{2}, \frac{1}{2}].$$

4.47. Find the Taylor series for $\cosh x$ in powers of x. Use the Taylor remainder formula to show that the series converges uniformly to $\cosh x$ on every interval $[-b, b]$.

4.48. Find the Taylor series for $\sinh(2x)$ in powers of x. Use the Taylor remainder formula to show that the series converges uniformly to $\sinh(2x)$ on every interval $[-b,b]$.

4.49. Find the Taylor series for $\cos x$ about $a = \pi/3$, i.e. in powers of $(x-\pi/3)$.

4.50. Prove the validity of Taylor's formula with remainder, Eq. (4.17), when $b < a$.
Hint: Consider the function $g(x) = f(a+b-x)$ over the interval $[b,a]$.

4.51. Find the binomial coefficients $b_k = \binom{\frac{1}{2}}{k}$ through b_3 in

$$\sqrt{1+y} = b_0 + b_1 y + b_2 y^2 + \cdots.$$

4.52. Consider the function $f(x) = \sqrt{x}$ on the interval $1 \le x \le 1+d$. Find values of d small enough that $t_2(x)$, the second-degree Taylor polynomial at $x=1$, approximates $f(x)$ on $[1,1+d]$ with an error of at most

(a) .1
(b) .01
(c) .001

4.53. Answer the question posed in Problem 4.52 for the third-degree Taylor polynomial t_3 in place of t_2.

4.54. Let s be a function with the following properties:

(a) s has derivatives of all orders.
(b) All derivatives of s lie between -1 and 1.
(c) $s^{(j)}(0) = \begin{cases} 0, & j \text{ even.} \\ (-1)^{(j-1)/2}, & j \text{ odd.} \end{cases}$

Determine a value of n so large that the nth-degree Taylor polynomial $t_n(x)$ approximates $s(x)$ with an error less than 10^{-3} on the interval $[-1,1]$. Determine the value of $s(0.7854)$ with an error less than 10^{-3}. What is the standard name for this function?

4.55. Let c be a function that has properties (a) and (b) of Problem 4.54 and satisfies

$$c^{(j)}(0) = \begin{cases} (-1)^{j/2}, & j \text{ even,} \\ 0, & j \text{ odd.} \end{cases}$$

Using a Taylor polynomial of appropriate degree, determine the value of $c(0.7854)$ with an error less than 10^{-3}. What is the standard name for this function?

4.56. Explain why there is no power series $|t| = \sum_{n=0}^{\infty} a_n t^n$.

4.57. In this problem, we rediscover Taylor's theorem for the case of polynomials. Let f be a polynomial of degree n, and a any constant, and set

$$g(x) = f(x+a) - xf'(x+a) + \frac{x^2}{2} f''(x+a) - \cdots + \frac{(-1)^n x^n}{n!} f^{(n)}(x+a).$$

(a) Evaluate $g(0)$ and $g(-a)$.
(b) Evaluate $g'(x)$ and simplify your result as much as possible.
(c) Conclude from part (b) the somewhat curious result that g is a constant function.
(d) Use the result of part (c) to express the relation between the values you found in part (a).

4.4 Approximating Derivatives

By definition, $f'(x)$ requires that we use a limiting process. That may be fine for human beings, but not for computers. The difference quotient

$$f_h(x) = \frac{f(x+h) - f(x)}{h}$$

can be computed once you know f, x, and h. But how good would such an approximation be? Let us look at some examples.

Example 4.18. If $f(x) = x^2$, then

$$f_h(x) = \frac{f(x+h) - f(x)}{h} = \frac{(x+h)^2 - x^2}{h} = \frac{x^2 + 2xh + h^2 - x^2}{h} = 2x + h.$$

For this function, replacing $f'(x)$ by $f_h(x)$ would lead to an error of only h. If we are willing to accept an error of say 0.00001, then we could approximate $f'(x)$ by $f_{0.00001}(x)$. Figure 4.20 shows the case in which $h = 0.07$.

Example 4.19. If $f(x) = x^3$, then

$$f_h(x) = \frac{f(x+h) - f(x)}{h} = \frac{(x+h)^3 - x^3}{h}$$

$$= \frac{x^3 + 3x^2h + 3xh^2 + h^3 - x^3}{h} = 3x^2 + 3xh + h^2.$$

For this function, replacing $f'(x)$ by $f_h(x)$ would introduce an error of $3xh + h^2$, an amount that depends on both h and x. This is illustrated in Fig. 4.21.

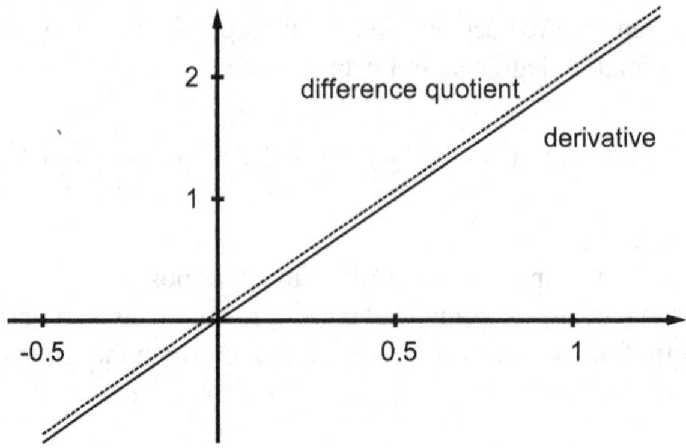

Fig. 4.20 Derivative $f'(x)$ and difference quotient $f_h(x) = \dfrac{f(x+h)-f(x)}{h}$ for $f(x) = x^2$ using $h = 0.07$

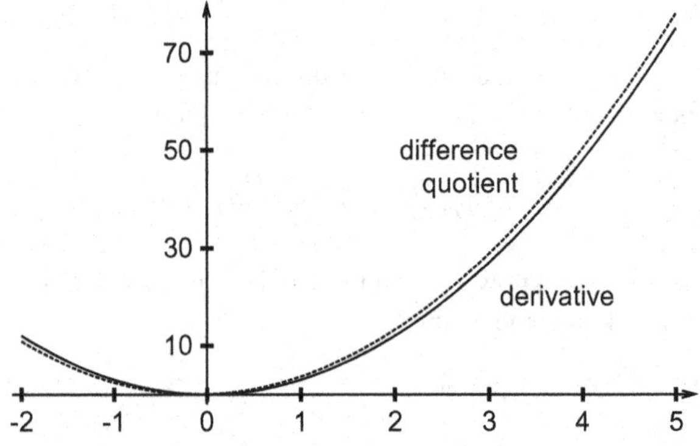

Fig. 4.21 Derivative $f'(x)$ and difference quotient $f_h(x) = \frac{f(x+h)-f(x)}{h}$ for $f(x) = x^3$ using $h = 0.2$

These examples lead to the concept of uniform differentiability.

Definition 4.8. A function f defined on an interval is called *uniformly differentiable* if given a tolerance $\varepsilon > 0$, there is a δ such that

$$\text{if } |h| < \delta, \text{ then } \left|\frac{f(x+h)-f(x)}{h} - f'(x)\right| < \varepsilon \text{ for every } x.$$

Example 4.20. The linear function $f(x) = mx + b$ is uniformly differentiable: the derivative is the constant function m. The difference quotient is given by

$$f_h(x) = \frac{m(x+h)+b-(mx+b)}{h} = m.$$

So it not only tends to $f'(x)$, but is equal to $f'(x)$ for all h and all x.

Example 4.21. We have shown in Sect. 3.3a that the exponential function e^x is differentiable at every x. We will now show that e^x is uniformly differentiable on each interval $[-c,c]$. We have

$$\frac{e^{x+h}-e^x}{h}-e^x = e^x\left(\frac{e^h-1}{h}-1\right).$$

Therefore, for every x in $[-c,c]$, this quantity is at most e^c times $\left(\frac{e^h-1}{h}-1\right)$, which does not depend on x, and which tends to zero as h tends to zero.

We claim the following theorem.

Theorem 4.13. *If f is uniformly differentiable on $[a,b]$, then f' is uniformly continuous on $[a,b]$.*

The proof is outlined in Problem 4.62 at the end of this section. The theorem has a converse, whose significance is that we may easily approximate derivatives:

Theorem 4.14. *If f' is uniformly continuous on $[a,b]$, then f is uniformly differentiable on $[a,b]$.*

Proof. We need to prove that $f'(x)$ and the difference quotient $f_h(x)$ differ by an amount that is small independently of x. More precisely, consider

$$\frac{f(x+h)-f(x)}{h}-f'(x).$$

According to the mean value theorem (Theorem 4.1), the difference quotient equals $f'(c)$ at some point c between x and $x+h$. So x and c differ by less than h, and we can rewrite the previous expression as

$$\frac{f(x+h)-f(x)}{h}-f'(x) = f'(c)-f'(x).$$

Since f' is uniformly continuous on $[a,b]$, given any tolerance ε, there is a precision δ such that if x and c are in $[a,b]$ and differ by less than δ, then $f'(c)$ and $f'(x)$ differ by less than ε. This proves that f is uniformly differentiable. □

Many of the functions we work with, such as polynomials, sine, cosine, exponential, and logarithm, have continuous derivatives, and are therefore uniformly differentiable on closed intervals. We give an example in Problem 4.63 of a differentiable function f for which f' is not continuous, and thus according to Theorem 4.13, f is not uniformly differentiable.

A Word of Caution about actually approximating $f'(x)$ by $f_h(x)$. When we asked how good an approximation $f_h(x) = \dfrac{f(x+h)-f(x)}{h}$ is to $f'(x)$, we assumed that we could perform the subtraction accurately. But when we subtract decimal approximations of numbers that are very close, we get very few digits of the difference correctly. Figure 4.22 shows the result of a computer program that attempted to calculate the difference quotient minus the derivative,

$$\frac{f(x+h)-f(x)}{h} - f'(x),$$

for $f(x) = x^2$ and $x = 1$. We know by algebra that

$$\frac{f(1+h)-f(1)}{h} - f'(1) = \frac{(1+h)^2-1^2}{h} - 2 = \frac{2h+h^2}{h} - 2 = h,$$

so the graph should be a straight line. But we see something quite different in Fig. 4.22.

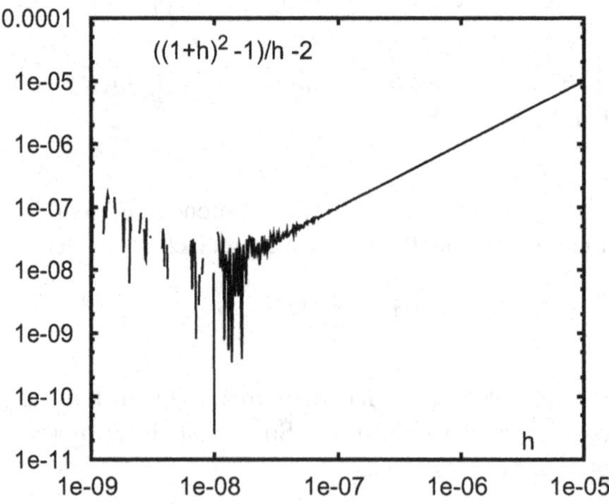

Fig. 4.22 The result of computing the difference quotient minus the derivative $\frac{(1+h)^2-1^2}{h} - 2$ is plotted for $10^{-9} \le h \le 10^{-5}$

Approximate Derivatives and Data. In experimental settings, functions are represented through tables of data, rather than by a formula. How can we compute f' when only tabular data are known? We give an application of the linear approximation theorem, Theorem 4.5, to the problem of approximating derivatives.

The difference quotient $\dfrac{f(x+h)-f(x)}{h}$ is asymmetric in the sense that it favors one side of the point x. By the linear approximation theorem,

$$f(x+h) = f(x) + f'(x)h + \frac{1}{2}f''(c_1)h^2 \quad \text{and} \quad f(x-h) = f(x) - f'(x)h + \frac{1}{2}f''(c_2)h^2,$$

where c_1 lies between x and $x+h$, and c_2 lies between x and $x-h$. Subtract and divide by $2h$ to get an estimate for the *symmetric* difference quotient

$$\frac{f(x+h)-f(x-h)}{2h} = f'(x) + \frac{1}{4}\big(f''(c_1) - f''(c_2)\big)h. \tag{4.23}$$

See Fig. 4.23, which illustrates a symmetric difference quotient. Both c_1 and c_2 differ by less than h from x. If f'' is continuous, then for small h, both $f''(c_1)$ and $f''(c_2)$ differ little from $f''(x)$. Thus we deduce that the symmetric difference quotient differs from $f'(x)$ by an amount sh, where $s = \frac{1}{4}\big(f''(c_1) - f''(c_2)\big)$ is small when h is small.

But we saw that the one-sided difference quotient differs from $f'(x)$ by $\frac{1}{2}f''(c_1)h$. We conclude that for twice differentiable functions and for small h, *the symmetric difference quotient is a better approximation of the derivative at x than the one-sided quotient.*

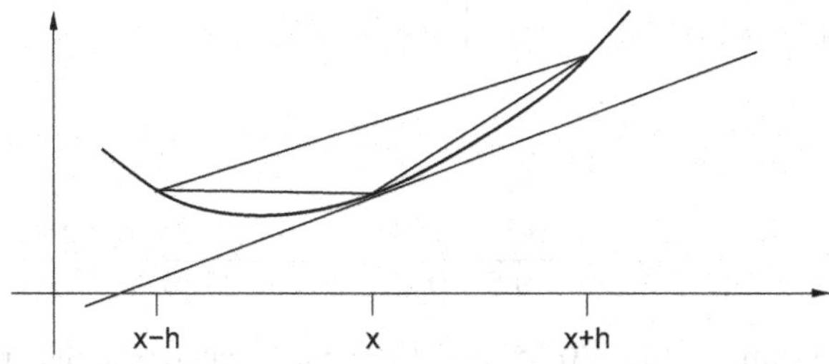

Fig. 4.23 The symmetric difference quotient is a better approximation to $f'(x)$ than one-sided quotients

Let us look at an example. Table 4.1 contains data for a function at eleven equidistant points between 0 and 1. Note that if we take $x+h = x_2$ and $x-h = x_1$ in Eq. (4.23), it becomes

$$\frac{f(x_2)-f(x_1)}{x_2-x_1} = f'\left(\frac{x_2+x_1}{2}\right) + \frac{1}{4}\big(f''(c_1) - f''(c_2)\big)\frac{x_2-x_1}{2}.$$

Table 4.1 shows estimates for f' at the midpoints $0.05, 0.15, 0.25, \ldots, 0.85, 0.95$ of the intervals using

$$f'\Big(\frac{x_1+x_2}{2}\Big) \approx \frac{f(x_2)-f(x_1)}{x_2-x_1}.$$

For example,

$$f'(0.35) = f'\Big(\frac{0.3+0.4}{2}\Big) \approx \frac{f(0.4)-f(0.3)}{0.4-0.3} = \frac{0.38941-0.29552}{0.1} = 0.9389$$

Table 4.1 Data in the $f(x)$ column, and approximate derivatives for the unknown function

x	$f(x)$	$\approx f'(x)$	$\approx f''(x)$	$\approx f'''(x)$	$\approx f''''(x)$
0	0				
0.05		0.9983			
0.1	0.09983		−0.100		
0.15		0.9883		−0.97	
0.2	0.19866		−0.197		( )
0.25		0.9686		−1	
0.3	0.29552		−0.297		( )
0.35		0.9389		−0.91	
0.4	0.38941		−0.388		( )
0.45		0.9001		−0.91	
0.5	0.47942		−0.479		( )
0.55		0.8522		( )	
0.6	0.56464		−0.565		( )
0.65		0.7957		( )	
0.7	0.64421		−0.643		( )
0.75		0.7314		( )	
0.8	0.71735		−0.717		( )
0.85		0.6597		( )	
0.9	0.78332		−0.782		
0.95		0.5815			
1	0.84147				

and

$$f'(0.45) = f'\left(\frac{0.4+0.5}{2}\right) \approx \frac{f(0.5)-f(0.4)}{0.5-0.4} = \frac{0.47942-0.38941}{0.1} = 0.9001.$$

Now using our estimates for $f'(0.35)$ and $f'(0.45)$ we can repeat the process to find estimates for f'' at $0.1, 0.2, \ldots, 0.9$. For example,

$$f''(0.4) \approx \frac{f'(0.45)-f'(0.35)}{0.45-0.35} \approx \frac{0.9001-0.9389}{0.1} = -0.388.$$

In Problem 4.60, we ask you to complete the table.

Problems

4.58. Consider the symmetric difference quotient graphed in Fig. 4.24. Use Taylor's theorem with remainder to show that the difference

$$\left|\frac{\sin(x+0.1)-\sin(x-0.1)}{0.2} - \cos x\right|$$

is less than 0.002 for all x. This is why the cosine appears to have been graphed in the figure.

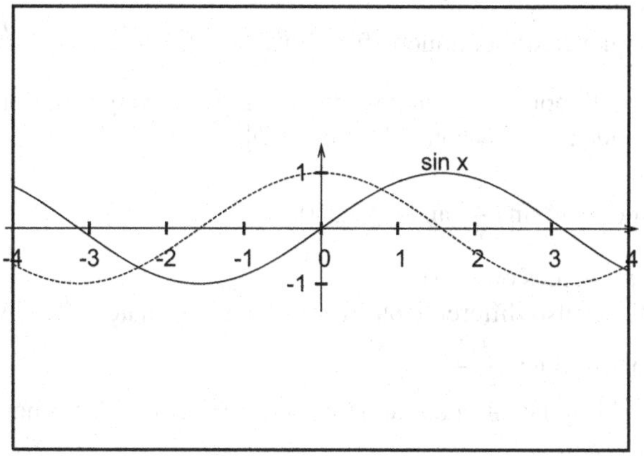

Fig. 4.24 The sine and a symmetric difference quotient $\dfrac{\sin(x+0.1)-\sin(x-0.1)}{0.2}$

4.59. Evaluate the one-sided difference quotient $\dfrac{f(x+h)-f(x)}{h}$ and the symmetric difference quotient $\dfrac{f(x+h)-f(x-h)}{2h}$ for the cases $f(x)=x^2$ and $f(x)=x^3$. If $x=10$ and $h=0.1$, by how much do these quotients differ from the derivative?

4.60.

(a) Use the approximation $f'''(\frac{x_1+x_2}{2}) \approx \dfrac{f''(x_2)-f''(x_1)}{x_2-x_1}$ to find estimates for f''' at the points 0.55, 0.65, 0.75, 0.85, which were left blank in Table 4.1.
(b) Use the approximation $f''''(\frac{x_1+x_2}{2}) \approx \dfrac{f'''(x_2)-f'''(x_1)}{x_2-x_1}$ to find estimates for f'''' at $x=0.2, 0.3, \ldots, 0.7, 0.8$.

4.61. Here we explore how to use approximate derivatives in a somewhat different way: to detect an isolated error in the tabulation of a smooth function. Suppose that in tabulating the data column that shows values of $f(x)$ in Table 4.1, a small error was made that interchanged two digits, $f(0.4)=0.38914$ instead of $f(0.4)=0.38941$.

(a) Recompute the table.
(b) Plot graphs of f, f', f'', f''' and f'''' for Table 4.1, and again for the recomputed table. What do you notice?

4.62. Assume that f is uniformly differentiable on $[a,b]$. Show that f' is continuous on $[a,b]$ by carrying out or justifying each of the following steps.

(a) Write down the definition of uniformly differentiable on $[a,b]$.
(b) Given any tolerance ε, explain why there is a positive integer n such that if $|h|<\frac{1}{n}$, then

$$\left|\frac{f(x+\frac{1}{n})-f(x)}{\frac{1}{n}} - f'(x)\right| < \varepsilon$$

for every x in $[a,b]$.

(c) Define a sequence of continuous functions $f_n(x) = \dfrac{f(x+\frac{1}{n}) - f(x)}{\frac{1}{n}}$. Explain why f_n is continuous on $[a,b]$ and show that f_n converges uniformly to f' on $[a,b]$. Conclude that f' is continuous on $[a,b]$.

4.63. Define $f(x) = x^2 \sin\left(\frac{1}{x}\right)$ and $f(0) = 0$.

(a) Find $f'(x)$ for values of $x \neq 0$.
(b) Verify that f is also differentiable at $x = 0$, and evaluate $f'(0)$, by considering the difference quotient $\dfrac{f(h) - f(0)}{h}$.
(c) Verify that f' is not continuous at 0 by showing that $f'(x)$ does not tend to $f'(0)$ as x tends to zero.
(d) Use Theorem 4.13 to argue that f is not uniformly differentiable.

Chapter 5
Applications of the Derivative

Abstract We present five applications of the calculus:

1. Atmospheric pressure in a gravitational field
2. Motion in a gravitational field
3. Newton's method for finding the zeros of a function
4. The reflection and refraction of light
5. Rates of change in economics

5.1 Atmospheric Pressure

If you have ever traveled in the mountains, you probably noticed that air pressure diminishes at higher altitudes. If you exert yourself, you get short of breath; if you cook, you notice that water boils at less than 100 °C. Our first application is to derive a differential equation to investigate the nature of air pressure as a function of altitude.

Let $P(y)$ be the air pressure [force/area] at altitude y above sea level. The force of air pressure at altitude y supports the weight of air above y. Imagine a column of air of unit cross section. The volume of air in the column between the altitudes y and $y+h$ (see Fig. 5.1) is h unit volumes. The weight of this air is $h\overline{\rho}g$, where h is the volume of the column, $\overline{\rho}$ is the average density [mass/volume] of air between altitudes y and $y+h$, and g is the acceleration due to gravity. This weight is supported by the force of air pressure at y minus the force of air pressure at $y+h$. Therefore,

$$h\overline{\rho}g = P(y) - P(y+h).$$

Dividing by h, we get

$$\overline{\rho}g = \frac{P(y) - P(y+h)}{h}.$$

P.D. Lax and M.S. Terrell, *Calculus With Applications*, Undergraduate Texts in Mathematics, 217
DOI 10.1007/978-1-4614-7946-8_5,

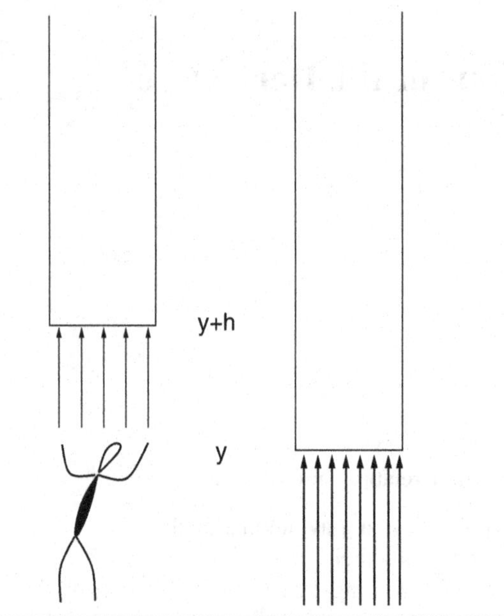

Fig. 5.1 The column of atmosphere above altitude y is heavier than that above $y+h$

As h tends to zero, the average density of the air, $\overline{\rho}$, tends to the density $\rho(y)$ at y, and the difference quotient on the right-hand side tends to $-P'(y)$, giving us the differential equation

$$\rho(y)g = -P'(y). \tag{5.1}$$

When gas is compressed, both its density and its pressure are increased. If we assume that air pressure and density are linearly related,

$$\rho = kP, \qquad k \text{ some constant},$$

and set this into the differential equation (5.1), we get

$$kgP(y) = -P'(y).$$

According to Theorem 3.11, the solutions of this equation are the exponential functions

$$P(y) = P(0)\mathrm{e}^{-kgy},$$

where $P(0)$ is atmospheric pressure at sea level.

So we have deduced that atmospheric pressure is an exponential function of altitude. Is there anything we can say about the constant k? We can determine the dimension of k from $k = \dfrac{\rho}{P}$. The dimension of k is then

$$\left[\frac{\text{density}}{\text{pressure}}\right] = \left[\frac{\frac{\text{mass}}{\text{volume}}}{\frac{\text{force}}{\text{area}}}\right].$$

Since force is equal to mass times acceleration, the dimension of k is

$$\left[\frac{\text{mass/length}^3}{\frac{\text{mass}\times\text{length}}{\text{time}^2}/\text{length}^2}\right] = \frac{\text{time}^2}{\text{length}^2} = \frac{1}{\text{velocity}^2}.$$

What is the velocity that appears in k? What velocity can one associate with the atmosphere? It turns out that $1/k$ is the square of the speed of *sound*. Let us check this value for k numerically using $P(y) = P(0)e^{-kgy}$ to calculate air pressure at Denver, Colorado, at altitude 1 mile. The speed of sound at sea level is approximately 1000 ft/s. Therefore,

$$k = 10^{-6} \ (\text{s/ft})^2, \qquad g = 32 \ \text{ft/s}^2, \qquad y = 5280 \ \text{ft},$$

and $kgy = 10^{-6}(32)(5280) = 0.169$. Since $e^{-0.169} = 0.844$, the air pressure formula gives

$$P(1 \text{ mile}) = 0.844\,P(0).$$

Atmospheric pressure at sea level is 14.7 psi (pounds per square inch). Using our formula, we get $(0.844)(14.7) = 12.4$ psi for air pressure at Denver. The measured value of atmospheric pressure in Denver is 12.1 psi, so our formula gives a good approximation.

Problems

5.1. Compare the atmospheric pressure at your city to the approximate value determined by the equation in this section.

5.2. While investigating atmospheric pressure, we assumed that air density is proportional to air pressure. In this problem, consider pressure in the ocean, where water density is nearly constant. You may ignore atmospheric pressure at the surface.

(a) Derive a differential equation for ocean pressure, similar to Eq. (5.1), in two different ways: In one equation, assume that y is measured from the surface down, and in the other, assume that y is measured from the bottom up. How do the resulting equations compare? What are the values of $P(0)$ in each case? Are there advantages of one equation over the other?

(b) Solve the "surface down" differential equation.

(c) Take the density of ocean water to be 1025 [kg/m^3], and atmospheric pressure 10^5 [N/m^2]. Divers use a rule of thumb that the pressure increases by one atmosphere for each 10 m of descent. Does this agree with the "surface down" pressure function you found in part (b)?

5.2 Laws of Motion

In this section, we see how calculus is used to derive differential equations of motion of idealized particles along straight lines. The beauty of these equations is their universality; we can use them to deduce how high we can jump on the surface of the Earth, as well as on the Moon. A *particle* is an idealization in physics, an indivisible point with no extent in space, so that a single coordinate gives the position. It has a *mass*, usually denoted by the letter m. In this simple case, the position of a particle is completely described by its distance y from an arbitrarily chosen point (the origin) on the line; y is taken to be positive if y lies to one side (chosen arbitrarily) of the origin and negative if the particle is located on the other side of the origin.

Since the particle moves, y is a function of time t. The derivative $y'(t)$ of this function is the *velocity* of the particle, a quantity usually denoted by v:

$$v = y' = \frac{dy}{dt}. \tag{5.2}$$

Note that v is positive if the y-coordinate of the particle increases during the motion.

The velocity of a particle may, of course, change as time changes. The rate at which it changes is called the *acceleration*, and is usually denoted by the letter a:

$$a = v' = \frac{dv}{dt} = \frac{d^2y}{dt^2}. \tag{5.3}$$

Newton's laws of motion relate the acceleration of a particle to its mass and the force acting on it as follows: A force f acting along the y-axis causes a particle of mass m to accelerate at the rate a, and

$$f = ma. \tag{5.4}$$

According to Eq. (5.4), a force acting along the y-axis is negative if it imparts a negative acceleration to a particle traveling along the y-axis. There is nothing mysterious about this negative sign. It merely means that the force is acting in the negative direction along the y-axis.

If a number of different forces act on a particle, as they do in most realistic situations, the effective force acting on the particle is the *sum* of the separate forces. For example, a body might be subject to the force of gravity f_g, the force of air resistance f_a, an electric force f_e, and a magnetic force f_m; the effective force f is then

$$f = f_g + f_a + f_e + f_m, \tag{5.5}$$

and the equations governing motions under this combination of forces are

$$y' = v, \qquad mv' = f. \tag{5.6}$$

There is a tremendous advantage in being able to synthesize the force acting on a particle from various simpler forces, each of which can be analyzed separately.

Although we can write down the equations of motion (5.6) governing particles subject to any combination of forces (5.5), we cannot, except in simple situations, write down formulas for the particle's position as a function of time. In this chapter, we show how to solve the simplest version of the problem when f is a constant due to gravity. In Chap. 10, we describe numerical methods to calculate the position of a particle as a function of time to a high degree of accuracy when the force is not constant. We also investigate equations of motion for a particle subject to a combination of forces.

Gravity. We illustrate how Newton's second law (5.4) can be used to describe motion in the specific situation in which the force is that of gravity at the surface of the Earth. According to a law that is again Newton's, the magnitude f_g of the force of gravity exerted on a particle of mass m is proportional to its mass:

$$f_g = gm. \tag{5.7}$$

The constant of proportionality g and the direction of the force depend on the gravitational pull of other masses acting on the particle. At a point near the surface of the Earth, the direction of the force is toward the center of the Earth, and the value of g is approximately

$$g = 9.81\,\mathrm{m/s^2}, \tag{5.8}$$

where m is meters and s is seconds. Near the surface of the Moon, the value of g is approximately

$$g = 1.6\,\mathrm{m/s^2}. \tag{5.9}$$

For the moment, let us stay near the Earth. Denote by y the distance from the surface of the Earth of a falling body, and denote by v the vertical velocity of this falling body. Since y was chosen to increase upward, and the force of gravity is downward, the force of gravity is $-gm$. Substituting this into Newton's law (5.4), we see that

$$-gm = ma,$$

where a is the acceleration of the falling body. Divide by m:

$$-g = a.$$

Recalling the definitions of velocity and acceleration, we can write $a = y''$, so Newton's law for a falling body is

$$y'' = -g. \tag{5.10}$$

We claim that all solutions of this equation are of the form

$$y = -\frac{1}{2}gt^2 + v_0 t + b, \tag{5.11}$$

where b and v_0 are constants. To see this, rewrite Eq. (5.10) as $0 = y'' + g = (y' + gt)'$, from which we conclude that $y' + gt$ is a constant; call it v_0. Then $y' + gt - v_0 = 0$. We rewrite this equation as $\left(y + \frac{1}{2}gt^2 - v_0 t\right)' = 0$. From this, we conclude that $y + \frac{1}{2}gt^2 - v_0 t$ is a constant; call it b. This proves that all solutions of Eq. (5.10) are of the form (5.11).

The significance of the constant v_0 is this: it is the particle's initial velocity; that is, $y' = v_0$ when $t = 0$. Similarly, b is the initial position of the particle, i.e., $y = b$ when $t = 0$. So we see that initial position and initial velocity can be prescribed arbitrarily, but thereafter, the motion is uniquely determined.

How High Can You Jump? Suppose that a student, let us call her Anya, idealized as a particle, can jump k meters vertically starting from a crouching position. How much force is exerted?

Denote by h Anya's height and by m her mass. Denote by f the force that her feet in crouching position exert on the ground. As long as Anya's feet are on the ground, the total upward force acting on her body is the force exerted by her feet minus the force of gravity:

$$f - gm.$$

According to Newton's law, Anya's motion, considered as the motion of a particle, is governed by Eq. (5.4),

$$my'' = f - gm,$$

where $y(t)$ is the distance of Anya's head from the ground at time t. We divide by m and rewrite the result as

$$y'' = p - g,$$

where $p = \frac{f}{m}$ is force per unit mass. As we have seen, all solutions of this equation are of the form (5.11):

$$y(t) = \frac{1}{2}(p-g)t^2 + v_0 t + b.$$

Setting $t = 0$, we get $y(0) = b$, the distance of Anya's head from the ground in crouching position. Differentiating y and setting $t = 0$, we get $y'(0) = v_0$. Since at the outset, the body is at rest, we have $v_0 = 0$. So

$$y(t) = \frac{1}{2}(p-g)t^2 + b. \tag{5.12}$$

The description (5.12) is valid until the time t_1 when Anya's feet leave the ground. That occurs when $y(t_1)$, the position of her head, equals her height h, that is, when $y(t_1) = h$. Setting this into Eq. (5.12), we get $\frac{1}{2}(p-g)t_1^2 = h - b$. Denote by c the difference between the position of Anya's head in the standing and crouching positions: $c = h - b$. Solving for t_1, the time at which Anya's feet leave the ground, we get

$$t_1 = \sqrt{\frac{2c}{p-g}}. \tag{5.13}$$

What is Anya's upward velocity v_1 at time t_1? Since velocity is the time derivative of position, we have $y'(t) = (p-g)t$. Using Eq. (5.13), we get

$$v_1 = y'(t_1) = \sqrt{2c(p-g)}. \tag{5.14}$$

After her feet leave the ground, the only force acting on Anya is gravity. So for $t > t_1$, the equation governing the position of her head is

$$y'' = -g.$$

The solutions of this equation are of the form (5.11). We rewrite it with t replaced by $t - t_1$ and b replaced by h:

$$y(t) = -\frac{1}{2}g(t-t_1)^2 + v_1(t-t_1) + h. \tag{5.15}$$

Here h is the position at time t_1. The greatest height reached by the trajectory (5.15) is time t_2, when the velocity is zero. Differentiating Eq. (5.15), we get

$$y'(t_2) = -g(t_2 - t_1) + v_1 = 0,$$

which gives $t_2 - t_1 = \frac{v_1}{g}$. Setting this into formula (5.15) for $y(t)$ gives

$$y(t_2) = -\frac{1}{2}g(t_2-t_1)^2 + v_1(t_2-t_1) + h = -\frac{v_1^2}{2g} + \frac{v_1^2}{g} + h = \frac{v_1^2}{2g} + h.$$

Using formula (5.14) for v_1, we get

$$y(t_2) = c\frac{p-g}{g} + h.$$

So the height of the jump $k = y(t_2) - h$ is given by

$$k = c\left(\frac{p}{g} - 1\right). \tag{5.16}$$

Using this relation, we can express the jumping force p per unit mass as

$$p = g\left(1 + \frac{k}{c}\right). \tag{5.17}$$

Notice that in order to be able to jump at all, the force per mass exerted has to be greater than the acceleration of gravity g.

Anya is rather tall, so we shall take her height h to be 2 m and b to be 1.5 m, making $c = h - b = 0.5$ m. We take the height of the jump k to be 0.25 m. Then by Eq. (5.17), $p = 1.5g$ (Fig. 5.2).

Let us see how far such a force would carry us on the Moon. The distance k_m jumped on the moon is given by formula (5.16), where g is replaced by g_m, the acceleration due to gravity on the Moon:

$$k_m = c\left(\frac{p}{g_m} - 1\right).$$

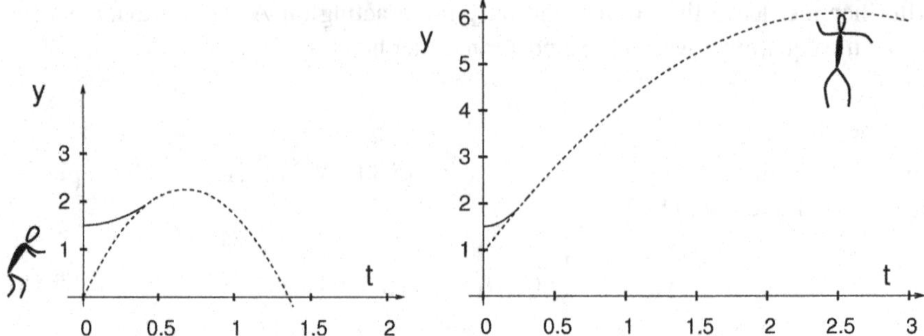

Fig. 5.2 Jumping on the surface of the Earth (*left*) and Moon (*right*) with the same force, plotting height as a function of time. The convex parabola indicates head position while the feet are pushing

Using $p = 1.5g$ and $c = 0.5$, we get

$$k_m = \frac{1}{2}\left(1.5\frac{g}{g_m} - 1\right). \tag{5.18}$$

Since $g = 9.8\ \mathrm{m/s^2}$, and $g_m = 1.6\ \mathrm{m/s^2}$, their ratio is 6.125. Setting these values into formula (5.18), we deduce that on the Moon, Anya can jump

$$k_m = \frac{1}{2}\big(1.5(6.125) - 1\big) = 4.1\,\mathrm{m}.$$

Problems

5.3. Suppose that the initial position and velocity of a particle subject to Earth's gravity are $y(0) = 0$ and $y'(0) = 10$ (m/s). Calculate position and velocity at time $t = 1$ and $t = 2$.

5.4. What is the largest value of $y(t)$ during the motions described in Problem 5.3?

5.5. Write Newton's law (5.4) as a differential equation for the position $y(t)$ of a particle of mass m in the following situation: (1) $y > 0$ is the distance down to a horizontal surface at $y = 0$. (2) There are two forces on the particle; one is downward due to constant gravitational acceleration g as we have discussed, and the other is a constant upward force f_{up}.

5.6. We have said that forces may be added, and that positions can be more difficult to find than forces. In this problem, a particle of mass m at position $y(t)$ moves in six different cases according to

$$\begin{aligned} my'' &= f, \\ y(0) &= 1, \\ y'(0) &= v, \end{aligned}$$

where v is 0 or 3, and f is 5 or 7 or $5+7$. Solve for the position functions $y(t)$ in the six cases. Is it ever true that positions can be added when the forces are added?

5.7. What theorem did we use to deduce that all solutions of Eq. (5.10) are of the form (5.11)?

5.3 Newton's Method for Finding the Zeros of a Function

In the preceding two sections, we applied calculus to problems in the physical sciences. In this section, we apply calculus to solving mathematical problems.

Many mathematical problems are of the following form: we are seeking a number, called "unknown" and denoted by, say, the letter z, which has some desirable property expressed in an equation. Such an equation can be written in the form

$$f(z) = 0,$$

where f is some function. Very often, additional restrictions are placed on the number z. In many cases, these restrictions require z to lie in a certain interval. So the task of "solving an equation" is really nothing but finding a number z for which a given function f vanishes, that is, where the value of f is zero. Such a number z is called a *zero* of the function f. In some problems, we are content to find *one* zero of f in a specified interval, while in other problems, we are interested in finding *all* zeros of f in an interval.

What does it mean to "find" a zero of a function? It means to devise a procedure that gives as good an approximation as desired, of a number z where the given function f vanishes. There are two ways of measuring the goodness of an approximation z_{approx}: one is to demand that z_{approx} differ from an exact zero z by, say, less than $\frac{1}{100}$, or $\frac{1}{1000}$, or 10^{-m}. Another way of measuring the goodness of an approximation is to insist that the value of f at z_{approx} be very small, say less than $\frac{1}{100}$, $\frac{1}{1000}$, or in general, less than 10^{-m}. Of course, these notions go hand in hand: if z_{approx} is close to the true zero z, then $f(z_{\text{approx}})$ will be close to $f(z) = 0$, provided that the function f is continuous.

In this section, we describe a method for finding approximations to zeros of functions f that are not only continuous but differentiable, preferably twice differentiable. The basic step of the method is this: starting with some fairly good approximation to a zero of f, we use the derivative to produce a much better one. If the approximation is not yet good enough, we repeat the basic step as often as necessary to produce an approximation that is good enough according to either of the two criteria mentioned earlier. There are two ways of describing the basic step, geometrically and analytically. We start with the geometric description.

Denote by z_{old} the starting approximation. We assume—and this is *crucial* for the applicability of this method—that $f'(z_{\text{old}}) \neq 0$. This guarantees that the line

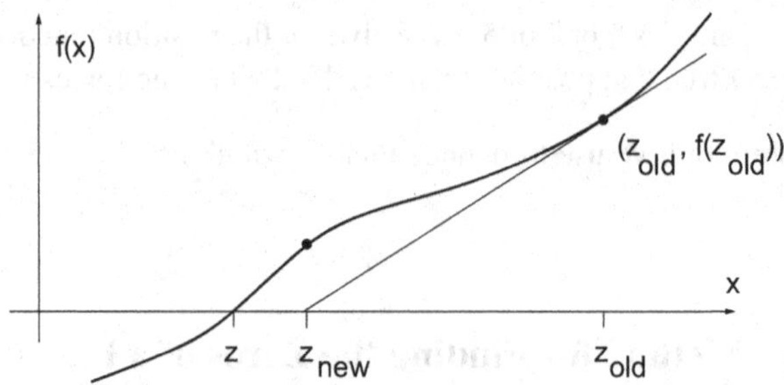

Fig. 5.3 Newton's method to approximate a root z

tangent to the graph of f at $(z_{\text{old}}, f(z_{\text{old}}))$ (see Fig. 5.3) is not parallel to the x-axis, and so intersects the x-axis at some point. This point of intersection is our new approximation z_{new}. We now calculate z_{new}. Since the slope of the tangent is $f'(z_{\text{old}})$, we have

$$f'(z_{\text{old}}) = \frac{f(z_{\text{old}}) - f(z_{\text{new}})}{z_{\text{old}} - z_{\text{new}}} = \frac{f(z_{\text{old}})}{z_{\text{old}} - z_{\text{new}}}.$$

From this relation we can determine z_{new}:

$$z_{\text{new}} = z_{\text{old}} - \frac{f(z_{\text{old}})}{f'(z_{\text{old}})}. \tag{5.19}$$

The rationale behind this procedure is this: if the graph of f were a straight line, z_{new} would be an exact zero of f. In reality, the graph of f is not a straight line, but if f is differentiable, its graph over a short enough interval is *nearly* straight, and so z_{new} can reasonably be expected to be a good approximation to the exact zero z, provided that the interval (z, z_{old}) is short enough.

We shall show at the end of this section that if z_{old} is a good enough approximation to a zero of the function f (in a sense made precise there), then z_{new} is a much better one, and that if we repeat the procedure, the resulting sequence of approximations will converge very rapidly to a zero of the function f. But first we give some examples.

The method described above has been devised, like so much else in calculus, by Newton and is therefore called *Newton's method.*

5.3a Approximation of Square Roots

Let $f(x) = x^2 - 2$. We seek a positive solution of

$$f(z) = z^2 - 2 = 0.$$

Let us see how closely we can approximate the exact solution, which is $z = \sqrt{2}$. For $f(x) = x^2 - 2$, we have $f'(x) = 2x$, so if z_{old} is an approximation to $\sqrt{2}$, Newton's recipe (5.19) yields

$$z_{\text{new}} = z_{\text{old}} - \frac{z_{\text{old}}^2 - 2}{2z_{\text{old}}} = \frac{z_{\text{old}}}{2} + \frac{1}{z_{\text{old}}}. \tag{5.20}$$

Notice that this relation is just Eq. (1.7) revisited. Let us take $z_{\text{old}} = 2$ as a first approximation to $\sqrt{2}$. Using formula (5.20), we get

$$z_{\text{new}} = 1.5.$$

We then repeat the process, with $z_{\text{new}} = 1.5$ now becoming z_{old}. Thus, we construct a sequence $z_1, z_2, \ldots$ of (hopefully) better approximations to $\sqrt{2}$ by choosing $z_1 = 2$ and setting

$$z_{n+1} = \frac{z_n}{2} + \frac{1}{z_n}.$$

The first six approximations are

$$\begin{aligned} z_1 &= 2.0 \\ z_2 &= 1.5 \\ z_3 &= 1.4166\ldots \\ z_4 &= 1.414215686\ldots \\ z_5 &= 1.414213562\ldots \\ z_6 &= 1.414213562\ldots \end{aligned}$$

Since z_5 and z_6 agree up to the first eight decimal places after the decimal point, we surmise that z_5 gives the first eight decimal places of $\sqrt{2}$ correctly. Indeed,

$$(1.41421356)^2 = 1.999999993\ldots$$

is very near and slightly less than 2, while

$$(1.41421357)^2 = 2.000000021\ldots$$

is very near, but slightly more than, 2. It follows from the intermediate value theorem that $z^2 = 2$ at some point between these numbers, i.e. that

$$1.41421356 < \sqrt{2} < 1.41421357.$$

When we previously encountered the sequence $\{z_n\}$, where it was constructed in a somewhat ad hoc fashion, the sequence was shown to converge.

5.3b Approximation of Roots of Polynomials

The beauty of Newton's method is its universality. It can be used to find the zeros of not only quadratic functions, but functions of all sorts.

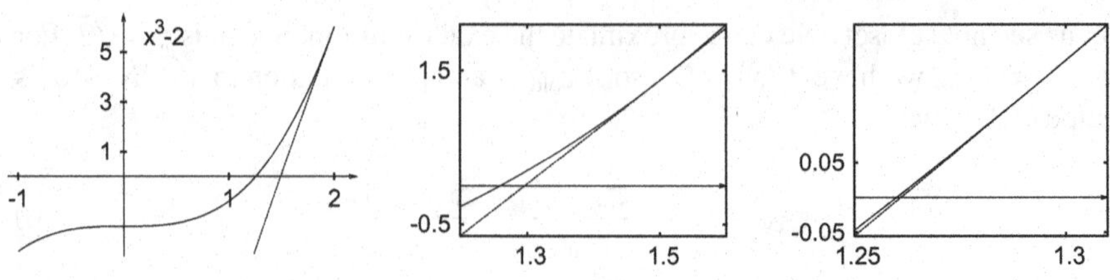

Fig. 5.4 Three steps of Newton's method to approximate a root of $f(x) = x^3 - 2$, drawn at different magnifications. The three tangent lines shown are at z_1, z_2, and z_3 of Example 5.1

Example 5.1. Let $f(x) = x^3 - 2$. We seek a sequence of approximations to a solution of

$$z^3 - 2 = 0, \tag{5.21}$$

i.e., to the number $\sqrt[3]{2}$. Since $f'(x) = 3x^2$, Newton's recipe gives the following sequence of approximations (Fig. 5.4):

$$z_{n+1} = z_n - \frac{z_n^3 - 2}{3z_n^2} = \frac{2z_n}{3} + \frac{2}{3z_n^2}. \tag{5.22}$$

Starting with $z_1 = 2$ as a first approximation, we have

$$\begin{aligned} z_1 &= 2.0 \\ z_2 &= 1.5 \\ z_3 &= 1.2962962\ldots \\ z_4 &= 1.2609322\ldots \\ z_5 &= 1.2599218\ldots \\ z_6 &= 1.2599210\ldots \end{aligned}$$

Since z_5 and z_6 agree up to the sixth digit after the decimal, we surmise that

$$\sqrt[3]{2} = 1.259921\ldots$$

Indeed, $(1.259921)^3 = 1.9999997\ldots$, while $(1.259922)^3 = 2.000004\ldots$, so that

$$1.259921 < \sqrt[3]{2} < 1.259922.$$

Next we find all zeros of

$$f(x) = x^3 - 6x^2 - 2x + 12. \tag{5.23}$$

Since f is a polynomial of degree 3, an odd number, $f(x)$ is very large and positive when x is very large and positive, and very large and negative when x is very large and negative. So by the intermediate value theorem, $f(x)$ is zero somewhere. To get a better idea where the zero, or zeros, might be located, we calculate the value of f at integers ranging from $x = -2$ to $x = 6$:

x	-2	-1	0	1	2	3	4	5	6
$f(x)$	-16	7	12	5	-8	-21	-28	-23	0

This table shows that f has a zero at $z=6$, and since the value of f at $x=-2$ is negative, at $x=-1$ positive, f has a zero in the interval $(-2,-1)$. Similarly, f has a zero in the interval $(1,2)$.

According to a theorem of algebra, if z is a zero of a polynomial, $x-z$ is a factor. Indeed, we can write our f in the factored form

$$f(x)=(x-6)(x^2-2).$$

This form for f shows that its other zeros are $z=\pm\sqrt{2}$, and that there are no others.

Let us ignore this exact knowledge of the zeros of f (which, after all, was due to a lucky accident). Let us see how well Newton's general method works in this case. The formula, for this particular function, reads

$$z_{n+1}=z_n-\frac{z_n^3-6z_n^2-2z_n+12}{3z_n^2-12z_n-2}.$$

Starting with $z_1=5$ as a first approximation to the exact root 6, we get the following sequence of approximations:

$$\begin{aligned} z_1 &= 5 \\ z_2 &= 6.76\ldots \\ z_3 &= 6.147\ldots \\ z_4 &= 6.007\ldots \\ z_5 &= 6.00001\ldots \end{aligned}$$

Similar calculations show that if we start with a guess z *close enough* to one of the other two zeros $\sqrt{2}$ and $-\sqrt{2}$, we get a sequence of approximations that converges *rapidly* to the exact zeros.

5.3c The Convergence of Newton's Method

How rapid is rapid, and how close is close? In the last example, starting with an initial guess that was off by 1, we obtained, after four steps of the method, an approximation that differs from the exact zero $z=6$ by 0.007.

Furthermore, perusal of the examples presented so far indicates that Newton's method works faster, the closer z_n gets to the zero! We shall analyze Newton's method to explain its extraordinary efficiency and also to determine its limitations.

Newton's method is based on a linear approximation. If there were no error in this approximation—i.e., if f had been a linear function—then Newton's method would have furnished in one step the exact zero of f. Therefore, in analyzing the error inherent in Newton's method, we must start with the deviation of the function f from its linear approximation. The deviation is described by the linear approximation theorem, Theorem 4.5:

$$f(x)=f(z_{\text{old}})+f'(z_{\text{old}})(x-z_{\text{old}})+\frac{1}{2}f''(c)(x-z_{\text{old}})^2, \tag{5.24}$$

where c is some number between z_{old} and x. Let us introduce for simplicity the abbreviations

$$f''(c)=s \qquad \text{and} \quad f'(z_{\text{old}})=m,$$

and let z be the exact zero of f. Then Eq. (5.24) yields for $x=z$,

$$f(z)=0=f(z_{\text{old}})+m(z-z_{\text{old}})+\frac{1}{2}s(z-z_{\text{old}})^2.$$

Divide by m and use Newton's recipe (5.19) to get

$$0=\frac{f(z_{\text{old}})}{m}+(z-z_{\text{old}})+\frac{1}{2}\frac{s}{m}(z-z_{\text{old}})^2=-z_{\text{new}}+z_{\text{old}}+(z-z_{\text{old}})+\frac{1}{2}\frac{s}{m}(z-z_{\text{old}})^2.$$

We can rewrite this as

$$z_{\text{new}}-z=\frac{1}{2}\frac{s}{m}(z_{\text{old}}-z)^2. \tag{5.25}$$

We are interested in finding out under what conditions z_{new} is a better approximation to z than z_{old}. Formula (5.25) is ideal for deciding this, since it asserts that $(z_{\text{new}}-z)$ is the product of $(z-z_{\text{old}})$ and $\frac{1}{2}\frac{s}{m}(z-z_{\text{old}})$. There is an improvement if and only if that second factor is less than 1 in absolute value, i.e., if

$$\frac{1}{2}\left|\frac{s}{m}\right||z-z_{\text{old}}|<1. \tag{5.26}$$

Suppose now that $f'(z)\neq 0$. Since f' is continuous, f' is bounded away from zero at all points close to z, and since f'' is continuous, s does not vary too much. Also, Eq. (5.26) holds if z_{old} is close enough to z. In fact, for z_{old} close enough, we have

$$\frac{1}{2}\left|\frac{s}{m}\right||z-z_{\text{old}}|<\frac{1}{2}. \tag{5.27}$$

If Eq. (5.27) holds, we deduce from Eq. (5.25) that

$$|z_{\text{new}}-z|\leq\frac{1}{2}|z_{\text{old}}-z|. \tag{5.28}$$

Now let z_1, z_2, … be a sequence of approximations generated by repeated applications of Newton's recipe. Suppose z_1 is so close to z that Eq. (5.27) holds for $z_{\text{old}}=z_1$ and for all z_{old} that are as close or closer to z than z_1. Then it follows from Eq. (5.28) that z_2 is closer to z than z_1 and, in general, that each z_{n+1} is closer to z than the previous z_n, and so Eq. (5.27) holds for all subsequent z_n. Repeated application of Eq. (5.28) shows that

$$|z_{n+1}-z| \le \frac{1}{2}|z_n - z| \le \left(\frac{1}{2}\right)^2 |z_{n-1}-z| \le \cdots \le \left(\frac{1}{2}\right)^n |z_1 - z|. \qquad (5.29)$$

This proves the following theorem.

Theorem 5.1. Convergence theorem for Newton's method. *Let f be twice continuously differentiable on an open interval and z a zero of f such that*

$$f'(z) \neq 0. \qquad (5.30)$$

Then repeated applications of Newton's recipe

$$z_{n+1} = z_n - \frac{f(z_n)}{f'(z_n)} \qquad (5.31)$$

yields a sequence of approximations z_1, z_2, ... that converges to z, provided that the first approximation z_1 is close enough to z.

A few comments are in order:

1. The proof that z_n tends to z is based on inequality (5.29), according to which $|z_{n+1}-z|$ is less than a constant times $\left(\frac{1}{2}\right)^n$. This is a gross overestimate; to understand the true rate at which z_n converges to z, we have to go back to relation (5.25). For z_{old} close to z, the numbers m and s differ little from $f'(z)$ and $f''(z)$ respectively, so that Eq. (5.25) asserts that $|z_{\text{new}}-z|$ is practically a constant multiple of $(z_{\text{old}}-z)^2$. Now if $|z_{\text{old}}-z|$ is small, its square is enormously small! To give an example, suppose that $\left|\frac{f''(z)}{2f'(z)}\right| \le 1$ and that $|z_{\text{old}}-z| \le 10^{-3}$. Then by Eq. (5.25), we conclude that

$$|z_{\text{new}}-z| \approx (z_{\text{old}}-z)^2 = 10^{-6}.$$

In words: If the first approximation lies within one-thousandth of an exact zero, and if $\left|\frac{f''(z)}{2f'(z)}\right| < 1$, then Newton's method takes us *in one step* to a new approximation that lies within one millionth of that exact zero.

Example 5.2. In Example 5.1, we have $\frac{1}{2}\frac{s}{m} < 1$ and

$$z_5 = \underline{1.2599218}6056593,$$
$$z_6 = \underline{1.25992104989}539,$$

where the underlined digits are correct, an increase of about twice as many in one step.

2. It is necessary to start close enough to z, not only to achieve rapid convergence, but to achieve convergence at all. Figure 5.5 shows an example in which Newton's method fails to get us any closer to a zero. The points z_{old} and z_{new} are chosen so

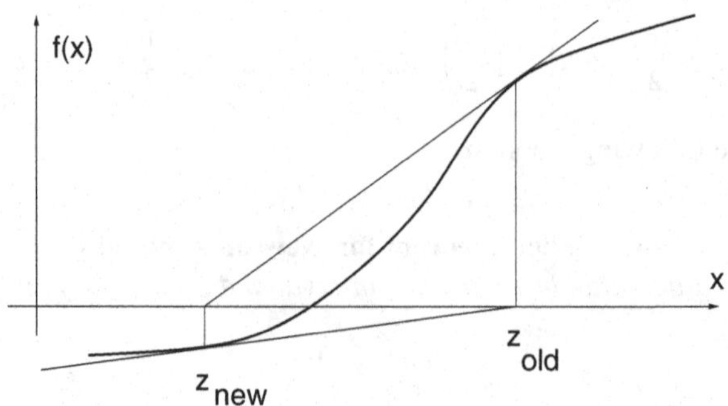

Fig. 5.5 Newton's method can fail by cycling

that the tangent to the graph of f at the point $(z_{\text{old}}, f(z_{\text{old}}))$ intersects the x-axis at z_{new}, and the tangent to the graph of f at $(z_{\text{new}}, f(z_{\text{new}}))$ intersects the x-axis at z_{old}. Newton's recipe brings us from z_{old} to z_{new}, then back to z_{old}, etc., without getting any closer to the zero between them.

3. Our analysis indicates difficulty with Newton's method when $f'(z) = 0$ at the zero z. Here is an example: the function $f(x) = (x-1)^2$ has a double zero at $z = 1$; therefore, $f'(z) = 0$. Newton's method yields the following sequence of iterates:

$$z_{n+1} = z_n - \frac{f(z_n)}{f'(z_n)} = z_n - \frac{(z_n - 1)^2}{2(z_n - 1)} = \frac{z_n + 1}{2}.$$

Subtracting 1 from both sides, we get $z_{n+1} - 1 = \dfrac{z_n - 1}{2}$. Using this relation repeatedly, we get

$$z_{n+1} - 1 = \frac{1}{2}(z_n - 1) = \frac{1}{4}(z_{n-1} - 1) = \cdots = \left(\frac{1}{2}\right)^n (z_1 - 1).$$

Thus z_n approaches the zero $z = 1$ at the rate of a constant times 2^{-n}, and not the super fast rate at which Newton's method converges when $f'(z) \neq 0$.

Problems

5.8. Use Newton's method to determine the first four digits after the decimal point of $3^{1/4}$ and of $\sqrt[3]{7}$.

Hint: Evaluating $c^{1/q}$ is equivalent to finding the zero of $z^q - c$.

5.9. Find all zeros of the following functions in the indicated domain:

(a) $f(x) = 1 + x^{1/3} - x^{1/2}$, $x \geq 0$. *Hint*: Try introducing $x = y^6$.
(b) $f(x) = x^3 - 3x^2 + 1$, $-\infty < x < \infty$.
(c) $f(x) = \dfrac{x}{x^2+1} + 1 - \sqrt{x}$, $x \geq 1$.

5.10. We claimed in the text that "if f had been a linear function, then Newton's method would have furnished in one step the exact zero of f." Show that this is true.

5.11. Show that you can find the largest value assumed by a function f in an interval $[a,b]$ by performing the following steps:

(a) Evaluate $f(x)$ at the endpoints a and b and at N equidistant points x_j between the endpoints, where N is to be specified. Set $x_0 = a$, $x_{N+1} = b$.
(b) If $f(x_j)$ is greater than both $f(x_{j-1})$ and $f(x_{j+1})$, there is a zero of $f'(x)$ in the interval (x_{j-1}, x_{j+1}). Use Newton's method to find such a zero, and denote it by z_j.
(c) Determine the largest of the values $f(z_j)$ and compare to the values of f at the endpoints.

Why is it important to select N sufficiently large?

5.12. Let z be a zero of the function f, and suppose that neither f' nor f'' is zero. Show that all subsequent approximations $z_2, z_3, \ldots$ generated by Newton's method are

(a) Greater than z if $f'(z)$ and $f''(z)$ have like signs,
(b) Less than z if $f'(z)$ and $f''(z)$ have opposite signs.

Verify the truth of these assertions for examples presented in this section.

5.13. In this exercise, we ask you to investigate the following method designed to obtain a sequence of progressively better approximations to a zero of a function f:

$$z_{\text{new}} = z_{\text{old}} - af(z_{\text{old}}).$$

Here a is a number to be chosen in some suitable way. Clearly, if z_{old} happens to be the exact root z, then $z_{\text{new}} = z_{\text{old}}$. The question is this: if z_{old} is a good approximation to z, will z_{new} be a better approximation, and how much better?

(a) Use this method to construct a sequence $z_1, z_2, \ldots$ of approximations to the positive root of

$$f(z) = z^2 - 2 = 0,$$

starting with $z_1 = 2$. Observe that

- For $a = \frac{1}{2}$, $z_n \to \sqrt{2}$, but the z_n are alternately less than and greater than $\sqrt{2}$.
- For $a = \frac{1}{3}$, $z_n \to \sqrt{2}$ monotonically.
- For $a = 1$, the sequence $(z_n\}$ diverges.

(b) Prove, using the mean value theorem, that

$$z_{\text{new}} - z = (1 - am)(z_{\text{old}} - z), \tag{5.32}$$

where m is the value of f' somewhere between z and z_{old}. Prove that if a is chosen so that

$$|1 - af'(z)| < 1,$$

then $z_n \to z$, provided that z_1 is taken close enough to z.
Can you explain your findings under (a) in light of formula (5.32)?

(c) What would be the best choice for a, i.e., one that would yield the most rapidly converging sequence of approximations?

(d) Since the ideal value of a in part (c) is not practical, try using some reasonable estimate of $1/f'$, perhaps $1/(\text{secant slope})$ in any interval where f changes sign, for the functions in Eqs. (5.21) and (5.23).

5.4 Reflection and Refraction of Light

Mathematics can be used in science to derive laws from basic principles. In this section, we show how to use calculus to derive the laws of reflection and refraction from Fermat's principle.

Fermat's principle: Among all possible paths PRQ connecting two points P and Q via a mirror, a ray of light travels along the one that takes the least time to traverse.

Flat Mirrors. We wish to calculate the path of a ray of light that is reflected from a flat mirror. The path of a reflected ray going from point P to point Q is pictured in Fig. 5.6. The ray consists of two straight line segments, one, the incident ray, leading from P to the point of reflection R, the other, the reflected ray, leading from R to Q.

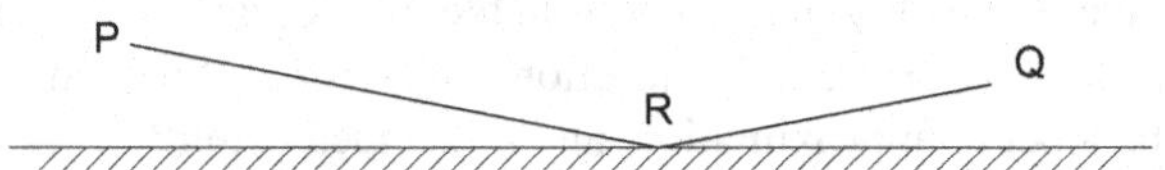

Fig. 5.6 A light ray reflects from a flat mirror

In a uniform medium like air, light travels with constant speed, so the time needed to traverse the path PRQ equals its length divided by the speed of light. So the path taken by the light ray will be the *shortest* path PRQ. We choose the mirror to be the axis of a Cartesian coordinate system. The coordinates of the point R are $(x,0)$. Denote the coordinates of the point P by (a,b), those of Q by (c,d). According to the Pythagorean theorem, the distances PR and RQ are

$$\ell_1(x) = PR = \sqrt{(x-a)^2 + b^2}, \qquad \ell_2(x) = RQ = \sqrt{(c-x)^2 + d^2}.$$

The total length ℓ of the path is

$$\ell(x) = \ell_1(x) + \ell_2(x) = \sqrt{(x-a)^2 + b^2} + \sqrt{(c-x)^2 + d^2}.$$

According to Fermat's principle, the coordinate x ought to minimize $\ell(x)$. The function ℓ is defined and differentiable for all real numbers x, not just those between a and c. For x large enough, positive or negative, $\ell(x)$ is also very large because $\ell(x) \geq |x-a| + |c-x|$. So by taking a large enough closed interval, ℓ assumes its maximum at an endpoint and its minimum at a point in the interior where $\ell'(x) = 0$.

Differentiating ℓ, we get

$$\ell'(x) = \frac{x-a}{\sqrt{(x-a)^2+b^2}} - \frac{c-x}{\sqrt{(c-x)^2+d^2}} = \frac{x-a}{\ell_1(x)} - \frac{c-x}{\ell_2(x)}.$$

Since $\ell'(x) = 0$, we have

$$\frac{x-a}{\ell_1(x)} = \frac{c-x}{\ell_2(x)}. \tag{5.33}$$

We ask you to check in Problem 5.14 that every x that satisfies Eq. (5.33) must be between a and c.

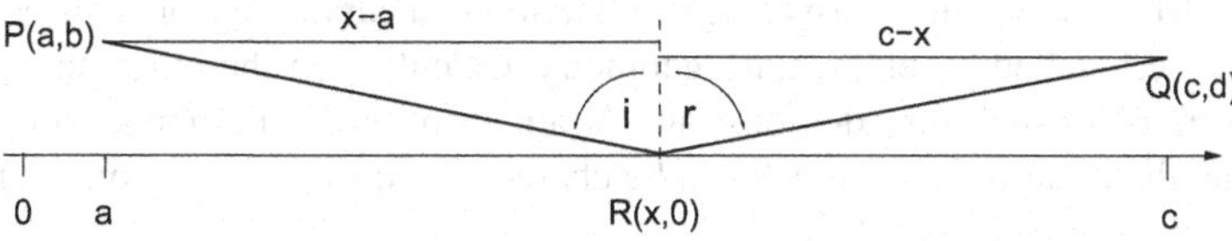

Fig. 5.7 Since the two right triangles are similar, $i = r$

The dashed line is perpendicular to the mirror at the point of reflection. The ratio $\frac{x-a}{\ell_1(x)}$ is the sine of the *angle of incidence*, defined as the angle i formed by the incident ray and the perpendicular to the mirror (Fig. 5.7). Similarly, $\frac{c-x}{\ell_2}$ is the sine of the *angle of reflection*, defined as the angle r formed by the reflected ray and the perpendicular to the mirror. Therefore,

$$\sin i = \sin r.$$

Since these angles are acute, this relation can be expressed by saying that *the angle of incidence equals the angle of reflection*. This is the celebrated *law of reflection*.

We now give a simple geometric derivation of the law of reflection from a plane mirror. See Fig. 5.8. We introduce the *mirror image* P' of the point P, so called because as will be apparent after this argument, it is the point you perceive if your eye is at Q. The mirror is the perpendicular bisector of the interval PP'. Then every point R of the mirror is equidistant from P and P',

$$PR = P'R,$$

so that

$$\ell(x) = PR + RQ = P'R + RQ.$$

The right side of this equation is the sum of two sides of the triangle $P'RQ$. According to the triangle inequality, that sum is at least as great as the third side:

$$\ell(x) \geq P'Q.$$

Equality holds in the triangle inequality only in the special case that the "triangle" is flat, that is, when P', R, and Q lie on a straight line. In that case, we see geometrically that the angle of incidence equals the angle of reflection.

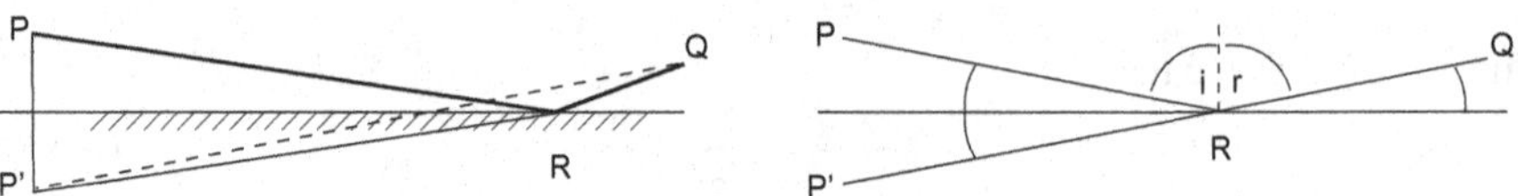

Fig. 5.8 A geometric argument to locate the path of shortest time

Curved Mirrors. We now turn to light reflection in a curved mirror. This case can no longer be handled by elementary geometry. Calculus, on the other hand, gives the answer, as we shall now demonstrate. Again according to Fermat's principle of least time, the point of reflection R can be characterized as that point on the mirror that minimizes the total length

$$\ell = PR + RQ.$$

We introduce Cartesian coordinates with R as the origin and the x-axis tangent to the mirror at R. See Fig. 5.9. In terms of these coordinates, the mirror is described by an equation of the form

$$y = f(x) \qquad \text{such that } f(0) = 0, \quad f'(0) = 0.$$

Denoting as before the coordinates of P by (a,b) and those of Q by (c,d), we see that the sum of the distances $\ell(x) = \ell_1(x) + \ell_2(x)$ from (c,d) to $(x, f(x))$ to (a,b) can be expressed as

$$\ell(x) = \sqrt{(x-a)^2 + (f(x)-b)^2} + \sqrt{(x-c)^2 + (f(x)-d)^2}.$$

We assumed that reflection occurs at $x = 0$. By Fermat's principle, the path from (a,b) to $(0,0)$ to (c,d) takes the least time and hence has the shortest total length ℓ. Therefore, $\ell'(0) = 0$. Differentiating, we get

$$\ell'(x) = \frac{(x-a) + f'(x)(f(x)-b)}{\ell_1(x)} + \frac{(x-c) + f'(x)(f(x)-d)}{\ell_2(x)}.$$

Since $f'(0) = 0$ at R, the value of $\ell'(x)$ at $x = 0$ simplifies to

$$\ell'(0) = \frac{-a}{\ell_1(0)} + \frac{-c}{\ell_2(0)} = \frac{-a}{\sqrt{a^2+b^2}} + \frac{-c}{\sqrt{c^2+d^2}} = 0.$$

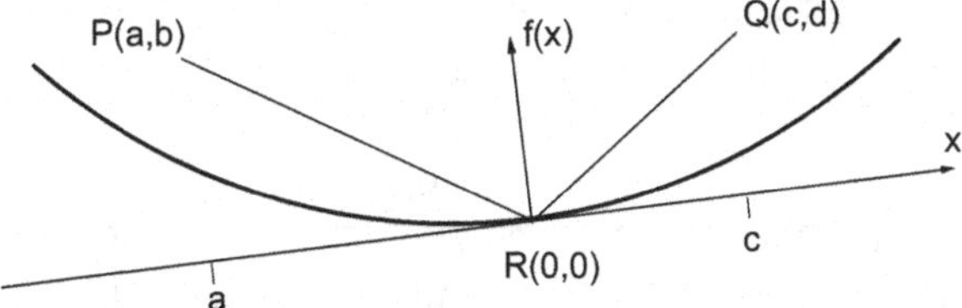

Fig. 5.9 The law of reflection also applies to a curved mirror

This equation agrees with the case of a straight mirror. We conclude as before that *the angle of incidence equals the angle of reflection*, except that in this case, these angles are defined as those formed by the rays with the line perpendicular to the tangent of the mirror at the point of reflection.

If we had not chosen the x-axis to be tangent to the mirror at the point of reflection, we would have had to use a fair amount of trigonometry to deduce the law of reflection from the relation $\ell'(x) = 0$. This shows that in calculus, as well as in analytic geometry, life can be made simpler by a wise choice of coordinate axes.

One difference between reflections from a straight mirror and from a curved mirror is that for a curved mirror, there may well be several points R that furnish a reflection. Only one of these is an absolute minimum, which points to an important modification of Fermat's principle: *Among all possible paths that connect two points P and Q via a mirror, light travels along those paths that take the least time to traverse compared to all nearby paths.* In other words, we do not care which one may be the absolute minimum; light seeks out those paths that are local, not absolute, minima. An observer located at P sees the object located at Q when he looks toward any of these points R; this can be observed in some funhouse mirrors as in Fig. 5.10.

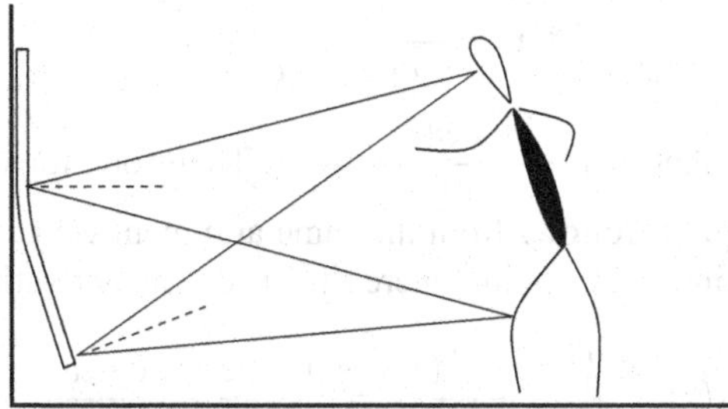

Fig. 5.10 The law of reflection. If the mirror is curved, you might see your knee in two places

Refraction of Light. We shall study the *refraction* of light rays, that is, their passage from one medium into another, in cases in which the propagation speed of light in the two media is different. A common example is the refraction at an air and water interface; see Fig. 5.11.

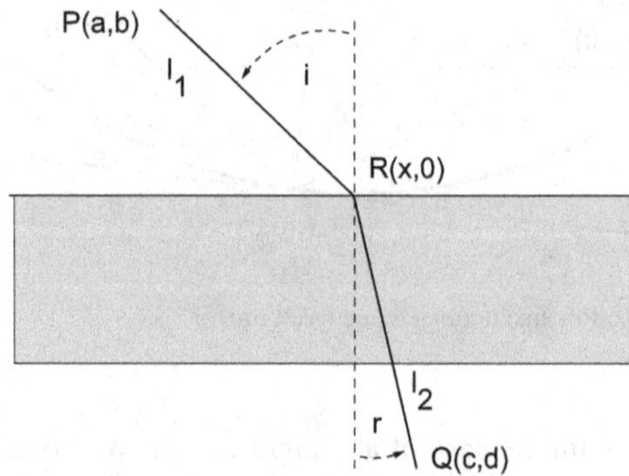

Fig. 5.11 Refraction of light traveling through air above and water below. On a straight path from P to Q, the light would spend more time in the water

We rely as before on Fermat's optical principle: among all possible paths PRQ, light travels along the one that takes the least time to traverse.

Denote by c_1 and c_2 the speed of light in air and water, respectively. The time it takes light to travel from P to R is $\frac{PR}{c_1}$, and from R to Q it takes $\frac{RQ}{c_2}$. The total time t is then

$$t = \frac{PR}{c_1} + \frac{RQ}{c_2}.$$

We introduce the line separating air and water as the x-axis. Denote as before the coordinates of P and of Q by (a,b) and (c,d) respectively, and the coordinate of R by $(x,0)$. Then

$$PR = \ell_1(x) = \sqrt{(x-a)^2 + b^2}, \qquad RQ = \ell_2(x) = \sqrt{(c-x)^2 + d^2},$$

and so

$$t(x) = \frac{\ell_1(x)}{c_1} + \frac{\ell_2(x)}{c_2}.$$

As before, we notice that $t(x) > \frac{|x-a|}{c_1} + \frac{|c-x|}{c_2}$. Therefore, for x large (positive or negative), $t(x)$ is large. It follows from the same argument we gave previously that $t(x)$ achieves its minimum at a point where $t'(x)$ is 0. The derivative is

$$t'(x) = \frac{\ell_1'(x)}{c_1} + \frac{\ell_2'(x)}{c_2} = \frac{x-a}{c_1\ell_1(x)} - \frac{c-x}{c_2\ell_2(x)}.$$

From the relation $t'(x) = 0$, we deduce that

$$\frac{c_2}{c_1}\frac{x-a}{\ell_1(x)} = \frac{c-x}{\ell_2(x)}.$$

As in Eq. (5.33), the ratios $\dfrac{x-a}{\ell_1(x)}$ and $\dfrac{c-x}{\ell_2(x)}$ can be interpreted geometrically (see Fig. 5.11) as the sine of the angle of incidence i and the sine of the angle of refraction r, respectively:

$$\frac{c_2}{c_1}\sin i = \sin r. \tag{5.34}$$

This is the *law of refraction*, named for the Dutch mathematician and astronomer Snell, and is often stated as follows: *when a light ray travels from a medium* 1 *into a medium* 2 *where the propagation speeds are* c_1 *and* c_2*, respectively, it is refracted so that the ratio of the sines of the angle of refraction to the angle of incidence is equal to the ratio* $\dfrac{c_2}{c_1}$ *of the propagation speeds*. The ratio $I = \dfrac{c_2}{c_1}$ is called the *index of refraction*. Since the sine function does not exceed 1, it follows from the law of refraction that $\sin r$ does not exceed the index of refraction I, i.e., from Eq. (5.34),

$$\sin r = I \sin i \le I. \tag{5.35}$$

The speed of light in water is less than that in air: $I = \dfrac{c_2}{c_1} < 1$. It follows from inequality (5.35) that r cannot exceed a critical angle r_{crit} defined by $\sin r_{\text{crit}} = I$. For water and air, the index of refraction is approximately $\dfrac{1}{1.33}$, so the critical angle is $\sin^{-1}\left(\dfrac{1}{1.33}\right) \approx 49°$. This means that an underwater observer who looks in a direction that makes an angle greater than 49° with the perpendicular cannot see points above the water, since such a refracted ray would violate the law of refraction. He sees instead reflections of underwater objects. (See Fig. 5.12.) This phenomenon, well known to snorkelers, is called total internal reflection.

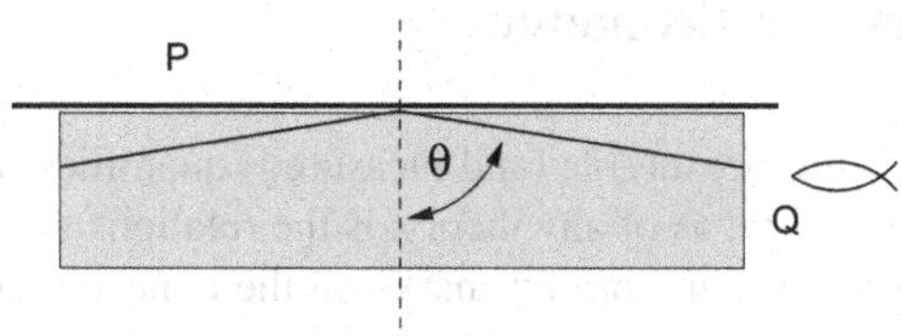

Fig. 5.12 A fish (or snorkeler) looking toward the surface with angle $\theta > 49°$ can't see an object at point P in the air

Problems

5.14. We showed in the derivation of the law of reflection that when $\ell(x)$ is at a minimum, $\dfrac{x-a}{\ell_1(x)} = \dfrac{c-x}{\ell_2(x)}$. Show that if x satisfies this equation, then x must be strictly between a and c.

5.15. The derivation of the law of reflection from Fermat's principle used calculus. Find all the places in the derivation in which knowledge of calculus was used.

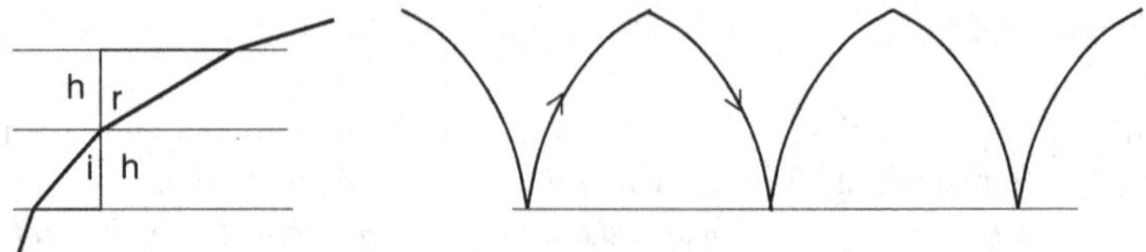

Fig. 5.13 *Left*: a light ray is refracted, in Problem 5.16. *Right*: with many thin layers, the ray may bend

5.16. On the left of Fig. 5.13, a ray has slope m_1 in the lower layer, and m_2 in the upper.

(a) If the light speeds are c_1 and c_2, use Snell's law to show that

$$\frac{c_2}{c_1}\frac{1}{\sqrt{m_1^2+1}} = \frac{1}{\sqrt{m_2^2+1}}.$$

(b) Suggest functions $c(y)$ and $y(x)$ for which $c(y)\sqrt{(y')^2+1}$ is constant, so that the graph of y produces one of the upward paths on the right of the figure. It is drawn to suggest repeated reflections between a mirror at the bottom and total internal reflection at the top.

5.5 Mathematics and Economics

Econometrics deals with measurable (and measured) quantities in economics. The basis of *econometric theory*, as of any theory, is the relations between such quantities. This section contains some brief remarks on the concepts of calculus applied to some of the functions that occur in economic theory.

Fixed and Variable Costs. Denote by $C(q)$ the total cost of producing q units of a certain commodity. Many ingredients make up the total cost; some, like raw materials needed, are *variable* and are dependent on the amount q produced. Others, like investment in a plant, are *fixed* and are independent of q. Now, $C(q)$ can be a rather complicated function of q, but it can be thought of as comprising two basic components, the *variable cost* $C_v(q)$ and the *fixed cost* $C_f(q) = F$, so that

$$C(q) = C_v(q) + F.$$

A manager who is faced with the decision whether to increase production has to know how much the additional production of h units will cost. The cost per additional unit is

$$\frac{C(q+h)-C(q)}{h}.$$

For reasonably small h, this is well approximated by $\frac{dC}{dq}$. This is called the *marginal cost of production*. Another function of interest is the *average cost function*

$$AC(q) = \frac{C(q)}{q} = \frac{C_v(q)+F}{q}.$$

Productivity. Let $G(L)$ be the amount of goods produced by a labor force of size L. A manager, in order to decide whether to hire more workers, wants to know how much additional goods will be produced by h additional laborers. The gain in production per laborer added is

$$\frac{G(L+h)-G(L)}{h}.$$

For reasonably small h, this is well approximated by $\frac{dG}{dL}$. This is called the *marginal productivity of labor*.

Demand. The consumer demand q for a certain product is a function of the price p of the product. The slope of the demand function, called the marginal demand, is the rate at which the demand changes given a change in price, $\frac{dq}{dp}$. The marginal demand is a measure of how responsive consumer demand is to a change in price. As you can see from the definition, the marginal demand depends on the units in which you measure quantity and price. For example, if you measure the quantity of oil in barrels rather than gallons, the marginal demand is $\frac{1}{42}$ as much, since there are 42 gallons to a barrel. Similarly with the price. Change the units from dollars to pesos, and the marginal demand will change depending on the exchange rate. Rather than specify units, economists define the *elasticity of demand*, ε, as

$$\varepsilon = \frac{p}{q}\frac{dq}{dp}.$$

First let us verify that ε is independent of a change in units. Suppose that the price is given as $P = kp$, and the quantity is given as $Q = cq$. Let the demand function be given by $q = f(p)$. Then $Q = cq = cf\left(\frac{P}{k}\right)$. By the chain rule,

$$\frac{dQ}{dP} = \frac{c}{k}f'\left(\frac{P}{k}\right) = \frac{c}{k}\frac{dq}{dp}.$$

It follows that ε is independent of units:

$$\varepsilon = \frac{P}{Q}\frac{dQ}{dP} = \frac{kp}{cq}\frac{dQ}{dP} = \frac{kp}{cq}\frac{c}{k}\frac{dq}{dp} = \frac{p}{q}\frac{dq}{dp}.$$

In Problem 5.18, we ask you to verify, using the chain rule, that an equivalent definition for the elasticity of demand is

$$\varepsilon = \frac{\mathrm{d}\log q}{\mathrm{d}\log p}.$$

Other Marginals. We give two further examples of derivatives in economics.

Example 5.3. Let $P(e)$ be the profit realized after the expense of e dollars. The added profit per dollar where h additional dollars are spent is

$$\frac{P(e+h)-P(e)}{h},$$

well approximated for small h by $\frac{\mathrm{d}P}{\mathrm{d}e}$, called the *marginal profit of expenditure*.

Example 5.4. Let $T(I)$ be the tax imposed on a taxable income I. The increase in tax per dollar on h additional dollars of taxable income is

$$\frac{T(I+h)-T(I)}{h}.$$

For moderate h and some values I, this is well approximated by $\frac{\mathrm{d}T}{\mathrm{d}I}$, called the *marginal rate of taxation*. However, T as prescribed by the tax code is only piecewise differentiable, in that the derivative $\frac{\mathrm{d}T}{\mathrm{d}I}$ fails to exist at certain points.

These examples illustrate two facts:

(a) The rate at which functions change is as interesting in economics, business, and finance as in every other kind of quantitative description.
(b) In economics, the rate of change of a function $y(x)$ is not called the derivative of y with respect to x but the marginal y of x.

Here are some examples of the uses to which the notion of derivative can be put in economic thinking. The managers of a firm would likely not hire additional workers when the going rate of pay exceeds the marginal productivity of labor, for the firm would lose money. Thus, declining productivity places a limitation on the size of a firm.

Actually, one can argue persuasively that efficiently run firms will stop hiring even before the situation indicated above is reached. The most efficient mode for a firm is one in which the cost of producing a unit of commodity is minimal. The cost of a unit commodity is

$$\frac{C(q)}{q}.$$

The derivative must vanish at the point where this is minimal. By the quotient rule, we get

$$\frac{q\frac{dC}{dq} - C(q)}{q^2} = 0,$$

which implies that at the point q_{max} of maximum efficiency,

$$\frac{dC}{dq}(q_{max}) = \frac{C(q_{max})}{q_{max}}. \tag{5.36}$$

In words: *At the peak of efficiency, the marginal cost of production equals the average cost of production.* The firm would still make more money by expanding production, but would not be as efficient as before, and so its relative position would be weakened.

Example 5.5. Let us see how this works in the simple case that the cost function is $C(q) = q^2 + 1$. Then the variable cost is $C_v(q) = q^2$, and the fixed cost is $C_f(q) = 1$. The average variable cost is $AC_v(q) = \frac{q^2}{q} = q$, the average fixed cost is $AC_f(q) = \frac{1}{q}$, the average cost is $AC(q) = \frac{q^2+1}{q} = q + \frac{1}{q}$, and the marginal cost is $C'(q) = 2q$. The average cost reaches its minimum when the average cost equals the marginal cost,

$$q + \frac{1}{q} = 2q,$$

which gives $q = 1$.

Equation (5.36) has the following geometric interpretation: The line connecting $(q_{max}, C(q_{max}))$ to the origin is tangent to the graph of C at $(q_{max}, C(q_{max}))$. Such a point does not exist for all functions, but does exist for functions $C(q)$ for which $\frac{C(q)}{q}$ tends to infinity as q tends to infinity. It has been remarked that the nonmonopolistic capitalistic system is possible precisely because the cost functions in capitalistic production have this property. We ask you for an example of this property in Problem 5.19.

We conclude this brief section by pointing out that to be realistic, economic theory has to take into account the enormous diversity and interdependence of economic activities. Any halfway useful models deal typically with functions of very many variables. These functions are not derived a priori from detailed theoretical considerations, but are empirically determined. For this reason, these functions are usually taken to be of very simple form, linear or quadratic; the coefficients are then determined by making a best fit to observed data. The fit that can be obtained in this way is as good as could be obtained by taking functions of more complicated forms. Therefore, there is no incentive or justification to consider more complicated functions. The mathematical theory of economics makes substantial use of statistical techniques for fitting linear and quadratic functions of many variables to recorded data and with maximizing or minimizing such functions when variables are subject to realistic restrictions.

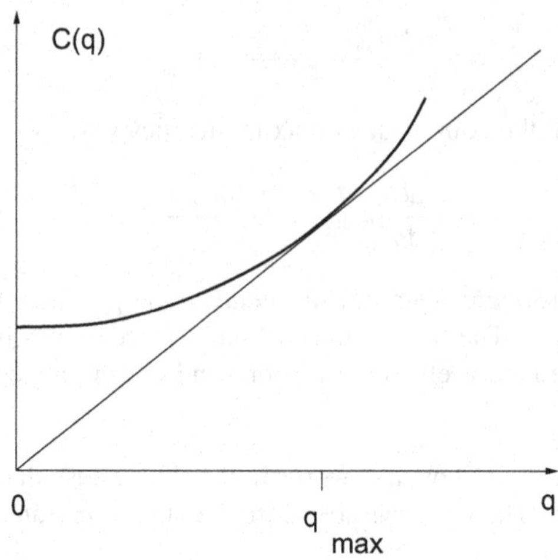

Fig. 5.14 Tangency at the peak of efficiency

Problems

5.17. Suppose you have two plants that have two different cost functions C_1 and C_2. You want to produce a total of Q units.

(a) Explain (including the meaning of q) why the total cost function C can be expressed as $C(q) = C_1(q) + C_2(Q-q)$.
(b) Show that the optimal division of production occurs when the marginal cost of production at plant 1 is *equal* to the marginal cost of production at plant 2.
(c) Suppose both costs are quadratic, $C_1(q) = aq^2$, $C_2(q) = bq^2$. Sketch a graph of C. If plant 2 is 20 % more expensive than plant 1, show that about 55 % of production should be done at plant 1.

If the costs were not equal, it would pay to switch some production from one plant to the other!

5.18. Elasticity of demand is defined as $\varepsilon = \dfrac{p}{q}\dfrac{dq}{dp}$. Use the chain rule to argue that an equivalent definition is $\varepsilon = \dfrac{d\log q}{d\log p}$.

5.19. Consider a cost function $C(q) = aq^k + b$, where a and b are positive. For what values of k is there a most efficient production level as illustrated in Fig. 5.14?

Chapter 6
Integration

Abstract The total amount of some quantity is an important and useful concept. We introduce the concept of the integral, the precise mathematical expression for total amount. The fundamental theorem of calculus tells us how the total amount is related to the rate at which that amount accumulates.

6.1 Examples of Integrals

We introduce the concept of the integral using three motivating examples of total amount: distance, mass, and area.

6.1a Determining Mileage from a Speedometer

At the beginning of Chap. 3, we investigated the relation between a car's odometer and speedometer. We showed that if the speedometer were broken, it would still be possible to determine the speed of the moving car from readings of the odometer and a clock. Now we investigate the inverse problem: how to determine the total mileage if we are given the speedometer readings at various times. We assume that we have at our disposal the *total record* of speedometer readings throughout the trip, i.e., that we know the speed of the car, f, as a function of time.

In the graph of f shown in Fig. 6.1, t is measured in hours, speed in miles per hour (mph). Our problem can be formulated as follows: *Given speed as function f of time, determine the distance covered during the time interval $[a,b]$.*

Let us denote the distance covered by

$$D(f,[a,b]).$$

P.D. Lax and M.S. Terrell, *Calculus With Applications*, Undergraduate Texts in Mathematics, 245
DOI 10.1007/978-1-4614-7946-8_6,

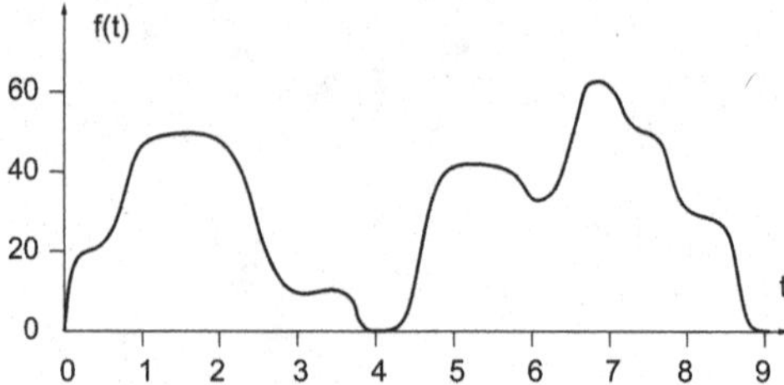

Fig. 6.1 Speed of a car

The notation emphasizes that D depends on f and $[a,b]$. How does D depend on $[a,b]$? Suppose we divide $[a,b]$ into two subintervals $[a,c]$ and $[c,b]$ that cover $[a,b]$ and do not overlap (Fig. 6.2).

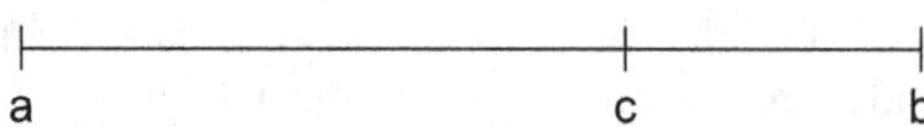

Fig. 6.2 The number c subdivides the interval

The distance covered during the total time interval $[a,b]$ is the sum of the distances covered during the intervals $[a,c]$ and $[c,b]$. This property is called the additivity property.

Additivity Property. For every c between a and b,

$$D(f,[a,b]) = D(f,[a,c]) + D(f,[c,b]). \tag{6.1}$$

How does distance D depend on the speed f? The distance covered by a car traveling with constant speed is

$$\text{distance} = (\text{speed})(\text{time}).$$

Suppose that between time a and b, the speed, f, is between m and M:

$$m \le f(t) \le M.$$

Two cars, one traveling with speed m the other with speed M, would cover the distances $m(b-a)$ and $M(b-a)$ during the time interval $[a,b]$. Our car travels a distance between these two. This property is called the lower and upper bound property.

Lower and Upper Bound Property. If $m \le f(t) \le M$ when $a \le t \le b$, then

$$m(b-a) \le D(f,[a,b]) \le M(b-a).$$

To make this example a bit more concrete, we use the data in Fig. 6.1 to find various lower and upper bounds for the distance traveled between $t = 2$ and $t = 7$.

From the graph, we see that the minimum speed during the interval was 0 mph and the maximum speed was 60 mph:

$$0 \le f(t) \le 60.$$

The time interval has length $7-2=5$ h. So we conclude that the distance traveled was between 0 and 300 miles,

$$0=(0)(5) \le D(f,[2,7]) \le (60)(5)=300.$$

This is not a very impressive estimate for the distance traveled. Let us see how to do better using additivity. We know that the time interval $[2,7]$ can be subdivided into two parts, $[2,5]$ and $[5,7]$, and that the total distance traveled is the sum of the distances traveled over these two shorter segments of the trip:

$$D(f,[2,7])=D(f,[2,5])+D(f,[5,7]).$$

The speed on $[2,5]$ is between 0 and 50 mph,

$$0 \le f(t) \le 50,$$

and the speed on $[5,7]$ is between 30 and 60 mph,

$$30 \le f(t) \le 60.$$

The length of $[2,5]$ is $5-2=3$, and the length of $[5,7]$ is $7-5=2$. The lower and upper bound property applied to each subinterval gives

$$(0)(3) \le D(f,[2,5]) \le (50)(3) \text{ and } (30)(2) \le D(f,[5,7]) \le (60)(2).$$

Adding these two inequalities together, we see that

$$60 \le D(f,[2,5])+D(f,[5,7]) \le 270 \quad \text{miles.}$$

Now recalling that $D(f,[2,7])=D(f,[2,5])+D(f,[5,7])$, we get

$$60 \le D(f,[2,7]) \le 270 \quad \text{miles,}$$

a better estimate for the distance traveled.

6.1b Mass of a Rod

Picture a rod of variable density along the x-axis, as in Fig. 6.3. Denote the density at position x by $f(x)$ in grams per centimeter. Let $R(f,[a,b])$ denote the mass of the portion of the rod between points a and b of the x axis-measured in centimeters.

How does R depend on $[a,b]$? If we divide the rod into two smaller pieces lying on $[a,c]$ and $[c,b]$, then the mass of the whole rod is the sum of the mass of the pieces,

$$R(f,[a,b]) = R(f,[a,c]) + R(f,[c,b]).$$

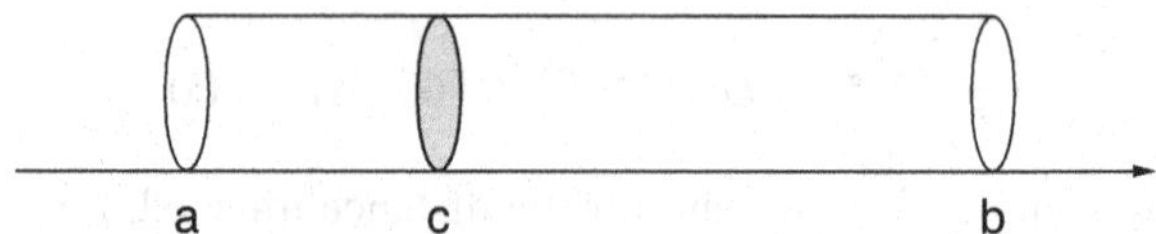

Fig. 6.3 A rod lying on a line between a and b

This property is the additivity property that we encountered in the distance example above.

How does the mass depend on f? If the density f is constant, then the mass is given by

$$\text{mass} = (\text{density})(\text{length}).$$

But our rod has variable density. If m and M are the minimum and maximum densities of the rod between a and b,

$$m \le f(x) \le M,$$

then the mass R of the $[a,b]$ portion of the rod is at least the minimum density times the length of the rod and is not more than the maximum density times the length of the rod:

$$m(b-a) \le R(f,[a,b]) \le M(b-a).$$

This is the lower and upper bound property that we encountered in the distance example.

Let us use the properties of additivity and lower and upper bounds to obtain various estimates for the mass of a particular rod. Suppose the rod lies along the x-axis between 1 and 5 cm and that its density at x is $f(x) = x$ grams per centimeter.

The greatest density occurs at $x = 5$ and is $f(5) = 5$, and the least density occurs at $x = 1$ and is $f(1) = 1$. The length of the rod is $5 - 1 = 4$ cm. So we conclude that

$$4 = (1)(4) \le R(f,[1,5]) \le (5)(4) = 20.$$

We can improve this estimate by subdividing the rod and using the properties of additivity and lower and upper bounds. This time, let us subdivide the rod into three shorter pieces as shown in Fig. 6.4.

By the additivity property, the mass of the rod between 1 and 5 is equal to the sum of the mass between 1 and 3 and the mass between 3 and 5. If we subdivide the

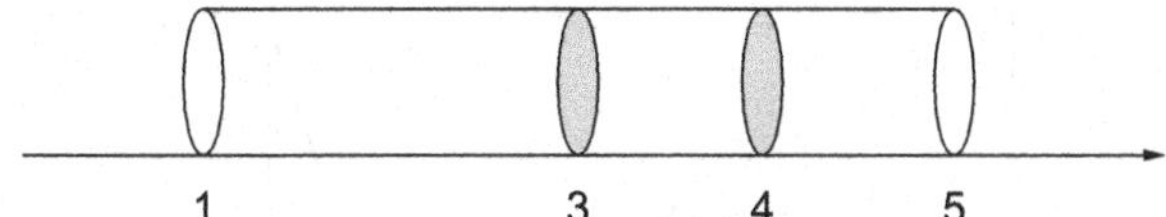

Fig. 6.4 A rod in three parts

interval $[3,5]$ at 4, then we see by additivity again that the mass between 3 and 5 is the sum of the mass between 3 and 4 and the mass between 4 and 5:

$$\begin{aligned} R(f,[1,5]) &= R(f,[1,3]) + R(f,[3,5]) \\ &= R(f,[1,3]) + R(f,[3,4]) + R(f,[4,5]). \end{aligned}$$

Next, we apply the lower and upper bound property to each segment to estimate their masses. The lower bounds for the density f on the three intervals are 1, 3, and 4, and the upper bounds are 3, 4, and 5, respectively. The lengths of the intervals are 2, 1, 1. By the lower and upper bound property, the masses of the three segments are

$$\begin{aligned} (1)(2) \le R(f,[1,3]) \le (3)(2), \\ (3)(1) \le R(f,[3,4]) \le (4)(1), \\ (4)(1) \le R(f,[4,5]) \le (5)(1). \end{aligned}$$

Now adding these three inequalities, we obtain

$$9 \le R(f,[1,3]) + R(f,[3,4]) + R(f,[4,5]) \le 15.$$

Recalling that mass has the additivity property

$$R(f,[1,5]) = R(f,[1,3]) + R(f,[3,4]) + R(f,[4,5]),$$

we obtain

$$9 \le R(f,[1,5]) \le 15,$$

a better estimate for the mass.

6.1c Area Below a Positive Graph

Let f be a function whose graph is shown in Fig. 6.5. We wish to calculate the area of the region contained between the graph of f, the x-axis, and the lines $x = a$ and $x = b$. Denote this area by

$$A(f,[a,b]).$$

How does A depend on $[a,b]$? For any c between a and b, subdivide $[a,b]$ into two subintervals $[a,c]$ and $[c,b]$, as on the left in Fig. 6.6. This subdivides the region

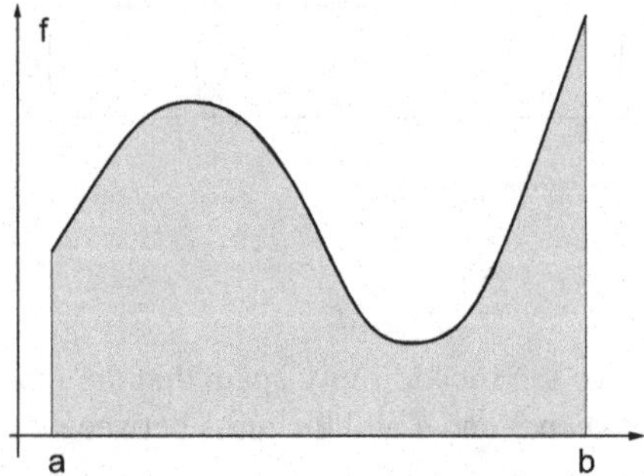

Fig. 6.5 The area of the shaded region is $A(f,[a,b])$

into two nonoverlapping regions. The area of the original region is the sum of the areas of the components. So A has the additive property,

$$A(f,[a,b]) = A(f,[a,c]) + A(f,[c,b]).$$

How does A depend on f? From the graph on the right in Fig. 6.6, we see that the values f takes on in $[a,b]$ lie between m and M:

$$m \le f(x) \le M \quad \text{for } x \text{ in } [a,b].$$

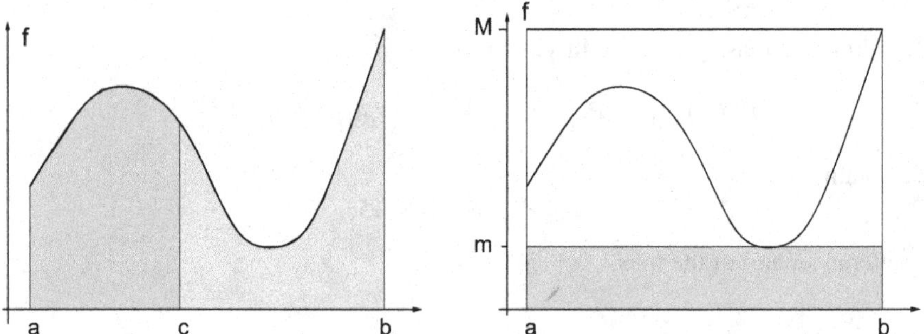

Fig. 6.6 *Left*: The interval subdivided. *Right*: Rectangles with heights m and M

Then, as Fig. 6.6 indicates, the region in question *contains* the rectangle with base $[a,b]$ and height m, and *is contained in* the rectangle with base $[a,b]$ and height M. Therefore, we conclude that

$$m(b-a) \le A(f,[a,b]) \le M(b-a). \tag{6.2}$$

That is, the lower and upper bound property holds.

Now just as we did in the examples of distance and mass, we look at a specific example. We estimate the area of the region bounded by the graph of $f(x) = x^2 + 1$, the x-axis, and the lines $x = -1$ and $x = 2$, as shown in Fig. 6.7. On $[-1,2]$, f is between 1 and 5, and so

$$3 = (1)(3) \leq A(x^2+1,[-1,2]) \leq (5)(3) = 15.$$

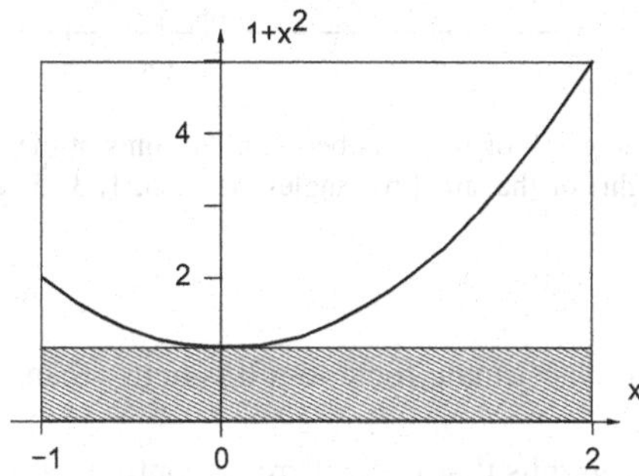

Fig. 6.7 The area below the graph of x^2+1 is between the areas of the smaller and larger rectangles. The height of the small rectangle is 1. The height of the large rectangle is 5

Since A is additive over intervals, we see that if we subdivide $[-1,2]$ into three intervals $[-1,-0.5]$, $[-0.5,1.5]$, and $[1.5,2]$, we get

$$A(x^2+1,[-1,2]) = A(x^2+1,[-1,-0.5]) + A(x^2+1,[-0.5,1.5]) + A(x^2+1,[1.5,2]).$$

Now on each of these subintervals, $f(x) = x^2 + 1$ takes on minimum and maximum values (see Fig. 6.8). So on $[-1,-0.5]$,

$$(1.25)(0.5) \leq A(x^2+1,[-1,-0.5]) \leq (2)(0.5).$$

On $[-.5,1.5]$,

$$(1)(2) \leq A(x^2+1,[-0.5,1.5]) \leq (3.25)(2);$$

and on $[1.5,2]$,

$$(3.25)(0.5) \leq A(x^2+1,[1.5,2]) \leq (5)(0.5).$$

Putting this all together, we get

$$\begin{aligned}&(1.25)(0.5) + (1)(2) + (3.25)(0.5)\\ &\leq A(x^2+1,[-1,2]) \leq (2)(0.5) + (3.25)(2) + (5)(0.5),\end{aligned}$$

or

$$4.25 \leq A(x^2+1,[-1,2]) \leq 10.$$

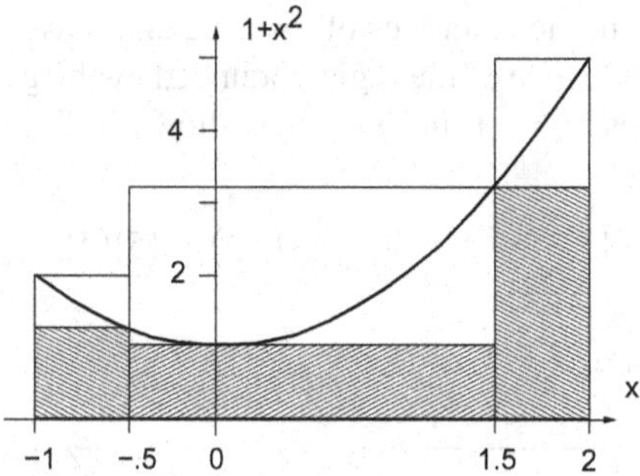

Fig. 6.8 The area below the graph of x^2+1 is between the sums of the areas of the smaller and larger rectangles. The heights of the small rectangles are 1.25, 1, 3.25. The heights of the large rectangles are 2, 3.25, 5

This is not a very accurate estimate for A, but it is better than our first estimate that A is between 3 and 15.

We can see from the graphs that if we were to continue to subdivide each of the intervals, the resulting estimates for A would become more accurate.

6.1d Negative Functions and Net Amount

So far, our function f has been positive, and in the case of density, that remains true. But signed distance D and area A also make sense for functions f which take on negative values.

The notion of positive and negative position of a car along a road can be defined in the same way as positive and negative numbers are defined on the number line: the starting point divides the road into two parts, one of which is arbitrarily labeled positive. Positions on the positive side are assigned a positive distance from the starting point, while positions on the negative side are assigned the negative of the distance from the starting point.

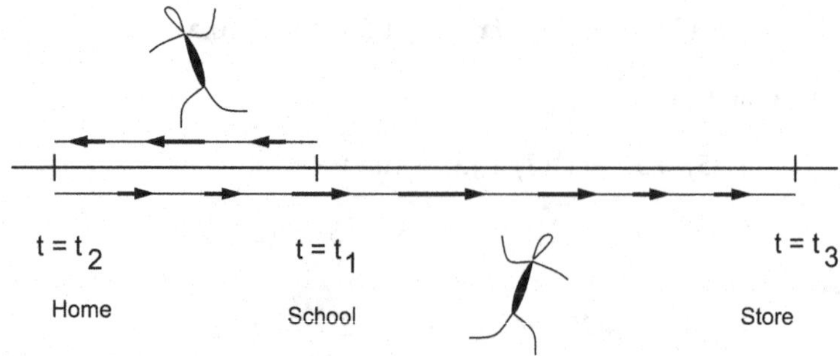

Fig. 6.9 A trip from school to home, then to the store

Velocity is defined as the derivative with respect to t of the position function as just defined. The change in position (or signed distance) traveled between two points in time is the ending position minus the starting position. On intervals over which velocity is always negative, the change in position (or signed distance) is negative. By arguments that are identical to those for positive velocity, we can see that signed distance D has both the additivity and the lower and upper bound properties. On intervals over which the velocity is at times positive and at times negative, we can see by additivity that D is the *net* distance: the sum of the signed distances traveled from the starting point. Fig. 6.9 shows an example.

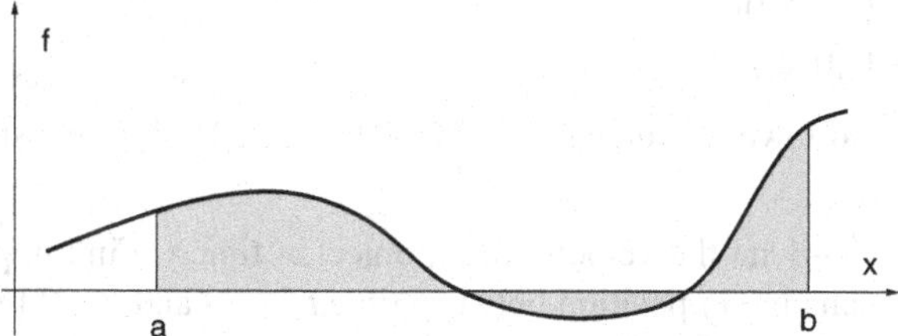

Fig. 6.10 A function f that has both positive and negative values

What should the interpretation of the area $A(f,[a,b])$ be when f takes on negative values, as pictured in Fig. 6.10? We propose to interpret the area above the x-axis as positive, and the area below the x-axis as negative, with the result that A will be defined as the algebraic sum of these positive and negative quantities. There is a reason for this interpretation: in many applications of area, the "underground" positions, i.e., those below the x-axis, have to be interpreted in a sense opposite to the portions aboveground. Only with this interpretation of positive and negative area is the lower and upper bound property (6.2) valid. This is particularly clear on intervals where f is negative.

We have shown that all three of the quantities distance D, rod mass R, and area A have the additive property with respect to $[a,b]$, and the lower and upper bound property with respect to f. We shall show in the next section that these two properties completely characterize D, R, and A. To put it sensationally, if you knew no more about D, R, and A than what you have learned so far, and if you were transported to a desert island, equipped only with pencil and paper, you could calculate the values of D, R, and A for any continuous function f on any interval $[a,b]$. The next section is devoted to explaining how.

Problems

6.1. Find a better estimate for the mass of the rod $R(x,[1,5])$ discussed in Sect. 6.1b, by

(a) Subdividing the rod into four subpieces of equal length,
(b) Subdividing the rod into eight subpieces of equal length.

6.2. Find better upper and lower estimates for the area $A(x^2+1,[-1,2])$ discussed in Sect. 6.1c, by subdividing $[-1,2]$ into six subintervals of equal length.

6.3. In this problem we explore signed area, or net area above the x-axis.

(a) Sketch the graph of $f(x) = x^2 - 1$ on the interval $[-3,2]$.

(b) Let $A(x^2-1,[a,b])$ be the signed area as described in Sect. 6.1d. Which of the following areas are clearly positive, clearly negative, or difficult to determine without some computation:

(i) $A(x^2-1,[-3,-2])$
(ii) $A(x^2-1,[-2,0])$
(iii) $A(x^2-1,[-1,0])$
(iv) $A(x^2-1,[0,2])$

(c) Find upper and lower estimates for $A(x^2-1,[-3,2])$ using five subintervals of equal length.

6.4. Let $f(t) = t^2 - 1$ be the velocity of an object at time t. Find upper and lower estimates for the change in position between time $t = -3$ and $t = 2$ by subdividing $[-3,2]$ into five subintervals of equal length.

6.2 The Integral

We have seen that all three quantities distance $D(f,[a,b])$, rod mass $R(f,[a,b])$, and area $A(f,[a,b])$ are additive with respect to the given interval and have the lower and upper bound property with respect to f. In this section, we show that using only these two properties, we can calculate D, R, and A with as great an accuracy as desired.

In other words, if f and $[a,b]$ are the same in each of the three applications, then the numbers $D(f,[a,b])$, $R(f,[a,b])$, and $A(f,[a,b])$ have the same value, even though D, R, and A have entirely different physical and geometric interpretations. Anticipating this result, we call this number *the integral of f over $[a,b]$* and denote it by

$$I(f,[a,b]).$$

The usual notation for the integral is

$$I(f,[a,b]) = \int_a^b f(t)\,\mathrm{d}t.$$

Example 6.1. The area $A(t,[0,b])$ is shown as the area of the large triangle in Fig. 6.11. In the integral notation it is written

$$\int_0^b t\,\mathrm{d}t,$$

and since it represents the area of a triangle of base b and height b, the value is $\frac{1}{2}b^2$. According to the additive property, if $0 < a < b$, then

$$\int_0^b t\,\mathrm{d}t = \int_0^a t\,\mathrm{d}t + \int_a^b t\,\mathrm{d}t.$$

We find by subtracting that

$$\int_a^b t\,\mathrm{d}t = \frac{b^2 - a^2}{2}.$$

This is the area of the shaded trapezoid in the figure.

We use the notation $I(f, [a,b])$ when we are developing the concept of the integral to emphasize that it is an operation whose *inputs* are *a function and an interval* and whose *output* is a *number*. The basic properties of the integral are (a) additivity with respect to the interval of integration, and (b) the lower and upper bound property with respect to the function being integrated.

Next we give an example to show how we can compute an integral using only the two basic properties.

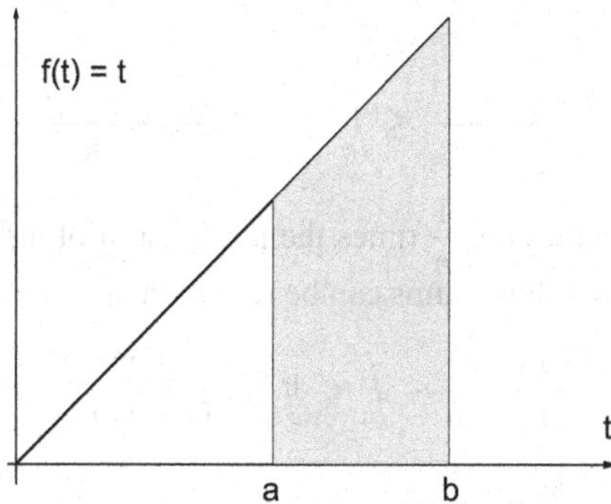

Fig. 6.11 The areas calculated in Example 6.1

The Integral of e^t on $[0,1]$: $\int_0^1 \mathrm{e}^t\,\mathrm{d}t$.

Let us look at what happens when we divide $[0,1]$ into three equal subintervals

$$0 < \frac{1}{3} < \frac{2}{3} < 1.$$

Since e^t is increasing, its lower bounds on these subintervals are

$$\mathrm{e}^0,\ \mathrm{e}^{1/3},\ \mathrm{e}^{2/3}.$$

Set $r = \mathrm{e}^{1/3}$. Then the lower bounds are $1,\ r,\ r^2$, and the upper bounds are $r,\ r^2,\ r^3$. The lower and upper bound property gives

$$(1)\left(\frac{1}{3}\right) \le \int_0^{1/3} \mathrm{e}^t \, \mathrm{d}t \le (r)\left(\frac{1}{3}\right),$$

$$(r)\left(\frac{1}{3}\right) \le \int_{1/3}^{2/3} \mathrm{e}^t \, \mathrm{d}t \le (r^2)\left(\frac{1}{3}\right),$$

$$(r^2)\left(\frac{1}{3}\right) \le \int_{2/3}^{1} \mathrm{e}^t \, \mathrm{d}t \le (r^3)\left(\frac{1}{3}\right).$$

Add these inequalities to get

$$\frac{1+r+r^2}{3} \le \int_0^{1/3} \mathrm{e}^t \, \mathrm{d}t + \int_{1/3}^{2/3} \mathrm{e}^t \, \mathrm{d}t + \int_{2/3}^{1} \mathrm{e}^t \, \mathrm{d}t \le \frac{r+r^2+r^3}{3}.$$

Then additivity gives

$$\frac{1+r+r^2}{3} \le \int_0^{1} \mathrm{e}^t \, \mathrm{d}t \le \frac{r+r^2+r^3}{3}.$$

Similarly, if the unit interval is divided into n equal parts, each of length $\frac{1}{n}$, we set $r = \mathrm{e}^{1/n}$. We get

$$\frac{1+r+\cdots+r^{n-1}}{n} \le \int_0^{1} \mathrm{e}^t \, \mathrm{d}t \le \frac{r+r^2+\cdots+r^n}{n}.$$

Both the left and right sides are $\frac{1}{n}$ times the partial sum of a geometric series (see Sect. 2.6a). We know that these sums can be rewritten as

$$\frac{1-r^n}{(1-r)n} \le \int_0^{1} \mathrm{e}^t \, \mathrm{d}t \le r\frac{1-r^n}{(1-r)n}. \tag{6.3}$$

Take $h = \frac{1}{n}$. Since $r = \mathrm{e}^{1/n}$, we have

$$\frac{1-r^n}{(1-r)n} = \frac{\mathrm{e}-1}{\frac{\mathrm{e}^h-1}{h}}.$$

As n tends to infinity, h tends to 0, and the limit $\lim_{h\to 0} \frac{\mathrm{e}^h-1}{h}$ is the derivative of e^x at $x = 0$, which equals 1. Therefore, in the limit as n tends to infinity, inequality (6.3) becomes $\mathrm{e}-1 \le \int_0^1 \mathrm{e}^t \, \mathrm{d}t \le \mathrm{e}-1$, and so

$$\int_0^1 \mathrm{e}^t \, \mathrm{d}t = \mathrm{e}-1. \tag{6.4}$$

6.2a The Approximation of Integrals

We show now how to determine the integral of any continuous function over any closed interval, using only the following two basic properties:

$$\text{If } a < c < b \text{ then } \int_a^b f(t)\,dt = \int_a^c f(t)\,dt + \int_c^b f(t)\,dt \tag{6.5}$$

and

$$\text{If } m \le f(t) \le M \text{ then } m(b-a) \le \int_a^b f(t)\,dt \le M(b-a). \tag{6.6}$$

By the additivity property in Eq. (6.5), we know that if

$$a < a_1 < b,$$

then

$$\int_a^b f(t)\,dt = \int_a^{a_1} f(t)\,dt + \int_{a_1}^b f(t)\,dt.$$

Similarly, divide $[a,b]$ into three intervals

$$a < a_1 < a_2 < b.$$

Then applying the additivity property twice, we get

$$\int_a^b f(t)\,dt = \int_a^{a_1} f(t)\,dt + \int_{a_1}^b f(t)\,dt \quad \text{and} \quad \int_{a_1}^b f(t)\,dt = \int_{a_1}^{a_2} f(t)\,dt + \int_{a_2}^b f(t)\,dt.$$

Therefore,

$$\int_a^b f(t)\,dt = \int_a^{a_1} f(t)\,dt + \int_{a_1}^{a_2} f(t)\,dt + \int_{a_2}^b f(t)\,dt.$$

Generally, if we divide $[a,b]$ into n intervals by

$$a < a_1 < a_2 < \cdots < a_{n-1} < b,$$

we find that repeated application of the additivity property gives

$$\int_a^b f(t)\,dt = \int_a^{a_1} f(t)\,dt + \int_{a_1}^{a_2} f(t)\,dt + \cdots + \int_{a_{n-1}}^b f(t)\,dt. \tag{6.7}$$

Set $a_0 = a$ and $a_n = b$. Since f is continuous on $[a,b]$, we know by the extreme value theorem that f has a minimum m and maximum M on $[a,b]$, and a minimum m_i and a maximum M_i on each of the subintervals, with $m \le m_i$ and $M_i \le M$. We can estimate each of the integrals on the right-hand side of Eq. (6.7) by the lower and upper bound property:

$$m_i(a_i - a_{i-1}) \le \int_{a_{i-1}}^{a_i} f(t)\,dt \le M_i(a_i - a_{i-1}).$$

By summing these estimates and using additivity, we get

$$\begin{aligned}&m_1(a_1-a_0)+m_2(a_2-a_1)+\cdots+m_n(a_n-a_{n-1})\\&\le\int_a^b f(t)\,\mathrm{d}t\le M_1(a_1-a_0)+M_2(a_2-a_1)+\cdots+M_n(a_n-a_{n-1}).\end{aligned}\tag{6.8}$$

Since each of the M_i is less than or equal to M, and $m \le m_i$, and since the sum of the lengths of the subintervals is the length $b-a$ of the entire interval, we see that inequality (6.8) is an improvement on the original estimate $m(b-a) \le \int_a^b f(t)\,\mathrm{d}t \le M(b-a)$. We saw examples of this kind of improvement in the distance, mass, and area examples.

Having a better estimate is good, but we would like to know that we can compute $\int_a^b f(t)\,\mathrm{d}t$ within any tolerance, no matter how small, through repeated uses of additivity and the lower and upper bound properties alone. We can achieve the tolerance we desire by making the difference between the left-hand side of Eq. (6.8), called the *lower sum*, and the right-hand side of Eq. (6.8), called the *upper sum*, as small as desired. For example, if the difference between them were less than $\frac{1}{1000}$, then we would know $\int_a^b f(t)\,\mathrm{d}t$ within that tolerance.

We recall now that every continuous function f on a closed interval is uniformly continuous. That is, given any tolerance $\varepsilon > 0$, there is a precision $\delta > 0$ such that if two points c and d in $[a,b]$ are closer than δ, then $f(c)$ and $f(d)$ differ by less than ε. So if we break up the interval $[a,b]$ into pieces $[a_{i-1},a_i]$ with length less than δ, the minimum m_i and the maximum M_i of f on $[a_{i-1},a_i]$ will differ by less than ε. The left and right sides of Eq. (6.8) differ by

$$(M_1-m_1)(a_1-a_0)+(M_2-m_2)(a_2-a_1)+\cdots+(M_n-m_n)(a_n-a_{n-1}),$$

which is less than

$$\varepsilon(a_1-a_0)+\varepsilon(a_2-a_1)+\cdots+\varepsilon(a_n-a_{n-1})=\varepsilon(a_n-a_0)=\varepsilon(b-a).$$

This shows that for a sufficiently fine subdivision of the interval $[a,b]$, the upper sum and the lower sum in Eq. (6.8) differ by less than $\varepsilon(b-a)$, so by Eq. (6.8), the upper and lower sums differ from $\int_a^b f(t)\,\mathrm{d}t$ by less than $\varepsilon(b-a)$. Since $b-a$ is fixed, we can choose ε so that $\varepsilon(b-a)$ is as small as we like.

So far, our descriptions of how to determine the integral of a continuous function on $[a,b]$ have relied heavily on finding the maximum and minimum on each subinterval. In general, it is not easy to find the absolute maximum and minimum values of a continuous function on a closed interval, even though we know they exist. We now give estimates of the integral of a function that are much easier to evaluate than the upper and lower bounds in Eq. (6.8), and that also approximate $\int_a^b f(t)\,\mathrm{d}t$ within any tolerance.

Definition 6.1. Choose any point t_i in the interval $[a_{i-1}, a_i]$, $i = 1, 2, \ldots, n$, and form the sum

$$I_{\text{approx}}(f, [a,b]) = f(t_1)(a_1 - a_0) + f(t_2)(a_2 - a_1) + \cdots + f(t_n)(a_n - a_{n-1}). \tag{6.9}$$

The sum I_{approx} is called an *approximate integral* of f on $[a,b]$ or a Riemann sum of f on $[a,b]$.

Example 6.2. To find an approximate integral of $f(t) = \sqrt{t}$ on the interval $[1,2]$ using the subdivision

$$1 < 1.3 < 1.5 < 2,$$

we need to choose a number from each subinterval. Let us choose them in such a way that their square roots are easy to calculate:

$$t_1 = 1.21 = (1.1)^2, \quad t_2 = 1.44 = (1.2)^2, \quad t_3 = 1.69 = (1.3)^2.$$

Then

$$\begin{aligned} I_{\text{approx}}(\sqrt{t}, [1,2]) &= \sqrt{t_1}(1.3 - 1) + \sqrt{t_2}(1.5 - 1.3) + \sqrt{t_3}(2 - 1.5) \\ &= (1.1)(0.3) + (1.2)(0.2) + (1.3)(0.5) = 1.22. \end{aligned}$$

Approximate integrals are easy to compute, but how close are they to $\int_a^b f(t)\,dt$? The value $f(t_i)$ lies between the minimum m_i and the maximum M_i of f on $[a_{i-1}, a_i]$. Therefore, $I_{\text{approx}}(f, [a,b])$ lies in the interval between

$$m_1(a_1 - a_0) + m_2(a_2 - a_1) + \cdots + m_n(a_n - a_{n-1})$$

and

$$M_1(a_1 - a_0) + M_2(a_2 - a_1) + \cdots + M_n(a_n - a_{n-1}).$$

That is, no matter how the t_i are chosen in each interval, each approximate integral lies within the same interval that contains $\int_a^b f(t)\,dt$. We saw that for continuous functions we can subdivide $[a,b]$ into subintervals such that the difference between the lower sum and the upper sum is less than $\varepsilon(b-a)$. Therefore, the exact and an approximate integral differ by an amount not greater than $\varepsilon(b-a)$. We state this result as the approximation theorem.

Theorem 6.1. Approximation theorem for the integral. *Suppose that for a continuous function f on $[a,b]$, $|f(c)-f(d)|<\varepsilon$ whenever $|c-d|<\delta$. Take a subdivision*

$$a = a_0 < a_1 < a_2 < \cdots < a_{n-1} < a_n = b, \tag{6.10}$$

where each length $(a_i - a_{i-1})$ is less than δ. Then every approximate integral

$$f(t_1)(a_1 - a_0) + \cdots + f(t_n)(a_n - a_{n-1})$$

differs from the exact integral $\int_a^b f(t)\,\mathrm{d}t$ by less than $\varepsilon(b-a)$.

Sometimes, the length $(a_i - a_{i-1})$ of the ith interval in (6.10) is denoted by $\mathrm{d}t_i$, so that the approximating sums are written as

$$I_{\text{approx}} = f(t_1)(a_1 - a_0) + \cdots + f(t_n)(a_n - a_{n-1}) = f(t_1)\mathrm{d}t_1 + \cdots + f(t_n)\mathrm{d}t_n.$$

One term of this sum is illustrated in Fig. 6.12. If we use the sum symbol, this sum may be abbreviated as

$$I_{\text{approx}}(f,[a,b]) = \sum_{i=1}^{n} f(t_i)\,\mathrm{d}t_i. \tag{6.11}$$

We use the classical notation

$$\int_a^b f(t)\,\mathrm{d}t$$

for the integral because of its resemblance to this formula.

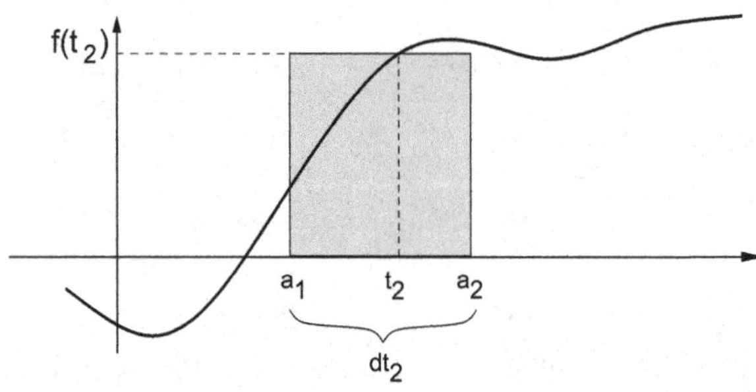

Fig. 6.12 Elements of an approximate integral

6.2b Existence of the Integral

We began our discussion of the integral with physical and geometric examples such as distance, mass, and area, in which it was reasonable to assume that there was a number, called the integral, that we were trying to estimate. For example, we found it reasonable to believe there is a number that can be assigned to the area of a planar region bounded by a nice boundary. But how do we know that a single number called area can be assigned to such a region in the first place?

In this section, we prove that for a continuous function on a closed interval, the approximate integrals converge to a limit as we refine the subdivision. We show that this limit does not depend on the particular sequence of subdivisions used. This limit is called the definite integral and is written $\int_a^b f(t)\,dt$.

Given a continuous function f on $[a,b]$ and given any tolerance ε, we can choose δ such that if two points s and t in $[a,b]$ differ by less than δ, then $f(s)$ and $f(t)$ differ by less than ε. Subdivide the interval $[a,b]$ as

$$a = a_0 < a_1 < a_2 < \cdots < a_{n-1} < a_n = b.$$

Then choose any point t_i in the ith subinterval, $a_{i-1} \le t_i \le a_i$, and form the approximate integral $I = \sum_{i=1}^{n} f(t_i)(a_i - a_{i-1})$. Then form another approximate integral $I' = \sum_{j=1}^{m} f(t'_j)(a'_j - a'_{j-1})$ using another subdivision

$$a = a'_0 < a'_1 < \cdots < a'_m = b$$

and points t'_j from each subinterval. Next, we show that if all the lengths of the subintervals are small enough,

$$a_i - a_{i-1} < \frac{1}{2}\delta \qquad \text{and} \qquad a'_j - a'_{j-1} < \frac{1}{2}\delta,$$

then the two approximate integrals differ from each other by less than $\varepsilon(b-a)$. That is,

$$|I - I'| < \varepsilon(b-a).$$

To see this, we form the common subdivision consisting of all the intersections of the a intervals and a' intervals that have positive length. We denote by s_{ij} the length of the intersection of $[a_{i-1}, a_i]$ with $[a'_{j-1}, a'_j]$. An example is indicated in Fig. 6.13. Many of the s_{ij} are zero, because most of the subintervals do not overlap.

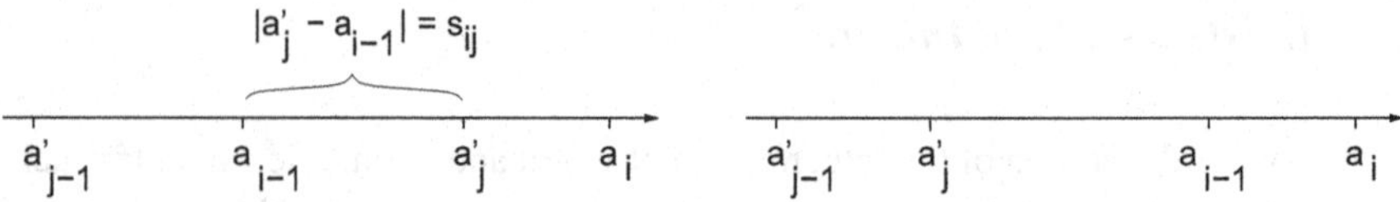

Fig. 6.13 *Left*: The subintervals intersect, with length s_{ij}. *Right*: They do not overlap, and $s_{ij} = 0$

We break up the sums I and I' as follows. The length $(a_i - a_{i-1})$ is the sum over j of the s_{ij} for which $[a_{i-1}, a_i]$ and $[a'_{j-1}, a'_j]$ intersect. Since the other s_{ij} are 0, we can sum over all i and j:

$$I = \sum_i f(t_i)(a_i - a_{i-1}) = \sum_{i,j} f(t_i) s_{ij}.$$

Similarly, the length $(a_j - a_{j-1})$ is the sum over i of the s_{ij} for which $[a_{i-1}, a_i]$ and $[a'_{j-1}, a'_j]$ intersect. So

$$I' = \sum_j f(t'_j)(a_j - a_{j-1}) = \sum_{i,j} f(t'_i) s_{ij}.$$

Therefore, $I - I' = \sum_{i,j}(f(t_i) - f(t'_j))s_{ij}$. The only terms of interest are those for which s_{ij} is nonzero, that is, for which $[a_{i-1}, a_i]$ and $[a'_{j-1}, a'_j]$ overlap. But the lengths of $[a_{i-1}, a_i]$ and $[a'_{j-1}, a'_j]$ have been assumed less than $\frac{1}{2}\delta$. Hence the points t_i and t'_j in each nonzero term differ by no more than δ. It follows that $f(t_i)$ and $f(t'_j)$ differ by less than ε. Therefore, by the triangle inequality,

$$|I - I'| \le \sum_{i,j} \varepsilon s_{ij} = \varepsilon(b - a). \tag{6.12}$$

This result enables us to make the following definition.

Definition 6.2. The integral of a continuous function on a closed interval. Take any sequence of subdivisions of $[a, b]$ with the following property: the length of the largest subinterval in the kth subdivision tends to zero as k tends to infinity. (For instance, we could take the kth subdivision to be the subdivision into k equal parts.) Denote by I_k any approximate integral using the kth subdivision.

Since f is a continuous function, given any tolerance $\varepsilon > 0$, there is a precision $\delta > 0$ such that the values of f differ by less than ε over any interval of length δ. Choose N so large that for $k > N$, each subinterval of the kth subdivision has length less than $\frac{1}{2}\delta$. It follows from (6.12) that for k and l greater than N, I_k and I_l differ by less than $\varepsilon(b - a)$. This proves the convergence of the sequence I_k.

The limit does not depend on our choice of the sequence of subdivisions. For given two such sequences, we can merge them into a single sequence, and the associated approximate integrals form a convergent sequence. This proves that the two sequences that were merged have the same limit.

This common limit is defined to be the integral $\int_a^b f(t)\,\mathrm{d}t$.

Other Integrable Functions. If f is not continuous on $[a,b]$, we may be able to define a continuous function g on $[a,b]$ by redefining f at finitely many points. If so, we say that f is integrable on $[a,b]$ and set $\int_a^b f(t)\,\mathrm{d}t = \int_a^b g(t)\,\mathrm{d}t$.

Example 6.3. To compute $\int_1^4 \frac{t^2-4}{t-2}\,\mathrm{d}t$, we notice that $\frac{t^2-4}{t-2}$ is not continuous on all of $[1,4]$ but is equal to $t+2$ for $t \neq 2$. So by redefining $\frac{t^2-4}{t-2}$ to be 4 at $t=2$, we obtain

$$\int_1^4 \frac{t^2-4}{t-2}\,\mathrm{d}t = \int_1^4 (t+2)\,\mathrm{d}t.$$

Example 6.4. To compute $\int_0^1 \frac{\sin t}{t}\,\mathrm{d}t$, we notice that $\frac{\sin t}{t}$ is not continuous at 0, because it is not defined at 0. We know that $\lim_{t\to 0} \frac{\sin t}{t} = 1$, so we define $g(0)=1$ and $g(t) = \frac{\sin t}{t}$ for $t \neq 0$. Then g is continuous on every closed interval $[a,b]$. In particular, $\int_0^1 \frac{\sin t}{t}\,\mathrm{d}t = \int_0^1 g(t)\,\mathrm{d}t$ is a number. We do not have an easy way to calculate this number, but using ten equal subintervals and the right-hand endpoints, you can see that it is approximately

$$\int_0^1 \frac{\sin t}{t}\,\mathrm{d}t \approx \sum_{n=1}^{10} \frac{\sin\left(\frac{n}{10}\right)}{\frac{n}{10}}\left(\frac{1}{10}\right) \approx 0.94.$$

Also, if f is not continuous on $[a,b]$ but is integrable on $[a,c]$ and $[c,b]$, then we say that f is integrable on $[a,b]$ and set $\int_a^b f(t)\,\mathrm{d}t = \int_a^c f(t)\,\mathrm{d}t + \int_c^b f(t)\,\mathrm{d}t$.

Example 6.5. Denote by $[x]$ the greatest integer that is less than or equal to x. See Fig. 6.14. Then $[x]$ is integrable on $[0,1]$, $[1,2]$, and $[2,3]$, and

$$\int_0^3 [x]\,\mathrm{d}x = \int_0^1 0\,\mathrm{d}x + \int_1^2 1\,\mathrm{d}x + \int_2^3 2\,\mathrm{d}x = 0 + 1(2-1) + 2(3-2) = 3.$$

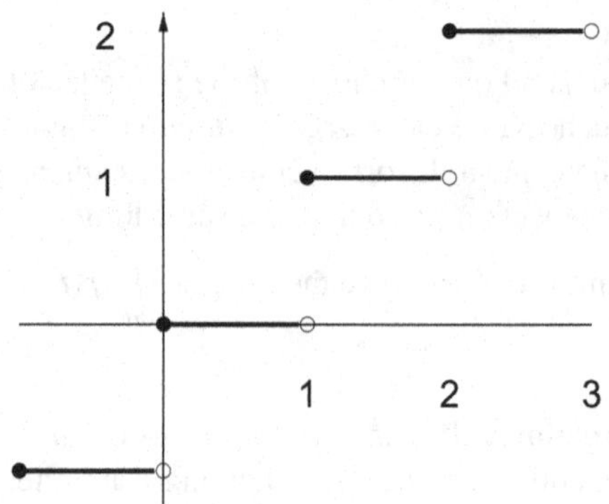

Fig. 6.14 Graph of the greatest integer function, in Example 6.5

The Properties of Integrals Revisited. We verify that the lower and upper bound property, as well as additivity property, are satisfied by Definition 6.2.

To check additivity, suppose $a < c < b$. Approximate $\int_a^b f(t)\,\mathrm{d}t$ using a subdivision of $[a,b]$ in which one point of the subdivision is c. This is permissible, because we have seen that any sequence of subdivisions is allowed, as long as the lengths of the subintervals tend to 0, and this can certainly be done while keeping c as one of the division points. Then for each such $I_{\text{approx}}(f,[a,b])$, we may separate the terms into two groups corresponding to the subintervals to the left of c and those to the right of c, and thus express $I_{\text{approx}}(f,[a,b])$ as a sum

$$I_{\text{approx}}(f,[a,b]) = I_{\text{approx}}(f,[a,c]) + I_{\text{approx}}(f,[c,b]).$$

As the lengths of the largest subintervals tend to zero, the approximate integrals tend to

$$\int_a^b f(t)\,\mathrm{d}t = \int_a^c f(t)\,\mathrm{d}t + \int_c^b f(t)\,\mathrm{d}t.$$

This verifies the additivity property (6.5). For the lower and upper bound property, suppose $m \le f(t) \le M$ on $[a,b]$. Then for every approximate integral, we have

$$\sum_i m(a_i - a_{i-1}) \le \sum_i f(t_i)(a_i - a_{i-1}) \le \sum_i M(a_i - a_{i-1}),$$

that is, $m(b-a) \le I_{\text{approx}}(f,[a,b]) \le M(b-a)$. In the limit, it follows that

$$m(b-a) \le \int_a^b f(t)\,\mathrm{d}t \le M(b-a),$$

which is the lower and upper bound property (6.6).

6.2c Further Properties of the Integral

We present some important properties of the integral.

Theorem 6.2. The mean value theorem for integrals. *If f is a continuous function on $[a,b]$, then there is a number c in $[a,b]$ for which*

$$\int_a^b f(t)\,dt = f(c)(b-a).$$

The number $f(c)$ is called the mean or average value of f on $[a,b]$.

Proof. By the extreme value theorem, f has a minimum value m and a maximum M on $[a,b]$. Then the lower and upper bound property gives

$$m \le \frac{1}{b-a}\int_a^b f(t)\,dt \le M.$$

Since a continuous function takes on all values between its minimum and maximum, there is a number c in $[a,b]$ for which

$$f(c) = \frac{1}{b-a}\int_a^b f(t)\,dt.$$

□

Example 6.6. The mean of $f(t) = t$ on $[a,b]$ is

$$\frac{1}{b-a}\int_a^b t\,dt = \frac{1}{b-a}\frac{b^2-a^2}{2} = \frac{a+b}{2} = f(c), \qquad \text{where } c = \frac{a+b}{2}.$$

In Problem 6.9, we encourage you to explore how the mean value of f on $[a,b]$ is related to the ordinary average taken over n numbers, and to the concept of a weighted average.

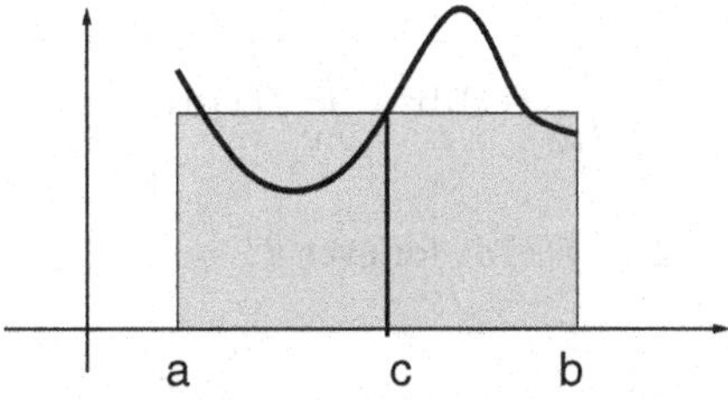

Fig. 6.15 The mean value of a positive function, illustrated in Example 6.7

Example 6.7. Take f positive. We interpret $\int_a^b f(t)\,\mathrm{d}t$ as the area under the graph of f in Fig. 6.15. Theorem 6.2 asserts that there is at least one number c between a and b such that the shaded rectangle shown has the same area as the region under the graph of f.

We make the following definition, which is occasionally useful in simplifying expressions.

Definition 6.3. When $a > b$ we define $\int_a^b f(t)\,\mathrm{d}t = -\int_b^a f(t)\,\mathrm{d}t$, and when $a = b$ we define $\int_a^b f(t)\,\mathrm{d}t = 0$.

Note that the additivity property holds for all numbers a, b, c within an interval on which f is continuous.

Example 6.8.

$$\int_1^3 f(t)\,\mathrm{d}t = \int_1^5 f(t)\,\mathrm{d}t + \int_5^3 f(t)\,\mathrm{d}t,$$

because this is merely a rearrangement of the earlier property

$$\int_1^3 f(t)\,\mathrm{d}t + \int_3^5 f(t)\,\mathrm{d}t = \int_1^5 f(t)\,\mathrm{d}t.$$

Even and Odd Functions. The integral of an odd function f, $f(-x) = -f(x)$, on an interval that is symmetric about 0 is zero. Consider approximate integrals for $\int_{-a}^0 f(x)\,\mathrm{d}x$ and $\int_0^a f(x)\,\mathrm{d}x$ for an odd function as in Fig. 6.16. We see that

$$\int_{-a}^0 f(x)\,\mathrm{d}x = -\int_0^a f(x)\,\mathrm{d}x.$$

Therefore, $\int_{-a}^a f(x)\,\mathrm{d}x = \int_{-a}^0 f(x)\,\mathrm{d}x + \int_0^a f(x)\,\mathrm{d}x = 0$ for odd f.

For an even function f, $f(-x) = f(x)$, as in the figure, we see that

$$\int_{-a}^0 f(x)\,\mathrm{d}x = \int_0^a f(x)\,\mathrm{d}x.$$

Therefore, $\int_{-a}^a f(x)\,\mathrm{d}x = 2\int_0^a f(x)\,\mathrm{d}x$ for even f.

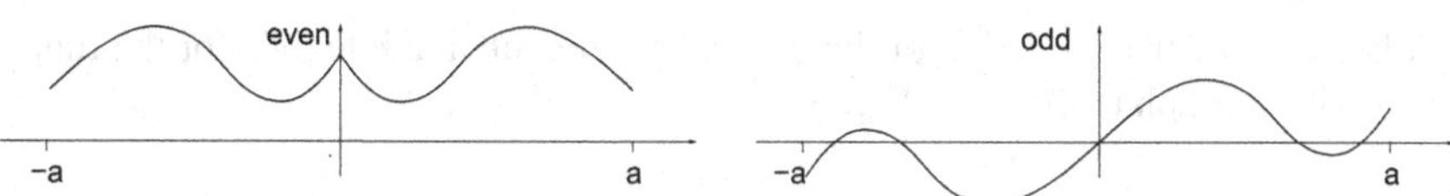

Fig. 6.16 Graphs of even and odd functions

Theorem 6.3. Linearity of the integral. *For any numbers* a, b, c_1, c_2 *and continuous functions* f_1 *and* f_2, *we have*

$$\int_a^b c_1 f_1(t) + c_2 f_2(t)\,\mathrm{d}t = c_1 \int_a^b f_1(t)\,\mathrm{d}t + c_2 \int_a^b f_2(t)\,\mathrm{d}t.$$

Proof. The approximate integrals satisfy

$$I_{\text{approx}}(c_1 f_1 + c_2 f_2, [a,b]) = c_1 I_{\text{approx}}(f_1, [a,b]) + c_2 I_{\text{approx}}(f_2, [a,b])$$

if we use the same subdivision and the same points t_i in each of the three sums. The limit of these relations then gives Theorem 6.3. □

Theorem 6.4. Positivity of the integral. *If* f *is a continuous function with* $f(t) \geq 0$ *on* $[a,b]$, *then* $\int_a^b f(t)\,\mathrm{d}t \geq 0$.

Proof. Each approximate integral consists of nonnegative terms, so the limit must be nonnegative. □

Example 6.9. If $f_1(t) \leq f_2(t)$ on $[a,b]$, then $\int_a^b f_1(t)\,\mathrm{d}t \leq \int_a^b f_2(t)\,\mathrm{d}t$. We see this by taking $f = f_2 - f_1$ in Theorem 6.4 and using the linearity of the integral, Theorem 6.3.

Problems

6.5. Calculate the approximate integral for the given function, subdivision of the interval, and choice of evaluation points t_i. For each problem, make a sketch of the graph of the function corresponding to the approximate integral.

(a) $f(t) = t^2 + t$ on $[1,3]$, using $1 < 1.5 < 2 < 3$ and $t_1 = 1.2$, $t_2 = 2$, $t_3 = 2.5$.
(b) $f(t) = \sin t$ on $[0,\pi]$, using $0 < \frac{\pi}{4} < \frac{\pi}{2} < \frac{3\pi}{4} < \pi$ and take the t_i to be the left endpoints of the subintervals.

6.6. Use an area interpretation of the integral to compute the integrals for the function f whose graph is shown in Fig. 6.17.

(a) $\int_0^1 f(t)\,\mathrm{d}t$

(b) $\int_1^4 f(t)\,\mathrm{d}t$

(c) $\int_0^4 f(t)\,\mathrm{d}t$

(d) $\int_1^6 f(t)\,\mathrm{d}t$

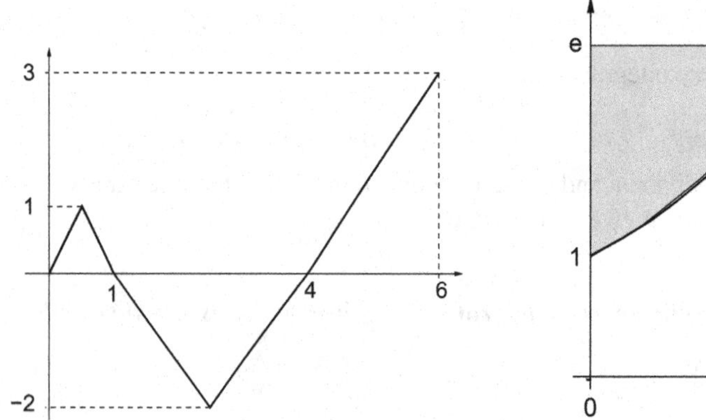

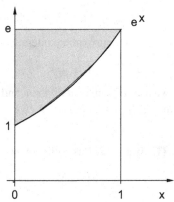

Fig. 6.17 *Left*: The graph of f in Problem 6.6. *Right*: The graph of e^x for Problem 6.7

6.7. Refer to Fig. 6.17.

(a) Use the result that $\int_0^1 e^t\,\mathrm{d}t = e - 1$ to find the area of the shaded region.

(b) Use a geometric argument to compute $\int_1^e \log t\,\mathrm{d}t$.

6.8. Use an area interpretation and properties of integrals to evaluate the following integrals.

(a) $\int_{-\pi}^{\pi} \sin(x^3)\,\mathrm{d}x$

(b) $\int_0^2 \sqrt{4-x^2}\,\mathrm{d}x$

(c) $\int_{-10}^{10} (e^{x^3} - e^{-x^3})\,\mathrm{d}x$

6.9. Let $T(x)$ be the temperature at point x on a rod, $0 \le x \le 15$.

(a) Use the measurements at six equally spaced points shown in Fig. 6.18. Make two estimates of the average temperature, first with an approximate integral using left endpoint values of T, and then using right endpoint values.

(b) Use the measurements taken at unequally spaced points in Fig. 6.18. Write two expressions to estimate the average temperature, first with an approximate integral using left endpoint values of T, and then using right endpoint values.

T	130	75	65	63	61	60
x	0	3	6	9	12	15

T	130	120	90	70	65	60
x	0	1	2	4	6	15

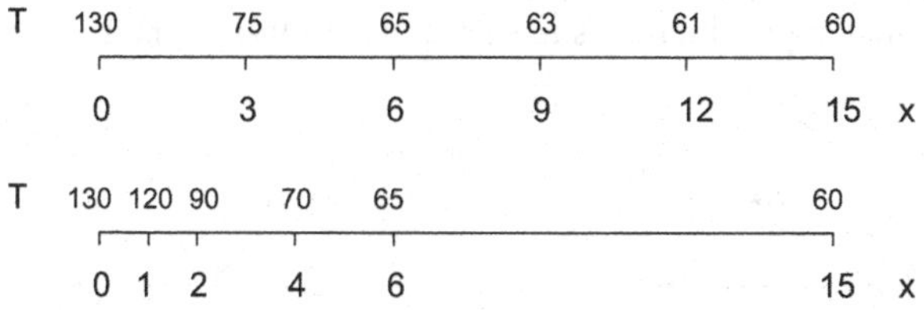

Fig. 6.18 Two sets of temperature measurements along the same rod in Problem 6.9

6.10. Express the limit

$$\lim_{n\to\infty} \frac{1+4+9+\cdots+(n-1)^2}{n^3}$$

as an integral of some function over some interval, and find its value.

6.11. Let k be some positive number. Consider the interval obtained from $[a,b]$ by stretching in the ratio $1:k$, i.e., $[ka,kb]$. Let f be any continuous function on $[a,b]$. Denote by f_k the function defined on $[ka,kb]$ obtained from f by stretching:

$$f_k(t) = f\left(\frac{t}{k}\right).$$

Using approximate integrals, prove that

$$\int_{ka}^{kb} f_k(t)\,\mathrm{d}t = k\int_a^b f(t)\,\mathrm{d}t$$

and make a sketch to illustrate this result. In Sect. 6.3, we shall prove this relation using the fundamental theorem of calculus.

Further properties of the integral are explored in the next problems. These properties can be derived here using approximate integrals. Later, in Sect. 6.3, we ask you to prove them using the fundamental theorem of calculus.

6.12. Let f be any continuous function on $[a,b]$. Denote by f_r the function obtained when f is shifted to the right by r. That is, f_r is defined on $[a+r,b+r]$ according to the rule

$$f_r(t) = f(t-r).$$

See Fig. 6.19. Prove that

$$\int_a^b f(x)\,dx = \int_{a+r}^{b+r} f_r(x)\,dx.$$

This property of the integral is called *translation invariance.*

Hint: Show that approximating sums are translation-invariant.

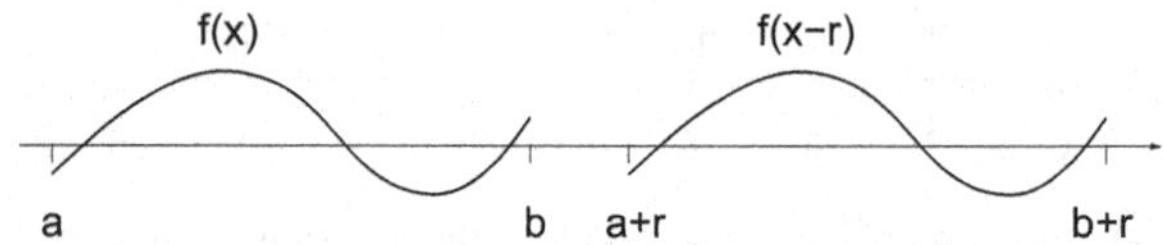

Fig. 6.19 Translation

6.13. For an interval $[a,b]$, the *reflected* interval is defined as $[-b,-a]$. If f is some continuous function on $[a,b]$, its *reflection*, denoted by f_-, is defined on $[-b,-a]$ as follows:

$$f_-(t) = f(-t).$$

The graph of f_- is obtained from the graph of f by reflection across the vertical axis; see Fig. 6.20. Prove that

$$\int_{-b}^{-a} f_-(t)\,dt = \int_a^b f(t)\,dt.$$

This property of the integral is called *invariance under reflection.*

Hint: Show that approximating sums are invariant under reflection.

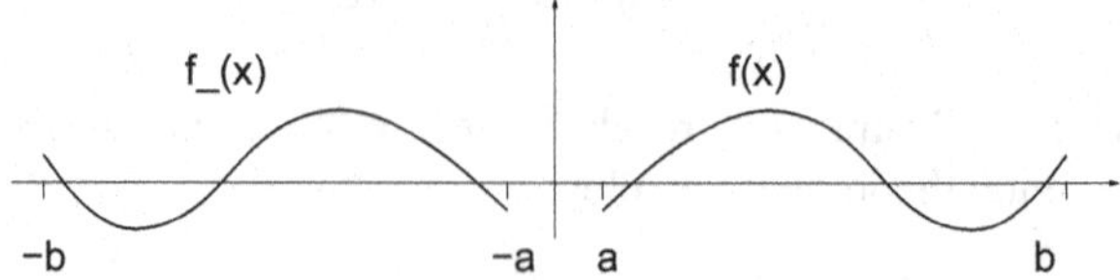

Fig. 6.20 Reflection

6.14. In this problem, you will evaluate $\int_0^1 \sqrt{t}\,dt = \frac{2}{3}$.

(a) Using $0 < \frac{1}{4} < \frac{1}{2} < 1$, verify that the upper and lower sums are

$$I_{\text{upper}} = \sqrt{\frac{1}{4}}\frac{1}{4} + \sqrt{\frac{1}{2}}\frac{1}{4} + \sqrt{1}\frac{1}{2}$$

and

$$I_{\text{lower}} = \sqrt{0}\frac{1}{4} + \sqrt{\frac{1}{4}}\frac{1}{4} + \sqrt{\frac{1}{2}}\frac{1}{2}.$$

(b) Let $0 < r < 1$, and use the subdivision $0 < r^3 < r^2 < r < 1$. Verify that the upper sum is

$$I_{\text{upper}} = \sqrt{r^3}r^3 + \sqrt{r^2}(r^2 - r^3) + \sqrt{r}(r - r^2) + \sqrt{1}(1 - r),$$

and find an expression for the lower sum.

(c) Write the upper sum for the subdivision $0 < r^n < r^{n-1} < \cdots < r^2 < r < 1$, recognize a geometric series in it, and check that

$$I_{\text{upper}} \to \frac{1 - r}{1 - r^{3/2}}$$

as n tends to infinity.

(d) Show that $\dfrac{1 - r}{1 - r^{3/2}}$ tends to $\frac{2}{3}$ as r tends to 1.

6.3 The Fundamental Theorem of Calculus

Earlier, we posed the following problem: determine the change of position, or *net distance*, D, of a moving vehicle during a time interval $[a,b]$ from knowledge of the *velocity* f of the vehicle at each instant of $[a,b]$. The answer we found in Sect. 6.1d was that D is the integral

$$D = I(f, [a,b])$$

of the velocity as a function of time over the interval $[a,b]$. This formula expresses the net distance covered during the whole trip. A similar formula holds, of course, for the net distance $D(t)$ up to time t. This formula is

$$D(t) = I(f, [a,t]),$$

where $[a,t]$ is the interval between the starting time and the time t.

In Sects. 3.1 and 6.1d, we discussed the *converse problem*: if we know the net distance $D(t)$ of a moving vehicle from its starting point to its position at time t, for all values of t, how can we determine its velocity as a function of time

t? The answer we found was that *velocity is the derivative* of D as a function of time:

$$f(t) = D'(t)$$

We also posed the problem early in this chapter of finding the mass of a rod R between points a and b from knowledge of its linear density f. We found that the mass of the rod is

$$R = I(f, [a,b]),$$

the integral of the density f over the interval $[a,b]$. A similar formula holds for the mass of the part of the rod up to the point x. This is

$$R(x) = I(f, [a,x]).$$

Again in Sect. 3.1 we discussed the converse problem: if we know the mass of the rod from one end to any point x, how do we find the linear density of the rod at x? The answer we found was that the linear density f at x is the derivative of the mass

$$f(x) = R'(x).$$

We can summarize these observations in the following words:

If a function F is defined to be the integral of f from a to x, then the derivative of F is f.

Omitting all qualifying phrases, we can express the preceding statement as an epigram.

Differentiation and integration are inverses of each other.

The argument presented in favor of this proposition was based on physical intuition. We proceed to give a purely mathematical proof.

Theorem 6.5. The fundamental theorem of calculus

(a) Let f be any continuous function on $[a,b]$. Then f is the derivative of some differentiable function. In fact, for x in $[a,b]$,

$$\frac{\mathrm{d}}{\mathrm{d}x}\left(\int_a^x f(t)\,\mathrm{d}t\right) = f(x). \tag{6.13}$$

(b) Let F be any function with a continuous derivative on $[a,b]$. Then

$$F(b) - F(a) = \int_a^b F'(t)\,\mathrm{d}t. \tag{6.14}$$

Proof. We first prove statement (a). Define a function

$$G(x) = \int_a^x f(t)\,dt.$$

Form the difference quotient

$$\frac{G(x+h) - G(x)}{h}.$$

We have to show that as h tends to zero, this quotient tends to $f(x)$. By definition of G,

$$G(x+h) = \int_a^{x+h} f(t)\,dt, \qquad \text{and} \qquad G(x) = \int_a^x f(t)\,dt.$$

By the additivity property of the integral,

$$\int_a^{x+h} f(t)\,dt = \int_a^x f(t)\,dt + \int_x^{x+h} f(t)\,dt.$$

This can be written as

$$G(x+h) = G(x) + \int_x^{x+h} f(t)\,dt.$$

The difference quotient is

$$\frac{G(x+h) - G(x)}{h} = \frac{1}{h}\int_x^{x+h} f(t)\,dt.$$

By the mean value theorem for integrals, Theorem 6.2, there is a number c between x and $x+h$ such that

$$\frac{1}{h}\int_x^{x+h} f(t)\,dt = f(c).$$

That is, as we see in Fig. 6.21, the area of a strip, divided by its width, is equal to the height of the strip at some point. Since f is continuous, $f(c)$ tends to $f(x)$ as h tends to 0. This proves that the difference quotient tends to $f(x)$. Therefore, the derivative of G is f. This concludes the proof of part (a).

We turn now to the proof of part (b). Since F' is continuous on $[a,b]$, we can define a function F_a by

$$F_a(x) = \int_a^x F'(t)\,dt, \quad a \le x \le b. \tag{6.15}$$

As we have shown in the proof of part (a), the derivative of F_a is F'. Therefore, the difference $F - F_a$ has derivative zero for every x in $[a,b]$. By Corollary 4.1 of the mean value theorem for derivatives, a function whose derivative is zero at every point of an interval is constant on that interval. Therefore,

$$F(x) - F_a(x) = \text{constant}$$

for every x. We evaluate the constant as follows. Let $x = a$. By the definition of F_a, $F_a(a) = 0$. Then $F(a) - F_a(a) = F(a) = \text{constant}$, and we get

$$F(x) = F_a(x) + F(a) \quad \text{on} \quad [a,b].$$

Setting $x = b$, it follows that

$$F(b) - F(a) = F_a(b) = \int_a^b F'(t)\,dt. \tag{6.16}$$

This completes a proof of the fundamental theorem of calculus. □

Here is another proof of part (b) of the fundamental theorem of calculus that uses the mean value theorem for derivatives more directly. Let

$$a = a_0 < a_1 < a_2 < \cdots < a_n = b$$

be any subdivision of the interval $[a,b]$. According to the mean value theorem for derivatives, in each subinterval $[a_{i-1}, a_i]$, there is a point t_i such that

$$F'(t_i) = \frac{F(a_i) - F(a_{i-1})}{a_i - a_{i-1}}.$$

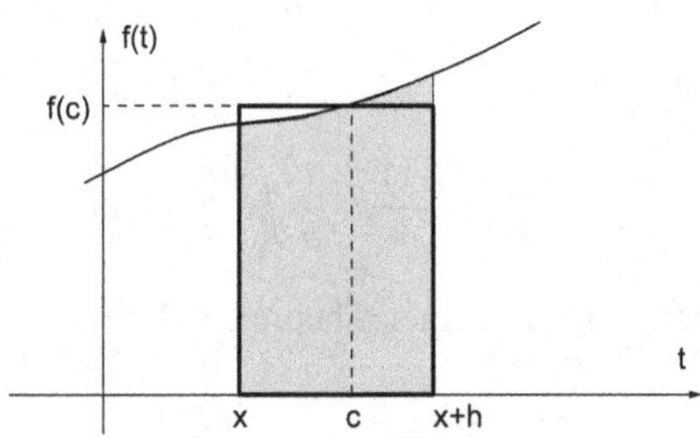

Fig. 6.21 The equation $\frac{1}{h}\int_x^{x+h} f(t)\,dt = f(c)$ illustrated

Therefore,

$$F'(t_i)(a_i - a_{i-1}) = F(a_i) - F(a_{i-1}).$$

Add these up for all i from 1 to n. We get

$$\begin{aligned}\sum_{i=1}^{n} F'(t_i)(a_i - a_{i-1}) &= F(a_1) - F(a_0) + F(a_2) - F(a_1) + \cdots + F(a_n) - F(a_{n-1}) \\ &= F(b) - F(a).\end{aligned}$$

The sum on the left is an approximation to the integral $\int_a^b F'(t)\,dt$. We have shown in Sect. 6.2b that the set of all approximations tends to the integral as the subdivision is refined. Our formula shows that no matter how fine the subdivision, *these particular* approximations are *exactly equal* to $F(b) - F(a)$. Therefore, the limit is $F(b) - F(a)$.

The fundamental theorem of calculus deserves its honorific name; it has at least two important uses. First and foremost, it is the *fundamental existence theorem* of analysis; it guarantees the existence of a function with a given derivative. Its second use lies in furnishing an exact method for evaluating the integral of any function we recognize to be the derivative of a known function.

We have shown how to deduce the fundamental theorem of calculus from the mean value theorem of differential calculus and the mean value theorem for integrals. The three can be related in one unifying statement. If F' is continuous on $[a,b]$, then for some c,

$$F'(c) = \frac{1}{b-a}\int_a^b F'(t)\,\mathrm{d}t = \frac{F(b)-F(a)}{b-a}.$$

We can express this relationship in words: the average of the instantaneous rates of change of F throughout an interval is equal to the average rate of change in F over the interval.

Notation. We sometimes denote $F(b)-F(a)$ by

$$F(b)-F(a) = [F(x)]_a^b = F(x)\Big|_a^b.$$

A function F whose derivative is f is called an *antiderivative* of f. One way to evaluate a definite integral on an interval is to find and evaluate an antiderivative F. We use the notation $\int f(x)\,\mathrm{d}x$ to denote an antiderivative of f. Often, $\int f(x)\,\mathrm{d}x = F(x)+C$ is used to denote all possible antiderivatives of f. Antiderivatives expressed in this way are called *indefinite* integrals. The constant C is called the constant of integration, and it can be assigned any value.

Evaluation of Some Integrals. Next, we illustrate how to use the fundamental theorem of calculus to evaluate the integral of a function that we recognize as a derivative.

Example 6.10. Since $(-\cos t)' = \sin t$, the fundamental theorem gives

$$\int_a^b \sin t\,\mathrm{d}t = -\cos t\Big|_a^b = -\cos b + \cos a.$$

Example 6.11. We know that $(\mathrm{e}^t)' = \mathrm{e}^t$. By the fundamental theorem, then,

$$\int_0^1 \mathrm{e}^t\,\mathrm{d}t = \mathrm{e}^1 - \mathrm{e}^0 = \mathrm{e} - 1,$$

in agreement with our computation of this integral in Sect. 6.2.

Example 6.12. Let $f(t) = t^c$, where c is any real number except -1. Then f is the derivative of $F(t) = \dfrac{t^{c+1}}{c+1}$. By the fundamental theorem,

$$\int_a^b t^c \,\mathrm{d}t = F(b) - F(a) = \frac{b^{c+1}}{c+1} - \frac{a^{c+1}}{c+1}.$$

In particular,

$$\int_a^b t \,\mathrm{d}t = \frac{t^2}{2}\Big|_a^b = \frac{b^2 - a^2}{2},$$

as we found in Sect. 6.2.

By now, you must have noticed that the key to evaluating integrals by the fundamental theorem lies in an ability to notice that the function f presented for integration is the derivative of another handy function F. How does one acquire the uncanny ability to find antiderivatives? It comes with the experience of differentiating many functions; in addition, the search for F can be systematized with the aid of a few basic techniques. These will be presented in the next chapter.

The Special Case $c = -1$. We showed in Chap. 3 that for every number c, positive or negative except -1, the function t^c, $t > 0$, is the derivative of $\frac{t^{c+1}}{c+1}$. In contrast, the function t^{-1} is the derivative of $\log t$. This seems very strange. The functions t^c change continuously with c as c passes through the value -1. Why is there such a drastic discontinuity at $c = -1$ of the antiderivative of the function?

We show now that the discontinuity is only apparent, not real. Let

$$F_c(t) = \int_1^t x^c \,\mathrm{d}x.$$

Using the fundamental theorem, we get

$$F_c(t) = \frac{t^{c+1} - 1}{c+1} \quad \text{for } c \neq -1, \qquad \text{and } F_{-1}(t) = \log t.$$

We shall show that as c tends to -1, F_c tends to F_{-1}. This is illustrated in Fig. 6.22. To prove this, we set $c = -1 + y$. Then

$$F_{y-1}(t) = \frac{t^y - 1}{y}.$$

Define the function g as $g(y) = t^y$, where t is some positive number. Note that $g(0)=1$. Using this function, we rewrite the previous relation as

$$F_{y-1}(t) = \frac{g(y) - g(0)}{y}.$$

The limit of this expression as y tends to zero is the derivative of g at $y = 0$. To evaluate that derivative, we write g in exponential form: $g(y) = \mathrm{e}^{y \log t}$. Using the chain rule, we get $\frac{\mathrm{d}g}{\mathrm{d}y} = g(y)\log t$. Since $g(0) = 1$, we have

$$\frac{dg}{dy}(0) = \log t.$$

In words: the limit of $F_{y-1}(t)$ as y tends to zero is $\log t$. Since $F_{-1}(t) = \log t$, this shows that $F_c(t)$ depends continuously on c at $c = -1$!

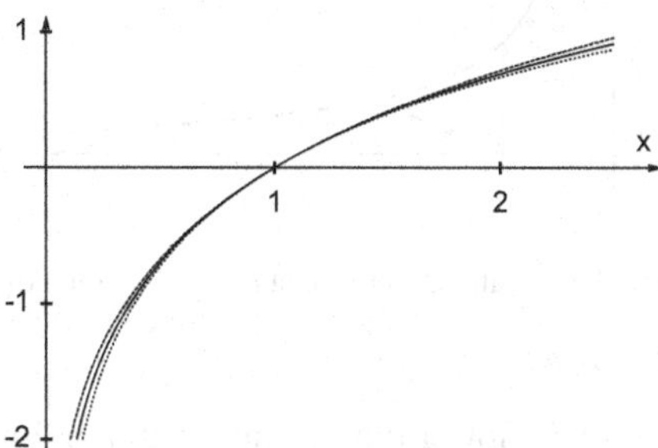

Fig. 6.22 The graphs of $\log x$ and $\frac{x^{c+1}-1}{c+1}$ for $c = -0.9$ and $c = -1.1$

In the next example, we illustrate how to use the fundamental theorem of calculus to construct a function with a particular derivative.

The Logarithm and Exponential Functions Redefined. Suppose we had not worked hard in Chaps. 1 and 2 to define e^x and $\log x$. Let us use the fundamental theorem part (a) to define a function $F(x)$ whose derivative is $\frac{1}{x}$. Let

$$F(x) = \int_1^x \frac{1}{t}\,dt$$

for any $x > 0$. See Fig. 6.23. Then by the fundamental theorem,

$$F'(x) = \frac{1}{x}.$$

Recall that the derivative of $\log x$ is also $\frac{1}{x}$. Two functions that have the same derivative on an interval differ by a constant, and since both are 0 at $x = 1$, we see that

$$\log x = \int_1^x \frac{1}{t}\,dt, \qquad x > 0. \tag{6.17}$$

Next we show that if we take Eq. (6.17) as the definition of the logarithm function, we can derive all properties of the logarithm from Eq. (6.17). The basic properties of $\log x$ are:

(a) $\log 1 = 0$,
(b) $\log x$ is an increasing function, and
(c) $\log(ax) = \log a + \log x$.

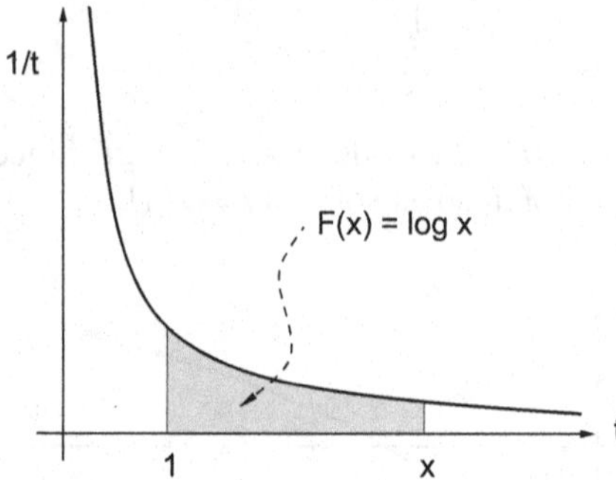

Fig. 6.23 For $x > 1$, $\log x$ can be visualized as the area under the graph of $\frac{1}{t}$

Part (a) follows because the lower limit of integration in Eq. (6.17) is 1. Part (b) follows because the derivative of $\log x$ is positive for x positive. For part (c), take the derivative of $\log(ax)$. Using the chain rule, we get

$$(\log ax)' = \frac{1}{ax}a = \frac{1}{x}.$$

This shows that $\log(ax)$ and $\log x$ have the same derivative on $(0,\infty)$. Therefore, they differ by a constant, $\log(ax) = C + \log x$. Setting $x = 1$, we see that C is $\log a$.

Since $\log x$ is increasing, we can define e^x to be the inverse of $\log x$:

$$\log e^x = x.$$

The basic properties of e^x are:

(a) $e^0 = 1$,
(b) $\left(e^x\right)' = e^x$, and
(c) $e^{a+x} = e^a e^x$.

Part (a) follows from property (a) of the log function. For part (b), note that $\log x$ is differentiable with continuous nonzero derivative. Its inverse, e^x, is then also differentiable. By the chain rule,

$$1 = x' = \left(\log e^x\right)' = \frac{1}{e^x}\left(e^x\right)'.$$

Multiply both sides by e^x to get $(e^x)' = e^x$. For part (c), we first verify that the derivative of $\frac{e^{a+x}}{e^x}$ is zero. By the quotient rule and chain rule,

$$\left(\frac{e^{a+x}}{e^x}\right)' = \frac{e^x e^{a+x} - e^{a+x}e^x}{(e^x)^2} = 0 \qquad \text{for all } x.$$

This means that $\frac{e^{a+x}}{e^x}$ is constant. Taking $x = 0$, we see that the constant is e^a.

Defining e^x as the inverse of $\log x$ is much, much simpler than the definition of e^x we used in Chap. 2: $\mathrm{e}^x = \lim_{n\to\infty}\left(1+\frac{x}{n}\right)^n$. Finding $(\mathrm{e}^x)'$ as the derivative of the inverse function of $\log x$ is much, much simpler than the discussion in Sect. 3.3a.

Many arguments are much easier once you know calculus. In Problem 6.22, we ask you to use calculus to find a much easier proof that $\left(1+\frac{1}{n}\right)^n$ is an increasing function of n than the one we gave in Sect. 1.4, which used no calculus.

Trigonometric Functions Redefined. In Sect. 2.4, we gave a geometric definition of trigonometric functions and then defined the inverse trigonometric functions from the trigonometric functions. We now show how their inverse functions can be defined independently, using integration. Then the trigonometric functions can be defined as *their* inverses. For $0 < x < 1$, let

$$F(x) = \int_0^x \frac{1}{\sqrt{1-t^2}}\,\mathrm{d}t. \tag{6.18}$$

Since $F'(x) = \frac{1}{\sqrt{1-x^2}}$ is positive, $F(x)$ is an increasing function of x for $0 < x < 1$, and therefore F has an inverse. We define this inverse to be $x = \sin t$, defined for $0 < t < p$, where $p = F(1)$ is defined as the limit of $F(x)$ as x tends to 1. All properties of the sine function can be deduced from this definition. In Sect. 7.3, we will see that $F(x)$ does have a limit as x approaches 1. To define the sine function to have the full domain we expect, we have to do a bit more work. But we see that $\sin t$ can be completely described without reference to triangles!

Problems

6.15. Use the fundamental theorem of calculus to calculate the derivatives.

(a) $\frac{\mathrm{d}}{\mathrm{d}x}\int_0^x t^3\,\mathrm{d}t$

(b) $\frac{\mathrm{d}}{\mathrm{d}x}\int_0^x t^3\mathrm{e}^{-t}\,\mathrm{d}t$

(c) $\frac{\mathrm{d}}{\mathrm{d}s}\int_{-2}^{s^2} x^3\mathrm{e}^{-x}\,\mathrm{d}x$

(d) $h'(4)$, if $h(x) = \int_1^x \sqrt{t}\cos\left(\frac{\pi}{t}\right)\mathrm{d}t$

6.16. Use the fundamental theorem to calculate the integrals.

(a) $\int_1^2 t^3\,\mathrm{d}t$

(b) $\int_0^b (x^3+5)\,\mathrm{d}x$

(c) $\displaystyle\int_0^1 \frac{1}{\sqrt{1+t}}\,\mathrm{d}t$

(d) $\displaystyle\int_0^7 \left(\cos t + (1+t)^{1/3}\right)\,\mathrm{d}t$

(e) $\displaystyle\int_a^b (t - \mathrm{e}^t)\,\mathrm{d}t$

6.17. Use the fundamental theorem to calculate the integrals.

(a) $\displaystyle\int_0^{\pi/4} \frac{1}{1+x^2}\,\mathrm{d}x$

(b) $\displaystyle\int_0^1 (x^2+2)^2\,\mathrm{d}x$

(c) $\displaystyle\int_1^4 \left(\frac{2}{\sqrt{x}} - \sqrt{x}\right)\,\mathrm{d}x$

(d) $\displaystyle\int_{-2}^{-1} (2 + 4t^{-2} - 8t^{-3})\,\mathrm{d}t$

(e) $\displaystyle\int_2^6 \left(2s + \frac{1}{s+1}\right)\,\mathrm{d}s$

6.18. Sketch the regions and find the areas.
(a) The region bounded by $y = \sqrt{x}$ and $y = \frac{1}{2}x$.
(b) The region bounded by $y = x^2$ and $y = x$.
(c) The region bounded by $y = \mathrm{e}^x$, $y = -x + 1$, and $x = 1$.

6.19. For the function $x(t) = \sin t$ defined as the inverse of F in equation (6.18), show that

$$\frac{\mathrm{d}x}{\mathrm{d}t} = \sqrt{1-x^2}.$$

6.20. Suppose f is an even function, i.e., $f(t) = f(-t)$, and set $g(x) = \displaystyle\int_0^x f(t)\,\mathrm{d}t$. Explain why g is odd, i.e., $g(x) = -g(-x)$.

6.21. Suppose g is a differentiable function and

$$F(x) = \int_a^{g(x)} f(t)\,\mathrm{d}t.$$

Explain why $F'(x) = f(g(x))g'(x)$.

6.22. Explain the following items, which prove that $\left(1+\dfrac{1}{n}\right)^n$ is an increasing sequence.

(a) For $x > 0$, $\left(1+\dfrac{1}{x}\right)^x = \mathrm{e}^{x\log\left(1+\frac{1}{x}\right)}$

(b) $\displaystyle\int_1^{1+1/x} \frac{1}{t}\,\mathrm{d}t > \frac{1}{x}$

(c) $\int_1^{1+1/x} \frac{1}{t}\,dt - \frac{1}{x+1} > 0$

(d) $\frac{d}{dx}\left(1+\frac{1}{x}\right)^x > 0$

(e) When n is a positive integer, $\left(1+\frac{1}{n}\right)^n$ is an increasing function of n.

6.23. Use the fundamental theorem to explain the following.

(a) $\frac{1}{4}((1+3^3)^4-(1+2^3)^4) = \int_2^3 3t^2(1+t^3)^3\,dt$

(b) $v(t_2)-v(t_1) = \int_{t_1}^{t_2} a(t)\,dt$ for the acceleration and velocity of a particle.

6.24. Work, or rework, Problems 6.11, 6.12, and 6.13 using the fundamental theorem.

6.4 Applications of the Integral

6.4a Volume

The volume of most regions in three-dimensional space is best described using integrals of functions of more than one variable, a topic in multivariable calculus. But some regions, such as solids of revolution and solids obtained by stacking thin slabs, can be expressed as integrals of functions of a single variable. For example, volumes of revolution can be expressed as quantities that satisfy the two basic properties of additivity and lower and upper bounds.

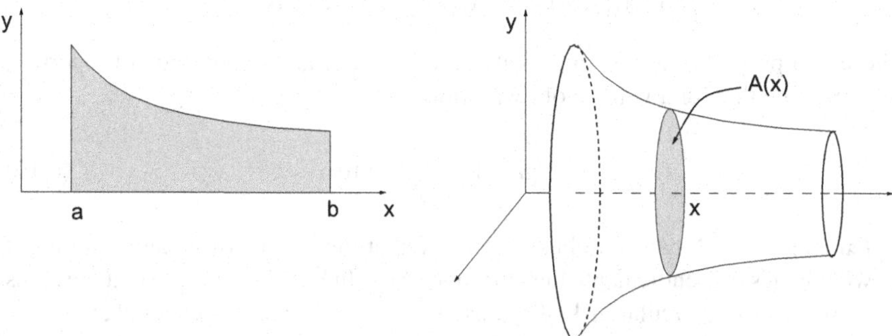

Fig. 6.24 *Left*: A planar region to revolve around the x-axis. *Right*: The solid of revolution. $A(x)$ is the shaded cross-sectional area of the solid at x

Let us write $V(A,[a,b])$ for the volume of a solid of revolution that is located in the region $a \le x \le b$, and where $A(x)$ is the cross-sectional area of the solid at each x. The solid is obtained by revolving a planar region as in Fig. 6.24 around the x-axis, and all cross sections are circular.

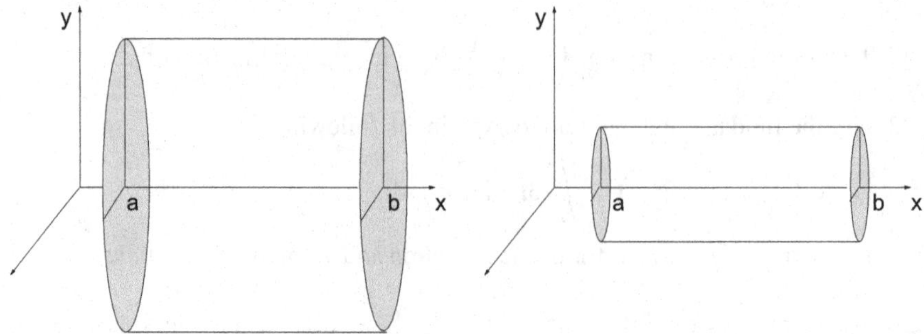

Fig. 6.25 If $m \le A(x) \le M$, the solid in Fig. 6.24 fits between cylinders with volumes $M(b-a)$ and $m(b-a)$

If M and m are the largest and smallest cross-sectional areas,

$$m \le A(x) \le M,$$

then the solid fits between two cylinders having cross-sectional areas m and M, and for the volumes of the three solids, we have (Fig. 6.25)

$$m(b-a) \le V(A,[a,b]) \le M(b-a).$$

Similarly, if we cut the object at c between a and b, we expect that the volumes of the two pieces add to the total:

$$V(A,[a,b]) = V(A,[a,c]) + V(A,[c,b]).$$

These two properties are the additivity and the lower and upper bound properties. So it must be that for a volume of revolution,

$$V(A,[a,b]) = \int_a^b A(x)\,\mathrm{d}x. \tag{6.19}$$

Example 6.13. A ball of radius r is centered at the origin. We imagine slicing it with planes perpendicular to the horizontal axis. In Fig. 6.26, we see that the cross section at x is a circular disk of radius $\sqrt{r^2-x^2}$, so the cross-sectional area is

$$A(x) = \pi(r^2-x^2).$$

Then the volume is

$$\int_{-r}^{r} A(x)\,\mathrm{d}x = \int_{-r}^{r} \pi(r^2-x^2)\,\mathrm{d}x = \pi\left[r^2x - \frac{x^3}{3}\right]_{-r}^{r} = 2\pi\left(r^3 - \frac{r^3}{3}\right) = \frac{4}{3}\pi r^3.$$

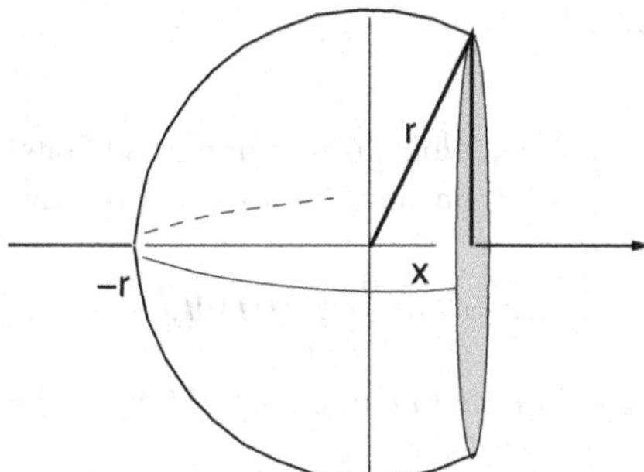

Fig. 6.26 By the Pythagorean theorem, the cross section of a ball is a disk of radius $\sqrt{r^2-x^2}$, in Example 6.13

Example 6.14. Consider the planar region bounded by the graph of $y=\dfrac{1}{x}$, the x-axis, and the lines $x=1$ and $x=2$. Generate a solid by revolving that region around the x-axis, as in Fig. 6.27. The volume is

$$\int_1^2 A(x)\,\mathrm{d}x,$$

where $A(x)$ is the cross-sectional area of the solid at x. The cross section is a disk of radius $\dfrac{1}{x}$, so $A(x)=\pi x^{-2}$. The volume is

$$\int_1^2 \pi x^{-2}\,\mathrm{d}x = -\pi x^{-1}\Big|_1^2 = \pi\left(-\frac{1}{2}+1\right) = \frac{\pi}{2}.$$

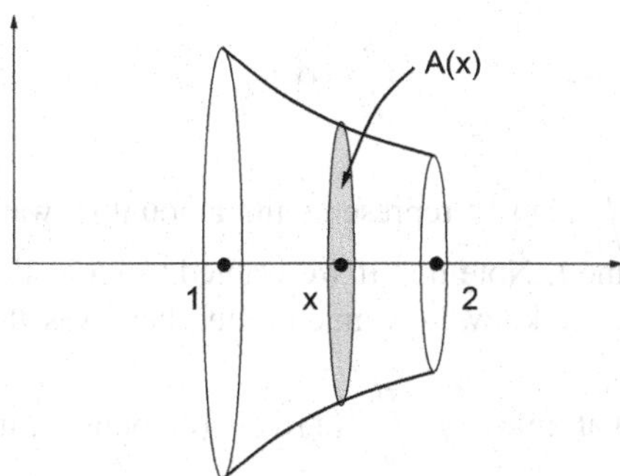

Fig. 6.27 The cross section of the solid in Example 6.14 is a disk of radius $\dfrac{1}{x}$

6.4b Accumulation

The fundamental theorem of calculus has two important consequences. One is that every continuous function f on an interval arises as the rate of change of some function:

$$f(x) = \frac{d}{dx}\int_a^x f(t)\,dt.$$

The other is that the integral of the rate of change F' equals the change in the function F between a and b:

$$F(b) - F(a) = \int_a^b F'(t)\,dt.$$

In this section, we look at ways to use the integral to answer the following question: How much?

Suppose we know that water is flowing into a pool at a rate $f(t)$ that varies continuously with time. How much water flows into the pool between time a and time b? We subdivide the interval of time into n very small subintervals,

$$a = a_0 < a_1 < \cdots < a_n = b,$$

and for each one, we find the rate at which water is flowing at some time t_i during that interval. The product of the rate $f(t_i)$ and the length of time $(a_i - a_{i-1})$ is a good estimate for the amount of water that entered the pool between times a_{i-1} and a_i. Summing all those estimates, we get the approximate integral $\sum_{i=1}^{n} f(t_i)(a_i - a_{i-1})$. We know that in the limit, such approximate integrals converge to $\int_a^b f(t)\,dt$. So the amount of water that accumulates in the pool between time a and time b is

$$\int_a^b f(t)\,dt.$$

The function $F(t) = \int_a^t f(\tau)\,d\tau$ represents the amount of water that accumulates between time a and time t. Note that if we wanted to know how much water is *in* the pool, we would need to know how much water there was at time a, and then

$$(\text{amount at time } t) = \int_a^t f(\tau)\,d\tau + (\text{amount at time } a).$$

6.4c Arc Length

Here is another example of how the integral is used to answer a "how much" question. Let f have a continuous derivative on $[a,b]$. The *arc length* of the graph of f from a to b is the least upper bound of the sum of lengths of line segments joining points on the graph. Let us see how to compute the arc length using an integral (Fig. 6.28).

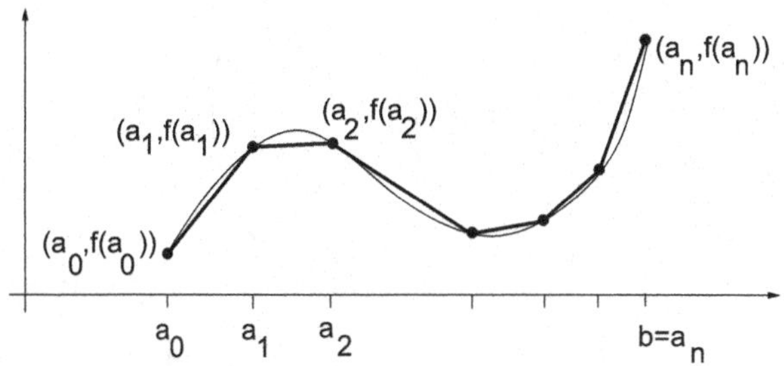

Fig. 6.28 The segments underestimate the arc length

Let

$$a = a_0 < a_1 < \cdots < a_{n-1} < a_n = b$$

be a subdivision of $[a,b]$. By the Pythagorean theorem, the length of the ith segment is

$$\sqrt{(a_i - a_{i-1})^2 + \big(f(a_i) - f(a_{i-1})\big)^2}.$$

By the mean value theorem for derivatives, there is a point t_i between a_{i-1} and a_i such that

$$f(a_i) - f(a_{i-1}) = f'(t_i)(a_i - a_{i-1})$$

and

$$\sqrt{(a_i - a_{i-1})^2 + \big(f(a_i) - f(a_{i-1})\big)^2} = \sqrt{1 + \big(f'(t_i)\big)^2}(a_i - a_{i-1}).$$

The length of the curve is approximately the sum of the lengths of the segments:

$$L \approx \sum_{i=1}^{n} \sqrt{1 + \big(f'(t_i)\big)^2}(a_i - a_{i-1}).$$

Since f' is continuous on $[a,b]$, $\sqrt{1+(f')^2}$ is continuous as well, and the approximate integrals $\sum_{i=1}^{n} \sqrt{1 + \big(f'(t_i)\big)^2}(a_i - a_{i-1})$ approach

$$\int_a^b \sqrt{1 + \big(f'(t)\big)^2}\,\mathrm{d}t,$$

the arc length of the curve from a to b.

Let us see how this formula works on a problem for which we already know the arc length.

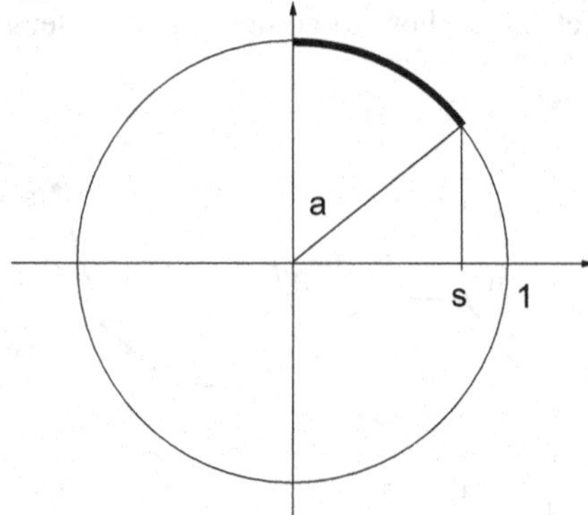

Fig. 6.29 In the first quadrant, the graph of $f(x) = \sqrt{1-x^2}$ is part of the unit circle. See Example 6.15

Example 6.15. According to our definition of the inverse sine function, the length of the heavy arc on the unit circle in Fig. 6.29 ought to be $a = \sin^{-1} s$. Let us check this against the arc-length formula. We have

$$f(x) = \sqrt{1-x^2}, \qquad f'(x) = \frac{-x}{\sqrt{1-x^2}},$$

and

$$\sqrt{1+(f')^2} = \sqrt{1+\frac{x^2}{1-x^2}} = \frac{1}{\sqrt{1-x^2}}.$$

So the arc-length formula gives

$$\int_0^s \frac{1}{\sqrt{1-x^2}}\,\mathrm{d}x = \sin^{-1} s,$$

in agreement with Sect. 3.4c.

Let us compute an arc length that we cannot compute geometrically.

Example 6.16. Find the arc length of the graph of $f(x) = \frac{2}{3}x^{3/2}$ from 0 to 1. We have $f'(x) = x^{1/2}$. Then

$$L = \int_0^1 \sqrt{1+(x^{1/2})^2}\,\mathrm{d}x = \int_0^1 \sqrt{1+x}\,\mathrm{d}x = \frac{2}{3}(1+x)^{3/2}\Big|_0^1 = \frac{2}{3}(2^{3/2}-1).$$

6.4d Work

The concept of work is readily visualized by the example of hoisting a load with the aid of a rope over a pulley. How much work is required depends on the weight of the load and on the difference between its initial height and the height to which it has to be hoisted. The following facts are suggested in our example by the intuitive notion of work:

(a) The work done is proportional to the *distance* through which the force acts.
(b) The amount of work done is proportional to the weight or force.

Accordingly, we *define* the work, W, done in elevating a load of weight f by the vertical distance h to be

$$W = fh.$$

According to Newton's theory, the dynamical effect of every force is the same. So we take $W = fh$ to be the work done by any force f acting through a distance h in the direction of the force (Fig. 6.30).

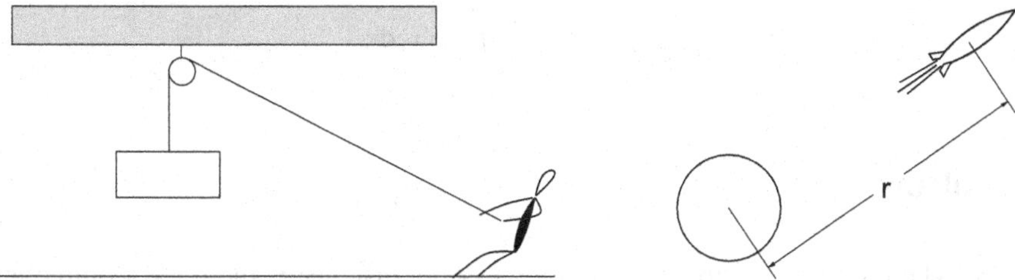

Fig. 6.30 *Left*: Work $W = fh$ to lift weight f through a distance h. *Right*: $W = \int f\,dr$ with a variable force

We show now that this formula for work is meaningful when h is negative, that is, when the displacement is in the opposite direction to the force. For let us lower the load to its original position. The total energy of the load has not changed, so the work done in lowering the load undoes the energy it gained from being raised, and so is a negative quantity.

How much work is done in moving an object through an interval $[a,b]$ against a variable force f, i.e., a force f whose magnitude differs at different points of $[a,b]$ and may even reverse its direction? In this case, f is a function defined on $[a,b]$; let us denote the work done by $W(f,[a,b])$.

What kind of function is W of $[a,b]$? Suppose $[a,b]$ is divided into two disjoint intervals,

$$a < c < b.$$

Since moving the object across $[a,b]$ means moving it first across $[a,c]$, then across $[c,b]$, it follows that the total work is the sum of the work done in accomplishing the separate tasks:

$$W(f,[a,b]) = W(f,[a,c]) + W(f,[c,b])$$

How does W depend on f? Clearly, if at every point of $[a,b]$ the force f stays below some value M, then the work done in pushing against f is less than the work done in pushing against a constant force of magnitude M. Likewise, if the force f is greater than m at every point of $[a,b]$, then pushing against f requires more work than pushing against a constant force of magnitude m. The work done by pushing against a constant force is given by $W = fh$. Thus, if the force f lies between the bounds

$$m \le f(x) \le M \text{ for } x \text{ in } [a,b],$$

then

$$m(b-a) \le W(f,[a,b]) \le M(b-a).$$

We recognize these as the additive and the lower and upper bound properties. These two properties characterize W as the integral

$$W(f,[a,b]) = \int_a^b f(x)\,\mathrm{d}x.$$

Problems

6.25. When a spring is stretched or compressed a distance x, the force required is kx, where k is a constant reflecting the physical properties of the spring. Suppose a spring requires a force of 2000 Newtons to compress it 4 mm. Verify that the spring constant is $k = 500{,}000$ N/m, and find the work done to compress the spring 0.004 m.

6.26. If at time t, oil leaks from a tank at a rate of $R(t)$ gallons per minute, what does $\int_3^5 R(t)\,\mathrm{d}t$ represent?

6.27. Water can be pumped from a tank at the rate of $2t + 10$ L/min. How long does it take to drain 200 L from the tank?

6.28. Find the volume of the solid obtained by revolving the graph of $\frac{2}{3}x^{3/2}$, for $0 \le x \le 1$, around the x-axis.

6.29. Consider the graph of $\frac{1}{x}$ on $[1,2]$.

(a) Set up but do not evaluate an integral for the arc length.
(b) Calculate two approximate integrals I_{approx} for the arc length, using ten subintervals, and taking the t_i at the left, respectively the right, endpoints.

6.30. During the space shuttle program, the shuttle orbiter had a mass of about 10^5 kg.

(a) When a body of mass m is close to the surface of the Earth, the force of gravity is essentially constant,
$$f = mg,$$
with a gravitational constant of $g = 9.8\,\text{m}/\text{s}^2$. Use the constant-force assumption $W = mgh$ to calculate the work done against gravity to lift the shuttle mass to a height of 50 m above the launch pad.

(b) When a mass is moved a great distance from the Earth, the force of gravity depends on the distance r from the center of the Earth,
$$f = \frac{GMm}{r^2}.$$
The radius of the Earth is about 6.4×10^6 m. Equating the two expressions for the weight of an object of mass m at the surface of the Earth, find GM.

(c) Calculate the work done against gravity to lift the shuttle orbiter to an altitude of 3.2×10^5 m.

6.31. We have calculated the volume of a solid of revolution as $\int_a^b A(x)\,dx$, where $A(x)$ is the cross-sectional area at x. We also know that every integral can be approximated with arbitrary accuracy by an approximate integral taken over a small enough subdivision:
$$\int_a^b A(x)\,dx \approx \sum_{i=1}^{n} A(x_i)\,dx_i.$$
We observe that each term $A(x_i)\,dx_i$ is the volume of a thin cylinder with thickness dx_i. Make a sketch to illustrate that the volume of the solid is well approximated by the volume of a stack of thin cylinders.

6.32. The density ρ of seawater changes with the depth. It is approximately 1025 kg/m^3 at the surface, and 1027 at 500 m depth.

(a) Assume that density is a linear function between 0 and 500 m. Find the mass of a column of water that has a uniform cross-sectional area of 1 m^2 and is located between 100 and 500 m deep.

(b) Assume also that $\rho = 1027$ from 500 to 800 m depth. Find the mass of the 1 m^2 column located between 100 and 700 m deep.

Chapter 7
Methods for Integration

Abstract In this chapter, we present techniques of integration and examples of how to use them.

7.1 Integration by Parts

The rules for differentiation in Sect. 3.2 specify how to express the derivatives of the sum and product of two functions in terms of the derivatives of those functions themselves. Using the fundamental theorem of calculus, we shall convert each of these rules into a rule for integration.

Linearity and Integration. Let f and g be the derivatives of F and G, respectively. The sum rule says that

$$(F+G)' = F' + G' = f + g.$$

Applying the fundamental theorem of calculus to the function $f+g$, we obtain

$$\int_a^b (f(t)+g(t))\,\mathrm{d}t = (F(b)+G(b)) - (F(a)+G(a)).$$

On the other hand,

$$\int_a^b f(t)\,\mathrm{d}t = F(b) - F(a) \text{ and } \int_a^b g(t)\,\mathrm{d}t = G(b) - G(a).$$

Comparing the first expression with the sum of the second two, we deduce the *sum rule* for integrals:

$$\int_a^b (f(t)+g(t))\,\mathrm{d}t = \int_a^b f(t)\,\mathrm{d}t + \int_a^b g(t)\,\mathrm{d}t.$$

P.D. Lax and M.S. Terrell, *Calculus With Applications*, Undergraduate Texts in Mathematics, 291
DOI 10.1007/978-1-4614-7946-8_7, © Springer Science+Business Media New York 2014

Similarly, if c is any constant, the constant multiple rule for derivatives says that $(cF)' = cF' = cf$. Applying the fundamental theorem of calculus to cf, we obtain

$$\int_a^b cf(t)\,\mathrm{d}t = \int_a^b (cF)'(t)\,\mathrm{d}t = cF(b) - cF(a) = c\big(F(b) - F(a)\big) = c\int_a^b f(t)\,\mathrm{d}t.$$

Combining these results, we obtain

$$\int_a^b (c_1 f(t) + c_2 g(t))\,\mathrm{d}t = c_1 \int_a^b f(t)\,\mathrm{d}t + c_2 \int_a^b g(t)\,\mathrm{d}t.$$

This rule is not new to us; we have already encountered it under the name of *linearity* in Theorem 6.3, where it was deduced from the linearity of approximating sums.

Integration by Parts. We now recall the product rule of differentiation:

$$(fg)' = f'g + fg'.$$

Integrate each side over an interval $[a,b]$ in which f' and g' are continuous, and apply linearity to obtain

$$\int_a^b (fg)'(t)\,\mathrm{d}t = \int_a^b f'(t)g(t)\,\mathrm{d}t + \int_a^b f(t)g'(t)\,\mathrm{d}t.$$

According to the fundamental theorem, we have

$$\int_a^b (fg)'(t)\,\mathrm{d}t = f(b)g(b) - f(a)g(a).$$

Subtraction then leads to the following result.

Theorem 7.1. Integration by parts *If f' and g' are continuous on $[a,b]$ then*

$$\begin{aligned}\int_a^b f'(t)g(t)\,\mathrm{d}t &= \int_a^b (fg)'(t)\,\mathrm{d}t - \int_a^b f(t)g'(t)\,\mathrm{d}t\\ &= f(b)g(b) - f(a)g(a) - \int_a^b f(t)g'(t)\,\mathrm{d}t.\end{aligned}$$

For indefinite integrals, we get the corresponding formula for intervals on which f' and g' are continuous:

$$\int f'(t)g(t)\,\mathrm{d}t = f(t)g(t) - \int f(t)g'(t)\,\mathrm{d}t.$$

Integration by parts is helpful if we know more about the integral on the right than about the integral on the left. "Knowing more" could mean knowing the exact value of the integral on the right, or it could mean that the integral on the right is easier to evaluate approximately than the one on the left. In the examples below, we illustrate both possibilities.

Example 7.1. To find

$$\int_a^b t\mathrm{e}^t\,\mathrm{d}t,$$

we consider whether the product $t\mathrm{e}^t = f'(t)g(t)$ ought to be viewed as giving $f'(t) = t$ or $f'(t) = \mathrm{e}^t$. If we try

$$f'(t) = t, \qquad g(t) = \mathrm{e}^t,$$

then we may take $f(t) = \frac{1}{2}t^2$. Integration by parts gives

$$\int_a^b t\mathrm{e}^t\,\mathrm{d}t = \left[\frac{1}{2}t^2\mathrm{e}^t\right]_a^b - \int_a^b \frac{1}{2}t^2\mathrm{e}^t\,\mathrm{d}t.$$

The new integration problem is no easier than the original one. On the other hand, if we take

$$f'(t) = \mathrm{e}^t, \qquad g(t) = t,$$

then we may take $f(t) = \mathrm{e}^t$ and $g'(t) = 1$. Integration by parts gives

$$\int_a^b t\mathrm{e}^t\,\mathrm{d}t = \mathrm{e}^b b - \mathrm{e}^a a - \int_a^b \mathrm{e}^t\,\mathrm{d}t = \mathrm{e}^b b - \mathrm{e}^a a - \mathrm{e}^b + \mathrm{e}^a.$$

This second choice for f' and g resulted in a very easy integral to evaluate.

The trick to applying integration by parts is to think ahead to the new integration problem that results from your choices for f' and g.

Example 7.2. To find $\displaystyle\int_2^3 \log x\,\mathrm{d}x$, we factor the integrand

$$\log x = (1)(\log x) = f'(x)g(x),$$

where $f'(x) = 1$ and $g(x) = \log x$. Using $f(x) = x$ and $g'(x) = \dfrac{1}{x}$, integration by parts gives

$$\begin{aligned}\int_2^3 \log x\,\mathrm{d}x &= 3\log 3 - 2\log 2 - \int_2^3 x\frac{1}{x}\,\mathrm{d}x\\ &= \log\left(\frac{3^3}{2^2}\right) - x\Big|_2^3 = \log\left(\frac{3^3}{2^2}\right) - 1.\end{aligned}$$

Example 7.3. To find $\displaystyle\int_0^1 \sqrt{x^2 - x^3}\,\mathrm{d}x$, factor the integrand:

$$\sqrt{x^2 - x^3} = (\sqrt{1-x})x = f'(x)g(x),$$

where $f'(x) = (1-x)^{1/2}$ and $g(x) = x$. Using $f(x) = -\frac{2}{3}(1-x)^{3/2}$ and $g'(x) = 1$, integration by parts gives

$$\int_0^1 x(1-x)^{1/2}\,dx = x\left(-\frac{2}{3}(1-x)^{3/2}\right)\Big|_0^1 - \int_0^1 -\frac{2}{3}(1-x)^{3/2}\,dx.$$

In evaluating $f(x)g(x)\big|_0^1$, we notice that f vanishes at one endpoint and g vanishes at the other. So $f(1)g(1) = 0,\ f(0)g(0) = 0$. Therefore,

$$\int_0^1 x(1-x)^{1/2}\,dx = -\int_0^1 -\frac{2}{3}(1-x)^{3/2}\,dx.$$

We recognize that $-\frac{2}{3}(1-x)^{3/2}$ is the derivative of $\frac{4}{15}(1-x)^{5/2}$. So, using the fundamental theorem of calculus, we find that $\displaystyle\int_0^1 -\frac{2}{3}(1-x)^{3/2}\,dx = -\frac{4}{15}$. Therefore, $\displaystyle\int_0^1 \sqrt{x^2 - x^3}\,dx = \frac{4}{15}$.

Example 7.4. To find an antiderivative $\int e^x \sin x\,dx$, we factor $e^x \sin x = f'(x)g(x)$, where $f'(x) = e^x$ and $g(x) = \sin x$. Using $f(x) = e^x$ and $g'(x) = \cos x$, integration by parts gives

$$\int e^x \sin x\,dx = e^x \sin x - \int e^x \cos x\,dx.$$

The new integral does not appear any easier than the original integral. Using integration by parts again, we factor the new integrand $e^x \cos x = f'(x)g(x)$, where $f'(x) = e^x$ and $g(x) = \cos x$. Using $f(x) = e^x$ and $g'(x) = -\sin x$, a second integration by parts gives

$$\int e^x \sin x\,dx = e^x \sin x - \left(e^x \cos x - \int e^x(-\sin x)\,dx\right)$$

$$= e^x \sin x - e^x \cos x - \int e^x \sin x\,dx.$$

Note that the last term on the right is our original integral. Therefore, we may solve to find an antiderivative

$$\int e^x \sin x\,dx = \frac{1}{2}e^x(\sin x - \cos x).$$

Example 7.5. To evaluate $\displaystyle\int_a^b \sin^2 t\,dt$, we use integration by parts. Take $f'(t) = \sin t$, $g(t) = \sin t$, $f(t) = -\cos t$, $g'(t) = \cos t$. Then

$$\int_a^b \sin t \sin t\,dt = [-\cos t \sin t]_a^b - \int_a^b (-\cos t)(\cos t)\,dt$$

$$= [-\cos t \sin t]_a^b + \int_a^b (1 - \sin^2 t)\,dt = [-\cos t \sin t + t]_a^b - \int_a^b \sin^2 t\,dt.$$

Solving for the integral, we get

$$\int_a^b \sin^2 t\,dt = \frac{1}{2}[-\cos t \sin t + t]_a^b.$$

In fact, we may extend the approach used in Example 7.5 to develop a reduction formula for integrating higher powers of $\sin t$. If $m = 3, 4, \ldots$, we integrate by parts, taking $f'(t) = \sin t$ and $g(t) = \sin^{m-1} t$:

$$\int_a^b \sin^m t\,dt = \int_a^b \sin t \sin^{m-1} t\,dt$$

$$= \left[-\cos t \sin^{m-1} t\right]_a^b - \int_a^b (-\cos t)\left((m-1)\sin^{m-2} t \cos t\right) dt$$

$$= \left[-\cos t \sin^{m-1} t\right]_a^b + (m-1)\int_a^b (1-\sin^2 t)\sin^{m-2} t\,dt$$

$$= \left[-\cos t \sin^{m-1} t\right]_a^b + (m-1)\int_a^b \sin^{m-2} t\,dt - (m-1)\int_a^b \sin^m t\,dt.$$

Solving for the integral of $\sin^m t$, we get

$$\int_a^b \sin^m t\,dt = \frac{1}{m}\left[-\cos t \sin^{m-1} t\right]_a^b + \frac{m-1}{m}\int_a^b \sin^{m-2} t\,dt. \tag{7.1}$$

Example 7.6. To evaluate the integral of $\sin^4 t$, we use the reduction formula (7.1) to get

$$\int_a^b \sin^4 t\,dt = \frac{1}{4}\left[-\cos t \sin^3 t\right]_a^b + \frac{3}{4}\int_a^b \sin^2 t\,dt.$$

According to Example 7.5, $\int_a^b \sin^2 t\,dt = \frac{1}{2}[-\cos t \sin t + t]_a^b$. Therefore,

$$\int_a^b \sin^4 t\,dt = \frac{1}{4}\left[-\cos t \sin^3 t\right]_a^b + \frac{3}{4}\frac{1}{2}[-\cos t \sin t + t]_a^b.$$

In Problem 7.8, we ask you to practice using integration by parts to derive the analogous reduction formula for integrating powers of the cosine.

The importance of integration by parts is not limited to these nearly miraculous cases of explicit integration; the examples below illustrate very different applications of the theorem.

7.1a Taylor's Formula, Integral Form of Remainder

We saw in Chap. 4 that if f is $n+1$ times continuously differentiable on an open interval containing a, we can form the nth Taylor polynomial $t_n(x)$ about a,

$$t_n(x) = \sum_{k=0}^{n} f^{(k)}(a)\frac{(x-a)^k}{k!},$$

for x in the interval. Theorem 4.10, Taylor's formula, shows that the remainder is given by

$$f(x) - t_n(x) = f^{(n+1)}(c)\frac{(x-a)^{n+1}}{(n+1)!}$$

for some c between a and x. Now we use integration by parts to obtain a different expression for the remainder,

$$f(x) - t_n(x) = \frac{1}{n!}\int_a^x (x-t)^n f^{(n+1)}(t)\,dt, \tag{7.2}$$

which we call the *integral form of the remainder*. First observe that when $n = 0$, this formula is

$$f(x) - f(a) = \int_a^x f'(t)\,dt, \tag{7.3}$$

a restatement of the fundamental theorem of calculus. Now factor the integrand $f'(t) = 1f'(t) = g'(t)f'(t)$, where $g(t) = t - x$. Integration by parts gives

$$\int_a^x f'(t)\,dt = \left[(t-x)f'(t)\right]_{t=a}^x - \int_a^x (t-x)f''(t)\,dt$$

$$= f'(a)(x-a) + \int_a^x (x-t)f''(t)\,dt.$$

Combining this with Eq. (7.3), we get

$$f(x) = f(a) + f'(a)(x-a) + \int_a^x (x-t)f''(t)\,dt, \tag{7.4}$$

which is Eq. (7.2) for $n = 1$. Now integrating by parts in the integral in Eq. (7.4), we have

$$\int_a^x (x-t)f''(t)\,dt = \left[-\frac{1}{2}(x-t)^2 f''(t)\right]_a^x + \int_a^x \frac{1}{2}(x-t)^2 f'''(t)\,dt$$

$$= f''(a)\frac{(x-a)^2}{2} + \int_a^x \frac{1}{2}(x-t)^2 f'''(t)\,dt.$$

Combining this with Eq. (7.4), we get

$$f(x) = f(a) + f'(a)(x-a) + f''(a)\frac{(x-a)^2}{2} + \int_a^x \frac{1}{2}(x-t)^2 f'''(t)\,dt,$$

which is Eq. (7.2) for $n = 2$. Continuing in this way, one can prove Eq. (7.2) by induction for all n.

Let us consider the function $f(x) = \log(1+x)$. The derivatives are

$$f'(x) = \frac{1}{1+x}, \quad f''(x) = \frac{-1}{(1+x)^2}, \quad f'''(x) = \frac{2}{(1+x)^3}, \quad f''''(x) = \frac{-3!}{(1+x)^4},$$

and so forth. For $n \geq 1$,

$$f^{(n)}(x) = (-1)^{n+1}\frac{(n-1)!}{(1+x)^n}, \quad f^{(n)}(0) = (-1)^{n+1}(n-1)!,$$

and the nth Taylor polynomial at $a = 0$ is

$$t_n(x) = \sum_{k=1}^{n}(-1)^{k+1}\frac{x^k}{k}.$$

For what values of x does $|\log(1+x) - t_n(x)|$ tend to zero? The integral form of the remainder (7.2) is

$$\log(1+x) - t_n(x) = \frac{1}{n!}\int_0^x (x-t)^n(-1)^{n+2}\frac{n!}{(1+t)^{n+1}}\,dt = (-1)^{n+2}\int_0^x \frac{(x-t)^n}{(1+t)^{n+1}}\,dt.$$

For $0 \leq x \leq 1$, we have $0 \leq t < x \leq 1$ and $1 \leq 1+t$. Therefore,

$$\left|(-1)^{n+2}\int_0^x \frac{(x-t)^n}{(1+t)^{n+1}}\,dt\right| \leq \int_0^x (x-t)^n\,dt = \frac{x^{n+1}}{n+1} \leq \frac{1}{n+1},$$

which tends to zero uniformly as n tends to infinity. So we have shown that for $0 \leq x \leq 1$,

$$\log(1+x) = x - \frac{x^2}{2} + \frac{x^3}{3} - \frac{x^4}{4} + \cdots,$$

the Taylor series for $\log(1+x)$. In particular, for $x = 1$,

$$\log 2 = x - \frac{1}{2} + \frac{1}{3} - \frac{1}{4} + \cdots.$$

In Problem 7.4, we ask you to prove the convergence of this Taylor series for $-1 < x < 0$.

7.1b Improving Numerical Approximations

We study the integral

$$\int_0^1 x^2\sqrt{1-x^2}\,dx. \tag{7.5}$$

We factor the integrand as follows:

$$x^2\sqrt{1-x^2} = (x\sqrt{1-x^2})x = f'g,$$

where

$$f(x) = -\frac{1}{3}(1-x^2)^{3/2}, \quad g(x) = x.$$

Note that the function f is 0 at $x=1$, and that g is 0 at $x=0$, so integrating by parts, we obtain

$$\int_0^1 x^2\sqrt{1-x^2}\,dx = \int_0^1 \frac{1}{3}(1-x^2)^{3/2}\,dx. \tag{7.6}$$

Seemingly our strategy of integrating by parts has merely reduced the original task of finding the value of an integral to the problem of finding the value of another integral. We shall now show that something useful has been achieved. The second integral is easier to approximate than the first one. To convince you of this, we plot the graphs of both integrands in Fig. 7.1.

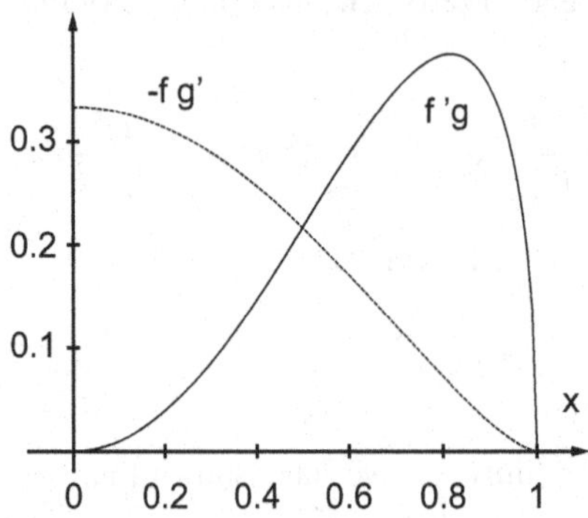

n	$I_{\text{mid}}(x^2\sqrt{1-x^2})$	$I_{\text{mid}}(\frac{1}{3}(1-x^2)^{3/2})$
1	0.21650635	0.21650635
2	0.21628707	0.19951825
5	0.20306231	0.19664488
10	0.19890221	0.19640021

Fig. 7.1 *Left*: Graphs of the integrands $f'(x)g(x) = x^2\sqrt{1-x^2}$ and $-f(x)g'(x) = \frac{1}{3}(1-x^2)^{3/2}$ in Eq. (7.6). *Right*: Approximate integrals using values at the midpoints of n subintervals

The derivative of the first function, $x^2\sqrt{1-x^2}$, is

$$2x\sqrt{1-x^2} - \frac{x^3}{\sqrt{1-x^2}},$$

which tends to minus infinity as x tends to the right endpoint $x=1$; this fact causes the graph of the function to have a vertical tangent at $x=1$. A function that has a large derivative is changing very fast. In contrast, the derivative of the second function, $\frac{1}{3}(1-x^2)^{3/2}$, is $-x\sqrt{1-x^2}$, which stays decently bounded throughout the whole interval $[0,1]$. A function with such a derivative changes slowly. We will see in Chap. 8 that *is easier to approximately evaluate the integral of a slowly changing function than the integral of one that changes quickly.*

The table in Fig. 7.1 lists, for various values of n and equal subdivisions of $[0,1]$, approximate integrals $I_{\text{mid}}(x^2\sqrt{1-x^2},[0,1])$ and $I_{\text{mid}}(\frac{1}{3}(1-x^2)^{3/2},[0,1])$ evaluated at the midpoint of each subinterval. We will show in Example 7.13 that the exact value of the integral is $\frac{\pi}{16} = 0.19634\ldots$. The table shows that for $n=10$, the approximation of the second integral is much closer to the exact value than the approximation of the first integral.

7.1c Application to a Differential Equation

Let $x(t)$ be any twice continuously differentiable function, and $[a,b]$ an interval at whose endpoints $x(t)$ is zero. Can one say something about the integral of the product xx'', that is, about $\int_a^b x(t)x''(t)\,dt$? To answer this question, we factor the integrand as $x''(t)x(t) = f'(t)g(t)$, where $f(t) = x'(t)$ and $g(t) = x(t)$. Since $g(t)$ is 0 at both endpoints, integration by parts gives

$$\int_a^b x''(t)x(t)\,dt = -\int_a^b x'(t)x'(t)\,dt = \int_a^b -(x'(t))^2\,dt. \tag{7.7}$$

Since the integrand on the right is negative, then so is the value of the integral on the right; this shows that the left side is negative. This demonstrates that integration by parts can sometimes reveal a quality of an integral such as negativity. There are some situations in which this is all we want to know.

Example 7.7. Suppose f_1 and f_2 are solutions of

$$f'' - f = 0$$

having the same value at two points, say $f_1(a) = f_2(a)$ and $f_1(b) = f_2(b)$. Take $x(t) = f_1(t) - f_2(t)$. Then $x'' - x = 0,\ \ x(a) = 0$, and $x(b) = 0$. Equation (7.7) becomes

$$\int_a^b (x(t))^2\,dt = -\int_a^b (x'(t))^2\,dt.$$

The left side is the integral of a square, and therefore is nonnegative; the right side is the integral of a square times -1 and is therefore nonpositive. This can be the case only if both sides are zero, but then $x = f_1 - f_2$ is zero for all values of t. Therefore, f_1 and f_2 are the same function. This furnishes another proof of Theorem 3.16 in Sect. 3.4d.

7.1d Wallis Product Formula for π

Let

$$W_n = \int_0^{\frac{\pi}{2}} \sin^n x\,dx.$$

Note that $W_0 = \frac{\pi}{2}$ and $W_1 = 1$. Using the reduction formula (7.1), we get

$$W_n = \int_0^{\frac{\pi}{2}} \sin^n x\,dx = \frac{1}{n}\left[-\cos x \sin x\right]_0^{\frac{\pi}{2}} + \frac{n-1}{n}\int_0^{\frac{\pi}{2}} \sin^{n-2} x\,dx,$$

and therefore

$$W_n = \frac{n-1}{n} W_{n-2} \qquad (n \geq 2). \tag{7.8}$$

Continuing in this fashion, and taking n even, $n = 2m$, we get

$$W_{2m} = \frac{2m-1}{2m} \frac{2m-3}{2m-2} \cdots \frac{1}{2} W_0,$$

while for $n = 2m+1$ odd, we get

$$W_{2m+1} = \frac{2m}{2m+1} \frac{2m-2}{2m-1} \cdots \frac{2}{3} W_1.$$

Since $W_0 = \frac{\pi}{2}$, we have found a relation between the even W_{2m} and the number π. Next, divide the even terms by the odd terms and use $W_1 = 1$. We get

$$\frac{W_{2m}}{W_{2m+1}} = \frac{(2m+1)(2m-1)(2m-1)(2m-3)\cdots(3)(1)}{(2m)(2m)(2m-2)(2m-2)\cdots(2)(2)} \frac{\pi}{2}.$$

Thus

$$\frac{\pi}{2} = \frac{(2m)(2m)(2m-2)(2m-2)\cdots(2)(2)}{(2m+1)(2m-1)(2m-1)(2m-3)\cdots(3)(1)} \frac{W_{2m}}{W_{2m+1}}.$$

Since

$$0 \leq \sin^{2m+1} x \leq \sin^{2m} x \leq \sin^{2m-1} x \qquad \text{for } 0 \leq x \leq \frac{\pi}{2},$$

we have $W_{2m+1} \leq W_{2m} \leq W_{2m-1}$. This implies, using $n = 2m+1$ in relation (7.8), that

$$1 = \frac{W_{2m+1}}{W_{2m+1}} \leq \frac{W_{2m}}{W_{2m+1}} \leq \frac{W_{2m-1}}{W_{2m+1}} = \frac{2m+1}{2m},$$

which tends to 1 as m tends to infinity. Therefore, we have proved the Wallis product formula

$$\frac{\pi}{2} = \lim_{m\to\infty} \frac{(2m)(2m)(2m-2)(2m-2)\cdots(4)(4)(2)(2)}{(2m+1)(2m-1)(2m-1)(2m-3)\cdots(5)(3)(3)(1)} = \frac{2}{1}\frac{2}{3}\frac{4}{3}\frac{4}{5}\frac{6}{5}\frac{6}{7}\cdots$$

Problems

7.1. Evaluate the following integrals.

(a) $\displaystyle\int_0^1 t^2 e^t \, dt$

(b) $\displaystyle\int_0^{\pi/2} t \cos t \, dt$

(c) $\displaystyle\int_0^{\pi/2} t^2 \cos t \, dt$

(d) $\displaystyle\int_0^1 x^3(1+x^2)^{1/2} \, dx$

7.2. Evaluate the following integrals.

(a) $\displaystyle\int_0^1 \sin^{-1} x \, dx$

(b) $\displaystyle\int_2^5 x \log x \, dx$

7.3. Evaluate the integral and the antiderivative.

(a) $\displaystyle\int_0^1 x \tan^{-1} x \, dx$

(b) $\displaystyle\int (\sin u + u \cos u) \, du$

7.4. Show that the Taylor series for $\log(1+x)$ converges for $-1 < x < 0$.

7.5. Suppose f_1 and f_2 are two solutions of

$$f'' - v(t)f = 0,$$

where v is a positive function, and suppose there are two points a and b where $f_1(a) = f_2(a)$, $f_1(b) = f_2(b)$. Use the integration by parts argument of Example 7.7 to show that f_1 and f_2 are the same function on $[a,b]$.

7.6. Find the antiderivatives.

(a) $\displaystyle\int x e^{-x} \, dx$

(b) $\displaystyle\int x e^{-x^2} \, dx$

(c) $\displaystyle\int x^3 e^{-x} \, dx$

(d) Express $\displaystyle\int x^2 e^{-x^2} \, dx$ in terms of $\displaystyle\int e^{-x^2} \, dx$.

7.7. Evaluate

(a) $\displaystyle\int_0^{\frac{\pi}{2}} \sin^2 t \, dt$

(b) $\displaystyle\int_0^{\frac{\pi}{2}} \sin^3 t \, dt$

7.8. Derive the following recurrence formula for powers $n = 2, 3, 4, \ldots$ of the cosine:

$$\int_a^b \cos^n t \, dt = \frac{1}{n} \left[\sin t \cos^{n-1} t\right]_a^b + \frac{n-1}{n} \int_a^b \cos^{n-2} t \, dt.$$

Use the formula to evaluate $\displaystyle\int_0^{\frac{\pi}{4}} \cos^2 t \, dt$ and $\displaystyle\int_0^{\frac{\pi}{4}} \cos^4 t \, dt$ exactly.

7.9. Find the antiderivatives ($x > 0$).

(a) $\int x^{-2}e^{-1/x}\,dx$

(b) Express $\int x^{-1}e^{-1/x}\,dx$ in terms of $\int e^{-1/x}\,dx$.

(c) $\int x^{-3}e^{-1/x}\,dx$

(d) $\int \left(x^{-2}e^{-1/x} + x^2e^{-x}\right)dx$

7.10. Define the number $B(n,m)$ by the integral

$$B(n,m) = \int_0^1 x^n(1-x)^m\,dx, \qquad n > 0,\ m > 0.$$

(a) Integrating by parts, show that

$$B(n,m) = \frac{n}{m+1}B(n-1,m+1).$$

(b) For positive integers n and m, show that repeated application of the recurrence relation derived in part (a) yields

$$B(n,m) = \frac{n!m!}{(n+m+1)!}.$$

7.11. Let K_m be an antiderivative $K_m(x) = \int x^m \sin x\,dx$ for $m = 0,1,2,\ldots$.

(a) Evaluate $K_1(x)$.
(b) Integrate by parts twice to show that

$$K_m(x) = -x^m\cos x + mx^{m-1}\sin x - m(m-1)K_{m-2}.$$

(c) Evaluate $K_0(x),\ K_2(x),\ K_4(x)$, and $\int_0^\pi x^4\sin x\,dx$.

(d) Evaluate $K_3(x)$.

7.2 Change of Variables in an Integral

We now recall the chain rule from Sect. 3.2b:

$$\left(F(g(t))\right)' = F'(g(t))g'(t).$$

Interpreting this as a statement about antiderivatives, we get

$$\int \left(F(g(t))\right)'\,dt = \int F'(g(t))g'(t)\,dt.$$

We now deduce an important formula for definite integrals.

Theorem 7.2. Change of variables *Let g be continuously differentiable on $[a,b]$, and let f be continuous on an interval that contains the range of g. Then*

$$\int_{g(a)}^{g(b)} f(u)\,\mathrm{d}u = \int_a^b f(g(t))\,g'(t)\,\mathrm{d}t.$$

Proof. By the fundamental theorem of calculus, f has an antiderivative F, so that $\dfrac{\mathrm{d}F(u)}{\mathrm{d}u} = f(u)$. Then by the chain rule,

$$\frac{\mathrm{d}}{\mathrm{d}t}\big(F(g(t))\big) = \frac{\mathrm{d}F}{\mathrm{d}u}(g)\frac{\mathrm{d}g}{\mathrm{d}t} = f(g(t))g'(t).$$

According to the fundamental theorem of calculus,

$$\int_a^b f(g(t))g'(t)\,\mathrm{d}t = F(g(b)) - F(g(a)).$$

Another application of the fundamental theorem gives

$$\int_{g(a)}^{g(b)} f(u)\,\mathrm{d}u = \int_{g(a)}^{g(b)} F'(u)\,\mathrm{d}u = F(g(b)) - F(g(a)).$$

Therefore

$$\int_{g(a)}^{g(b)} f(u)\,\mathrm{d}u = \int_a^b f(g(t))\,g'(t)\,\mathrm{d}t.$$

□

We show how to use the change of variables formula, also known as *substitution*, through some examples.

Example 7.8. Consider the integral

$$\int_0^{2\pi} 2t\cos(t^2)\,\mathrm{d}t.$$

Let $u = g(t) = t^2$, as in Fig. 7.2. Then $g'(t) = 2t$. When $t = 0,\ \ u = g(0) = 0$. When $t = 2\pi,\ \ u = g(2\pi) = (2\pi)^2$. By the change of variables theorem,

$$\int_{t=0}^{t=2\pi} 2t\cos(t^2)\,\mathrm{d}t = \int_{u=0}^{u=(2\pi)^2} \cos u\,\mathrm{d}u = \sin u\Big|_0^{(2\pi)^2} = \sin\big((2\pi)^2\big).$$

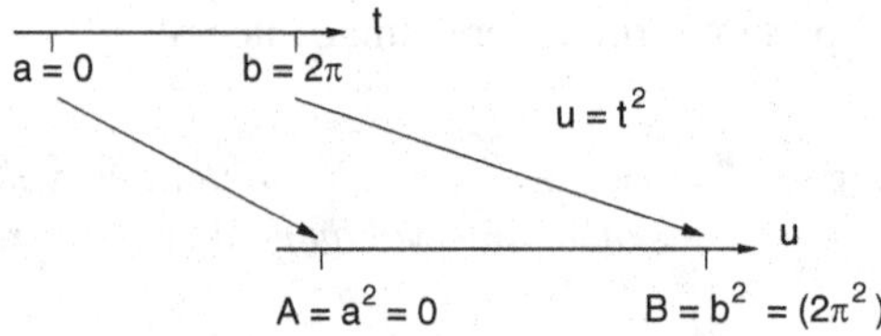

Fig. 7.2 The change of variable $u = t^2$ in Example 7.8

Change of variables works for finding antiderivatives as well, i.e., if we write $g(t) = u,\ g'(t) = \dfrac{du}{dt}$, then

$$\int f(g(t))g'(t)\,dt = \int f(u)\frac{du}{dt}\,dt = \int f(u)\,du.$$

Here are some examples.

Example 7.9. To find an antiderivative $\int \frac{1}{t}\log t\,dt \quad (t > 0)$, let $u = \log t$. Then $\dfrac{du}{dt} = \dfrac{1}{t}$, and

$$\int \log t\left(\frac{1}{t}\right)dt = \int u\frac{du}{dt}\,dt = \int u\,du = \frac{1}{2}u^2 = \frac{1}{2}(\log t)^2.$$

Let us check our answer:

$$\left(\frac{1}{2}(\log t)^2\right)' = \frac{1}{2}2(\log t)\frac{1}{t}.$$

Example 7.10. To find an antiderivative $\int \frac{x}{3}(x^2+3)^{1/2}\,dx$, let $u = x^2+3$. Then $\dfrac{du}{dx} = 2x$ and $du = 2x\,dx$. Solving for $x\,dx$, we get $x\,dx = \frac{1}{2}du$ and $\frac{1}{3}x\,dx = \frac{1}{6}du$.

Therefore,

$$\int (x^2+3)^{1/2}\frac{x}{3}\,dx = \int \frac{1}{6}u^{1/2}\,du = \frac{1}{6}\frac{2}{3}u^{3/2} = \frac{1}{9}(x^2+3)^{3/2}.$$

Let us check our answer:

$$\left(\frac{1}{9}(x^2+3)^{3/2}\right)' = \frac{1}{9}\frac{3}{2}(x^2+3)^{1/2}(2x) = \frac{x}{3}(x^2+3)^{1/2}.$$

Example 7.11. To find an antiderivative $\int \cos(2t)\,dt$, let $u = 2t$. Then $\dfrac{du}{dt} = 2$, $du = 2\,dt$, and $\frac{1}{2}du = dt$. Then

$$\int \cos(2t)\,dt = \frac{1}{2}\int \cos u\,du = \frac{1}{2}\sin u = \frac{1}{2}\sin(2t).$$

Often, the change of variables formula is used in the opposite way, that is, by replacing the variable of integration u by a function of t, as in the following example.

Example 7.12. The area of a circle of radius r is represented by the integral

$$4\int_0^r \sqrt{r^2-u^2}\,\mathrm{d}u.$$

Using the change of variables $u = g(t) = r\sin t$, we obtain $g'(t) = r\cos t$. See Fig. 7.3. To determine the endpoints of integration, we notice that $u = r$ when $t = \frac{\pi}{2}$ and that $u = 0$ when $t = 0$. By the change of variables theorem, then,

$$4\int_0^r \sqrt{r^2-u^2}\,\mathrm{d}u = 4\int_0^{\frac{\pi}{2}} \sqrt{r^2-r^2\sin^2 t}\,r\cos t\,\mathrm{d}t = 4\int_0^{\frac{\pi}{2}} r^2\cos^2 t\,\mathrm{d}t.$$

We use the trigonometric identity $\cos^2 t = \frac{1}{2}(1+\cos(2t))$. Then this is equal to

$$4r^2\int_0^{\frac{\pi}{2}} \frac{1}{2}(1+\cos(2t))\,\mathrm{d}t = 2r^2\left[t+\frac{1}{2}\sin(2t)\right]_0^{\frac{\pi}{2}} = \pi r^2.$$

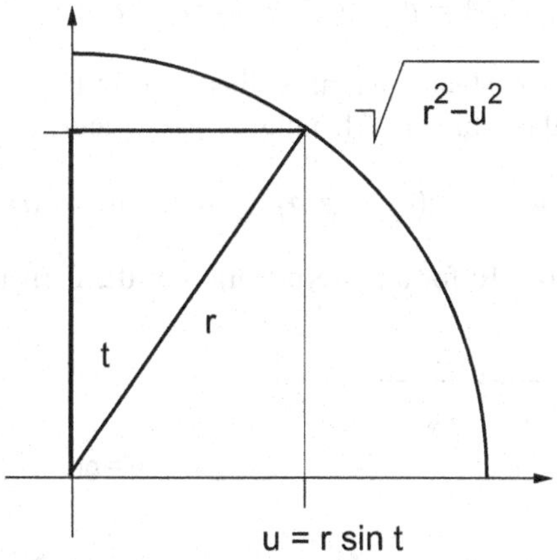

Fig. 7.3 The graph of $f(u) = \sqrt{r^2-u^2}$ on $[0,r]$, and the change of variables $u = r\sin t$ in Example 7.12

Example 7.13. Previously, we stated that the exact value of the integral (7.5) is $\int_0^1 x^2\sqrt{1-x^2}\,\mathrm{d}x = \frac{\pi}{16}$. If we use the substitution $x = \sin t$, $0 \le t \le \frac{\pi}{2}$, we will see why:

$$\int_0^1 x^2\sqrt{1-x^2}\,\mathrm{d}x = \int_0^{\frac{\pi}{2}} \sin^2 t\sqrt{1-\sin^2 t}(\cos t)\,\mathrm{d}t = \int_0^{\frac{\pi}{2}} \sin^2 t(1-\sin^2 t)\,\mathrm{d}t$$

$$= \int_0^{\frac{\pi}{2}} \sin^2 t\,\mathrm{d}t - \int_0^{\frac{\pi}{2}} \sin^4 t\,\mathrm{d}t = W_2 - W_4,$$

where the W_n are defined in Sect. 7.1d. Since $W_2 = \frac{\pi}{4}$ and $W_4 = \frac{3}{4}W_2$, we conclude that

$$\int_0^1 x^2\sqrt{1-x^2}\,\mathrm{d}x = \frac{\pi}{4} - \frac{3}{4}\frac{\pi}{4} = \frac{\pi}{16}.$$

In the last two examples, we simplified integrals involving $\sqrt{a^2-x^2}$ using the substitution $x = a\sin t$. This is called a trigonometric substitution. In Problem 7.14, we guide you through two other useful substitutions, $x = a\tan t$ for integrals involving $\sqrt{a^2+x^2}$, and $x = a\sec t$ for integrals involving $\sqrt{x^2-a^2}$.

Geometric Meaning of the Change of Variables. The change of variables formula has a geometric meaning that is not revealed by the somewhat formal proof we presented. Here is another, more revealing, proof, which we give in the special case that the new variable $u = g(t)$ in Theorem 7.2 is an increasing function.

Proof. Subdivide the interval $[a,b]$ on the t-axis into subintervals as

$$a = a_0 < a_1 < a_2 < \cdots < a_n = b.$$

The function g maps the interval $[a,b]$ on the t-axis into the interval $[A,B]$ on the u-axis, creating a subdivision

$$A = u_0 < u_1 < \cdots < u_n = B,$$

where $u_i = g(a_i)$. A typical subinterval is illustrated in Fig. 7.4. According to the mean value theorem, the subinterval lengths $u_j - u_{j-1}$ and $a_j - a_{j-1}$ are related by

$$u_j - u_{j-1} = g(a_j) - g(a_{j-1}) = g'(t_j)(a_j - a_{j-1})$$

for some t_j. (See Sect. 3.1c for an interpretation of the derivative as a "stretching factor.")

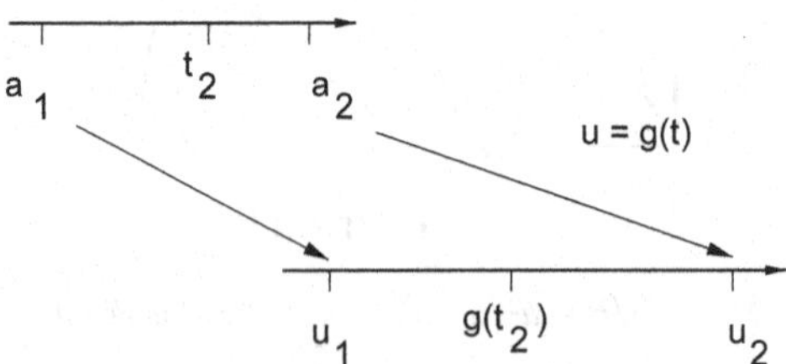

Fig. 7.4 The interval $[a_1,a_2]$ is mapped to $[u_1,u_2]$ by the change of variables $u = g(t)$

To show that

$$\int_A^B f(u)\,\mathrm{d}u = \int_a^b f(g(t))g'(t)\,\mathrm{d}t,$$

we look at sums approximating the integrals. On the left we choose to evaluate f at the points $g(t_j)$:

$$\sum_{j=1}^{n} f(g(t_j))(u_j - u_{j-1}).$$

On the right we evaluate $f \circ g$ at the points t_j:

$$\sum_{j=1}^{n} f(g(t_j))g'(t_j)(a_j - a_{j-1}).$$

Since g is an increasing function and t_j is between a_{j-1} and a_j, it follows that $g(t_j)$ is between u_{j-1} and u_j. With these choices, the two sums are equal. Since these sums approximate the integrals as closely as we wish, it follows that the integrals are equal too. □

As we have seen, the formula for changing variables in an integral holds out the possibility of transforming an integral into another one that can be expressed in terms of known functions. But it also can be used to give us insight into integrals more generally.

Example 7.14. Let f be any continuous function, and let g be the linear function

$$g(t) = kt + c,$$

where k and c are constants. We get by a change of variables that

$$\int_{ka+c}^{kb+c} f(x)\,\mathrm{d}x = \int_a^b f(kt+c)k\,\mathrm{d}t = k\int_a^b f(kt+c)\,\mathrm{d}t.$$

Notice that this embodies two rules:

(a) Setting $k = 1$, we obtain the translation invariance of the integral, and
(b) Setting $c = 0$, we obtain the effect of stretching on integrals.

Thus we see that the change of variables theorem is a powerful generalization of the simple rules we previously found.

Improving an Estimate by Change of Variables. Often after a change of variables, the transformed integral cannot be *evaluated* explicitly but is easier to *estimate* than the original integral. Here is an example:

We know how to evaluate the integral

$$I = \int_1^4 \frac{1}{1+x}\mathrm{d}x$$

exactly. It is $\left[\log(1+x)\right]_1^4 = \log\left(\frac{5}{2}\right) = 0.9162\ldots$. But suppose we did not know the antiderivative and we performed a quick estimate. On $[1,4]$, $\dfrac{1}{1+x}$ is between $\frac{1}{2}$ and $\frac{1}{5}$. Using the estimate of the integral in terms of the upper and lower bounds of the integrand, we get

$$0.6 = \frac{1}{5}(4-1) \le \int_1^4 \frac{1}{1+x}\mathrm{d}x \le \frac{1}{2}(4-1) = 1.5.$$

Now let us change variables. Set $x = t^2$. Then

$$I = \int_1^4 \frac{1}{1+x}\,\mathrm{d}x = \int_1^2 \frac{1}{1+t^2}\frac{\mathrm{d}x}{\mathrm{d}t}\,\mathrm{d}t = \int_1^2 \frac{2t}{1+t^2}\,\mathrm{d}t.$$

Now the length of the interval of integration is $2-1=1$. The new integrand is a decreasing function of t. Its maximum, taken on at $t=1$, is 1, and its minimum, taken on at $t=2$, is 0.8. Therefore, according to the lower and upper bound property, I lies between these limits:

$$0.8 \le I \le 1.$$

Notice how much narrower these bounds are than those we had previously.

The most important use by far of the formula for changing variables in an integral comes from frequent encounters with integrals where the integrand is not handed down from on high to be evaluated, but is a function that arises in another part of a larger problem. It often happens that it is necessary, or convenient, to change variables in the other part of the problem; it is then necessary to perform that change of variable in the integral, too.

Problems

7.12. Use the substitution $t = 2\sin\theta$ similar to that of Example 7.12 to evaluate

$$\int_0^1 \sqrt{4-t^2}\,\mathrm{d}t,$$

and use a similar sketch to help determine the limits of integration.

7.13. Use the substitution $u = 1+x^2$ to evaluate the integral

$$\int_1^2 \left(1+x^2\right)^{3/2} x\,dx.$$

7.14. Evaluate $\displaystyle\int_0^3 \frac{1}{9+x^2}\,\mathrm{d}x$ using the substitution $x = 3\tan t$.

7.15. Use a change of variables to evaluate the following integrals.

(a) $\displaystyle\int_0^1 \frac{t}{t^2+1}\,\mathrm{d}t$

(b) $\displaystyle\int_0^1 \frac{t}{(t^2+1)^2}\,\mathrm{d}t$

(c) $\displaystyle\int_0^1 \frac{1}{(t^2+1)^2}\,\mathrm{d}t$ *Hint*: let $t = \tan u$.

(d) $\displaystyle\int_{-1}^1 x^2 e^{x^3}\,\mathrm{d}x$

(e) $\displaystyle\int_{-1}^1 \frac{2t+3}{t^2+9}\,\mathrm{d}t$

(f) $\int_0^1 \sqrt{2+t^2}\,dt$ *Hint*: Let $t = \sqrt{2}\sinh u$. You may also need to refer to Problem 3.66.

7.16. Express the area bounded by the ellipse

$$\frac{x^2}{a^2} + \frac{y^2}{b^2} = 1$$

as a definite integral. Make a change of variables that converts this integral into one representing the area of a circle, and then evaluate it.

7.17. Evaluate

$$\int_0^1 \sqrt{1+\sqrt{x}}\,dx.$$

Hint: Let $\sqrt{x} = t$.

7.18. (a) Find $\frac{d}{dx}\log(\sec x + \tan x)$. For what values of x is your result valid?
(b) Explain which, if any, of the following integrals have been defined:

$$\int_1^{3/2} \sec x\,dx, \quad \int_1^{\pi/2} \sec x\,dx \quad \int_{\pi/2}^{2} \sec x\,dx \quad \int_1^{2} \sec x\,dx$$

7.19. Let g be a function whose derivative on the interval $[a,b]$ is negative. Prove that in this case, we have the following formula for changing variables in an integral:

$$\int_{g(b)}^{g(a)} f(x)\,dx = \int_a^b f(g(t))|g'(t)|\,dt.$$

7.20. Use the change of variable $x = \sin^2(\theta)$ to express the integral

$$\int_0^{\pi/2} \sin^{2n+1}(\theta)\cos^{2m+1}(\theta)\,d\theta$$

in terms of the numbers $B(n,m)$ in Problem 7.10.

7.21. Use the change of variables theorem to prove:

(a) If f is any continuous function on $[a,b]$ and if $f_r(x)$ is defined on $[a+r, b+r]$ according to the rule $f_r(x) = f(x-r)$, then

$$\int_a^b f(x)\,dx = \int_{a+r}^{b+r} f_r(x)\,dx.$$

(b) If f is continuous on $[a,b]$ and its reflection, denoted by f_-, is defined on $[-b,-a]$ as $f_-(x) = f(-x)$, then

$$\int_a^b f(x)\,dx = \int_{-b}^{-a} f_-(x)\,dx.$$

7.3 Improper Integrals

In this section, we study "improper" integrals. The first kind of improper integral arises when the interval of integration becomes infinite, e.g., if one or both endpoints of $[a,b]$ tend to plus or minus infinity. The second kind of improper integral arises when a function becomes unbounded at some point of the interval of integration. We shall see that an improper integral of either type may or may not be meaningful.

Example 7.15. Consider the integral $\int_1^b \frac{1}{x^2}\,\mathrm{d}x$. The function $\frac{1}{x^2}$ is continuous on $(0,\infty)$ and is the derivative of $-\frac{1}{x}$, so for $b>0$, the integral can be evaluated with the aid of the fundamental theorem of calculus:

$$\int_1^b \frac{1}{x^2}\,\mathrm{d}x = -\frac{1}{b} - (-1) = -\frac{1}{b} + 1.$$

Let b tend to infinity. Then $\frac{1}{b}$ tends to zero, so we get the result

$$\lim_{b\to\infty}\int_1^b \frac{1}{x^2}\,\mathrm{d}x = \lim_{b\to\infty}\left(-\frac{1}{b}+1\right) = 1.$$

The limit on the left is denoted, not unreasonably, by $\int_1^\infty \frac{1}{x^2}\,\mathrm{d}x$.

More generally, we make the following definition.

Definition 7.1. Whenever f is continuous on $[a,b]$ and $\int_a^b f(x)\,\mathrm{d}x$ tends to a limit as b tends to infinity, that is, whenever

$$\lim_{b\to\infty}\int_a^b f(x)\,\mathrm{d}x$$

exists, that limit is denoted by

$$\int_a^\infty f(x)\,\mathrm{d}x.$$

Such an integral is called an *improper integral*, and the function f is said to be *integrable* on $[a,\infty)$, and the integral is said to *converge*. If the limit does not exist, we say that the function f is not integrable on $[a,\infty)$ and that the improper integral *diverges*.

Example 7.16. Let $n > 1$ be any number, and $b > 0$. We get from the fundamental theorem of calculus that

$$\int_1^b \frac{1}{x^n}\,\mathrm{d}x = -\frac{1}{n-1}\frac{1}{b^{n-1}} + \frac{1}{n-1}.$$

Letting b tend to infinity, we deduce that $\frac{1}{b^{n-1}}$ tends to zero and

$$\int_1^\infty \frac{1}{x^n}\,\mathrm{d}x = \frac{1}{n-1} \qquad (n > 1),$$

so the integral converges.

Example 7.17. For $n < 1$, $\frac{1}{b^{n-1}}$ tends to infinity as b does, so

$$\int_1^\infty \frac{1}{x^n}\,\mathrm{d}x = \lim_{b\to\infty}\int_1^b \frac{1}{x^n}\,\mathrm{d}x = \lim_{b\to\infty}\left(-\frac{1}{n-1}\frac{1}{b^{n-1}} + \frac{1}{n-1}\right)$$

does not exist. Therefore,

$$\int_1^\infty \frac{1}{x^n}\,\mathrm{d}x \qquad \text{divergesfor } n < 1.$$

Example 7.18. Next, we show that the improper integral $\int_1^\infty \frac{1}{x}\,\mathrm{d}x$ does not converge. For $b > 1$, we have $\int_1^b \frac{1}{x}\,\mathrm{d}x = \log b - \log 1 = \log b$. The logarithm increases without bound, so the limit $\lim_{b\to\infty}\int_1^b \frac{1}{x}\,\mathrm{d}x$ does not exist, and

$$\int_1^\infty \frac{1}{x}\,\mathrm{d}x \qquad \text{diverges.}$$

Here is another way to see that the improper integral $\int_1^\infty \frac{1}{x}\,\mathrm{d}x$ diverges, a way that does not rely on our knowledge of the logarithm function. Recall from the lower and upper bound property of the integral that $\int_a^b f(x)\,\mathrm{d}x \geq (b-a)m$, where m is the minimum of f in $[a,b]$. Apply this to the integral $\int_a^{2a} \frac{1}{x}\,\mathrm{d}x$. The minimum of $\frac{1}{x}$ occurs at the right endpoint of the interval of integration, so

$$\int_1^2 \frac{1}{x}\,\mathrm{d}x > (2-1)\frac{1}{2} = \frac{1}{2}, \qquad \int_2^4 \frac{1}{x}\,\mathrm{d}x > (4-2)\frac{1}{4} = \frac{1}{2},$$

and so forth. This, together with the additivity of integrals, allows us to write the integral of $\frac{1}{x}$ over the interval $[1,2^k]$ as the following sum:

$$\int_1^{2^k} \frac{1}{x}\,\mathrm{d}x = \int_1^2 \frac{1}{x}\,\mathrm{d}x + \int_2^4 \frac{1}{x}\,\mathrm{d}x + \int_4^8 \frac{1}{x}\,\mathrm{d}x + \cdots + \int_{2^{k-1}}^{2^k} \frac{1}{x}\,\mathrm{d}x$$

$$> \frac{1}{2} + \frac{1}{2} + \frac{1}{2} + \cdots + \frac{1}{2} = \frac{k}{2}.$$

See Fig. 7.5. This estimate shows that as k tends to infinity, $\int_1^{2^k} \frac{1}{x}\,\mathrm{d}x$ also tends to infinity, and therefore does not have a limit. Thus the function $\frac{1}{x}$ is not integrable on $[1,\infty)$.

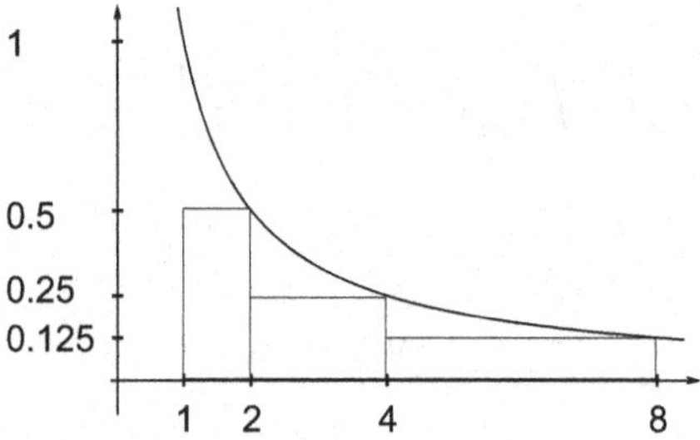

Fig. 7.5 The rectangles under the graph of $\frac{1}{x}$ each have area $\frac{1}{2}$

The Comparison Theorem. Now we present a very useful criterion for deciding which functions are integrable on $[a,\infty)$ and which are not. It is based on the monotonicity of the integral, that is, on the fact that functions with greater values have greater integrals. We start by examining what it means that the limit

$$\lim_{b\to\infty} \int_a^b f(x)\,\mathrm{d}x$$

exists. The meaning is that for b large enough, the value of the integral on $[a,b]$ depends very little on b. More precisely, for any choice of tolerance $\varepsilon > 0$ there is an N that depends on ε such that

$$\left| \int_a^{b_1} f(x)\,\mathrm{d}x - \int_a^{b_2} f(x)\,\mathrm{d}x \right| < \varepsilon$$

for every b_1 and b_2 larger than N, that is, for b_1 and b_2 large enough. For $a < b_1 < b_2$,

$$\int_a^{b_2} f(x)\,\mathrm{d}x = \int_a^{b_1} f(x)\,\mathrm{d}x + \int_{b_1}^{b_2} f(x)\,\mathrm{d}x.$$

Therefore, the above difference of integrals can be rewritten as a single integral over $[b_1, b_2]$, and we need

$$\left|\int_{b_1}^{b_2} f(x)\,dx\right| < \varepsilon \tag{7.9}$$

for b_1 and b_2 large enough.

Theorem 7.3. Comparison theorem for improper integrals. *Suppose f and g are continuous and*

$$|f(x)| \le g(x)$$

on $[a,\infty)$. If g is integrable on $[a,\infty)$, then so is f. The same result holds if $[a,\infty)$ is replaced by $(-\infty,a]$.

Proof. Since $|f(x)| \le g(x)$,

$$\left|\int_{b_1}^{b_2} f(x)\,dx\right| \le \int_{b_1}^{b_2} |f(x)|\,dx \le \int_{b_1}^{b_2} g(x)\,dx.$$

Since g is integrable from a to ∞, then according to (7.9),

$$\int_{b_1}^{b_2} g(x)\,dx < \varepsilon$$

for b_1, b_2 large enough. Then it follows also that

$$\left|\int_{b_1}^{b_2} f(x)\,dx\right| < \varepsilon.$$

This means that f is integrable on $[a,\infty)$. □

We can also obtain the result for intervals $(-\infty,a]$. This theorem can be used as an integrability criterion for f, or a nonintegrability criterion for g. We illustrate with a few examples.

Example 7.19. Is the function $\frac{1}{1+x^2}$ integrable on $[1,\infty)$? To answer this question quickly, observe that

$$\frac{1}{1+x^2} < \frac{1}{x^2}.$$

We have seen that the function $\frac{1}{x^2}$ is integrable on $[1,\infty)$. Therefore, it follows from the comparison theorem that $\frac{1}{1+x^2}$ is also integrable on $[1,\infty)$.

We know how to evaluate $\int_1^\infty \frac{1}{1+x^2}\,dx$ exactly, so let us redo the last example.

Example 7.20. We have

$$\int_1^\infty \frac{1}{1+x^2}\,\mathrm{d}x = \lim_{t\to\infty}\int_1^t \frac{1}{1+x^2}\,\mathrm{d}x$$

$$= \lim_{t\to\infty}\left[\tan^{-1}x\right]_1^t = \lim_{t\to\infty}\left(\tan^{-1}t - \tan^{-1}1\right) = \frac{\pi}{2} - \frac{\pi}{4} = \frac{\pi}{4}.$$

The comparison theorem gives us a tool that allows us sometimes to quickly determine the convergence or divergence of an improper integral. But it does not tell us the value of a convergent integral.

Example 7.21. Is the function $\dfrac{x}{1+x^2}$ integrable on $[1,\infty)$? Since $x^2 \ge 1$ on the interval in question, we have the inequality

$$\frac{x}{1+x^2} \ge \frac{x}{x^2+x^2} = \frac{1}{2x} \quad \text{for} \quad x \ge 1.$$

The function $\dfrac{1}{2x}$ is just half the function $\dfrac{1}{x}$. We have seen that $\displaystyle\int_1^b \frac{1}{x}\,\mathrm{d}x$ tends to infinity with b. Therefore, it follows from the comparison theorem that

$$\int_1^b \frac{x}{1+x^2}\,\mathrm{d}x \ge \frac{1}{2}\int_1^b \frac{1}{x}\,\mathrm{d}x,$$

so $\displaystyle\int_1^\infty \frac{x}{1+x^2}\,\mathrm{d}x$ diverges.

Example 7.22. We evaluate, using the change of variables $u = 1+x^2$,

$$\int_1^b \frac{x}{1+x^2}\,\mathrm{d}x = \frac{1}{2}\int_2^{1+b^2} \frac{1}{u}\,\mathrm{d}u = \frac{1}{2}\left(\log\left(1+b^2\right) - \log 2\right).$$

This increases without bound as b tends to infinity. This shows that $\displaystyle\int_1^\infty \frac{x}{1+x^2}\,\mathrm{d}x$ diverges, as we saw in Example 7.21 by a simple comparison.

Setting up a simple comparison can be less onerous than evaluating the integral directly, as we see in the next example.

Example 7.23. Is $\dfrac{\sin x}{x^2}$ integrable on $(-\infty,-1]$? Since $\left|\dfrac{\sin x}{x^2}\right| \le \dfrac{1}{x^2}$, and since we saw that $\dfrac{1}{x^2}$ is integrable on $(-\infty,-1]$, so is $\dfrac{\sin x}{x^2}$.

Example 7.24. Is the function e^{-x^2} integrable on $[1,\infty)$? Since $x < x^2$ for $1 < x$, it follows that $0 \le \mathrm{e}^{-x^2} \le \mathrm{e}^{-x}$. The function e^{-x} is integrable on $[1,\infty)$, since

$\int_1^b e^{-x}\,dx = e^{-1} - e^{-b}$, which tends to e^{-1} as b tends to infinity. According to the comparison theorem,

$$\int_1^\infty e^{-x^2}\,dx \qquad \text{converges.}$$

It can be shown that

$$\int_0^\infty e^{-x^2}\,dx = \frac{1}{2}\sqrt{\pi}. \tag{7.10}$$

The convergence or divergence of improper integrals can be used as a criterion for the convergence or divergence of infinite series.

Theorem 7.4. Integral test for the convergence of a series.
Let $f(x)$ be a positive decreasing continuous function on $[1,\infty)$.

(a) Suppose f is integrable on $[1,\infty)$, and let $\sum_{n=1}^{\infty} a_n$ be an infinite series whose terms satisfy the inequalities

$$|a_n| \le f(n).$$

Then the series $\sum_{n=1}^{\infty} a_n$ converges.

(b) Let $\sum_{n=1}^{\infty} a_n$ be a convergent infinite series whose terms satisfy the inequalities

$$a_n \ge f(n).$$

Then f is integrable on $[1,\infty)$.

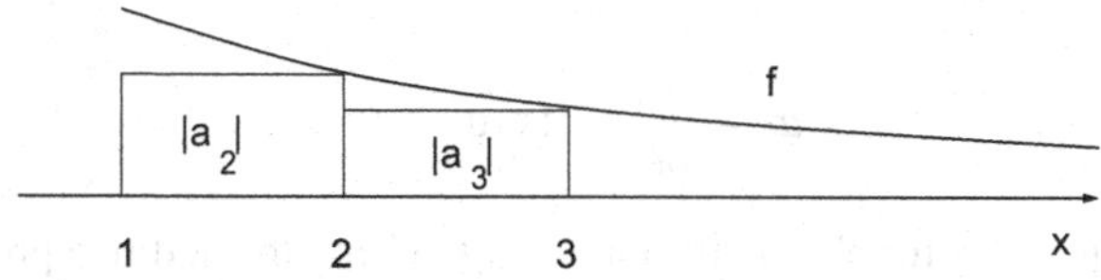

Fig. 7.6 The idea of the proof of Theorem 7.4, part (a). The numbers $|a_n|$ may be viewed as areas

Proof. Since $f(x)$ is assumed to be decreasing, its minimum on the interval $[n-1,n]$ occurs at the right endpoint n. By the lower bound property of the integral (Fig.7.6),

$$f(n) \le \int_{n-1}^{n} f(x)\,dx.$$

Using the hypothesis that $|a_n| \le f(n)$, we conclude, therefore, that

$$|a_n| \le \int_{n-1}^{n} f(x)\,dx.$$

Add these inequalities for all n between j and k. Using the additive property of the integral, we get

$$|a_{j+1}|+\cdots+|a_k| \le \int_j^{j+1} f(x)\,\mathrm{d}x+\cdots+\int_{k-1}^{k} f(x)\,\mathrm{d}x = \int_j^k f(x)\,\mathrm{d}x.$$

The function f is assumed to be integrable on $[1,\infty)$, so the right-hand side is less than ε for j and k large enough. Therefore,

$$|a_{j+1}+\cdots+a_k| \le |a_{j+1}|+\cdots+|a_k| \le \varepsilon$$

for j, k large enough. This means that the jth and kth partial sums of $\sum_{n=1}^{\infty} a_n$ differ by very little for j and k large enough, and the series $\sum_{n=1}^{\infty} a_n$ converges (see Theorem 1.20). This proves part (a).

To prove part (b), we observe that since f is decreasing, its maximum on the interval $(n,n+1)$ is reached at n. By the upper bound property of the integral (Fig. 7.7),

$$f(n) \ge \int_n^{n+1} f(x)\,\mathrm{d}x.$$

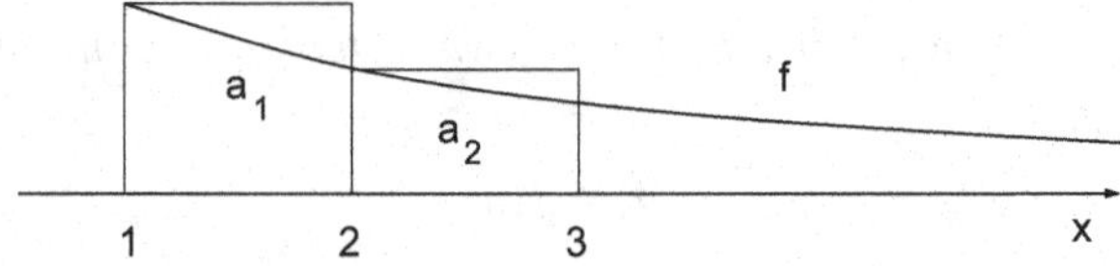

Fig. 7.7 The idea of the proof of Theorem 7.4, part (b). The numbers a_n are viewed as areas

Using the assumption that $a_n \ge f(n) \ge 0$, we conclude that

$$a_n \ge \int_n^{n+1} f(x)\,\mathrm{d}x \ge 0.$$

Add these inequalities for all n between j and k. Using the additive property of the integral, we get

$$a_j+\cdots+a_{k-1} \ge \int_j^{j+1} f(x)\,\mathrm{d}x+\cdots+\int_{k-1}^{k} f(x)\,\mathrm{d}x = \int_j^k f(x)\,\mathrm{d}x \ge 0.$$

We have assumed that the series $\sum_{n=1}^{\infty} a_n$ converges, and hence its jth and kth partial sums differ by very little for j and k large enough. This means that no matter how small ε is, we have

$$a_j+\cdots+a_{k-1} \le \varepsilon$$

for j, k large enough. It follows then that

$$\left| \int_j^k f(x)\,\mathrm{d}x \right| \le \varepsilon.$$

This implies that f is integrable on $[1,\infty)$. This completes the proof of the integral test for convergence. □

The integral test is enormously useful in applications. Here are some examples.

Example 7.25. $f(x) = \dfrac{1}{x^p}$, $p > 1$. In this case, we have seen that f is integrable on $[1,\infty)$. Set

$$a_n = f(n) = \frac{1}{n^p}.$$

It follows from part (a) of the integral test for convergence that

$$\sum_{n=1}^{\infty} \frac{1}{n^p} = 1 + \frac{1}{2^p} + \frac{1}{3^p} + \cdots$$

converges for $p > 1$. This sum is the celebrated zeta function $\zeta(p)$ of number theory.

Example 7.26. We saw in Example 1.21 that the series

$$\sum_{n=1}^{\infty} \frac{1}{n}$$

diverges. Here is a new argument using the integral test. Take $f(x) = \dfrac{1}{x}$, and set $a_n = f(n) = \dfrac{1}{n}$. If the series converged, then according to part (b) of the integral test, the function $f(x) = \dfrac{1}{x}$ would be integrable over $[1,\infty)$, but we have seen that it is not.

Unbounded Integrand. We turn now to another class of integrals also called "improper," with the feature that the integrand is not bounded on the closed interval of integration.

Example 7.27. Consider the integral $\displaystyle\int_a^1 \frac{1}{\sqrt{x}}\,\mathrm{d}x$, where $a > 0$. The function $\dfrac{1}{\sqrt{x}}$ is the derivative of $2\sqrt{x}$, so the fundamental theorem of calculus yields $\displaystyle\int_a^1 \frac{1}{\sqrt{x}}\,\mathrm{d}x = 2 - 2\sqrt{a}$. Now let a tend to zero. Then $\sqrt{a}$ also tends to zero, and

$$\lim_{a\to 0} \int_a^1 \frac{1}{\sqrt{x}}\,\mathrm{d}x = 2.$$

The limit on the left is denoted, not unreasonably, by $\displaystyle\int_0^1 \frac{1}{\sqrt{x}}\,\mathrm{d}x$.

Although the function $\dfrac{1}{\sqrt{x}}$ becomes unbounded as x tends to zero, its integral over $[0,1]$ exists. More generally, we make the following definition.

Definition 7.2. Let f be a function defined on a half-open interval $(a,b]$ such that f is continuous on every subinterval $[a+h,b]$, with $h>0$, but not on the interval $[a,b]$ itself. The function f is called *integrable on* $(a,b]$ if the limit

$$\lim_{h\to 0}\int_{a+h}^{b} f(x)\,\mathrm{d}x$$

exists. This limit is denoted by $\displaystyle\int_a^b f(x)\,\mathrm{d}x$ and is called an improper integral. If the limit does not exist, we say that f is not integrable on $(a,b]$, or that the integral *diverges*.

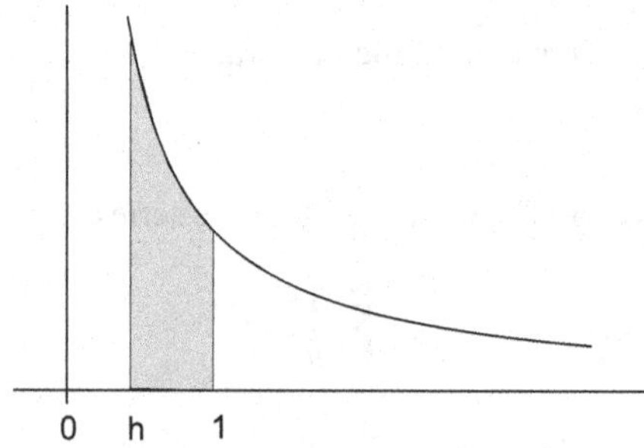

Fig. 7.8 The graph of $\frac{1}{x}$ for Example 7.28. The shaded area tends to infinity as $h\to 0$

Example 7.28. Consider $\displaystyle\int_0^1 \frac{1}{x}\,\mathrm{d}x$. The function $\dfrac{1}{x}$ is not defined at the left endpoint $x=0$ of the interval of integration. We add a small number $0<h<1$ to the left endpoint, and we integrate over the modified interval $[h,1]$. The fundamental theorem yields

$$\int_h^1 \frac{1}{x}\,\mathrm{d}x = \log 1 - \log h = \log\left(\frac{1}{h}\right).$$

As $h\to 0$, $\log\left(\frac{1}{h}\right)$ tends to infinity, so the integral diverges. See Fig. 7.8.

Example 7.29. Consider the integral $\displaystyle\int_0^1 \frac{1}{x^p}\,\mathrm{d}x$, where $p\neq 1$. Figure 7.9 shows graphs of some of the integrands. Since the function $\dfrac{1}{x^p}$ is the derivative of $\dfrac{x^{1-p}}{1-p}$, it follows from the fundamental theorem that for $h>0$,

$$\int_h^1 \frac{1}{x^p}\,\mathrm{d}x = \frac{1}{1-p} - \frac{h^{1-p}}{1-p}.$$

When $p < 1$, h^{1-p} tends to zero as h does. Therefore, the improper integral

$$\int_0^1 \frac{1}{x^p}\,dx \text{ is equal to } \frac{1}{1-p} \qquad \text{for all } 0 < p < 1.$$

When $p > 1$, h^{1-p} tends to infinity. Therefore,

$$\int_0^1 \frac{1}{x^p}\,dx \qquad \text{diverges if } p > 1.$$

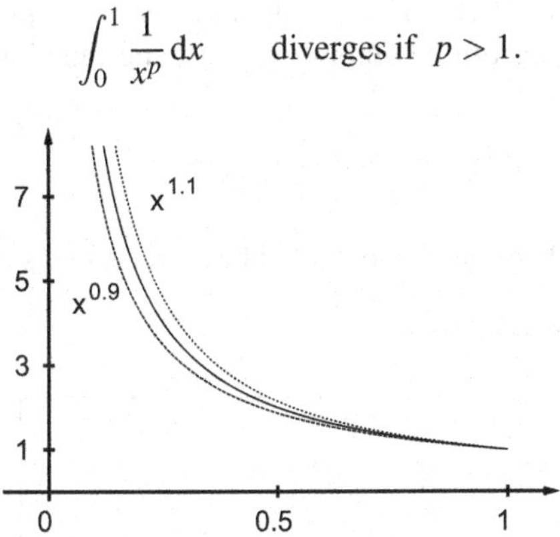

Fig. 7.9 The graphs of $\frac{1}{x^p}$ for $p = 0.9$, 1, and 1.1, for Example 7.29. As $h \to 0$, the area between $[h, 1]$ and the graph converges if $p < 1$ and diverges if $p \geq 1$

Example 7.30. Consider $\int_{-1}^1 \frac{-1}{x^2}\,dx$. Since the integrand is the derivative of $\frac{1}{x}$, it is tempting to use the fundamental theorem of calculus, giving $\left[\frac{1}{x}\right]_{-1}^1 = 2$. However, the theorem cannot be applied, because $\frac{-1}{x^2}$ is not continuous on $[-1, 1]$. The right way to approach such an integral is to evaluate the two improper integrals $\int_{-1}^0 \frac{-1}{x^2}\,dx$ and $\int_0^1 \frac{-1}{x^2}\,dx$. If they exist, then their sum is the answer. Otherwise, the integral diverges. In fact, we know that the second of these integrals diverges, so the integral

$$\int_{-1}^1 \frac{-1}{x^2}\,dx \qquad \text{does not exist.}$$

The comparison criterion also applies to improper integrals. If $|f(x)| \leq g(x)$ on $(a, b]$, and if g is integrable on $(a, b]$, then f is also integrable on $(a, b]$.

Example 7.31. Consider the improper integral

$$\int_0^1 \frac{1}{\sqrt{x + x^2}}\,dx.$$

The integrand satisfies the inequality $\frac{1}{\sqrt{x+x^2}} < \frac{1}{\sqrt{x}} = \frac{1}{x^{\frac{1}{2}}}$. Since we have seen in Example 7.29 that $\frac{1}{x^{\frac{1}{2}}}$ is integrable on $(0, 1]$, the integral converges.

Example 7.32. The function $f(x) = \dfrac{\sin x}{x}$ is not defined at $x = 0$. But because

$$\lim_{x \to 0} \frac{\sin x}{x} = 1,$$

we can extend the definition of f to be 1 at $x = 0$. Then f is continuous at 0. (See the graph of f in Fig. 7.10.) This is called a *continuous extension* of f, and we make these extensions often without renaming the function. So from this point of view, we say that

$$\int_0^1 \frac{\sin x}{x}\,\mathrm{d}x$$

is a proper integral because the integrand has a continuous extension that is integrable.

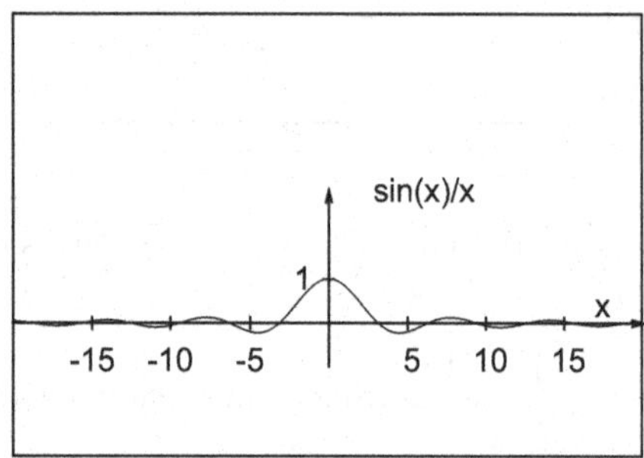

Fig. 7.10 The graph of $\dfrac{\sin x}{x}$ in Example 7.32

A Change of Variables. Next we show how the introduction of $z = \dfrac{1}{x}$ as a new variable of integration can change an improper integral into a proper one, sometimes exchanging an interval of infinite length for a finite one. According to the change of variable formula for integrals,

$$\int_a^b f(x)\,\mathrm{d}x = \int_{1/b}^{1/a} f\left(\frac{1}{z}\right)\frac{1}{z^2}\,\mathrm{d}z.$$

Suppose that the function $f(x)$ is such that as x tends to infinity, $x^2 f(x)$ tends to a finite limit L. Then the change of variable leads to the function $f\left(\dfrac{1}{z}\right)\dfrac{1}{z^2}$, which tends to the same limit L as z tends to zero. That is, $f\left(\dfrac{1}{z}\right)\dfrac{1}{z^2}$ has a continuous extension to the value L at $z = 0$. It follows that as b tends to infinity, the integral $\displaystyle\int_a^b f(x)\,\mathrm{d}x$ tends to the proper integral

$$\int_0^{1/a} f\left(\frac{1}{z}\right)\frac{1}{z^2}\,dz.$$

The next example makes use of such a change of variable.

Example 7.33. We consider $\int_1^{\infty} \frac{1}{1+x^2}\,dx$. The change of variables $x = \frac{1}{z}$ gives

$$\lim_{b\to\infty} \int_1^b \frac{1}{1+x^2}\,dx = \lim_{b\to\infty} \int_{1/b}^1 \frac{1}{1+\left(\frac{1}{z}\right)^2}\frac{1}{z^2}\,dz = \lim_{b\to\infty} \int_{1/b}^1 \frac{1}{z^2+1}\,dz.$$

As b tends to infinity, the integral on the right tends to the perfectly proper integral (Fig. 7.11)

$$\int_0^1 \frac{1}{z^2+1}\,dz = \tan^{-1} 1 - \tan^{-1} 0 = \frac{\pi}{4}.$$

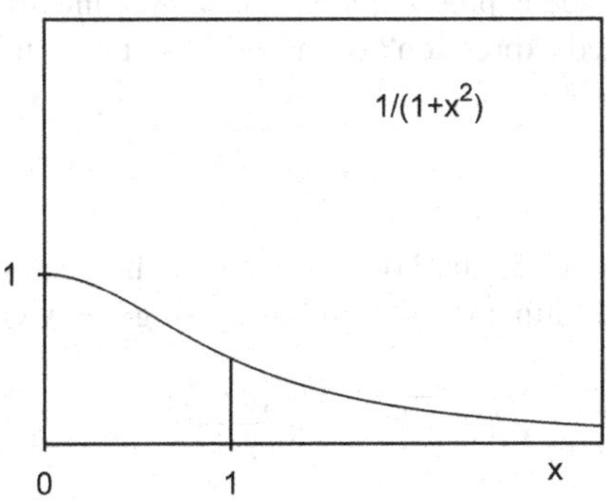

Fig. 7.11 The graph of $\frac{1}{1+x^2}$. The area over $[0,1]$ is equal to the area over $[1,\infty)$. See Example 7.33

Estimation of $n!$. We have shown in Sect. 4.1 that the exponential function e^x grows faster than any power of x as x tends to infinity. It follows that $x^n e^{-x}$ goes to zero faster than any negative power of x, no matter how large n is. It follows that the function $x^n e^{-x}$ is integrable from 0 to infinity (Fig. 7.12).

We show now how to evaluate, using integration by parts, the improper integral

$$\int_0^{\infty} x^n e^{-x}\,dx,$$

where n is a positive integer. Write the integrand as $f'(x)g(x)$, where $f(x) = -e^{-x}$ and $g(x) = x^n$. We integrate from 0 to b and then let b tend to infinity. The product

$f(x)g(x)$ is zero at $x=0$, and its value at $x=b$ tends to zero as b tends to infinity. Since $g'(x)=nx^{n-1}$, we get

$$\int_0^\infty x^n e^{-x}\,dx = \lim_{b\to\infty}\left(\left[-e^{-x}x^n\right]_0^b + \int_0^b nx^{n-1}e^{-x}\,dx\right) = n\int_0^\infty x^{n-1}e^{-x}\,dx.$$

Repeating this procedure, we obtain $\int_0^\infty x^n e^{-x}\,dx = n(n-1)\int_0^\infty x^{n-2}e^{-x}\,dx$. After n integrations by parts, we obtain

$$\int_0^\infty x^n e^{-x}\,dx = n!\int_0^\infty e^{-x}\,dx.$$

For every t, $\int_0^t e^{-x}\,dx = -e^{-t}-(-1)$, which tends to 1 as t tends to infinity. Therefore, we have shown that

$$\int_0^\infty x^n e^{-x}\,dx = n!. \tag{7.11}$$

One is tempted to think of formula (7.11) as expressing a complicated integral on the left by the simple expression $n!$. But actually, it is the other way around; (7.11) expresses the complicated expression $n!$ by the simple integral on the left! Why do we call $n!$ a complicated expression? Because it is not easy to estimate how large it is for large n. We can replace all the factors $1,2,3,\ldots$ by n and conclude that

$$n! < n^n,$$

but this is a very crude estimate. We can obtain a sharper estimate by pairing the factors k and $n-k$ and estimating their product using the A-G inequality:

$$\sqrt{k(n-k)} \le \frac{k+(n-k)}{2} = \frac{n}{2}.$$

Squaring this, we get $k(n-k)\le(n/2)^2$. In the product $n!$ we have $n/2$ pairs k and $n-k$, so applying the above inequality to all such pairs, we get that

$$n! < \left(\frac{n}{2}\right)^n. \tag{7.12}$$

This is an improvement by a factor of 2^n over our previous crude estimate, but as we shall see, this is still a very crude estimate.

As we shall show, the integral on the left in Eq. (7.11) is much easier to estimate, since most of the contribution to an integral comes from around the point where the integrand reaches its maximum. Our first task is to locate the maximum of $x^n e^{-x}$. This is easily accomplished by looking at the derivative of the integrand. Using the product rule, we get

$$\frac{d}{dx}\left(x^n e^{-x}\right) = nx^{n-1}e^{-x} - x^n e^{-x}.$$

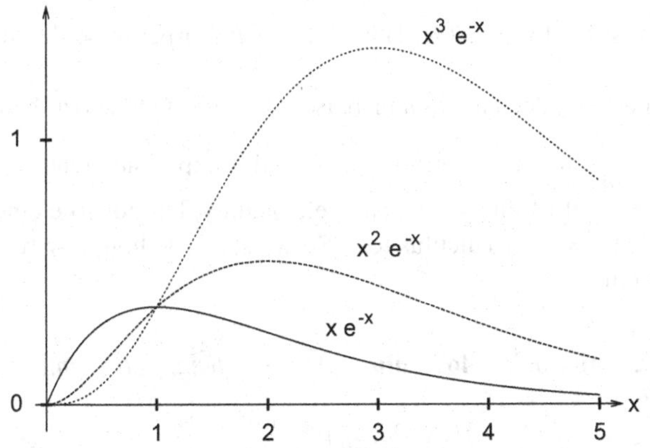

Fig. 7.12 Graphs of $x^n e^{-x}$ for $n = 1,2,3$

We can write the right side as

$$(n-x)x^{n-1}e^{-x}.$$

This formula shows that the derivative of $x^n e^{-x}$ is positive for x less than n, and negative for x greater than n. This shows that $x^n e^{-x}$ reaches its maximum at $x = n$; its value there is $n^n e^{-n}$. Let us factor it out:

$$x^n e^{-x} = \left(\frac{n}{e}\right)^n \left(\frac{x}{n}\right)^n e^{n-x}.$$

So we can write the integral for $n!$ as

$$n! = \left(\frac{n}{e}\right)^n \int_0^\infty \left(\frac{x}{n}\right)^n e^{n-x}\, dx.$$

We make the change of variable $x = ny$, $dx = n\,dy$ and get

$$n! = \left(\frac{n}{e}\right)^n \int_0^\infty y^n e^{n(1-y)} n\, dy = n\left(\frac{n}{e}\right)^n \int_0^\infty \left(ye^{1-y}\right)^n dy. \tag{7.13}$$

We denote the integral on the right in Eq. (7.13) by $d(n)$ and rewrite (7.13) as

$$n! = \frac{n^{n+1}}{e^n} d(n), \tag{7.14}$$

where

$$d(n) = \int_0^\infty \left(ye^{1-y}\right)^n dy.$$

What can we say about this integral? The integrand is the nth power of ye^{1-y}, a function that is equal to 1 at $y = 1$ and is less than 1 everywhere else. This shows that the integrand is a decreasing function of n for every y; therefore, so is its integral

$d(n)$. This shows that formula (7.14) gives a much sharper upper bound for $n!$ than Eq. (7.12).

How fast does $d(n)$ decrease as n increases? It turns out that $d(n)$ behaves asymptotically like $\sqrt{\frac{2\pi}{n}}$; that is, the ratio of $d(n)$ to this expression tends to 1 as n tends to infinity. The proof of this statement is elementary, but not so elementary that it belongs to an introductory calculus text. So we state, without any further ado, the following theorem.

Theorem 7.5. Stirling's formula. *As n tends to infinity, $n!$ is asymptotic to*

$$\sqrt{2\pi n}\left(\frac{n}{e}\right)^n,$$

in the sense that the ratio of $n!$ to this expression tends to 1.

Problems

7.22. Two sequences a_n and b_n are said to be asymptotic to each other if their ratio $\frac{a_n}{b_n}$ tends to 1 as n tends to infinity. If a_n and b_n are asymptotic to each other, does their difference $a_n - b_n$ tend to 0?

7.23. Suppose a_n and b_n are positive asymptotic sequences.

(a) Are na_n and nb_n asymptotic to each other? How about na_n and $\sqrt{1+n^2}b_n$?
(b) Find $\lim_{n\to\infty}(\log a_n - \log b_n)$.

7.24. We stated Stirling's formula, that $\sqrt{2\pi n}\left(\frac{n}{e}\right)^n \sim n!$. Use this to find the limit of

$$\log(n!) - \left(n+\frac{1}{2}\right)\log n + n$$

as n tends to infinity.

7.25. Use the integral test to determine which of these series converge.

(a) $\sum_{n=1}^{\infty}\frac{1}{n^2}$
(b) $\sum_{n=1}^{\infty}\frac{1}{n^{1.2}}$
(c) $\sum_{n=2}^{\infty}\frac{1}{n\log n}$
(d) $\sum_{n=1}^{\infty}\frac{1}{n^{9/10}}$

7.26. Which of the functions below are integrable on $[0,\infty)$?

(a) $\dfrac{1}{\sqrt{x}(1+x)}$

(b) $\dfrac{x}{1+x^2}$

(c) $\dfrac{\sqrt{x+1}-\sqrt{x}}{1+x}$

(d) $\dfrac{1}{x+x^4}$

Hint: Use the comparison theorem. Treat the two endpoints separately.

7.27. Let $f(x)=\dfrac{p(x)}{q(x)}$, where $q(x)$ is a polynomial of degree $n\geq 2$, and suppose that $q(x)$ is nonzero for $x\geq 1$. Let $p(x)$ be a polynomial of degree $\leq n-2$.

(a) Show that the introduction of $z=x^{-1}$ as a new variable of integration turns the improper integral $\int_1^\infty f(x)\,\mathrm{d}x$ into a proper one.

(b) Use this method to show that $\displaystyle\int_1^\infty \frac{1}{1+x^3}\,\mathrm{d}x=\int_0^1 \frac{z}{z^3+1}\,\mathrm{d}z.$

7.28. Integrate by parts to turn the improper integral $\displaystyle\int_0^1 \frac{1}{\sqrt{x+x^2}}\,\mathrm{d}x$ into a proper one.

7.29. Integrate by parts to verify that the improper integral $\displaystyle\int_1^\infty \frac{\sin x}{x}\,\mathrm{d}x$ converges. Then use the following items to verify that $\displaystyle\int_1^\infty \frac{|\sin x|}{x}\,\mathrm{d}x$ does not converge:

(a) Explain why $\displaystyle\int_1^{n\pi} \frac{|\sin x|}{x}\,\mathrm{d}x \geq \sum_{k=2}^{n}\int_{(k-1)\pi}^{k\pi} \frac{|\sin x|}{x}\,\mathrm{d}x.$

(b) Explain why $\displaystyle\int_{(k-1)\pi}^{k\pi} \frac{|\sin x|}{x}\,\mathrm{d}x \geq \frac{1}{k\pi}\int_{(k-1)\pi}^{k\pi} |\sin x|\,\mathrm{d}x.$

(c) Verify that $\displaystyle\int_{(k-1)\pi}^{k\pi} |\sin x|\,\mathrm{d}x = 2.$

(d) Why do these facts imply that $\displaystyle\int_1^\infty \frac{|\sin x|}{x}\,\mathrm{d}x$ diverges?

7.30. Show that the following improper integrals exist, and evaluate them.

(a) $\displaystyle\int_1^\infty \frac{\mathrm{d}x}{x^{1.0001}}$

(b) $\displaystyle\int_0^4 \frac{\mathrm{d}x}{\sqrt{4-x}}$

(c) $\displaystyle\int_0^1 \frac{\mathrm{d}x}{x^{0.9999}}$

(d) $\displaystyle\int_0^1 \frac{\mathrm{d}x}{x^{2/3}}$

7.31. Show that the integral $\int_s^\infty \frac{1}{x\log x}\,\mathrm{d}x$ diverges, where $s>1$.

The next two problems indicate an algebraic method that is useful for some integrals.

7.32.(a) Bring to a common denominator $\frac{A}{x+2}+\frac{B}{x-2}$.

(b) Write $\frac{3x-4}{x^2-4}$ as a sum $\frac{3x-4}{x^2-4}=\frac{A}{x+2}+\frac{B}{x-2}$.

(c) Evaluate $\int_3^4 \frac{3x-4}{x^2-4}\,\mathrm{d}x$.

7.33. Evaluate the integral $\int_2^\infty \frac{1}{y-y^2}\,\mathrm{d}y$.

7.34. Show that $\int_0^\infty t^n \mathrm{e}^{-pt}\,\mathrm{d}t = \frac{n!}{p^{n+1}}$ for positive p by repeated integration by parts.

7.35. Use integration by parts twice to show that

$$\int_0^\infty \sin(at)\mathrm{e}^{-pt}\,\mathrm{d}t = \frac{a}{a^2+p^2} \qquad \text{and} \qquad \int_0^\infty \cos(at)\mathrm{e}^{-pt}\,\mathrm{d}t = \frac{p}{a^2+p^2}.$$

7.4 Further Properties of Integrals

7.4a Integrating a Sequence of Functions

We have seen that polynomial functions are very easy to integrate, and that many important functions can be represented by Taylor series. We would like to know whether we may evaluate the integral of f on $[a,b]$ by integrating its Taylor series term by term instead. The next theorem answers this question.

Theorem 7.6. Convergence theorem for integrals. *If a sequence of functions f_n converges uniformly on an interval to f, then for any numbers a and b in the interval, the sequence of integrals of f_n converges to the integral of f:*

$$\lim_{n\to\infty}\int_a^b f_n(t)\,\mathrm{d}t = \int_a^b f(t)\,\mathrm{d}t.$$

Proof. The f_n converge uniformly to f on an interval containing $[a,b]$. This means that for every $\varepsilon>0$, $f_n(t)$ and $f(t)$ differ by less than ε for all t on $[a,b]$, provided that n is large enough.

Choose N so large that $|f_n(t)-f(t)|<\varepsilon$ for $n\ge N$, $a\le t\le b$. By the properties of integrals, we have

$$\left| \int_a^b f_n(t)\,dt - \int_a^b f(t)\,dt \right| \le \int_a^b |f_n(t) - f(t)|\,dt \le \varepsilon(b-a)$$

for all $n \ge N$. So $\lim_{n\to\infty} \int_a^b f_n(t)\,dt = \int_a^b f(t)\,dt$. □

Example 7.34. We know that $\sum_{n=0}^{\infty} \frac{x^{2n+1}}{(2n+1)!} = x - \frac{x^3}{3!} + \frac{x^5}{5!} - \cdots$ converges uniformly to $\sin x$ on every interval $[-c,c]$. Therefore, the series

$$\sum_{n=0}^{\infty} \frac{(x^2)^{2n+1}}{(2n+1)!} = x^2 - \frac{x^6}{3!} + \frac{x^{10}}{5!} - \cdots$$

converges uniformly to $\sin(x^2)$ on every interval $[a,b]$. By Theorem 7.6,

$$\int_0^1 \sin(x^2)\,dx = \left[\frac{x^3}{3} - \frac{x^7}{3!7} + \frac{x^{11}}{5!11} - \cdots \right]_0^1 = \frac{1}{3} - \frac{1}{3!7} + \frac{1}{5!11} - \cdots.$$

Theorem 7.6 may also be used to generate power series that converge to a given function, as in the next examples.

Example 7.35. We know from Sect. 2.6a that the geometric series

$$1 - t + t^2 - t^3 + \cdots$$

converges uniformly to $\frac{1}{1+t}$ on every interval $[-c,c]$, where $0 < c < 1$. According to Theorem 7.6, we may integrate from 0 to any number x in $(-1,1)$. Integrating from 0 to x is a new way to obtain the power series for $\log(1+x)$:

$$x - \frac{x^2}{2} + \frac{x^3}{3} - \frac{x^4}{4} + \cdots = \log(1+x),$$

with uniform convergence on every $[-c,c]$, where $0 < c < 1$.

Example 7.36. The sequence of polynomials

$$f_n(t) = \sum_{k=0}^{n} (-1)^k t^{2k} = 1 - t^2 + t^4 - t^6 + \cdots + (-1)^n t^{2n}$$

converges uniformly to $\frac{1}{1-(-t^2)} = \frac{1}{1+t^2}$ in every interval $[-c,c]$ where $0 < c < 1$. The $f_n(t)$ are in fact the Taylor polynomials of $f(t) = \frac{1}{1+t^2}$ at $a = 0$. By Theorem 7.6, we can write for $-1 < x < 1$,

$$\int_0^x \frac{1}{1+t^2}\,dt = \lim_{n\to\infty} \int_0^x \left(1 - t^2 + t^4 - t^6 + \cdots + (-1)^n t^{2n}\right) dt$$

$$= \lim_{n\to\infty}\left(x - \frac{x^3}{3} + \frac{x^5}{5} - \cdots + (-1)^n \frac{x^{2n+1}}{2n+1}\right) = \sum_{n=0}^{\infty}(-1)^n \frac{x^{2n+1}}{2n+1}.$$

By the fundamental theorem of calculus, $\int_0^x \frac{1}{1+t^2}\,dt = \tan^{-1}x$, and we get $\tan^{-1}x = \sum_{n=0}^{\infty}(-1)^n \frac{x^{2n+1}}{2n+1}$ for $|x| < 1$.

Theorem 7.6 is a great convenience when it applies. Happily, as we saw in Chap. 2, when Taylor polynomials converge, they do so uniformly on $[a,b]$, so opportunities to use Theorem 7.6 abound. The next example, however, demonstrates that strange things can happen if a sequence of functions converges pointwise but not uniformly on $[a,b]$.

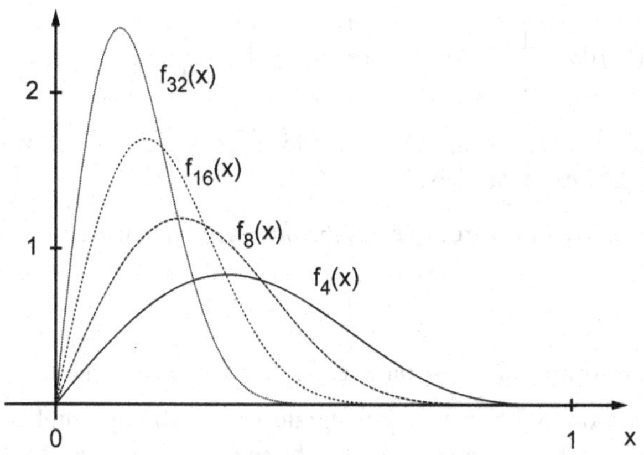

Fig. 7.13 The functions $f_n(x)$ in Example 7.37 converge to 0 pointwise, but their integrals do not converge to 0

Example 7.37. Let $f_n(x) = nx(1-x^2)^n$ on $[0,1]$. For each $n = 1,2,3,\ldots$, the function f_n is continuous and

$$\int_0^1 f_n(x)\,dx = n\int_0^1 x(1-x^2)^n\,dx = \frac{n}{n+1}\left(-\frac{1}{2}\right)\left[(1-x^2)^{n+1}\right]_0^1 = \frac{1}{2}\frac{n}{n+1}.$$

Therefore,

$$\lim_{n\to\infty}\int_0^1 f_n(x)\,dx = \frac{1}{2}\lim_{n\to\infty}\frac{n}{n+1} = \frac{1}{2}.$$

Now let us investigate $f_n(x)$ as n tends to infinity for each x in $[0,1]$. At $x = 0$, $f_n(0) = 0$ for all n. For $0 < x \le 1$, the number $b = 1 - x^2$ is less than 1, and we have $\lim_{n\to\infty} f_n(x) = x\lim_{n\to\infty} nb^n$. This limit is 0, because according to Theorem 2.10 on exponential growth, the reciprocals $\frac{b^{-n}}{n}$ tend to infinity. Therefore $f_n(x)$ tends to 0 for all x in the interval $[0,1]$, but

$$\int_0^1 \lim_{n\to\infty} f_n(x)\,\mathrm{d}x = \int_0^1 0\,\mathrm{d}x = 0 \neq \frac{1}{2}.$$

The limit of the integrals is not equal to the integral of the limit. A careful look at the graphs in Fig. 7.13 gives some insight into why this example does not contradict Theorem 7.6.

7.4b Integrals Depending on a Parameter

A very important extension of the convergence theorem for integrals deals with functions that depend on an additional parameter. We introduce the idea of integrals that depend on a parameter through an example.

Consider gas in a tube of unit cross-sectional area. The tube lies along the x-axis. Denote the linear density (mass per length) of the gas at x by $\rho(x)$. Then the mass M of the gas contained in the portion $[a,b]$ of the tube is given by the integral

$$M = \int_a^b \rho(x)\,\mathrm{d}x.$$

Suppose that the gas flows, and its linear density ρ changes with time t. Then the mass of the gas in the tube is a function of time. Let $\rho[t](x)$ be the linear density at x at time t. Then the mass in the tube at time t is

$$M[t] = \int_a^b \rho[t](x)\,\mathrm{d}x.$$

If the linear density at each point is a continuous function of time, we would expect the mass in the tube to depend continuously on time as well. This leads to a definition and a theorem (Fig.7.14).

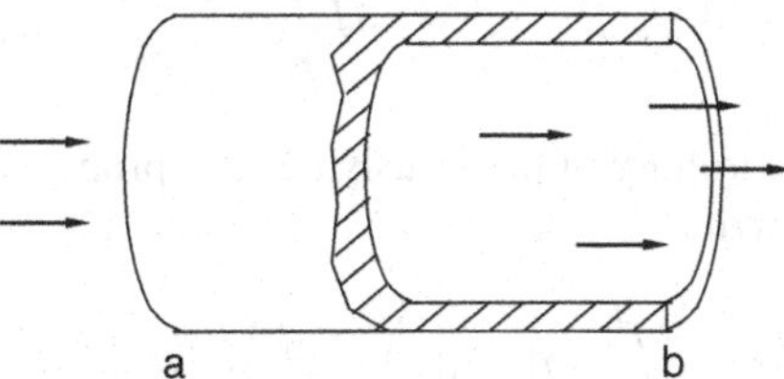

Fig. 7.14 Gas flowing in a tube has varying density ρ. M is the mass between points a and b

We say that a one-parameter family of functions ρ on $[a,b]$ *depends continuously on the parameter* t if given any tolerance ε, there is a δ such that when t and s differ by less than δ, $\rho[t](x)$ and $\rho[s](x)$ differ by less than ε at all points x of $[a,b]$.

If a function ρ depends on a parameter t, so does its integral over a fixed interval $[a,b]$. We indicate this dependence explicitly by $I(\rho[t],[a,b]) = I(t)$. In many situations, it is very important to learn how $I(t)$ varies with t. The following theorem is a variant of the convergence theorem 7.6.

Theorem 7.7. *Suppose that a one-parameter family of functions ρ depends continuously on the parameter t on the interval $[a,b]$. Then the integral of ρ from a to b depends continuously on t.*

So for the gas in the tube, the theorem tells us that if the linear density at x depends continuously on t, then the mass of gas in the tube depends continuously on time. Next, we want to know whether we can determine the rate of change in the mass with respect to time if we know the rate of change of linear density with respect to time at each x in $[a,b]$. To do so, we need the following definition.

Definition 7.3. We say that a one-parameter family of functions $\rho[t]$ depends differentiably on the parameter t if for every t, the difference quotients

$$\frac{\rho[t+h](x) - \rho[t](x)}{h}$$

tend uniformly on the interval $[a,b]$ to a limit function as h tends to 0. We denote this limit function by $\dfrac{\mathrm{d}\rho}{\mathrm{d}t}$.

The following theorem shows how to differentiate $M(t)$ with respect to t:

Theorem 7.8. Differentiation theorem for the integral. *Suppose that $\rho[t]$ depends differentiably on the parameter t. Then $\int_a^b \rho[t](x)\,\mathrm{d}x$ depends differentiably on t, and*

$$\frac{\mathrm{d}}{\mathrm{d}t}\int_a^b \rho[t](x)\,\mathrm{d}x = \int_a^b \frac{\mathrm{d}\rho}{\mathrm{d}t}(x)\,\mathrm{d}x.$$

Proof. This result is a corollary of previously derived properties: using the linearity of the integral, we can write

$$\frac{1}{h}\left(\int_a^b \rho[t+h](x)\,\mathrm{d}x - \int_a^b \rho[t](x)\,\mathrm{d}x\right) = \int_a^b \left(\frac{\rho[t+h](x) - \rho[t](x)}{h}\right)\mathrm{d}x.$$

Since we have assumed that the difference quotients tend uniformly to $\dfrac{\mathrm{d}\rho}{\mathrm{d}t}$, we conclude from the convergence theorem for integrals that the right side tends to the integral of $\dfrac{\mathrm{d}\rho}{\mathrm{d}t}$. The left side tends to $\dfrac{\mathrm{d}}{\mathrm{d}t}\int_a^b \rho[t](x)\,\mathrm{d}x$. So Theorem 7.8 follows. □

A colloquial way of stating Theorem 7.8 is this:

The derivative of an integral with respect to a parameter is the integral of the derivative of the integrand.

Theorem 7.8 can be regarded as an extension of the rule that the derivative of a sum is the sum of the derivatives. This result is enormously useful.

In the gas example, how does mass M change with time? If ρ depends differentiably on the parameter t, then

$$\frac{dM}{dt} = \int_a^b \frac{d\rho}{dt}\,dx.$$

Factorial of a Noninteger. In Sect. 7.3, we expressed $n!$ as an integral

$$n! = \int_0^\infty x^n e^{-x}\,dx.$$

This formula allows us to define $n!$ for positive noninteger n.

Problems

7.36. Use Theorem 4.12 to express the integrand in

$$\sin^{-1} x = \int_0^x (1-t^2)^{-1/2}\,dt$$

as a binomial series. Integrate this series term by term to produce a series for the inverse sine function.

7.37. Show that for all real positive n, $n! = n(n-1)!$.

7.38. Make a change of variables in Eq. (7.10) to show that

$$\int_0^\infty e^{-ty^2}\,dy = \frac{1}{2}\sqrt{\pi}t^{-1/2}.$$

Differentiate both sides with respect to t to find expressions for the integrals

$$\int_0^\infty y^2 e^{-y^2}\,dy, \qquad \int_0^\infty y^4 e^{-y^2}\,dy.$$

7.39. Make a change of variables in one of the integrals of Problem 7.38 to calculate the factorial $\left(\frac{1}{2}\right)!$.

7.40. Consider the sequence of functions $g_n(x) = n^2 x(1-x^2)^n$ on $[0,1]$. Show that the $g_n(x)$ converge to 0 pointwise but that the integrals $\int_0^1 g_n(x)\,dx$ tend to infinity. Compare a sketch of the graphs of the g_n with those in Fig. 7.13.

7.41. The density of gas in a tube of unit cross section is $\rho(x) = 1 + x^2 + t$ at time t, and $0 < x < 10$. Calculate

(a) The mass $M = \int_0^{10} \rho(x)\,dx$ in the tube.
(b) The rate of change $\dfrac{dM}{dt}$ using $\dfrac{d}{dt}\int_0^{10} \rho(x)\,dx$.
(c) The rate of change $\dfrac{dM}{dt}$ using $\int_0^{10} \dfrac{d\rho}{dt}\,dx$.
(d) If gas enters the tube at rate R [mass/time] at the left end $x = 0$, and no mass enters or leaves by any other means, what is the relation between R and $\int_0^{10} \dfrac{d\rho}{dt}\,dx$?

Chapter 8
Approximation of Integrals

Abstract In this chapter, we explore different ways to approximate $\int_a^b f(t)\,dt$ and ask the question, "How good are they?"

8.1 Approximating Integrals

In Sect. 6.2, we defined the integral $\int_a^b f(t)\,dt$ by approximating it with sums of the form $\sum_{j=1}^{n} f(t_j)(a_j - a_{j-1})$, where $a = a_0 < a_1 < a_2 < \cdots < a_n = b$ is a subdivision of $[a,b]$, and t_j is a point in $[a_{j-1}, a_j]$. The basic observation behind this approximation was that if $[a,b]$ is a short interval, and t a point of $[a,b]$, then $f(t)(b-a)$ is a good approximation to the integral of f over $[a,b]$.

We now turn to the practical application of the approximation theorem, i.e., to calculate integrals approximately. Among all possible approximation formulas, we single out three classes for numerical study: those in which the function is evaluated at the left endpoint, the midpoint, or the right endpoint of each subinterval. We denote these by $I_{\text{left}}(f,[a,b])$, $I_{\text{mid}}(f,[a,b])$, and $I_{\text{right}}(f,[a,b])$. See Fig. 8.1, where just one subinterval is shown.

Example 8.1. We know that $\int_0^1 t\,dt = \frac{1}{2}$. Let us compare approximate integrals of $f(t) = t$ on $[0,1]$ to the exact value. In the case of $n = 2$ equal subintervals (see Fig. 8.2), we obtain

$$\begin{aligned}
I_{\text{left}}(t,[0,1]) &= f(0)(0.5) + f(0.5)(0.5) = 0 + 0.5(0.5) = 0.25,\\
I_{\text{mid}}(t,[0,1]) &= f\left(\frac{0+0.5}{2}\right)(0.5) + f\left(\frac{0.5+1}{2}\right)(0.5) = 0.25(0.5) + 0.75(0.5) = 0.5,\\
I_{\text{right}}(t,[0,1]) &= f(0.5)(0.5) + f(1)(0.5) = 0.5(0.5) + 1(0.5) = 0.75.
\end{aligned}$$

P.D. Lax and M.S. Terrell, *Calculus With Applications*, Undergraduate Texts in Mathematics, DOI 10.1007/978-1-4614-7946-8_8,

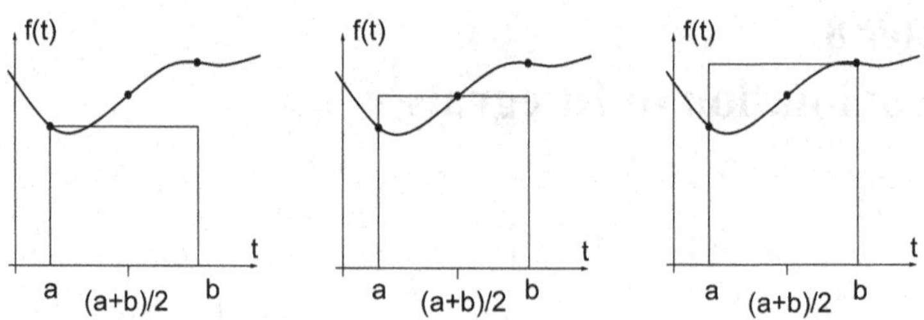

Fig. 8.1 To approximate $\int_a^b f(t)\,dt$ using only one subinterval, we may evaluate f at the left, middle, or right point of $[a,b]$

Computing other cases of n equal subintervals in a similar manner gives the results shown in the table. We suggest that you use your calculator or computer to generate such tables yourself. Observe that the midpoint rule gives the exact answer for $\int_0^1 t\,dt$, while I_{left} and I_{right} deviate from the exact value.

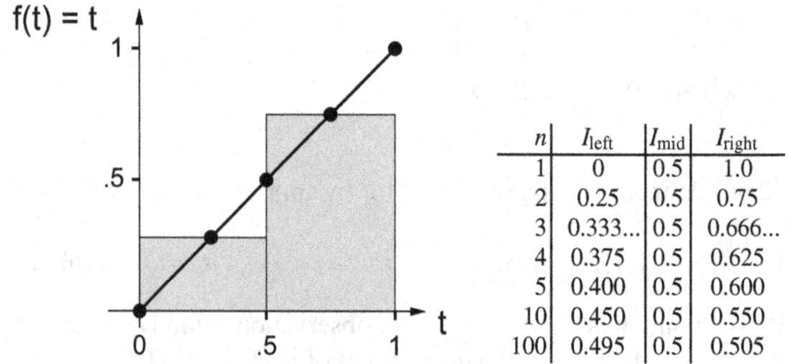

n	I_{left}	I_{mid}	I_{right}
1	0	0.5	1.0
2	0.25	0.5	0.75
3	0.333...	0.5	0.666...
4	0.375	0.5	0.625
5	0.400	0.5	0.600
10	0.450	0.5	0.550
100	0.495	0.5	0.505

Fig. 8.2 Approximate integrals for $f(t) = t$ on $[0,1]$. The rectangles shown correspond to the midpoint rule

Example 8.2. The exact value of $\int_0^1 t^2\,dt$ is $\frac{1}{3}$. Compare that to the approximate integrals in the table, where we again use equal subintervals. Again, you should create your own table using your calculator or computer.

n	I_{left}	I_{mid}	I_{right}
1	0.0000000	0.2500000	1.0000000
5	0.2400000	0.3300000	0.4400000
10	0.2850000	0.3325000	0.3850000
100	0.3283500	0.3333250	0.3383500

Notice that for all n listed, I_{mid} gives a better approximation to the exact value than either I_{left} or I_{right}.

Example 8.3. We can compute the exact value

$$\int_0^1 \frac{t}{1+t^2}\,dt = \left[\frac{1}{2}\log(1+t^2)\right]_0^1 = \frac{1}{2}\log 2 = 0.34657359027\ldots.$$

Compare this to the approximate integrals given in the table.

n	I_{left}	I_{mid}	I_{right}
1	0.0	0.4	0.5
5	0.2932233...	0.3482550...	0.3932233...
10	0.3207392...	0.3469911...	0.3707392...
100	0.3440652...	0.3465777...	0.3490652...

Observe that I_{mid} again provides a consistently better approximation to the value of the integral than either I_{left} or I_{right}.

8.1a The Midpoint Rule

In Examples 8.1–8.3, we saw that picking the point t to be the midpoint of the subintervals was better than picking one of the endpoints. We repeat the definition:

Definition 8.1. The *midpoint rule* is the approximate integral

$$I_{\text{mid}}(f,[a,b]) = f\left(\frac{a+b}{2}\right)(b-a).$$

We use a sum of such terms when the interval is subdivided.

We show now that the midpoint rule gives the exact value of the integral when f is a linear function. First, the midpoint rule gives the exact value of the integral for a constant function $f(t)=k$: $f\left(\frac{a+b}{2}\right)(b-a) = k(b-a)$, which is equal to the integral of f on $[a,b]$. What about the function $f(t)=t$? By the fundamental theorem of calculus,

$$\int_a^b t\,dt = \left[\frac{1}{2}t^2\right]_a^b = \frac{b^2-a^2}{2} = \frac{a+b}{2}(b-a).$$

This is exactly the midpoint rule, since $f\left(\frac{a+b}{2}\right) = \frac{a+b}{2}$ for $f(t)=t$.

For any linear function $f(t)=mt+k$, we have by linearity of the integral that

$$\int_a^b (mt+k)\,dt = m\int_a^b t\,dt + \int_a^b k\,dt$$

$$= m\frac{a+b}{2}(b-a) + k(b-a) = \left(m\frac{a+b}{2}+k\right)(b-a).$$

This proves that the midpoint rule gives the exact value of the integral of any linear function.

8.1b The Trapezoidal Rule

Our introductory examples showed that using either endpoint gives a less-accurate approximation than that produced using the midpoint. What about using the average of f at both endpoints? We call this the trapezoidal rule. Figure 8.3 suggests the origin of the name.

Definition 8.2. The *trapezoidal rule* is the approximate integral

$$I_{\text{trap}}(f,[a,b]) = \frac{1}{2}(f(a)+f(b))(b-a).$$

We use a sum of such terms when the interval is subdivided.

We show now that the trapezoidal rule, like the midpoint rule, gives the exact value of the integral when $f(t) = mt + k$, a linear function. By the linearity property again,

$$\int_a^b (mt+k)\,dt = m\int_a^b t\,dt + k\int_a^b 1\,dt.$$

Therefore, it suffices to test the function $f(t) = t$ and the constant function 1. Setting $f(t) = t$ into the rule gives $\frac{1}{2}(a+b)(b-a)$, which is the exact value of the integral $\int_a^b t\,dt$. Setting $f(t) = 1$ into the rule gives $\frac{1}{2}(1+1)(b-a)$, which is the exact value of the integral $\int_a^b 1\,dt$. Therefore, the trapezoidal rule is exact when $f(t) = mt + k$.

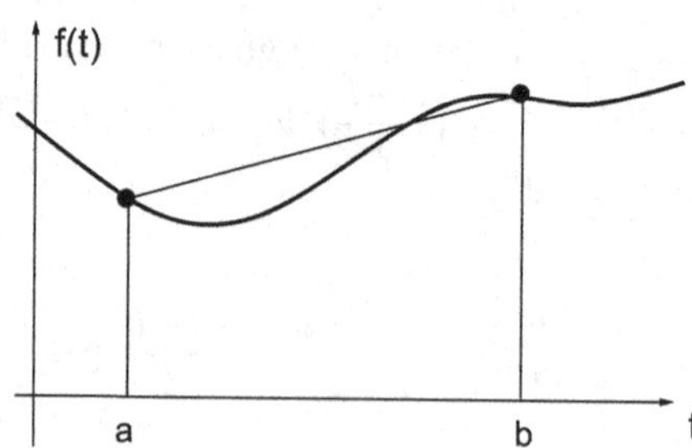

Fig. 8.3 The area of the trapezoid is $\frac{f(a)+f(b)}{2}(b-a)$

Let us see how well these rules work for the function $f(t) = t^2$. To make the formulas not too messy, we take $a = 0$. Since t^2 is the derivative of $\frac{1}{3}t^3$, the integral of t^2 over $[0,b]$ is $\frac{1}{3}b^3$. The midpoint and trapezoidal rules give respectively

$$I_{\text{mid}}(t^2,[0,b]) = \left(\frac{1}{2}b\right)^2 b = \frac{1}{4}b^3$$

and

$$I_{\text{trap}}(t^2,[0,b]) = \frac{1}{2}(0^2+b^2)b = \frac{1}{2}b^3.$$

Neither of these is the exact value. How much do they deviate from the exact value?

$$\int_0^b t^2\,\mathrm{d}t - I_{\text{mid}}(t^2,[0,b]) = b^3\left(\frac{1}{3}-\frac{1}{4}\right) = \frac{1}{12}b^3,$$

and

$$\int_0^b t^2\,\mathrm{d}t - I_{\text{trap}}(t^2,[0,b]) = b^3\left(\frac{1}{3}-\frac{1}{2}\right) = -\frac{1}{6}b^3.$$

In words: for the function t^2, the trapezoidal rule approximation is twice as far from the exact value of the integral as the midpoint rule approximation is from the exact value, and in the opposite direction. It follows from this that for $f(t)=t^2$, the combination

$$\frac{2}{3}I_{\text{mid}}(f,[a,b]) + \frac{1}{3}I_{\text{trap}}(f,[a,b]) \tag{8.1}$$

gives the exact value of the integral. Since for linear functions, both I_{mid} and I_{trap} give the exact value of the integral, it follows from linearity of the integral (Theorem 6.3) that this new combination gives the exact value of the integral for all quadratic functions integrated over $[0,b]$. We ask you to write out the details of this in Problem 8.4. This new approximation is called Simpson's rule, denoted by $I_S(f,[a,b])$.

So far, we have assumed that the lower endpoint of the interval of integration is zero. As we point out in Problem 6.12, any interval of integration $[a,c]$ can be reduced to this case by replacing the function $f(t)$ to be integrated by $f(t+a)$, and the interval of integration by $[0,c-a]$. We point out another way to show that $\frac{2}{3}I_{\text{mid}}(t^2,[a,b])+\frac{1}{3}I_{\text{trap}}(t^2,[a,b])$ is exactly $\int_a^b t^2\,\mathrm{d}t$ without assuming that the lower endpoint is zero in Problem 8.6.

Problems

8.1. Compute the values of I_{left}, I_{right}, and I_{mid} in the following cases:

(a) $f(x)=x^3$, $[1,2]$, $n=1,2,4$.
(b) $f(x)=\sqrt{1-x^2}$, $[0,\frac{1}{\sqrt{2}}]$, $n=1,2,4$.
(c) $f(x)=\dfrac{1}{1+x^2}$, $[0,1]$, $n=1,2,4$.

8.2. When a drug is administered once every 24 h, the concentration of the drug in the blood, $c(t)$ (micrograms per milliliter), varies over time. Measurements are taken of the concentration at several times during the 24 h period from 72 to 96 h after the first dose, and an average is calculated using the trapezoidal rule:

$$\frac{1}{24}\int_{72}^{96} c(t)\,\mathrm{d}t \approx \frac{1}{24}I_{\text{trap}}.$$

Some clinical decisions are based on this average value, which is expected to be a good estimate of the steady-state concentration during repeated use of the drug. Carry out this approximation to find the average, assuming that the data available are as in Fig. 8.4.

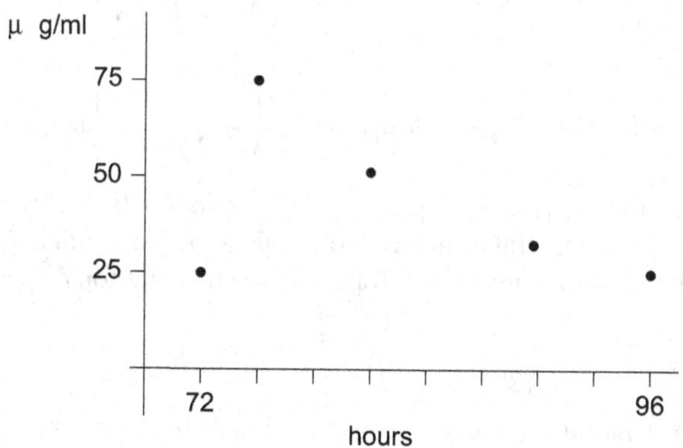

Fig. 8.4 Measurements of drug concentration in Problem 8.2

8.3. Use a calculator or computer to evaluate approximations to the integrals using the midpoint rule, using partitions into n equal subintervals, $n = 1,\ 5,\ 10,\ 100$. Compare your results to the exact value of the integrals.

(a) $\displaystyle\int_0^1 \frac{x}{\sqrt{1+x^2}}\,dx$ (b) $\displaystyle\int_0^1 \frac{1}{1+x^2}\,dx$ (c) $\displaystyle\int_0^2 \frac{1}{(1+x)^2}\,dx$

8.4. We showed that Simpson's rule is exact for $f(t) = t^2$, and that the midpoint and trapezoidal rules are exact for $mt+k$.

(a) Show that $I_S(kt^2, [a,b]) = \displaystyle\int_a^b kt^2\,dt$ for every number k.
(b) Show that Simpson's rule is exact for every quadratic function.

8.5. Suppose that the function f is convex over $[a,b]$, i.e., that its second derivative is positive there. Show that

(a) $\displaystyle\int_a^b f(x)\,dx \geq I_{\text{mid}}(f, [a,b])$ (b) $\displaystyle\int_a^b f(x)\,dx \leq I_{\text{trap}}(f, [a,b])$.

Can you give a geometric interpretation of these inequalities?

8.6. Using one subinterval $[a,b]$, show the following, which prove that Simpson's rule gives exactly $\displaystyle\int_a^b t^2\,dt$ on every interval.

(a) $I_{\text{mid}}(t^2, [a,b]) = \frac{1}{4}(-a^2b + ab^2 + b^3 - a^3)$.
(b) $I_{\text{trap}}(t^2, [a,b]) = \frac{1}{2}(a^2b - ab^2 + b^3 - a^3)$.

(c) Conclude that $\int_a^b t^2\,dt = \frac{2}{3}I_{\text{mid}}(t^2,[a,b]) + \frac{1}{3}I_{\text{trap}}(t^2,[a,b])$.
(d) Why does this also prove the case in which many subintervals are used?

8.7. Explain each of the following steps, which prove an error estimate for the midpoint rule.

(a) $I_{\text{mid}}(f,[-h,h]) - \int_{-h}^{h} f(x)\,dx = K(h) - K(-h)$, where $K(h) = hf(0) - \int_0^h f(x)\,dx$.
(b) $K(h)-K(-h)$ is an odd function, and Taylor's theorem gives

$$K(h)-K(-h) = 0+0+0+\left(K'''(c_2)+K'''(-c_2)\right)\frac{h^3}{3!} = \left(-f''(c_2)-f''(-c_2)\right)\frac{h^3}{6}$$

for some c_2 in $[-h,h]$ depending on h.
(c) Therefore, if $[a,b]$ is subdivided into n equal parts, and M_2 is an upper bound for $|f''|$ on $[a,b]$, then

$$\left|I_{\text{mid}}(f,[a,b]) - \int_a^b f(x)\,dx\right| \le n\frac{2}{6}M_2\left(\frac{b-a}{2n}\right)^3 = \frac{1}{24}M_2(b-a)\left(\frac{b-a}{n}\right)^2.$$

8.2 Simpson's Rule

We observed at the end of Sect. 8.1b that the combination of the midpoint rule and the trapezoid rule, formula (8.1), called Simpson's rule, gives the exact integral for quadratic integrands f. In this section, we explore the application of Simpson's rule to the integration of arbitrary smooth functions.

Definition 8.3. *Simpson's rule* is the approximate integral $\frac{2}{3}I_{\text{mid}} + \frac{1}{3}I_{\text{trap}}$, that is,

$$I_S(f,[a,b]) = \left(\frac{1}{6}f(a) + \frac{2}{3}f\left(\frac{a+b}{2}\right) + \frac{1}{6}f(b)\right)(b-a).$$

We use a sum of such terms when the interval is subdivided.

By design, Simpson's rule gives the exact value of the integral of f when f is a quadratic polynomial. Let us see how good an approximation it gives to the integrals of some other functions.

Example 8.4. Let $f(t) = t^3$. Since t^3 is the derivative of $\frac{1}{4}t^4$, the integral of t^3 over the interval $[a,b]$ is $\frac{1}{4}(b^4-a^4)$. Simpson's rule on a single subinterval gives

$$\left(\frac{1}{6}a^3 + \frac{2}{3}\left(\frac{a+b}{2}\right)^3 + \frac{1}{6}b^3\right)(b-a)$$

$$= \left(\frac{1}{6}a^3 + \frac{a^3 + 3a^2b + 3ab^2 + b^3}{12} + \frac{1}{6}b^3\right)(b-a)$$

$$= \frac{1}{4}(a^3 + a^2b + ab^2 + b^3)(b-a) = \frac{1}{4}(b^4 - a^4).$$

Note that this is exactly right.

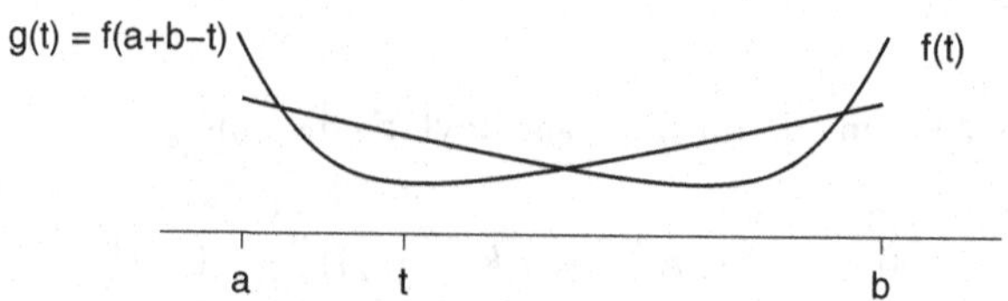

Fig. 8.5 Graphs of a function f and of the reflection $g(t) = f(a+b-t)$

So rather surprisingly, Simpson's rule gives the exact value of the integral for $f(t) = t^3$. Here is a reason why: For a function $f(t)$ defined on the interval $[a,b]$, we define $g(t) = f(a+b-t)$. The graphs of f and g are reflections of each other across the midpoint, $\frac{a+b}{2}$, as in Fig. 8.5. Therefore, the integrals of f and g over $[a,b]$ are equal:

$$\int_a^b g(t)\,dt = \int_a^b f(t)\,dt.$$

So it follows from the linearity of the integral that

$$\int_a^b (f(t) + g(t))\,dt = 2\int_a^b f(t)\,dt. \tag{8.2}$$

By reflection, $g(a) = f(b),\ g(b) = f(a)$, and $g(\frac{a+b}{2}) = f(\frac{a+b}{2})$. It follows that

$$I_S(f,[a,b]) = \left[\frac{1}{6}f(a) + \frac{2}{3}f(\frac{a+b}{2}) + \frac{1}{6}f(b)\right](b-a)$$

$$= \left[\frac{1}{6}g(b) + \frac{2}{3}g(\frac{a+b}{2}) + \frac{1}{6}g(a)\right](b-a) = I_S(g,[a,b]).$$

Since Simpson's rule depends linearly on f, it follows that

$$I_S(f+g,[a,b]) = 2I_S(f,[a,b]). \tag{8.3}$$

Now take $f(t)$ to be t^3. Then $g(t) = (a+b-t)^3$. The sum

$$f(t) + g(t) = t^3 + (a+b-t)^3$$

is a quadratic function, because the cubic term cancels out. Since Simpson's rule gives the exact value of the integral for quadratic functions, the left side of (8.2) equals the left side of (8.3) when $f(t) = t^3$. But then the right sides are also equal:

$$2\int_a^b t^3\,\mathrm{d}t = 2I_S(t^3,[a,b]).$$

It follows by linearity that for every cubic polynomial, Simpson's rule furnishes the exact value of the integral. We have proved the following theorem.

Theorem 8.1. Simpson's rule for cubic polynomials. *For any cubic polynomial $f(t)$ and any subdivision of $[a,b]$, Simpson's rule gives the exact value of $\int_a^b f(t)\,\mathrm{d}t$.*

Example 8.5. For $f(t) = t^4$ and $[0,c]$, the exact integral is $\int_0^c t^4\,\mathrm{d}t = \frac{1}{5}c^5$. However, Simpson's rule gives

$$\left(\frac{1}{6}(0) + \frac{2}{3}\left(\frac{c}{2}\right)^4 + \frac{1}{6}(c^4)\right)c = \left(\frac{1}{24} + \frac{1}{6}\right)c^5 = \frac{5}{24}c^5.$$

At last we have found a case in which the rule fails to give the exact value. The failure is not excessive. The relative error, $\frac{\frac{5}{24} - \frac{1}{5}}{\frac{1}{5}}$, is about 4.1%.

How well does Simpson's rule work for other functions? That depends on how closely these functions can be approximated by cubic polynomials. Taylor's theorem, Theorem 4.10, gives the estimate

$$f(b) = f(a) + f'(a)(b-a) + f''(a)\frac{(b-a)^2}{2} + f'''(a)\frac{(b-a)^3}{6} + R,$$

where $R = f''''(c)\frac{(b-a)^4}{24}$ for some number c between a and b. This estimate can be used to judge the accuracy of Simpson's rule for four-times differentiable functions.

We have been considering Simpson's method on a single interval $[a,b]$. To achieve better accuracy, we divide the given interval of integration $[a,b]$ into n short subintervals of length $h = \frac{b-a}{n}$. Define $a_j = a + j\frac{b-a}{2n}$ and use the subdivision

$$a_0 < a_2 < a_4 < \cdots < a_{2n}.$$

The a_j with j odd are the midpoints of these intervals. We apply Simpson's rule to each interval $[a_{2(j-1)}, a_{2j}]$ for $j = 1,\ldots,n$, and add the approximations. We get

$$I_S(f,[a,b]) = \sum_{j=1}^{n} \left(\frac{1}{6} f(a_{2(j-1)}) + \frac{2}{3} f(a_{2j-1}) + \frac{1}{6} f(a_{2j}) \right) h$$

$$= \left(\frac{1}{6} f(a_0) + \frac{2}{3} f(a_1) + \frac{1}{3} f(a_2) + \frac{2}{3} f(a_3) + \cdots + \frac{2}{3} f(a_{2n-1}) + \frac{1}{6} f(a_{2n}) \right) \frac{b-a}{n}.$$

Example 8.6. We evaluate the integral $\int_0^1 \frac{s}{1+s^2}\,\mathrm{d}s$ using Simpson's rule applied without subdivision. This gives the approximate value

$$\left(\frac{1}{6}(0) + \frac{2}{3} \left(\frac{\frac{1}{2}}{1+\frac{1}{4}} \right) + \frac{1}{6} \left(\frac{1}{2} \right) \right) (1-0) = \frac{7}{20} = 0.35$$

for the integral. We display the results of approximating this integral by Simpson's rule but with n equal subdivisions.

n	1	5	10	100
I_S	0.35...	0.346577...	0.3465738...	0.3465735903...

The exact value is $\frac{1}{2}\log 2 = 0.34657359027\ldots$; Simpson's rule with five subintervals gives five digits correctly.

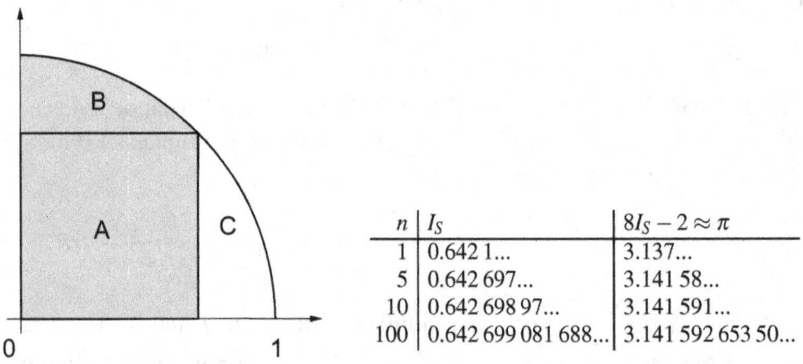

n	I_S	$8I_S - 2 \approx \pi$
1	0.642 1...	3.137...
5	0.642 697...	3.141 58...
10	0.642 698 97...	3.141 591...
100	0.642 699 081 688...	3.141 592 653 50...

Fig. 8.6 *Left*: The shaded area is expressed as an integral in Example 8.7. *Right*: output from Simpson's rule

Example 8.7. Let $f(s) = \sqrt{1-s^2}$ and set $I = \int_0^{1/\sqrt{2}} \sqrt{1-s^2}\,\mathrm{d}s$. The geometric meaning of the integral I is the area of the shaded region shown in Fig. 8.6, the sum of the area of a square A of side length $\frac{1}{\sqrt{2}}$ and the area of a piece marked B. The quarter circle is the union of A, B, and C. Since B and C are congruent, and the area of A is $\frac{1}{2}$, we get that

$$\text{area of quarter circle } = \frac{\pi}{4} = \text{area}(A) + \text{area}(B) + \text{area}(C) = \frac{1}{2} + 2\,\text{area}(B).$$

Hence $\pi = 2 + 8\,\text{area}(B)$. But $I = \text{area}(A) + \text{area}(B) = \frac{1}{2} + \text{area}(B)$, so that $\text{area}(B) = I - \frac{1}{2}$. Thus $\pi = 2 + 8(I - \frac{1}{2}) = 8I - 2$. The table in Fig. 8.6 lists various approximations to I and to $8I - 2$ using Simpson's rule with a partition into n equal subintervals.

Observe that for $n = 10$, we get the first six digits of π correctly.

8.2a An Alternative to Simpson's Rule

We describe here an approximation scheme of integrals that is as good as Simpson's rule for functions that have continuous fourth derivatives. We start again with Taylor's approximation of f to fourth order,

$$f(x) = f(0) + f'(0)x + \frac{1}{2}f''(0)x^2 + \frac{1}{6}f'''(0)x^3 + R,$$

where the remainder R is equal to $\frac{1}{24}f''''(c)x^4$ for some c between 0 and x. Integrate with respect to x over $[-h, h]$:

$$\int_{-h}^{h} f(x)\,\mathrm{d}x = 2f(0)h + \frac{1}{3}f''(0)h^3 + \int_{-h}^{h} R(x)\,\mathrm{d}x. \tag{8.4}$$

We also have by Taylor's approximation of f'' to second order that

$$f''(x) = f''(0) + f'''(0)x + R_2,$$

where $R_2 = \dfrac{1}{2}f''''(c_2)x^2$ for some c_2 between 0 and x. Integrating with respect to x over $[-h, h]$, we get

$$\int_{-h}^{h} f''(x)\,\mathrm{d}x = 2f''(0)h + \int_{-h}^{h} R_2\,\mathrm{d}x. \tag{8.5}$$

To eliminate the $f''(0)$ term in (8.4), multiply (8.5) by $\frac{1}{6}h^2$ and subtract from (8.4), giving

$$\int_{-h}^{h} f(x)\,\mathrm{d}x - \frac{1}{6}h^2\int_{-h}^{h} f''(x)\,\mathrm{d}x = 2f(0)h + \int_{-h}^{h}\left(R - \frac{1}{6}h^2R_2\right)\mathrm{d}x.$$

The integral of f'' can be expressed by f', so we get

$$\int_{-h}^{h} f(x)\,\mathrm{d}x = 2f(0)h + \frac{1}{6}(f'(h) - f'(-h))h^2 + \int_{-h}^{h}\left(R - \frac{1}{6}h^2R_2\right)\mathrm{d}x.$$

The integral on the right is less than a constant times h^5.

Divide the interval $[a,b]$ into n equal parts of width $w = 2h = \dfrac{b-a}{n}$, and apply the previous formula to each subinterval. To use the formula in intervals other than $[-h,h]$, we use the translation invariance of the integral (see Problem 6.12). This gives on any interval $[c,d]$,

$$\int_c^d f(x)\,\mathrm{d}x = f\left(\frac{c+d}{2}\right)(d-c) + \frac{1}{6}\left(\frac{d-c}{2}\right)^2\left(f'(d) - f'(c)\right) + \text{error}.$$

If we denote the subdivision of $[a,b]$ by $a_0 < m_1 < a_2 < m_2 < a_4 < \cdots < a_n$ and use the approximation on each $[a_{2(j-1)}, a_{2j}]$, then f is evaluated only at the midpoints m_j. When we sum over all subintervals, the f' terms cancel except at the ends $a_0 = a$ and $a_n = b$. This gives the following approximation formula:

$$\int_a^b f(x)\,\mathrm{d}x = w\sum_{j=1}^{n} f(m_j) + \frac{1}{24}w^2\left(f'(b) - f'(a)\right) + \text{error},$$

where the error is bounded by a constant times $w^4(b-a)$.

Definition 8.4. We denote this alternative to Simpson's rule by

$$I_A(f,[a,b]) = w\sum_{j=1}^{n} f(m_j) + \frac{1}{24}w^2\left(f'(b) - f'(a)\right),$$

where the m_j are the midpoints of n subintervals of length $w = \dfrac{b-a}{n}$.

Example 8.8. Using Simpson's rule and the alternative, we evaluate

$$\int_0^1 \sqrt{2+s^2}\,\mathrm{d}s.$$

The exact value of this integral can be expressed in terms of hyperbolic functions, as we have suggested in Problem 7.15. Its first eight digits are

$$1.5245043\ldots.$$

The following table gives approximations with partition into n equal subintervals. "Count" is the number of function evaluations.

n	I_S	I_S count	I_A	I_A count	error $\int_0^1 - I_A$
1	1.524 377 3...	3	1.524 056 2...	3	.0004480...
2	1.524 495 9...	5	1.524 474 9...	4	.0000294...
4	1.524 503 8...	9	1.524 502 5...	6	.0000018...
8	1.524 504 3...	17	1.524 504 2...	10	.0000001...

Thus we see that with only four subdivisions, both rules give correctly the first six digits of the integral. The last column agrees with the h^4 error prediction, because with each doubling of n, the error is reduced by a factor of about 2^4.

Problems

8.8. Use a calculator or computer to evaluate approximations to the integrals using Simpson's rule, employing partitions into n equal subintervals, $n = 1,\ 5,\ 10,\ 100$. Compare your results to the exact value of the integrals and to the result of Problem 8.3.

(a) $\displaystyle\int_0^1 \frac{x}{\sqrt{1+x^2}}\,dx$ (b) $\displaystyle\int_0^1 \frac{1}{1+x^2}\,dx$ (c) $\displaystyle\int_0^2 \frac{1}{(1+x)^2}\,dx$

8.9. Calculate I_A for $\displaystyle\int_0^1 t^4\,dt$ with two subintervals. Verify that the result is about 0.1 % too large.

8.10. We showed by the change of variables $z = \dfrac{1}{x}$ that $\displaystyle\int_1^\infty \frac{1}{1+x^2}\,dx = \int_0^1 \frac{1}{1+z^2}\,dz$.

(a) Approximate the integral on the right by Simpson's rule, $n = 2$.

(b) Use the equation above to evaluate approximately $\displaystyle\int_{-\infty}^\infty \frac{1}{1+x^2}\,dx = \pi$. How close is your approximation?

8.11. Looking at Fig. 8.6, we see that $\displaystyle\int_0^1 \sqrt{1-s^2}\,ds = \frac{\pi}{4}$. Approximate, using Simpson's rule, the integral on the left. You will observe that even if a large number of subintervals is used, the approximation to $\frac{\pi}{4}$ is poor. Can you explain why Simpson's rule works so poorly in this case?

Chapter 9
Complex Numbers

Abstract We develop the properties of the number system called the complex numbers. We also describe derivatives and integrals of some basic functions of complex numbers.

9.1 Complex Numbers

Most people first encounter complex numbers as solutions of quadratic equations $x^2+bx+c=0$, that is, zeros z of the function $f(x)=x^2+bx+c$.

Example 9.1. Take as an example the equation $x^2+1=0$. The quadratic formula for the roots gives $z=\pm\sqrt{-1}$. There is no real number $\sqrt{-1}$. We introduce a new number $\mathrm{i}=\sqrt{-1}$.

Definition 9.1. A complex number z is defined as the sum of a real number x and a real multiple y of i,

$$z=x+\mathrm{i}y,$$

where i denotes a square root of minus one, $\mathrm{i}^2=-1$. The number $x=\mathrm{Re}\,(z)$ is called the *real part* of z, and the real number $y=\mathrm{Im}\,(z)$ is called its *imaginary part*. A complex number whose imaginary part is zero is called (naturally enough) real. A complex number whose real part is zero is called *purely imaginary*. The *complex conjugate* of z is $\overline{z}=x-\mathrm{i}y$.

You might be wondering how solving an equation such as $r^2+1=0$ might arise in calculus. Consider the differential equation

$$y''+y=0.$$

We know that $\sin x$ and $\cos x$ solve the equation. But what about $y=\mathrm{e}^{rx}$? The second derivative of $y=\mathrm{e}^{rx}$ is r^2 times y, so $y=\mathrm{e}^{rx}$ solves the equation

P.D. Lax and M.S. Terrell, *Calculus With Applications*, Undergraduate Texts in Mathematics, DOI 10.1007/978-1-4614-7946-8_9,

$$y'' + y = (r^2+1)e^{rx} = 0$$

if r solves $r^2+1=0$. We will see in this chapter that e^{ix}, $\sin x$, and $\cos x$ are related, and we will see in Chap. 10 that functions of complex numbers help us solve many useful differential equations. Complex numbers also have very practical applications; they are used to analyze alternating current circuits.[1]

9.1a Arithmetic of Complex Numbers

We now describe a natural way of doing arithmetic with complex numbers. To add complex numbers, we add their real and imaginary parts separately:

$$(x+iy)+(u+iv) = x+u+i(y+v).$$

Similarly, for subtraction, $(x+iy)-(u+iv) = x-u+i(y-v)$. To multiply complex numbers, we use the distributive law:

$$(x+iy)(u+iv) = xu+iyu+xiv+iyiv.$$

Rewrite xi as ix and yi as iy, since we are assuming that multiplication of real numbers and i is commutative. Then, since $i^2=-1$, we can write the product above as

$$(xu-yv)+i(yu+xv).$$

Example 9.2. Multiplication includes squaring:

$$(-i)^2 = -1, \quad (3-i)^2 = 9-6i+i^2 = 8-6i, \quad (5i)^2 = -25.$$

It is easy to divide a complex number by a real number r: $\dfrac{x+iy}{r} = \dfrac{x}{r}+i\dfrac{y}{r}$. To express the quotient of two complex numbers $\dfrac{x+iy}{u+iv}$ as a complex number in the form $s+it$, multiply the numerator and denominator by the complex conjugate[2] $\overline{u+iv} = u-iv$. We get

$$\frac{x+iy}{u+iv} = \frac{(x+iy)(u-iv)}{(u+iv)(u-iv)} = \frac{xu+yv+i(yu-xv)}{u^2+v^2} = \frac{xu+yv}{u^2+v^2}+i\frac{yu-xv}{u^2+v^2}.$$

Notice that the indicated division by u^2+v^2 can be carried out unless both u and v are zero. In that case, the divisor $u+iv$ is zero, so we do not expect to be able to carry out the division.

---

[1] For electrical engineers, the letter i denotes current and nothing but current. So they denote the square root of -1 by the letter j.

[2] For physicists, the conjugate is denoted by an asterisk: $(u+iv)^* = u-iv$.

Example 9.3. One handy example of a quotient is the reciprocal of i,

$$\frac{1}{\mathrm{i}} = \frac{1}{\mathrm{i}}\frac{(-\mathrm{i})}{(-\mathrm{i})} = \frac{-\mathrm{i}}{1} = -\mathrm{i}.$$

Example 9.4. Division is needed almost every time you solve an equation. Let us try to solve for a if

$$\mathrm{i}(\mathrm{i}+a) = 2a+1.$$

Use properties of complex arithmetic and collect like terms to get $\mathrm{i}^2 - 1 = (2-\mathrm{i})a$. Next divide both sides by $2-\mathrm{i}$:

$$a = \frac{-2}{2-\mathrm{i}} = \frac{-2}{2-\mathrm{i}}\frac{2+\mathrm{i}}{2+\mathrm{i}} = \frac{-4-2\mathrm{i}}{4+1} = -\frac{4}{5} - \frac{2}{5}\mathrm{i}.$$

Addition and multiplication of complex numbers are defined so that the associative, commutative, and distributive rules of arithmetic hold. For complex numbers v, w, and z, we have

- Associativity rules: $v+(w+z) = (v+w)+z$ and $v(wz) = (vw)z$,
- Commutativity rules: $v+w = w+v$ and $vw = wv$,
- Distributivity rule: $v(w+z) = vw+vz$.
- The numbers $0 = 0+0\mathrm{i}$ and $1 = 1+\mathrm{i}0$ are distinguished, inasmuch as adding 0 and multiplying by 1 do not alter a number.

Rules for the Conjugate of a Complex Number. In the case of complex numbers, we have the additional operation of conjugation. The following rules concerning conjugation are very useful. They can be verified immediately using the rules of arithmetic for complex numbers. We already encountered some of the rules of conjugation when we explained how to divide by a complex number.

- Symmetry: The conjugate of the conjugate is the original number: $\overline{\overline{z}} = z$.
- Additivity: The conjugate of a sum is the sum of the conjugates: $\overline{z+w} = \overline{z}+\overline{w}$.
- The sum of a complex number and its conjugate is real:

$$z+\overline{z} = 2\mathrm{Re}(z). \tag{9.1}$$

- The conjugate of a product is the product of the conjugates: $\overline{zw} = \overline{z}\,\overline{w}$.
- The product of a complex number $z = x+\mathrm{i}y$ and its conjugate is

$$z\overline{z} = x^2+y^2. \tag{9.2}$$

  The number $z\overline{z}$ is real and nonnegative.

The Absolute Value of a Complex Number. Since $z\overline{z}$ is real and nonnegative, we can make the following definition.

Definition 9.2. The *absolute value*, $|z|$, of $z = x + \mathrm{i}y$ is the nonnegative square root of $z\overline{z}$:

$$|z| = \sqrt{z\overline{z}} = \sqrt{x^2 + y^2}. \tag{9.3}$$

For z real, this definition coincides with the absolute value of a real number. Next, we see that the absolute values of the real and imaginary parts of $z = x + \mathrm{i}y$ are always less than or equal to $|z|$.

$$|\mathrm{Re}\,(z)| = |x| = \sqrt{x^2} \le \sqrt{x^2 + y^2} = |z| \tag{9.4}$$

and

$$|\mathrm{Im}\,(z)| = |y| = \sqrt{y^2} \le \sqrt{x^2 + y^2} = |z|.$$

Example 9.5. To find $|z|$ for $z = 3 + 4\mathrm{i}$, we have

$$|z| = \sqrt{z\overline{z}} = \sqrt{(3+4\mathrm{i})(3-4\mathrm{i})} = \sqrt{3^2 + 4^2} = 5.$$

Therefore, $|3 + 4\mathrm{i}| = 5$.

The absolute value of z is also called the modulus of z or the magnitude of z.

Example 9.6. There are infinitely many complex numbers with the same absolute value. Let

$$z = \cos\theta + \mathrm{i}\sin\theta,$$

where θ is any real number. Then $z\overline{z} = \cos^2\theta + \sin^2\theta = 1$. So

$$|z| = \sqrt{1} = 1.$$

We show below that this extension of the notion of absolute value to complex numbers has some familiar properties.

Theorem 9.1. Properties of absolute value

(a) Positivity: $|0| = 0$*, and if* $z \neq 0$ *then* $|z| > 0$.
(b) Symmetry: $|\overline{z}| = |z|$.
(c) Multiplicativity: $|wz| = |w||z|$.
(d) Triangle inequality: $|w + z| \le |w| + |z|$.

Proof. Positivity and symmetry follow directly from the definition, as we ask you to check in Problem 9.8.

For multiplicativity we use the properties of complex numbers and conjugation along with the definition of absolute value to get

$$|wz|^2 = wz\overline{wz} = wz\overline{w}\,\overline{z} = w\overline{w}z\overline{z} = |w|^2|z|^2 = (|w||z|)^2,$$

and so $|wz| = |w||z|$.

The proof of the triangle inequality again begins with the definition of absolute value and then uses the additive property and the distributive rule to get

$$|w+z|^2 = (w+z)(\overline{w+z}) = (w+z)(\overline{w}+\overline{z})$$
$$= w\overline{w}+w\overline{z}+z\overline{w}+z\overline{z} = |w|^2+w\overline{z}+z\overline{w}+|z|^2.$$

Observe that $w\overline{z}$ and $z\overline{w}$ are conjugates of each other. According to (9.1), their sum is equal to twice the real part

$$w\overline{z}+z\overline{w} = 2\mathrm{Re}\,(w\overline{z}).$$

As we saw in (9.4), the real part of a complex number does not exceed its absolute value. Therefore,

$$w\overline{z}+z\overline{w} \le 2|w\overline{z}|.$$

Now by multiplicativity and symmetry, $2|w\overline{z}| = 2|w||\overline{z}| = 2|w||z|$. Therefore,

$$|w+z|^2 = |w|^2+2\mathrm{Re}\,(w\overline{z})+|z|^2 \le |w|^2+2|w||z|+|z|^2 = (|w|+|z|)^2.$$

Therefore,

$$|w+z| \le |w|+|z|.$$

□

Example 9.7. We verify the triangle inequality for $w = 1-2\mathrm{i}$ and $z = 3+4\mathrm{i}$. Adding, we get $w+z = 4+2\mathrm{i}$. Taking absolute values, we have

$$|w| = \sqrt{5}, \qquad |z| = \sqrt{25}, \qquad |w+z| = \sqrt{20},$$

and $\sqrt{20} \le \sqrt{5}+\sqrt{25}$. So we see that $|w+z| \le |w|+|z|$.

Absolute Value and Sequence Convergence. Just as with sequences of real numbers, absolute values help us define what it means for numbers to be close, and hence what it means for a sequence of complex numbers to converge. A sequence of complex numbers $\{z_n\} = \{z_1, z_2, \ldots, z_n, \ldots\}$ is said to converge to z if for any tolerance $\varepsilon > 0$, $|z_n - z|$ is less than ε for all z_n far enough out in the sequence.

A sequence of complex numbers z_n is a *Cauchy sequence* if $|z_n - z_m|$ can be made less than any prescribed tolerance for all n and m large enough. A sequence $z_n = x_n + \mathrm{i}y_n$ of complex numbers gives rise to two sequences of real numbers: the real parts $x_1, x_2, \ldots$ and the imaginary parts $y_1, y_2, \ldots$. How is the behavior of $\{z_n\}$ related to the behavior of these sequences of real and imaginary parts $\{x_n\}$ and $\{y_n\}$? For example, suppose $\{z_n\}$ is a Cauchy sequence. According to (9.4),

$$|x_n - x_m| = |\mathrm{Re}(z_n - z_m)| \le |z_n - z_m|$$

and

$$|y_n - y_m| = |\mathrm{Im}(z_n - z_m)| \le |z_n - z_m|.$$

This shows that $\{x_n\}$ and $\{y_n\}$ are Cauchy sequences of real numbers, and hence converge to limits x and y. The sequence $\{z_n\}$ converges to $z = x + \mathrm{i}y$, because according to the triangle inequality,

$$|z - z_n| = |x - x_n + \mathrm{i}(y - y_n)| \le |x - x_n| + |y - y_n|,$$

and this sum on the right will be as small as desired whenever n is large enough.

On the other hand, suppose x_n and y_n are Cauchy sequences of real numbers. Then the sequence of complex numbers

$$z_n = x_n + \mathrm{i}y_n$$

is a Cauchy sequence of complex numbers, because

$$|z_n - z_m| = \sqrt{(x_n - x_m)^2 + (y_n - y_m)^2}$$

can be made as small as desired for all n and m large enough. Take n and m so large that *both* of $|x_n - x_m|$ and $|y_n - y_m|$ are less than ε. Then

$$|z_n - z_m| < \sqrt{\varepsilon^2 + \varepsilon^2} = \sqrt{2}\varepsilon.$$

In summary, the convergence of a sequence of complex numbers boils down to the convergence of the sequences of their real and imaginary parts. As a result, much of the work we did in Chap. 1 to prove theorems about convergence of sequences can be extended to complex numbers.

Example 9.8. Suppose $z_n = x_n + \mathrm{i}y_n$ is a Cauchy sequence tending to $z = x + \mathrm{i}y$. Then $x_n \to x$ and $y_n \to y$. Therefore, the sequence

$$z_n^2 = x_n^2 - y_n^2 + 2\mathrm{i}x_n y_n \quad \text{tends to} \quad x^2 - y^2 + 2\mathrm{i}xy.$$

That is, $z_n^2 \to z^2$.

Alternatively, without mentioning x_n or y_n, we have

$$|z^2 - z_n^2| = |z + z_n||z - z_n|.$$

The factor $|z + z_n|$ is nearly $|z + z|$ for n large, and the factor $|z - z_n|$ tends to zero. Therefore, z_n^2 tends to z^2.

9.1b Geometry of Complex Numbers

We now present a *geometric representation* of complex numbers. This turns out to be a very useful way of thinking about complex numbers, just as it is useful to think of the real numbers as points of the number line. The complex numbers are conveniently represented as points in a plane, called the *complex number plane*.

To each point (x,y) of the Cartesian plane we associate the complex number $x+\mathrm{i}y$. The horizontal axis consists of real numbers. It is called the *real axis*. The vertical axis consists of purely imaginary numbers and is called the *imaginary axis*. See Fig. 9.1.

The complex conjugate has a simple geometric interpretation in the complex plane. The complex conjugate of $x+\mathrm{i}y$, $x-\mathrm{i}y$, is obtained from $x+\mathrm{i}y$ by *reflection across the real axis*. The geometric interpretation of the absolute value $\sqrt{x^2+y^2}$ of $x+\mathrm{i}y$ is very striking: it is the *distance* of $x+\mathrm{i}y$ *from the origin*. To visualize geometrically the sum of two complex numbers z and w, move the coordinate system rigidly and parallel to itself, i.e., without rotating it, so that the origin ends up where the point z was originally located. Then the point w will end up where the point $z+w$ was located in the original coordinate system. It follows from this geometric description of addition that the four points 0, w, z, $w+z$ are the vertices of a parallelogram; in particular, it follows that the distance of w to $w+z$ equals the distance from 0 to z. See on the left in Fig. 9.2.

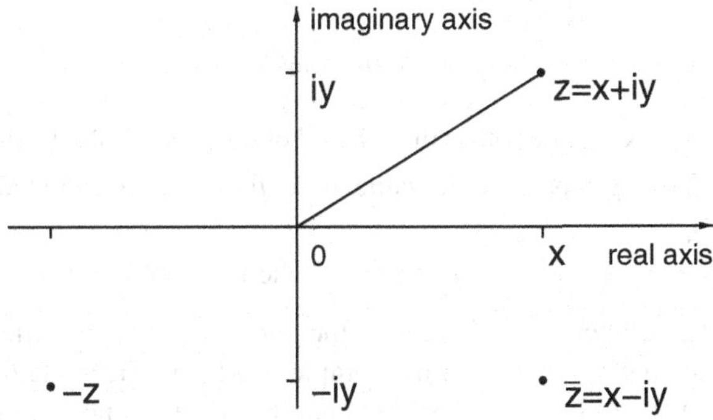

Fig. 9.1 Some complex numbers

Now consider the triangle whose vertices are 0, w, and $w+z$, as on the right in Fig. 9.2. The length of the side 0 to $w+z$ is $|w+z|$, the length of the side 0 to w is $|w|$, and using the parallelogram interpretation, the length of the side w to $w+z$ is $|z|$. According to a famous inequality of geometry, the length of any one side of a triangle does not exceed the sum of the lengths of the other two; therefore,

$$|w+z| \leq |w|+|z|.$$

This is precisely the triangle inequality and is the reason for its name. Our earlier proof of this inequality made no reference to geometry, so it may be regarded as a proof of a theorem about triangles with the aid of complex numbers! In Problems 9.16 and 9.17 we shall give further examples of how to prove geometric results with the aid of complex numbers.

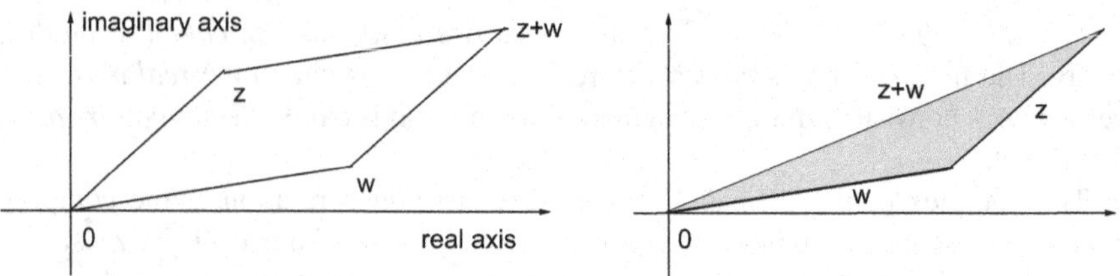

Fig. 9.2 *Left*: addition of complex numbers. *Right*: the triangle inequality illustrated

We have seen that addition of complex numbers can be visualized using Cartesian coordinates. Next, we see how multiplication of complex numbers can be visualized using *polar coordinates*.

Suppose p is a complex number with absolute value 1. Such a point lies on the unit circle. The Cartesian coordinates of p are $(\cos\theta,\ \sin\theta)$, where θ is the radian measure of the angle made by the real axis and the ray from 0 through p. So the complex number p with $|p| = 1$ is

$$p = \cos\theta + \mathrm{i}\sin\theta. \tag{9.5}$$

Let z be any complex number other than 0, and denote its absolute value by $r = |z|$. Define $p = \dfrac{z}{r}$. Then p has absolute value 1, so p can be represented in the form (9.5). Therefore,

$$z = r(\cos\theta + \mathrm{i}\sin\theta), \quad \text{where} \quad r = |z|. \tag{9.6}$$

See Fig. 9.3. The numbers (r,θ) are called the *polar coordinates* of the point (x,y), and (9.6) is called the *polar form* of the complex number. The angle θ is called the *argument* of z, denoted by $\arg z$; it is the angle between the positive real axis and the ray connecting the origin to z. In (9.6), we may replace θ by θ plus any integer multiple of 2π.

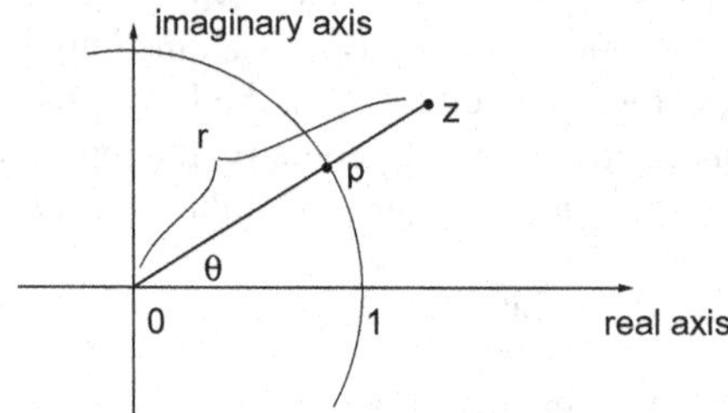

Fig. 9.3 Polar coordinates (r,θ) for z. The case $|z| > 1$ is drawn

Let z and w be a pair of complex numbers. Represent each in polar form,

$$z = r(\cos\theta + \mathrm{i}\sin\theta), \qquad w = s(\cos\phi + \mathrm{i}\sin\phi).$$

Multiplying these together, we get that

$$zw = rs(\cos\theta + \mathrm{i}\sin\theta)(\cos\phi + \mathrm{i}\sin\phi) \tag{9.7}$$
$$= rs\big((\cos\theta\cos\phi - \sin\theta\sin\phi) + \mathrm{i}(\cos\theta\sin\phi + \sin\theta\cos\phi)\big).$$

We use the addition laws for the cosine and sine, Eq. (3.17), to rewrite the product formula (9.7) in a particularly simple form. Recalling that r denotes $|z|$ and s denotes $|w|$, we find that

$$zw = |z||w|[\cos(\theta+\phi) + \mathrm{i}\sin(\theta+\phi)]. \tag{9.8}$$

This formula gives the polar form of the product zw. It is a symbolic statement of the following theorem.

Theorem 9.2. Multiplication rule for complex numbers in polar form

(a) The absolute value of the product zw is the product of the absolute values of its factors.
(b) The argument of the product zw is the sum of the arguments of its factors z and w.

In symbols:

$$|zw| = |z||w|, \tag{9.9}$$

$$\arg(zw) = \arg z + \arg w + 2\pi n, \tag{9.10}$$

where n is 0 or 1.

Example 9.9. Let $z = w = -i$, where $\arg z = \arg w = \dfrac{3\pi}{2}$. Then $zw = -1$, $\arg(zw) = \arg(-1) = \pi$, and $\arg z + \arg w = 3\pi$. Therefore, in this case,

$$\arg z + \arg w = \arg(zw) + 2\pi.$$

Square Root of a Complex Number. The complex number z^2 has twice the argument of z, and its absolute value is $|z|^2$. This suggests that to find square roots, we must halve the argument and form the square root of the absolute value.

Let us use these properties to locate a square root of i. The square root of 1 is 1, and half of $\dfrac{\pi}{2}$ is $\dfrac{\pi}{4}$. This describes the number

$$z = \cos\left(\frac{\pi}{4}\right) + \mathrm{i}\sin\left(\frac{\pi}{4}\right) = \frac{1}{\sqrt{2}} + \mathrm{i}\frac{1}{\sqrt{2}} = \frac{1+\mathrm{i}}{\sqrt{2}}.$$

Let us verify that $z^2 = \mathrm{i}$:

$$\left(\frac{1+\mathrm{i}}{\sqrt{2}}\right)^2 = \frac{1+2\mathrm{i}+\mathrm{i}^2}{2} = \mathrm{i}.$$

It works. Next we show that when we have z in polar form, it is possible to take powers and roots of z easily.

De Moivre's Theorem. By the multiplication rule, if

$$z = r(\cos\theta + i\sin\theta),$$

then

$$z^2 = r^2(\cos 2\theta + i\sin 2\theta),$$
$$z^3 = r^3(\cos 3\theta + i\sin 3\theta),$$

and for every positive integer n,

$$\big(r(\cos\theta + i\sin\theta)\big)^n = r^n\big(\cos(n\theta) + i\sin(n\theta)\big). \tag{9.11}$$

This result is known as de Moivre's theorem. The reciprocal is

$$\frac{1}{r(\cos\theta + i\sin\theta)} = r^{-1}(\cos\theta - i\sin\theta),$$

because $r(\cos\theta + i\sin\theta)\frac{1}{r}(\cos\theta - i\sin\theta) = \frac{r}{r}(\cos^2\theta + \sin^2\theta) = 1$. Similarly, we ask you to verify in Problem 9.10 that for $n = 1, 2, 3, \ldots$,

$$\text{if } z = r(\cos\theta + i\sin\theta) \qquad \text{then } z^{-n} = r^{-n}\big(\cos(n\theta) - i\sin(n\theta)\big).$$

After our successful polar representation of positive and negative integer powers of z, we tackle the problem of rational powers, $z^{p/q}$. Since $z^{p/q} = (z^{1/q})^p$, we first settle the problem of finding the qth roots of z, for $q = 2, 3, 4, \ldots$.

By analogy to (9.11), we shall tentatively represent a qth root of z by

$$w_1 = r^{1/q}\left(\cos\left(\frac{\theta}{q}\right) + i\sin\left(\frac{\theta}{q}\right)\right).$$

Indeed, this is a qth root of z, for its qth power is

$$(w_1)^q = (r^{1/q})^q\left(\cos\left(q\frac{\theta}{q}\right) + i\sin\left(q\frac{\theta}{q}\right)\right) = r(\cos\theta + i\sin\theta) = z,$$

but it is not the only qth root. Another root, different from w_1, is the number

$$w_2 = r^{1/q}\left(\cos\left(\frac{\theta + 2\pi}{q}\right) + i\sin\left(\frac{\theta + 2\pi}{q}\right)\right).$$

You may also check, using the periodicity of the sine and cosine functions, that the q numbers

$$w_{k+1} = r^{1/q}\left(\cos\left(\frac{\theta + 2k\pi}{q}\right) + i\sin\left(\frac{\theta + 2k\pi}{q}\right)\right), \qquad k = 0, 1, \ldots, q-1, \tag{9.12}$$

yield the q distinct qth roots of z, if $z \neq 0$. If $z = 0$, they are all equal to zero.

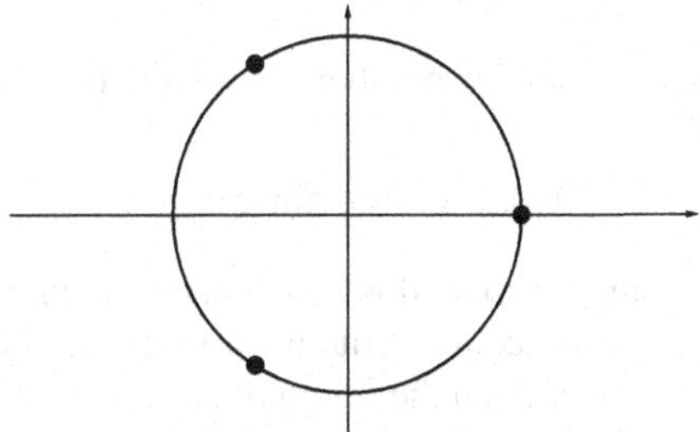

Fig. 9.4 The cube roots of 1, given in Example 9.10. All three lie on the unit circle

Example 9.10. The three cube roots of $1 = \cos 0 + \mathrm{i}\sin 0$ are

$$\cos\frac{0}{3} + \mathrm{i}\sin\frac{0}{3} = 1, \quad \cos\frac{2\pi}{3} + \mathrm{i}\sin\frac{2\pi}{3} = -\frac{1}{2} + \mathrm{i}\frac{\sqrt{3}}{2},$$

and

$$\cos\frac{4\pi}{3} + \mathrm{i}\sin\frac{4\pi}{3} = -\frac{1}{2} - \mathrm{i}\frac{\sqrt{3}}{2},$$

which correspond to taking $k = 0$, 1, and 2 in (9.12). See Fig. 9.4.

We end this section on the geometry of complex numbers by showing how to use products of complex numbers to find the area A of a triangle with vertices at three complex numbers in the plane. We take the case in which one of the vertices of the triangle is the origin. We lose no generality by this assumption, for the area of the triangle with vertices p, q, r is the same as the area of the translated triangle $0, q-p, r-p$.

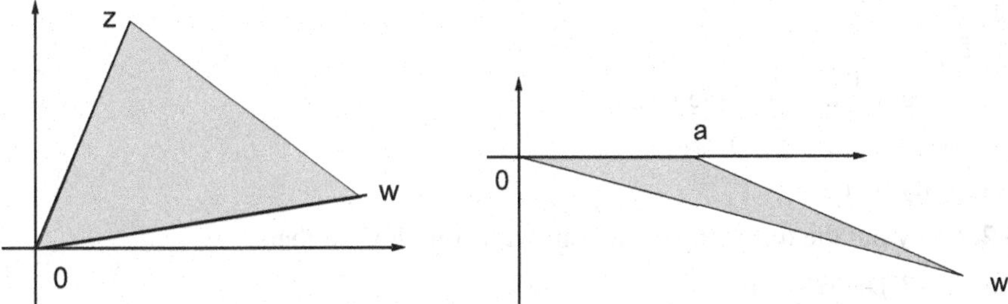

Fig. 9.5 *Left*: we prove that the area of the triangle is $\frac{1}{2}|\mathrm{Im}(\bar{z}w)|$. *Right*: the height of this triangle is $|\mathrm{Im}(w)|$

Denote by a a positive real number, and by w a complex number. The triangle whose vertices are 0, a, and w has base whose length is a and whose altitude is the absolute value of the imaginary part of w. See Fig. 9.5. Therefore, the area of the triangle with vertices 0, a, w is

$$A(0,a,w) = \frac{1}{2}a|\mathrm{Im}(w)| = \frac{1}{2}|\mathrm{Im}(aw)|.$$

Now let z and w denote any complex numbers. We claim that the area of the triangle with vertices 0, z, w is

$$A(0,z,w) = \frac{1}{2}\left|\operatorname{Im}(\overline{z}w)\right|. \tag{9.13}$$

In the case that z is real and positive, this agrees with the first case. Now consider arbitrary $z \neq 0$. Let p denote any complex number of absolute value 1. Multiplication by p is equivalent to rotation around the origin; therefore, the triangle with vertices 0, pz, pw has the same area as the triangle with vertices 0, z, w:

$$A(0,z,w) = A(0,pz,pw).$$

Choose $p = \dfrac{\overline{z}}{|z|}$. Then $pz = |z|$ is real, so the area is given by the formula

$$A(0,pz,pw) = \frac{1}{2}\left|\operatorname{Im}(\overline{p}\overline{z}pw)\right| = \frac{1}{2}\left|\operatorname{Im}(\overline{z}w)\right|.$$

Example 9.11. We find the area of the triangle with vertices $(0,1)$, $(2,3)$, and $(-5,7)$. Translate so that the first vertex is at the origin, giving $(0,0)$, $(2,2)$, and $(-5,6)$, or the complex numbers 0, $z = 2+2\mathrm{i}$, and $w = -5+6\mathrm{i}$. Then

$$A(0,z,w) = \frac{1}{2}|\operatorname{Im}(\overline{z}w)| = \frac{1}{2}\left|\operatorname{Im}\big((2-2\mathrm{i})(-5+6\mathrm{i})\big)\right| = 11.$$

Problems

9.1. For each of the numbers $z = 2+3\mathrm{i}$ and $z = 4-\mathrm{i}$, calculate the following:

(a) $|z|$
(b) $\overline{z}$
(c) Express $\dfrac{1}{z}$ and $\dfrac{1}{\overline{z}}$ in the form $a+b\mathrm{i}$.
(d) Verify that $z+\overline{z} = 2\operatorname{Re}(z)$.
(e) Verify that $z\overline{z} = |z|^2$.

9.2. Carry out the following operations with complex numbers:

(a) $(2+3\mathrm{i})+(5-4\mathrm{i})$
(b) $(3-2\mathrm{i})-(8-7\mathrm{i})$
(c) $(3-2\mathrm{i})(4+5\mathrm{i})$
(d) $\dfrac{3-2\mathrm{i}}{4+5\mathrm{i}}$
(e) Solve $2\mathrm{i}z = \mathrm{i}-4z$

9.3. Express $z = 4+(2+\mathrm{i})\mathrm{i}$ in the form $x+\mathrm{i}y$, where x and y are real, and then find the conjugate $\overline{z}$.

9.4. Give an example to show that the ray from 0 to $\mathrm{i}z$ is 90 degrees counterclockwise from the ray from 0 to z, unless z is 0.

9.5. Find the absolute values of the following complex numbers:

(a) $3+4\mathrm{i}$
(b) $5+6\mathrm{i}$
(c) $\dfrac{3+4\mathrm{i}}{5+6\mathrm{i}}$
(d) $\dfrac{1+\mathrm{i}}{1-\mathrm{i}}$

9.6. For each item, describe geometrically the set of complex numbers that satisfy the given condition.

(a) $\mathrm{Im}\,(z)=3$
(b) $\mathrm{Re}\,(z)=2$
(c) $2<\mathrm{Im}\,(z)\le 3$
(d) $|z|=1$
(e) $|z|=0$
(f) $1<|z|<2$

9.7. Show that $\mathrm{Im}\,z=\dfrac{z-\overline{z}}{2\mathrm{i}}$ and $\mathrm{Re}\,z=\dfrac{z+\overline{z}}{2}$.

9.8. Verify the positivity and symmetry properties of absolute value $|z|$, as listed in Theorem 9.1.

9.9. Verify the identity $(a^2+b^2)(c^2+d^2)=(ac-bd)^2+(ad+bc)^2$. Then explain how this proves the property $|z||w|=|zw|$ of absolute value.

9.10. Verify that the reciprocal of $z^n=r^n\big(\cos\theta+\mathrm{i}\sin\theta\big)^n$ is

$$z^{-n}=r^{-n}\big(\cos(n\theta)-\mathrm{i}\sin(n\theta)\big).$$

9.11. Prove that for complex numbers z_1 and z_2,

(a) $|z_1-z_2|^2=|z_1|^2+|z_2|^2-2\mathrm{Re}\,(z_1\overline{z_2})$,
(b) $\big||z_1|-|z_2|\big|\le|z_1-z_2|$.

9.12.(a) Show that for any pair of complex numbers z and w,

$$|z+w|^2+|z-w|^2=2|z|^2+2|w|^2.$$

(b) Using the parallelogram interpretation of the addition of complex numbers, deduce from (a) that the sum of the squares of the diagonals of a parallelogram equals the sum of the squares of its four sides.

9.13. Find all three cube roots of -1 and represent each in the complex plane.

9.14.(a) Show that the two square roots of i are

$$\frac{1}{\sqrt{2}}(1+\mathrm{i}),\ -\frac{1}{\sqrt{2}}(1+\mathrm{i}),$$

and sketch each in the complex plane.

(b) Verify that the two square roots of i are fourth roots of -1. Then find the other two fourth roots of -1. Sketch all four.

9.15. Suppose a triangle has vertices 0, $a = a_1 + ia_2$, and $b = b_1 + ib_2$. Derive from the area formula (9.13) that the area of the triangle is

$$\frac{1}{2}|a_1b_2 - a_2b_1|.$$

9.16. Verify the following.

(a) The argument of $\overline{w}$ is the negative of the argument of w.
(b) The argument of $z\overline{w}$ is the difference of the arguments of z and w.
(c) The argument of $\frac{z}{w}$ is the difference of the arguments of z and w.
(d) The ray from 0 to z is perpendicular to the ray from 0 to w if and only if the number $z\overline{w}$ is purely imaginary.
(e) Let p be a point on the unit circle. Prove that the ray connecting p to the point 1 on the real axis is perpendicular to the ray connecting p to the point -1.

9.17. Let p and q be complex numbers of absolute value 1. Verify the following.

(a) $(p-1)^2\overline{p}$ is real.
(b) $\left((p-1)(\overline{q}-1)\right)^2\overline{p}q$ is real.
(c) Prove the angle-doubling theorem: If p and q lie on the unit circle, then the angle β between the rays from the origin to p and q is twice the angle α between the rays connecting the point $z = 1$ to p and q. In Fig. 9.6, $\beta = 2\alpha$.

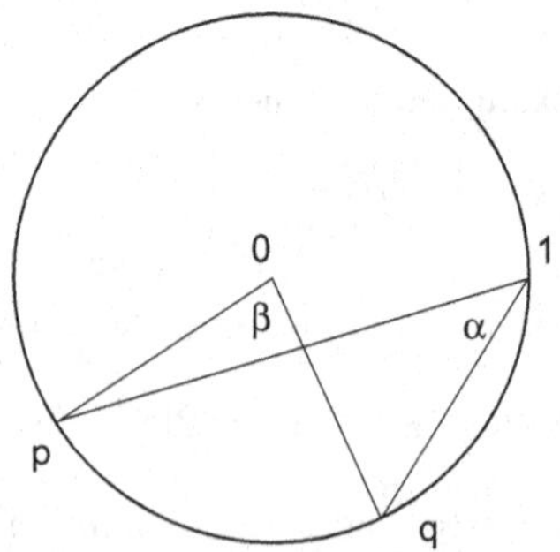

Fig. 9.6 The angle at 0 is twice the angle at 1, in Problem 9.17

9.18. Show that the sequence of complex numbers

$$1,\ 1+z,\ 1+z+z^2,\ 1+z+z^2+z^3, \ldots$$

is a Cauchy sequence, provided that $|z|$ is less than 1. What is the limit of the sequence?

9.19. We explore roots of 1.

(a) Find all zeros of the function $w(x) = x^4 + x^3 + x^2 + x + 1$ by first observing that the function $(x-1)w(x) = x^5 - 1$ is 0 whenever $w(x)$ is 0.
(b) Sketch all n of the nth roots of 1 in the complex plane.
(c) Let r be the nth root of 1 with smallest nonzero argument. Verify that the remaining nth roots are $r^2, r^3, \ldots, r^n$. Explain why the $n-1$ numbers

$$r,\ r^2,\ r^3,\ \ldots,\ r^{n-1}$$

are the roots of $x^{n-1} + x^{n-2} + \cdots + x + 1 = 0$.

9.20. Generalize the geometric series argument used in Problem 1.56 to show that for each complex number z, the partial sums

$$s_n = 1 + z + \frac{z^2}{2!} + \cdots + \frac{z^n}{n!}$$

are a Cauchy sequence.

9.2 Complex-Valued Functions

In this section we discuss the relationship of complex numbers to the concept of function. So far in this text, numbers have entered the concept of function in two places: as *input* and *output*. In the first part of this section, we show how simple it is to replace real numbers by complex ones as output if the input is kept real; such functions are called complex-valued functions with real input. In the second part, we discuss some very special but important functions whose input is complex, and whose output is also complex, i.e., complex-valued functions with complex input.

Definition 9.3. A *complex-valued function of a real variable* means a function

$$f(t) = p(t) + iq(t),$$

where p and q are real-valued functions. The variable t is real, in some interval that is the domain of p, q and f.

The theory of complex-valued functions is simple, because it can be reduced at one stroke to the theory of real-valued functions. There are two ways of going about it. The first is to observe that everything (well, almost everything) that has been said about real-valued functions makes sense when carried over to complex-valued functions. "Everything" includes the following notions:

(a) The concept of function itself.
(b) The operations of adding and multiplying functions and forming the reciprocal of a function (with the usual proviso that the function should not be zero).

(c) The concept of a continuous function.
(d) The concept of a differentiable function and its derivative.
(e) Higher derivatives.
(f) The integral of a function over an interval.

9.2a Continuity

Let us review the concept of continuity. The intuitive meaning of the continuity of a function f is that to determine $f(t)$ approximately, approximate knowledge of t is sufficient. The precise version was discussed in Definition 2.3 back in Chapter 2. Both the intuitive and the precise definitions make sense for complex-valued functions:

- Intuitive: A complex-valued function $f(t)$ is continuous at t if approximate knowledge of t suffices to determine approximate knowledge of $f(t)$.
- Precise: A complex-valued function f is uniformly continuous on an interval I if given any tolerance $\varepsilon > 0$, there is a precision $\delta > 0$ such that for any pair of numbers t and s in I that differ by less than δ, $f(t)$ and $f(s)$ differ by less than ε.

Sums and products of uniformly continuous complex-valued functions are uniformly continuous, and so is the reciprocal of a uniformly continuous function that is never zero.

Example 9.12. Let $f(t) = 3 + it$. To check uniform continuity of f we may write

$$f(t) - f(s) = (3 + it) - (3 + is) = i(t - s).$$

Then

$$|f(t) - f(s)| = |i(t - s)| = |t - s|.$$

If t and s are within δ, then the complex numbers $f(t)$ and $f(s)$ are as well. Therefore, f is uniformly continuous on every interval.

9.2b Derivative

We turn to the concept of differentiation. The function $f(t)$ is differentiable at t if

$$f_h(t) = \frac{f(t+h) - f(t)}{h}$$

tends to a limit as h tends to zero. This limit is called the derivative of f at t, and is denoted by $f'(t)$. We can separate the difference quotient into its real and imaginary parts:

$$\frac{p(t+h) + iq(t+h) - p(t) - iq(h)}{h} = p_h(t) + iq_h(t).$$

It follows that $f = p + iq$ is differentiable at t if and only if its real and imaginary parts are differentiable at t.

The usual rules for differentiating sums, products, and reciprocals of differentiable functions hold, i.e.,

$$\begin{aligned}(f+g)' &= f' + g',\\ (fg)' &= fg' + f'g,\\ \left(\frac{1}{f}\right)' &= -\frac{f'}{f^2}.\end{aligned}$$

You are urged to consult Chap. 3 to convince yourself that the proofs offered there retain their validity for complex-valued functions as well.

Example 9.13. Suppose $f(t) = z_1 t + z_2$, where z_1 and z_2 are arbitrary complex numbers. Then $f'(t) = z_1$.

Example 9.14. If $f(x) = \dfrac{1}{x+\mathrm{i}}$, then the rule for differentiating the reciprocal of a function yields

$$f'(x) = -\frac{1}{(x+\mathrm{i})^2}.$$

Example 9.15. Again let $f(x) = \dfrac{1}{x+\mathrm{i}}$. Let us find $f'(x)$ a different way. Separate f into its real and imaginary parts:

$$f(x) = \frac{1}{x+\mathrm{i}} = \frac{1}{x+\mathrm{i}}\frac{x-\mathrm{i}}{x-\mathrm{i}} = \frac{x-\mathrm{i}}{x^2+1} = \frac{x}{x^2+1} - \frac{\mathrm{i}}{x^2+1} = a(x) + ib(x).$$

Differentiating, we get

$$a'(x) = \frac{(x^2+1)1 - x(2x)}{(x^2+1)^2} = \frac{1-x^2}{(x^2+1)^2}, \qquad b'(x) = \left(-\frac{1}{x^2+1}\right)' = \frac{2x}{(x^2+1)^2}.$$

To see that the real and imaginary parts of $f'(x)$, as computed in Example 9.14, are $a'(x)$ and $b'(x)$ of Example 9.15, we write

$$f'(x) = -\frac{1}{(x+\mathrm{i})^2} = \frac{-1}{(x+\mathrm{i})^2}\frac{(x-\mathrm{i})^2}{(x-\mathrm{i})^2} = \frac{-x^2+2x\mathrm{i}+1}{(x^2+1)^2} = \frac{1-x^2}{(x^2+1)^2} + \mathrm{i}\frac{2x}{(x^2+1)^2}.$$

Example 9.16. We use the quotient rule for differentiating $f(x) = \dfrac{x}{x^2+\mathrm{i}}$, and obtain

$$f'(x) = \frac{(x^2+\mathrm{i})x' - x(x^2+\mathrm{i})'}{(x^2+\mathrm{i})^2} = \frac{\mathrm{i}-x^2}{(x^2+\mathrm{i})^2}.$$

Example 9.17. Now consider $f(x) = (x+\mathrm{i})^2$. If we carry out the indicated squaring, we can split f into its real and imaginary parts:

$$f(x) = x^2 + 2\mathrm{i}x - 1 = x^2 - 1 + 2\mathrm{i}x = a(x) + \mathrm{i}b(x).$$

Differentiating $a(x) = x^2 - 1$ and $b(x) = 2x$, we get $a'(x) = 2x$ and $b'(x) = 2$, so that

$$f'(x) = a'(x) + \mathrm{i}b'(x) = 2x + 2\mathrm{i}.$$

The function $f(x) = (x+\mathrm{i})^2 = (x+\mathrm{i})(x+\mathrm{i})$, when differentiated using the product rule, yields

$$f'(x) = 1(x+\mathrm{i}) + (x+\mathrm{i})1 = 2(x+\mathrm{i}),$$

the same answer we got before.

Next we consider the chain rule. Suppose $g(t)$ is a real-valued function and f is a complex-valued function defined at all values taken on by g. Then we can form their composition $f \circ g$, defined as $f(g(t))$. If f and g are both differentiable, so is the composite, and the derivative of the composite is given by the usual chain rule. The proof is similar to the real-valued case.

Example 9.18. Let $f(x) = \dfrac{1}{x+\mathrm{i}}$ and $g(t) = t^2$. By Example 9.14, $f'(x) = -(x+\mathrm{i})^{-2}$. The derivative of $f \circ g$ can be calculated by the chain rule:

$$\left(\frac{1}{t^2+\mathrm{i}}\right)' = (f \circ g)'(t) = f'(g(t))g'(t) = -(g(t)+\mathrm{i})^{-2}g'(t) = \frac{1}{(t^2+\mathrm{i})^2}2t.$$

9.2c Integral of Complex-Valued Functions

Splitting a complex-valued function into its real and imaginary parts is suitable for defining the integral of a complex-valued function.

Definition 9.4. For $f = p + \mathrm{i}q$, where p and q are continuous real-valued functions on $[a,b]$, we set

$$\int_a^b f(t)\,\mathrm{d}t = \int_a^b p(t)\,\mathrm{d}t + \mathrm{i}\int_a^b q(t)\,\mathrm{d}t.$$

The properties of integrals of complex-valued functions follow from the definition and the corresponding properties of integrals of real-valued functions that are continuous on the indicated intervals:

- *Additivity:* $\displaystyle\int_a^c f(t)\,\mathrm{d}t + \int_c^b f(t)\,\mathrm{d}t = \int_a^b f(t)\,\mathrm{d}t.$

- *Linearity:* For a complex constant k, $\int_a^b kf(t)\,\mathrm{d}t = k\int_a^b f(t)\,\mathrm{d}t$ and

$$\int_a^b (f(t)+g(t))\,\mathrm{d}t = \int_a^b f(t)\,\mathrm{d}t + \int_a^b g(t)\,\mathrm{d}t.$$

- *Fundamental theorem:* $\dfrac{\mathrm{d}}{\mathrm{d}x}\int_a^x f(t)\,\mathrm{d}t = f(x)$ and $\int_a^b F'(t)\,\mathrm{d}t = F(b)-F(a)$.

Recall that the integrals of p and q in Definition 9.4 are defined as the limits of approximating sums $I_{\text{approx}}(p,[a,b])$ and $I_{\text{approx}}(q,[a,b])$. Let us use the same subdivision of $[a,b]$ and choices of the t_j and define

$$I_{\text{approx}}(f,[a,b]) = I_{\text{approx}}(p,[a,b]) + \mathrm{i}I_{\text{approx}}(q,[a,b]).$$

We can conclude that $I_{\text{approx}}(f,[a,b])$ tends to $\int f(t)\,\mathrm{d}t$. That is, the integral of a complex-valued function can be defined in terms of approximating sums. This bears out our original contention that most of the theory we have developed for real-valued functions applies verbatim to complex-valued functions. Another property of the integral of real-valued functions is the following.

- *Upper bound property:* If $|f(t)| \le M$ for every t in $[a,b]$, then $\left|\int_a^b f(t)\,\mathrm{d}t\right| \le M(b-a)$.

This inequality, too, remains true for complex-valued functions, and for the same reason: the analogous estimate holds for the approximating sums, with absolute value as defined for complex numbers.

9.2d Functions of a Complex Variable

Next we consider complex-valued functions $f(z)=w$ of a complex variable z. Do such functions have derivatives? In Chap. 3, we introduced the derivative in two ways:

- the rate at which the value of the function changes;
- the slope of the line tangent to the graph of the function.

For functions of a complex variable, the geometric definition as slope is no longer available, but the rate of change definition is still meaningful.

Definition 9.5. A complex-valued function $f(z)$ of a complex variable z is differentiable at z if the difference quotients

$$\frac{f(z+h)-f(z)}{h}$$

tend to a limit as the complex number h tends to zero. The limit is called the derivative of f at z and is denoted by $f'(z)$.

Example 9.19. Let $f(z)=z^2$. Then

$$\frac{f(z+h)-f(z)}{h}=\frac{(z+h)^2-z^2}{h}=\frac{2zh+h^2}{h}=2z+h.$$

As the complex number h tends to 0, $2z+h$ tends to $2z$. So $f'(z)=2z$.

Theorem 9.3. *Every positive integer power of z,*

$$z^m \qquad (m=1,2,3,\ldots),$$

is a differentiable function of the complex variable z, and its derivative is mz^{m-1}.

Proof. According to the binomial theorem, valid for complex numbers,

$$(z+h)^m=z^m+mz^{m-1}h+\cdots+h^m. \tag{9.14}$$

All the terms after the first two have h raised to the power 2 or higher. So

$$\frac{(z+h)^m-z^m}{h}=mz^{m-1}+\cdots,$$

where each of the terms after the plus sign contains h as a factor. It follows that as h tends to zero, the difference quotient above tends to mz^{m-1}. This concludes the proof. □

Since sums and constant multiples of differentiable functions are differentiable, it follows that every polynomial $p(z)$ is a differentiable function of z. Next we show that we can extend Newton's method to find complex roots of polynomials.

Newton's Method for Complex Roots. Newton's method for estimating a real root of a real function $f(x)=0$ relied on finding a root of the linear approximation to f at a previous estimate. We carry this idea into the present setting of complex roots.

Suppose z_{old} is an approximate root of $p(z) = 0$. Denote by h the difference between the exact zero z of p and z_{old}:

$$h = z - z_{\text{old}}.$$

Using Eq. (9.14), we see that

$$0 = p(z) = p(z_{\text{old}} + h) = p(z_{\text{old}}) + p'(z_{\text{old}})h + \text{error}, \tag{9.15}$$

where the error is less than a constant times $|h|^2$. Let the new approximation be

$$z_{\text{new}} = z_{\text{old}} - \frac{p(z_{\text{old}})}{p'(z_{\text{old}})}, \tag{9.16}$$

as we did in Sect. 5.3a. We use (9.15) to express

$$p(z_{\text{old}}) = -p'(z_{\text{old}})h - \text{error}.$$

Setting this in (9.16), we get

$$z_{\text{new}} = z_{\text{old}} + h + \frac{\text{error}}{p'(z_{\text{old}})}.$$

Since h was defined as $z - z_{\text{old}}$, we can rewrite this relation as

$$z_{\text{new}} - z = \frac{\text{error}}{p'(z_{\text{old}})}.$$

If $p'(z_{\text{old}})$ is bounded away from zero (which is a basic assumption for Newton's method to work), we can rewrite this relation as the inequality

$$|z - z_{\text{new}}| < (\text{constant})h^2 = (\text{constant})|z - z_{\text{old}}|^2.$$

If z_{old} is so close to the exact root z that the quantity $(\text{constant})|z - z_{\text{old}}|$ is less than 1, then the new approximation is closer to the exact root than the old one. Repeating this process produces a sequence of approximations that converge to the exact root with extraordinary rapidity.

Example 9.20. We have seen that the three cube roots of 1 are 1, $\frac{1+\sqrt{3}\mathrm{i}}{2}$, and $\frac{1-\sqrt{3}\mathrm{i}}{2}$. Let us see what Newton's method produces. For $f(z) = z^3 - 1$, Newton's iteration is

$$z_{\text{new}} = z - \frac{f(z)}{f'(z)} = z - \frac{z^3 - 1}{3z^2}.$$

Table 9.1 shows the results starting from three different initial states, and these are sketched in Fig. 9.7.

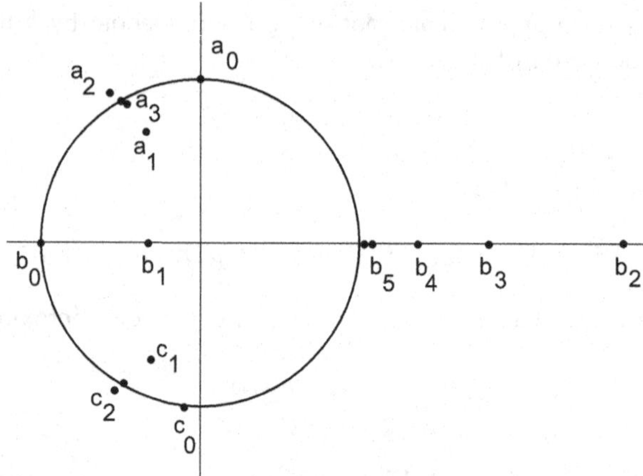

Fig. 9.7 Newton's method used to approximate three roots of $f(z) = z^3 - 1$, starting from $a_0 = \mathrm{i}$, $b_0 = -1$, and $c_0 = -0.1 - \mathrm{i}$. See Example 9.20

Table 9.1 Newton's method applied in Example 9.20 to $z^3 - 1 = 0$, from three starting values

n	a_n	b_n	c_n
0	i	−1	−0.1 − i
1	−0.33333 + 0.66667i	−0.33333	−0.39016 − 0.73202i
2	−0.58222 + 0.92444i	2.7778	−0.53020 − 0.89017i
3	−0.50879 + 0.86817i	1.8951	−0.50135 − 0.86647i
4	−0.50007 + 0.86598i	1.3562	−0.50000 − 0.86602i
5	−0.50000 + 0.86603i	1.0854	−0.50000 − 0.86603i
6	−0.50000 + 0.86603i	1.0065	−0.50000 − 0.86603i
7	−0.50000 + 0.86603i	1.0000	−0.50000 − 0.86603i

9.2e The Exponential Function of a Complex Variable

We now turn to the function $C(t) = \cos t + \mathrm{i}\sin t$. Its image in the complex plane lies on the unit circle. The function C is differentiable, and $C'(t) = -\sin t + \mathrm{i}\cos t$. A moment's observation discloses that

$$C' = \mathrm{i}C.$$

We now recall that for a real, $\mathrm{e}^{at} = P(t)$ satisfies the differential equation

$$P' = aP.$$

This suggests how to define $\mathrm{e}^{\mathrm{i}t}$.

Definition 9.6.

$$e^{it} = \cos t + i \sin t.$$

The functional equation of the exponential function suggests that

$$e^{x+iy} = e^x e^{iy};$$

combining this with Definition 9.6, we arrive at the definition of the exponential of a complex number.

Definition 9.7. If $x,\ y$ are real numbers, we define

$$e^{x+iy} = e^x\big(\cos y + i \sin y\big).$$

The exponential function as defined above has complex input, complex output, and it has all the usual properties of the exponential function, as we shall now show.

Theorem 9.4. *For all complex numbers z and w,*

$$e^{z+w} = e^z e^w.$$

Proof. Set $z = x + iy$ and $w = u + iv$. Then by definition, we get

$$e^z e^w = e^x(\cos y + i \sin y)e^u(\cos v + i \sin v).$$

Using the functional equation of the exponential function for real inputs and arithmetic properties of complex numbers yields

$$\begin{aligned} e^z e^w &= e^{x+u}(\cos y + i \sin y)(\cos v + i \sin v) \\ &= e^{x+u}\big((\cos y \cos v - \sin y \sin v) + i(\cos y \sin v + \sin y \cos v)\big). \end{aligned}$$

Similarly, using the addition formulas for sine and cosine, we conclude that

$$e^z e^w = e^{x+u}(\cos(y+v) + i \sin(y+v)) = e^{x+u+i(y+v)}.$$

We conclude that $e^z e^w = e^{z+w}$, as asserted. □

Theorem 9.5. Differential equation. *For every complex number c,*

$$P(t) = e^{ct}$$

satisfies

$$P'(t) = cP(t).$$

Proof. Set $c = a + ib$, where a and b are real. By the definition of complex exponentials,

$$e^{ct} = e^{at}(\cos(bt) + i\sin(bt)).$$

The derivative of this is, by the product rule,

$$ae^{at}(\cos(bt) + i\sin(bt)) + e^{at}(-b\sin(bt) + ib\cos(bt)).$$

This is the same as $ae^{ct} + ibe^{ct} = (a + ib)e^{ct} = ce^{ct}$, as was to be proved. □

Example 9.21. Let us show that $y = e^{it}$ satisfies the differential equation

$$y'' + y = 0.$$

We have $y' = ie^{it}$, so $y'' = i^2 e^{it} = -e^{it} = -y$ and $y'' + y = 0$.

Theorem 9.6. Series representation. *For every complex number z,*

$$e^z = 1 + z + \frac{z^2}{2} + \cdots + \frac{z^n}{n!} + \cdots.$$

Proof. We saw in Problem 1.55 of Chap. 1 that the theorem is true in the case of $z = 1$. Certainly it is true when $z = 0$. We proved it for all real z in Sect. 4.3a, and in Problem 9.20, we suggested a method to show that the partial sums are a Cauchy sequence. For the sake of simplicity, we prove this now only for pure imaginary z. Assume $z = ib$, where b is real. Substituting $z = ib$ into the series on the right, we get

$$1 + ib - \frac{b^2}{2} + \cdots + \frac{(ib)^n}{n!} + \cdots.$$

The powers i^n have period 4, i.e., we have

$$i^0 = 1, \qquad i^1 = i, \qquad i^2 = -1, \qquad i^3 = -i,$$

and then the pattern is repeated. This shows that the terms of even order are real, and the odd ones are purely imaginary. Next, we assume that it is valid to rearrange the terms of this series:

$$\begin{aligned}
&1 + ib - \frac{b^2}{2} + \cdots + \frac{(ib)^n}{n!} + \cdots \\
&= \left(1 - \frac{b^2}{2} + \frac{b^4}{24} - \cdots + \frac{(-1)^m b^{2m}}{(2m)!} + \cdots\right) \\
&+ i\left(b - \frac{b^3}{6} + \frac{b^5}{120} + \cdots + \frac{(-1)^m b^{2m+1}}{(2m+1)!} + \cdots\right).
\end{aligned}$$

The real and imaginary parts are $\cos b$ and $\sin b$; see Sect. 4.3a and Eq. (4.20). So our series is

$$1+ib-\frac{b^2}{2}+\cdots+\frac{(ib)^n}{n!}+\cdots.=\cos b+\mathrm{i}\sin b,$$

which is e^{ib}, as was to be proved. □

Set $t=2\pi$ in the definition of the exponential function, and observe that

$$\mathrm{e}^{\mathrm{i}2\pi}=\cos(2\pi)+\mathrm{i}\sin(2\pi)=1.$$

More generally, since $\cos(2\pi n)=1,\ \sin(2\pi n)=0$,

$$\mathrm{e}^{\mathrm{i}2\pi n}=\cos(2\pi n)+\mathrm{i}\sin(2\pi n)=1$$

for every integer n.

Now set $t=\pi$ in the definition. Since $\cos\pi=-1$, $\sin\pi=0$, we get that

$$\mathrm{e}^{\mathrm{i}\pi}=\cos\pi+\mathrm{i}\sin\pi=-1.$$

This can be rewritten in the form[3] $\mathrm{e}^{\mathrm{i}\pi}+1=0$.

The Derivative of the Exponential Function of a Complex Variable. We have shown in Theorem 9.3 that polynomials $f(z)$ are differentiable for complex z. We shall show now that so is the function e^z.

Theorem 9.7. *The function* e^z *is differentiable, and its derivative is* e^z.

Proof. We have to show that the difference quotient

$$\frac{\mathrm{e}^{z+h}-\mathrm{e}^z}{h}$$

tends to e^z as h tends to 0. Using the functional equation for the exponential function, $\mathrm{e}^{z+h}=\mathrm{e}^z\mathrm{e}^h$, we can write the difference quotient as

$$\mathrm{e}^z\frac{\mathrm{e}^h-1}{h}.$$

So what has to be proved is that $\dfrac{\mathrm{e}^h-1}{h}$ tends to 1 as h tends to zero. Decompose h into its real and imaginary parts: $h=x+\mathrm{i}y$. Then

$$\mathrm{e}^h-1=\mathrm{e}^x(\cos y+\mathrm{i}\sin y)-1.$$

[3] It is worth pointing out to those interested in number mysticism that this relation contains the most important numbers and symbols of mathematics: $0,1,\mathrm{i},\pi,\mathrm{e},+$, and $=$.

As h approaches 0, both x and y approach 0 as well. Use the linear approximations of the exponential and trigonometric functions for x and y near zero:

$$e^x = 1 + x + r_1, \quad \cos y = 1 + r_2, \quad \sin y = y + r_3,$$

where the remainders r_1, r_2, and r_3 are less than a constant times $x^2 + y^2 = |h|^2$. Using these approximations, we get

$$\begin{aligned} e^h - 1 &= (1 + x + r_1)\big(1 + r_2 + i(y + r_3)\big) - 1 \\ &= x + iy + xiy + \text{remainder} = h + ixy + \text{remainder}, \end{aligned}$$

where the absolute value of the remainder is less than a constant times $|h|^2$. According to the A-G inequality, it is also true that the absolute value $|ixy|$ is equal to

$$|ixy| = |x||y| \le \frac{1}{2}\left(|x|^2 + |y|^2\right) = \frac{1}{2}|h|^2.$$

Therefore, $e^h - 1 = h + \text{remainder}$, where the absolute value of the remainder is less than a constant times $|h|^2$. Dividing by h, we see that

$$\frac{e^h - 1}{h} = 1 + \frac{\text{remainder}}{h}$$

tends to 1, as claimed. □

So far, all the functions $f(z)$ that we have considered, i.e., polynomials $p(z)$ and the exponential e^z, have been differentiable with respect to z. However, there are simple functions that are not differentiable. Here is an example.

Example 9.22. Let $f(z) = \overline{z}$. Then

$$\lim_{h \to 0} \frac{f(z+h) - f(z)}{h} = \lim_{h \to 0} \frac{\overline{z+h} - \overline{z}}{h} = \lim_{h \to 0} \frac{\overline{h}}{h}$$

if the limit exists. For real $h = x + 0i \ne 0$, we have $\overline{h} = h$, so $\dfrac{\overline{h}}{h} = 1$. For imaginary $h = 0 + yi \ne 0$, we have $\overline{h} = -yi$ and $\dfrac{\overline{h}}{h} = -1$. Therefore, the limit does not exist, and f is not differentiable.

Problems

9.21. Differentiate the following complex functions of a real variable t:

(a) $e^t + i \sin t$

(b) $\dfrac{1}{t - i} + \dfrac{1}{t + i}$

(c) $\mathrm{i}e^{t^2}$
(d) $\mathrm{i}\sin t + \dfrac{1}{t+3+\mathrm{i}}$

9.22. Differentiate the following complex functions of a complex variable z:

(a) $2\mathrm{i} - z^2$
(b) $z^3 - z + 5e^z$

9.23. Express $\cos t$ and $\sin t$ in terms of $e^{\mathrm{i}t}$.

9.24. Compute the complex integrals:

(a) $\displaystyle\int_0^1 e^{\mathrm{i}s}\,ds$
(b) $\displaystyle\int_0^{\pi/2} (\cos s + \mathrm{i}\sin s)\,ds$

9.25. Since we have defined e^z for every complex z, we can extend the definition of the hyperbolic cosine to every complex number. Write that definition, and then verify the identity $\cosh(\mathrm{i}t) = \cos t$ for all real numbers t.

9.26. For a real and positive and z complex, write a definition of a^z by expressing a as $e^{\log a}$. Show that $a^{z+w} = a^z a^w$.

9.27. Find the value of the integral $\displaystyle\int_0^\infty e^{\mathrm{i}kx-x}\,dx$, k real.

9.28. In this exercise we ask you to test experimentally Newton's method (9.16) for finding complex roots of an algebraic equation $p(z) = 0$. Take for p the cubic polynomial

$$p(z) = z^3 + z^2 + z - \mathrm{i}.$$

(a) Write a computer program that constructs a sequence $z_1, z_2, \ldots$ according to Newton's method, stopping when both the real and imaginary parts of z_{n+1} and z_n differ by less than 10^{-6}, or when n exceeds 30.
(b) Show that $p(z) \neq 0$ if $|z| > 2$ or if $|z| < \frac{1}{2}$.
Hint: Use the triangle inequality in the form

$$|a-b| \geq |a| - |b|$$

to show that $|z^3 + z^2 + z - \mathrm{i}| \geq 1 - |z^3| - |z^2| - |z|$, which is positive if $|z| < \frac{1}{2}$. Similarly, show that if $|z| > 2$, then the z^3 term is the most important to consider, because its absolute value is greater than that of the sum of the other three terms.
(c) Starting with the first approximation

$$z_0 = 0.35 + 0.35\mathrm{i},$$

construct the sequence z_n by Newton's method, and determine whether it converges to a solution of $p(z) = 0$.
(d) Find all solutions of $p(z) = 0$ by starting with different choices for z.

Chapter 10
Differential Equations

Abstract The laws of the exact sciences are often formulated as differential equations, that is, equations connecting functions and their derivatives. In this chapter, we present examples from three different fields: mechanical vibrations, population growth, and chemical reactions.

10.1 Using Calculus to Model Vibrations

Most people realize that *sound*—its generation, transmission, and perception—is a vibration. For this reason alone, vibration is a very important subject. But vibrations are more general and pervasive than mere sound, and they constitute one of the fundamental phenomena of physics. The reason is the mechanical stability of everyday objects from bells and horns to the basic constituents of matter. Mechanical stability means that when an object is distorted by an outside force, it springs back into its original shape when released. This is accomplished by restoring forces inherent in any object. Restoring forces work in a peculiar way: they not only bring the object back to its original shape, but they tend to overcorrect and distort it in the opposite direction. This is again overcorrected, and so ad infinitum, leading to a vibration around an equilibrium. In this section, we shall explain this process in simple situations, as an example of the application of calculus.

10.1a Vibrations of a Mechanical System

The fundamental concepts of one-dimensional mechanics are particle, mass, position, velocity, acceleration, and force.

The *position* of a particle along a line is specified by a single real number x. Since the position of the particle changes in time, it is a function of the time t. The derivative of position with respect to t is the velocity of the particle, denoted usually

P.D. Lax and M.S. Terrell, *Calculus With Applications*, Undergraduate Texts in Mathematics,
DOI 10.1007/978-1-4614-7946-8_10,

by $v(t)$. The derivative of velocity with respect to time is called *acceleration*, and is denoted by $a(t)$:

$$x' = v, \quad v' = x'' = a.$$

The mass of the particle, denoted by m, does not change throughout the motion. Newton's law of motion says that

$$f = ma,$$

where f is the total force acting on the particle, m the mass, and a the acceleration. To put teeth into Newton's law, we have to be able to calculate the total force acting on the particle. According to Newton, the total force acting on a particle (in the direction of increasing x) is the sum of all the various forces acting on it. In this section, we shall deal with two kinds of forces: *restoring forces* and *frictional forces*. We shall describe them in the following specific context.

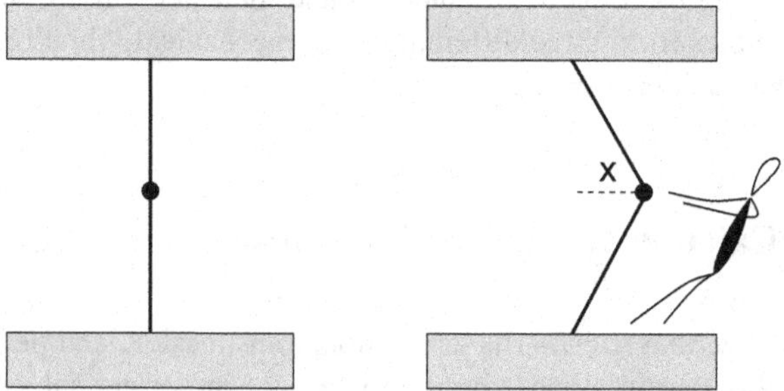

Fig. 10.1 *Left*: position of static equilibrium. *Right*: mass displaced by distance x

Imagine a piece of elastic string (rubber band, elastic wire) placed in a vertical position with its endpoints fastened and a mass attached to its middle. In this position the mass is at rest. Now displace the mass to one side (Fig. 10.1). In this position, the elastic string exerts a force on the mass. It is clear, to anyone who ever shot paper clips with a rubber band, that the restoring force depends on position, and

(a) *the force acts in the direction opposite to the displacement, tending to restore the mass to its previous position;*
(b) *the greater the magnitude of the displacement, the greater the magnitude of the force.*
A force with these two properties is called a *restoring force*, written f_{re}. The graph of a typical restoring force is shown in Fig. 10.2. Many restoring forces, such as the one exerted by a rubber band, have yet a third property, symmetry:
(c) *Displacements by the same magnitude but in opposite directions generate restoring forces f_{re} that are equal in magnitude but opposite in direction.*

We turn next to describing the force of friction. Friction can be caused by various mechanisms, one of which is air resistance. As anyone who has ever bicycled at high speed knows, the frictional force depends on velocity, and

(a) *the force of air resistance acts in the direction opposite to the direction of motion*;
(b) *the greater the velocity, the greater the force of resistance.*

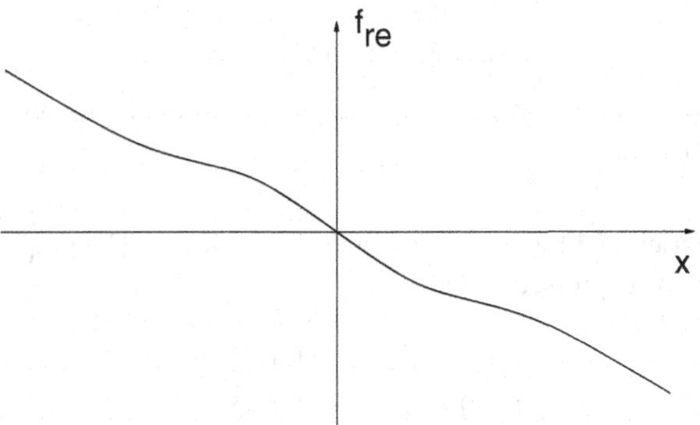

Fig. 10.2 A restoring force f_{re}, as a function of displacement

Any force with these two properties is called a *frictional force*, written f_{fr}. The graph of a typical frictional force is shown in Fig. 10.3. This graph displays yet another property common to most frictional forces, their symmetry:

(c) *The magnitude of the frictional force f_{fr} depends only on the magnitude of the velocity.*

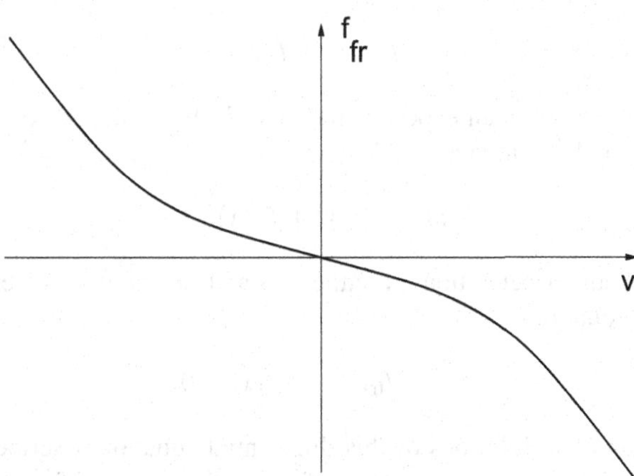

Fig. 10.3 A frictional force f_{fr}, as a function of velocity

In order to turn the verbal descriptions of these two kinds of forces into mathematical descriptions, we regard the restoring force $f_{\rm re}$ as a function of the position x. The properties (a)–(c) can be expressed as follows in the language of functions:

$$f_{\rm re}(x) = \begin{cases} <0 & \text{for } x > 0, \\ =0 & \text{for } x = 0, \\ >0 & \text{for } x < 0, \end{cases}$$

and

$$f_{\rm re}(x) \text{ is a decreasing function of } x.$$

The symmetry property of $f_{\rm re}(x)$ can be expressed in the following way: $f_{\rm re}$ is an *odd* function of x, i.e.,

$$f_{\rm re}(-x) = -f_{\rm re}(x).$$

We regard the frictional force $f_{\rm fr}$ as a function of velocity. The properties of a frictional force can be expressed as follows:

$$f_{\rm fr}(v) = \begin{cases} <0 & \text{for } v > 0, \\ =0 & \text{for } v = 0, \\ >0 & \text{for } v < 0, \end{cases}$$

and

$$f_{\rm fr}(v) \text{ is a decreasing function of } v.$$

Assumption (c), that the magnitude of friction depends only on the magnitude of velocity, together with assumption (a) implies that $f_{\rm fr}$ is an odd function, i.e.,

$$f_{\rm fr}(-v) = -f_{\rm fr}(v).$$

The total force f is the sum of the individual forces:

$$f = f_{\rm fr} + f_{\rm re}.$$

The additivity of forces is an experimental fact. With this decomposition of force, Newton's law may be written as

$$ma = f_{\rm fr}(v) + f_{\rm re}(x). \tag{10.1}$$

Since velocity v and acceleration a are the first and second derivatives of x, this is the differential equation

$$mx'' - f_{\rm fr}(x') - f_{\rm re}(x) = 0 \tag{10.2}$$

for x as function of t. Solutions of this differential equation describe all possible motions of a particle subject to a restoring force and a frictional force.

In the rest of Sect. 10.1, we shall study the behavior of solutions of Eq. (10.2) for various kinds of restoring force and frictional force. The basic fact is that if we prescribe the initial position and velocity of a particle, then the motion of the particle is completely determined for all time by the differential equation. We call this basic result the *uniqueness theorem*:

Theorem 10.1. Uniqueness. *Denote by $x(t)$ and $y(t)$ two solutions of the differential equation (10.2) that are equal at some time s and whose first derivatives are equal at time s:*

$$x(s) = y(s), \qquad x'(s) = y'(s).$$

Then $x(t)$ and $y(t)$ are equal for all $t \geq s$.

The steps of the proof of this theorem are outlined in Problem 10.21 at the end of Sect. 10.1, where we ask you to justify each step.

Example 10.1. We have encountered the following differential equations in previous chapters:

$$\text{(a) } x' = x, \quad \text{(b) } x'' - x = 0\,, \quad \text{(c) } x'' + x = 0.$$

Which of these are examples of Eq. (10.2) governing vibrations of a simple mechanical system?

(a) The equation $x' = x$ can be rewritten $0 + x' - x = 0$, but it is not an example of Eq. (10.2), because there is no second-order term mx''.

(b) $x'' - x = 0$ looks promising. Take mass $m = 1$, friction $f_{\text{fr}}(x') = 0$, and restoring force $f_{\text{re}}(x) = x$. This seems to fit the form. However, in our model we made the assumption that f_{re} is decreasing and odd. Since this $f_{\text{re}}(x)$ is not decreasing, $x'' - x = 0$ is not an example of Eq. (10.2).

(c) $x'' + x = 0$ looks very much like case (b), except that now the restoring force $f_{\text{re}}(x) = -x$ is decreasing and odd. Therefore, the equation $x'' + x = 0$ is an example of the equation $mx'' - f_{\text{fr}}(x') - f_{\text{re}}(x) = 0$ governing vibrations of a simple mechanical system.

10.1b Dissipation and Conservation of Energy

This section will be devoted to the mathematics of extracting information about solutions of the differential equation $mx'' - f_{\text{fr}}(x') - f_{\text{re}}(x) = 0$. It is remarkable how much we can deduce about the solutions without knowing the frictional force $f_{\text{fr}}(v)$ or the restoring force $f_{\text{re}}(x)$ explicitly, but knowing only that they both are decreasing odd functions.

We start with a trick. Multiply the equation by v, obtaining

$$mva - vf_{\text{fr}}(v) - vf_{\text{re}}(x) = 0.$$

Since the sign of $f_{\text{fr}}(v)$ is opposite to that of v, it follows that $-vf_{\text{fr}}(v)$ is positive except when $v = 0$. By dropping this positive term, we convert the equality into the inequality

$$mva - vf_{\text{re}}(x) \leq 0. \tag{10.3}$$

Recalling that acceleration is the derivative of velocity, we can rewrite the term mva as mvv'. We recognize this as the derivative of $\frac{1}{2}mv^2$:

$$mva = \frac{\mathrm{d}}{\mathrm{d}t}\left(\frac{1}{2}mv^2\right). \tag{10.4}$$

Recalling that v is the derivative of x, we can rewrite the term $vf_{\text{re}}(x)$ as $x'f_{\text{re}}(x)$.

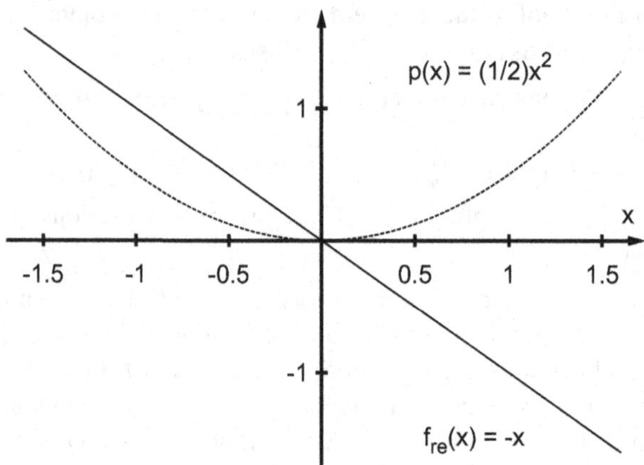

Fig. 10.4 Graphs of the restoring force $f_{\text{re}}(x) = -x$ and the potential energy $p(x) = \frac{1}{2}x^2$ for the equation $x'' + x = 0$ in Example 10.1

Let us introduce the function $p(x)$ as the integral of $-f_{\text{re}}$,

$$p(x) = -\int_0^x f_{\text{re}}(y)\,\mathrm{d}y.$$

By the fundamental theorem of calculus, the derivative of p is $-f_{\text{re}}$,

$$\frac{\mathrm{d}}{\mathrm{d}x}p(x) = -f_{\text{re}}(x), \tag{10.5}$$

and by definition,

$$p(0) = 0.$$

The derivative of $p(x)$ with respect to x is $-f_{\text{re}}(x)$, which is positive for x positive, and negative for x negative. (See Figs. 10.2 and 10.4.) According to the monotonicity criterion, this means that p is increasing for $x > 0$ and decreasing for $x < 0$. Since $p(0) = 0$, it follows that $p(x)$ is positive for all $x \neq 0$. Using the chain rule and Eq. (10.5), we can express the derivative of $p(x(t))$ with respect to t as

$$\frac{\mathrm{d}}{\mathrm{d}t}p(x(t)) = x'(t)\frac{\mathrm{d}p}{\mathrm{d}x} = -x'(t)f_{\mathrm{re}}(x(t)) = -vf_{\mathrm{re}}(x). \tag{10.6}$$

Substituting this result and expression (10.4) for the first and second terms into inequality (10.3) we obtain

$$\frac{\mathrm{d}}{\mathrm{d}t}\left(\frac{1}{2}mv^2 + p(x)\right) \leq 0. \tag{10.7}$$

According to the monotonicity criterion, a function whose derivative is less than or equal to zero is decreasing. We use "decreasing" to mean "nonincreasing." So we conclude that the function

$$\frac{1}{2}mv^2 + p(x)$$

decreases with time. This function, and both terms appearing in it, have physical meaning: the quantity $\frac{1}{2}mv^2$ is called *kinetic energy*, and the quantity $p(x)$ is called *potential energy*. The sum of kinetic and potential energies is called the *total energy*. In this terminology, we have derived the following.

Law of Decrease of Energy. *The total energy of a particle moving under the influence of a restoring force and a frictional force decreases with time.*

Suppose there is no frictional force, i.e., f_{fr} is zero. Then the energy inequality (10.7) becomes an equality: $\frac{\mathrm{d}}{\mathrm{d}t}\left(\frac{1}{2}mv^2 + p(x)\right) = 0$. A function whose derivative is zero for all t is a constant, E, so we have derived the following.

Law of Conservation of Energy. *In the absence of friction, the total energy of a particle moving under the influence of a restoring force does not change with time.*

$$\frac{1}{2}mv^2 + p(x) = E.$$

From Eq. (10.2) governing vibrations of a simple mechanical system, we have derived energy laws for the particle in the presence, and in the absence, of friction. When there is no friction, the total mechanical energy does not change with time. When there is friction, the total *mechanical* energy decreases. That energy is not lost but is turned into heat energy.

10.1c Vibration Without Friction

Next, we turn our attention to the study of the motion of a particle subject to a restoring force in the absence of friction. That is, the forces satisfy

$$mx'' - f_{\mathrm{re}}(x) = 0. \tag{10.8}$$

In Example 10.1, we saw that the differential equation $x'' - (-x) = x'' + x = 0$ is an example of Eq. (10.8), and we showed in Sect. 3.4b that all solutions of $x'' + x = 0$ are of the form $x(t) = u\cos t + v\sin t$, where u and v are arbitrary constants. These functions have period 2π:

$$x(t+2\pi) = x(t).$$

Now we show that *every* function $x(t)$ satisfying Eq. (10.8) is periodic. We start with a qualitative description of the motion determined by Eq. (10.8) and the law of conservation of energy that we derived from it in the last section:

$$\frac{\mathrm{d}}{\mathrm{d}t}\left(\frac{1}{2}mv^2 + p(x)\right) = 0, \quad p(x) = -\int_0^x f_{\mathrm{re}}(y)\,\mathrm{d}y, \quad \frac{1}{2}mv^2 + p(x) = E.$$

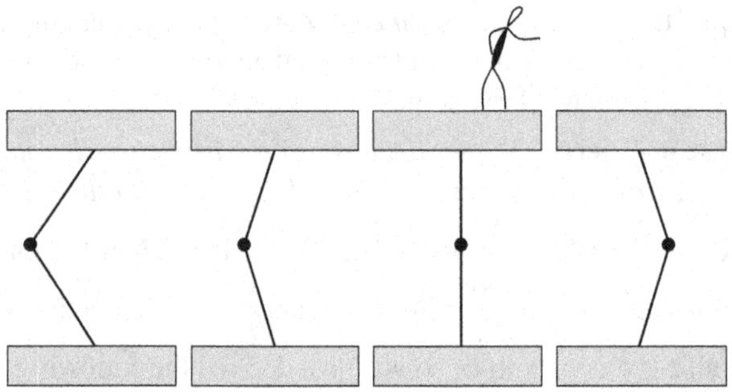

Fig. 10.5 At *left*, $x = -b$, $v = 0$, and $p = E$. Then x is shown between $-b$ and 0, with v positive and $p < E$. Then $x = 0$, $p = 0$, and $\frac{1}{2}mv^2 = E$. At *right*, the particle has almost reached the point $x = b$ halfway through the cycle

Suppose we start the motion at time $t = 0$ by displacing the particle to the position $x = -b$, $b > 0$, and holding it there until we let it go, so that initially its velocity is zero. See Fig. 10.5. The total energy imparted thereby to the system is $p(-b) = E$. On being released, the restoring force starts moving the particle toward the position $x = 0$. For negative x, $p(x)$ decreases with x. It follows then from the law of conservation of energy that the kinetic energy, $\frac{1}{2}mv^2$, increases. Since v^2 is increasing, the particle gains speed during this phase of the motion. The potential energy reaches its minimum at $x = 0$. As soon as the particle swings past $x = 0$, its potential energy starts increasing, and its kinetic energy decreases accordingly. This state of affairs persists until the particle reaches the position $x = b$. At this point, its potential energy equals $p(b)$. Since p is the integral of an odd function, p is an even function, and $p(b) = p(-b) = E$ is the total energy. Therefore, at point b, the kinetic energy $\frac{1}{2}mv^2$ is zero, and b is the right endpoint of the interval through which the particle moves. On reaching $x = b$, the particle is turned around by the restoring force and describes a similar motion from right to left until it returns to its original position $x = -b$. Its velocity at this time $t = T$ is zero, so everything is just as it was at the beginning of the motion. Therefore, according to Theorem 10.1 in the previous section, the *same* pattern is repeated all over again. Such motion is called *periodic*, and the time T taken by the particle to return to its original position is called the *period* of the motion. The mathematical expression of periodicity is

$$x(t+T) = x(t),$$

and the graph of such a period-T function is shown in Fig. 10.6. In fact, due to the assumption that $f_{\rm re}$ is odd, the position $x = b$ occurs at exactly $t = \frac{1}{2}T$, for the motions to left and right are mirror images of each other.

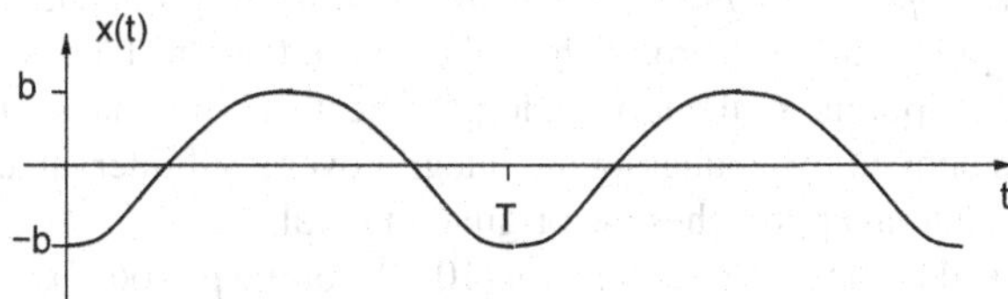

Fig. 10.6 The motion repeats with period T

We now turn from this qualitative description of motion to a quantitative description, which we also shall deduce from the law of conservation of energy. Using the energy equation $\frac{1}{2}mv^2 + p(x) = E$, we can express v as function of x:

$$v = \sqrt{\frac{2}{m}(E - p(x))}.$$

In the first phase of the motion, $0 \le t \le \frac{1}{2}T$, x is an increasing function of time. Therefore, $v = x'$ is positive, so that the positive square root is to be taken. Since $x(t)$ is strictly monotonic during this interval, we can express t as a function of x. According to the rule for differentiating the inverse of a function, the derivative of t with respect to x is

$$\frac{dt}{dx} = \frac{1}{\frac{dx}{dt}} = \frac{1}{v}.$$

Using the formula above for v, we deduce that

$$\frac{dt}{dx} = \sqrt{\frac{m}{2(E - p(x))}}.$$

According to the fundamental theorem of calculus, t is the integral with respect to x of $\frac{dt}{dx}$:

$$t(y_2) - t(y_1) = \int_{y_1}^{y_2} \sqrt{\frac{m}{2(E - p(x))}}\, dx. \tag{10.9}$$

The integral on the right expresses the time it takes for the particle to move from position y_1 to position y_2 during the first phase of the motion. Take, in particular, $y_1 = -b$ and $y_2 = b$. These positions are reached at $t = 0$ and $t = \frac{1}{2}T$, respectively. Therefore, $\frac{1}{2}T - 0 = \int_{-b}^{b} \sqrt{\frac{m}{2(E - p(x))}}\, dx$, and multiplying by 2, we get

$$T = \int_{-b}^{b} \sqrt{\frac{2m}{E - p(x)}}\, dx. \tag{10.10}$$

We have seen that the energy conservation $\frac{1}{2}mv^2 + p(x) = E$ at times $t = 0$ and $t = \frac{1}{2}T$ gives $E = p(-b) = p(b)$. This shows that as x approaches $-b$ or b, the difference $E - p(x)$ tends to zero. This makes the integrand tend to infinity as x approaches the endpoints. In the terminology of Sect. 7.3, this integral is improper, and therefore is defined by evaluating the integral over a subinterval and taking the limit as the subinterval approaches the original interval.

We show now that the improper integral (10.10) for the period T converges. According to the mean value theorem, the function in the denominator of Eq. (10.10) is

$$E - p(x) = E - p(b) - p'(c)(x - b)$$

for some c between b and x. Since $E = p(b)$, this gives

$$E - p(x) = -f_{\text{re}}(c)(b - x).$$

For x slightly less than b, that is, near the upper limit in integral (10.10), this is a positive multiple of $(b - x)$, because $-f_{\text{re}}(c)$ is nearly $-f_{\text{re}}(b) > 0$. Therefore,

$$\sqrt{\frac{2m}{E - p(x)}} \le \frac{\text{const}}{\sqrt{b - x}}.$$

The integrand is similarly bounded near $-b$, the lower bound of integration. As we have seen in Example 7.27 of Sect. 7.3, such a function is integrable. In other words, the period T is well defined by the integral (10.10).

We have been able to deduce quite a bit about the function $x(t)$ from the fact that it satisfies $mx'' - f_{\text{re}}(x) = 0$. First, we showed that $x(t)$ is periodic. Second, we showed that the period is a number $T = \int_{-b}^{b} \sqrt{\frac{2m}{E - p(x)}}\, dx$ that depends on the initial displacement b of the particle and on the restoring force f_{re}, since $p(x) = -\int_0^x f_{\text{re}}(y)\, dy$. Next, we look at the specific cases in which the restoring forces are linear functions.

10.1d Linear Vibrations Without Friction

Suppose that the restoring force is a *differentiable* function of x. According to the basic tenet of differential calculus, *a differentiable function can be well approximated over a short interval by a linear function.* We have seen earlier that the motion is confined to the interval $-b \le x \le b$, where $-b$ is the initial displacement. For small b, $f_{\text{re}}(x)$ can be well approximated over $[-b,b]$ by a linear function (Fig. 10.7). It is reasonable to expect that if we replace the true restoring force by its linear approximation over the small interval $[-b,b]$, the characteristic properties of motions with small displacements will not change drastically.

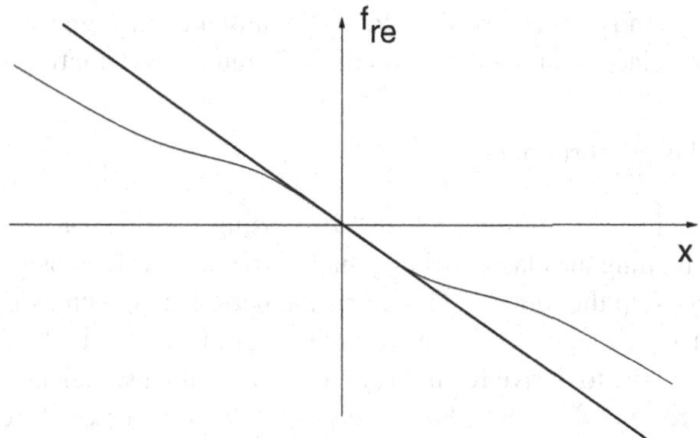

Fig. 10.7 Linearized restoring force

In this section, we study vibration under a *linear* restoring force

$$f_{\text{re}}(x) = -kx.$$

The positive constant k measures the stiffness of the elastic medium exerting the force, i.e., the larger k is, the greater the resistance to the displacement. For this reason, k is called the *stiffness* constant. The corresponding potential

$$p(x) = -\int_0^x f_{\text{re}}(s)\,\mathrm{d}s = -\int_0^x -ks\,\mathrm{d}s = \frac{1}{2}kx^2$$

is quadratic. Let us substitute it into formula (10.10) for the period of the motion. Using the fact that $E = p(b)$, we get

$$T = \int_{-b}^{b} \sqrt{\frac{2m}{E - p(x)}}\,\mathrm{d}x = \int_{-b}^{b} \sqrt{\frac{4m}{kb^2 - kx^2}}\,\mathrm{d}x.$$

Performing the change of variable $x = by$, we get

$$T = \int_{-b}^{b} \sqrt{\frac{4m}{kb^2 - kx^2}}\,\mathrm{d}x = \int_{-1}^{1} \sqrt{\frac{4m}{kb^2 - k(by)^2}}\,b\,\mathrm{d}y = 2\sqrt{\frac{m}{k}} \int_{-1}^{1} \frac{\mathrm{d}y}{\sqrt{1-y^2}}.$$

We recall that the function $\dfrac{1}{\sqrt{1-y^2}}$ is the derivative of $\sin^{-1} y$, so

$$\int_{-1}^{1} \frac{1}{\sqrt{1-y^2}}\, dy = \sin^{-1} 1 - \sin^{-1}(-1) = \frac{\pi}{2} - (-\frac{\pi}{2}) = \pi.$$

Substituting this into the above formula for T, we obtain

$$T = 2\pi\sqrt{\frac{m}{k}}. \tag{10.11}$$

This remarkable formula shows how the period of the motion depends on the data:

(a) The period is independent of the size of the initial displacement, provided that the initial displacement is small enough to warrant approximating $f_{\rm re}$ by a linear function.

(b) The period is proportional to $\sqrt{\dfrac{m}{k}}$.

What does our physical intuition tell us? Increasing the mass m slows down the motion, and tightening the elastic string, which is the same as increasing the stiffness constant k, speeds up the motion. Therefore, the period is an increasing function of m and a decreasing function of k; this is evident from formula (10.11).

We show now how to derive formula (10.11) using dimensional analysis. In a linear restoring force $f_{\rm re}(x) = -kx$, the dimension of the number k is force per length, which is equal to

$$\frac{(\text{mass})(\text{acceleration})}{\text{length}} = \frac{(\text{mass})\frac{\text{length}}{(\text{time})^2}}{\text{length}} = \frac{\text{mass}}{(\text{time})^2}.$$

The only way to build a number whose dimension is time out of the two numbers m and k is $\sqrt{\dfrac{m}{k}}$. Therefore, the period T must be a constant multiple of $\sqrt{\dfrac{m}{k}}$. Calculus is needed only to nail down that constant as 2π.

A periodic motion is often called a vibration. Any portion of such motion lasting a full period is called a *cycle*. The number of cycles per unit time is called *frequency*, i.e.,

$$\text{frequency} = \frac{1}{\text{period}} = \frac{1}{2\pi}\sqrt{\frac{k}{m}}.$$

The most striking manifestation of vibration is caused by the pressure waves transmitted through the air to the ears of a nearby auditor who perceives them as sound. The pitch of the sound is determined by the number of pressure pulses per unit time reaching the eardrum, and this number is the frequency of the vibrating source of the sound. When struck by a hammer, piece of metal vibrates. We know from everyday observation that the pitch of the sound generated does *not* depend on how hard the metal has been struck, although the loudness of the sound does. On the other hand, the sound generated by a plucked rubber band has a twangy quality, indicating that the pitch changes as the displacement changes. We conclude that the elastic force that acts in metal when slightly displaced from equilibrium is a linear function of displacement, while the force exerted by a rubber band is a nonlinear function of displacement.

10.1e Linear Vibrations with Friction

We now turn to the study of motion with friction. We shall restrict our study to motions for which displacement x and velocity v are relatively small, so small that both f_{re} and f_{fr} are so well approximated by linear functions that we might as well take them to be linear, i.e., $f_{\text{re}} = -kx$ and $f_{\text{fr}} = -hv$, where $k > 0$ and $h > 0$. Newton's equation becomes

$$mx'' + hx' + kx = 0, \tag{10.12}$$

where the constant h is called the *friction* constant. Such a differential equation whose coefficients m, k, and h are constants has a solution of the form e^{rt}. Substituting e^{rt} and its first and second derivatives re^{rt}, r^2e^{rt} into Eq. (10.12), we get $mr^2e^{rt} + hre^{rt} + ke^{rt} = 0$, and factoring out the exponential yields $(mr^2 + hr + k)e^{rt} = 0$. Since the exponential factor is never zero, the sum in the parentheses must be zero:

$$mr^2 + hr + k = 0. \tag{10.13}$$

Our efforts have led to a solution e^{rt} of Eq. (10.12) for each root of Eq. (10.13).

This is a quadratic equation for r, whose solutions are

$$r_{\pm} = -\frac{h}{2m} \pm \frac{\sqrt{h^2 - 4mk}}{2m}.$$

There are two cases, depending on the sign of the quantity under the square root.

- Case I: $h^2 - 4mk$ negative, or $h < 2\sqrt{mk}$.
- Case II: $h^2 - 4mk$ nonnegative, or $2\sqrt{mk} \le h$.

In Case I, the roots are complex, while in Case II, they are real. We first consider Case I.

Case I, $h < 2\sqrt{mk}$. Denote by w the real quantity

$$\frac{1}{2m}\sqrt{4mk - h^2} = w.$$

Then the roots can be written as

$$r_{\pm} = -\frac{h}{2m} \pm \mathrm{i}w.$$

This gives two complex-valued solutions,

$$x_-(t) = e^{r_- t} = e^{(-\frac{h}{2m} - iw)t} \quad \text{and} \quad x_+(t) = e^{r_+ t} = e^{(-\frac{h}{2m} + iw)t}.$$

We have seen in Chap. 9 that $e^{a+ib} = e^a(\cos b + i \sin b)$. Using this with $a = -\frac{h}{2m}$ and $b = w$, and again with $b = -w$, we can express

$$x_\pm(t) = e^{r_\pm t} = e^{-\frac{h}{2m}t}(\cos wt \pm i \sin wt).$$

We have shown in Theorem 9.3 that complex exponentials satisfy $(e^{rt})' = re^{rt}$, so the functions x_+ and x_- are solutions of Eq. (10.12). We ask you to verify in Problem 10.11 that sums and complex multiples of these solutions are also solutions. As a result,

$$\frac{1}{2}\big(x_+(t) + x_-(t)\big) = e^{-\frac{h}{2m}t}\cos wt \quad \text{and} \quad \frac{1}{2i}\big(x_+(t) - x_-(t)\big) = e^{-\frac{h}{2m}t}\sin wt$$

are solutions. These functions are the product of a trigonometric and an exponential function. The trigonometric function is periodic, with period $\frac{2\pi}{w}$, and the exponential function tends to 0 as t tends to infinity. The exponential function diminishes by the factor $e^{-\frac{h}{2m}}$ per unit time. This is called the *decay rate* of $x(t)$. Such motion is called a *damped vibration.*

By applying the linearity principle for real solutions of a differential equation (see Problem 10.9), we see that every combination of the form

$$x(t) = e^{-\frac{h}{2m}t}(A\cos wt + B\sin wt),$$

where A and B are constants, is also a solution.

Example 10.2. Consider the equation

$$x'' + \frac{1}{2}x' + \frac{17}{16}x = 0. \tag{10.14}$$

Solving $r^2 + \frac{1}{2}r + \frac{17}{16} = 0$, we obtain $r = -\frac{1}{4} + i$, so $e^{-\frac{1}{4}t}\big(\cos t + i \sin t\big)$ is a complex solution. The functions $e^{-\frac{1}{4}t}\cos t$ and $e^{-\frac{1}{4}t}\sin t$ are both real solutions to Eq. (10.14), and so are all linear combinations of them. The particular linear combination

$$x(t) = -e^{-\frac{1}{4}t}\Big(\cos t + \frac{1}{4}\sin t\Big)$$

is graphed in Fig. 10.8. This solution has initial values $x(0) = -1$ and $x'(0) = 0$.

Case II, $2\sqrt{mk} \le h$. The special case of equal roots, $h = 2\sqrt{mk}$, will be discussed in Problem 10.13. For $h > 2\sqrt{mk}$, the roots

$$r_\pm = -\frac{h}{2m} \pm \frac{\sqrt{h^2 - 4mk}}{2m}$$

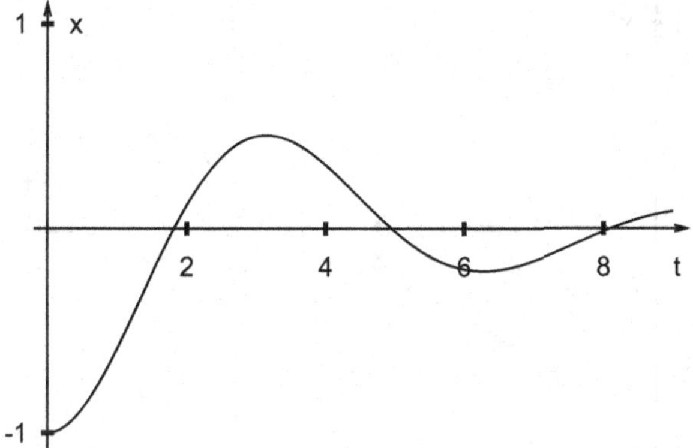

Fig. 10.8 A graph of the damped vibration $x(t) = -e^{-t/4}(\cos t + \frac{1}{4}\sin t)$ in Example 10.2

are both real, and they furnish two distinct real exponential solutions, e^{r_+t} and e^{r_-t}. According to the principle of linearity, every combination of them,

$$x(t) = A_+e^{r_+t} + A_-e^{r_-t},$$

is also a solution. We would like to choose the constants A_+ and A_- so that the initial displacement is $x(0) = -b$ and the initial velocity is $x'(0) = v(0) = 0$. The desired values of A_+ and A_- have to satisfy

$$\begin{aligned} x(0) &= A_+ + A_- = -b, \\ v(0) &= r_+A_+ + r_-A_- = 0. \end{aligned}$$

Since r_+ and r_- are unequal, A_+ and A_- are easily determined from these relations.

Example 10.3. We consider the equation

$$x'' + \frac{3}{2}x' + \frac{1}{2}x = 0$$

with initial displacement $x(0) = -1$. The roots of $r^2 + \frac{3}{2}r + \frac{1}{2} = 0$ are $r_- = -1$ and $r_+ = -\frac{1}{2}$. We need to solve $x(0) = A_+ + A_- = -1$ and $-\frac{1}{2}A_+ - A_- = 0$. Adding, we obtain $A_+ = -2$, then $A_- = 1$. Our solution is

$$x(t) = -2e^{-\frac{1}{2}t} + e^{-t}.$$

This is graphed for $t > 0$ in Fig. 10.9.

Both roots r_+ and r_- in Case II are negative. Consequently, both exponentials tend to zero as t tends to infinity. Of the two negative roots, r_- has the greater magnitude:

$$|r_-| > |r_+|.$$

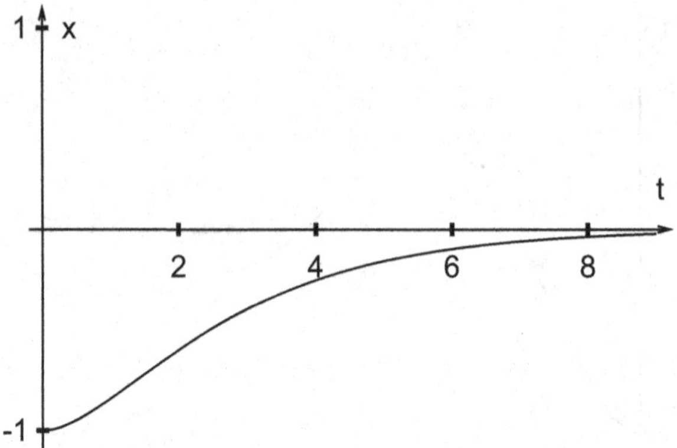

Fig. 10.9 The overdamped vibration $x(t) = \mathrm{e}^{-t} - 2\mathrm{e}^{-t/2}$ in Example 10.3

It is also true that $|A_+| > |A_-|$, and that for $t > 0$, $\mathrm{e}^{r_+t} > \mathrm{e}^{r_-t}$. Hence $|A_+\mathrm{e}^{r_+t}|$ is always greater than $|A_-\mathrm{e}^{r_-t}|$. As t tends to infinity, the first term becomes very much greater than the second. This shows that the decay of $x(t)$ is governed by the decay rate of the first term. That decay rate is e^{r_+}.

Rates of Decay. The difference between Case I and Case II is that in Case I, the force of friction is not strong enough to prevent the particle from swinging back and forth, although it does diminish the magnitude of successive swings. In Case II, friction is so strong compared to the restoring force that it slows down the particle to such an extent that it never swings over to the other side (except possibly in the rare case where $h = 2\sqrt{mk}$). This motion is called *overdamped*.

In both Case I and Case II, motion decays to zero as t tends to infinity. We now investigate the rates of this decay, respectively $\mathrm{e}^{-\frac{h}{2m}}$ and e^{r_+}. The logarithms of these decay rates are called *coefficients of decay* and are denoted by the symbol ℓ. We have the following formula for ℓ:

$$\ell(h) = \begin{cases} -\frac{h}{2m} & \text{for } h < 2\sqrt{mk}, \text{ Case I damped,} \\ \frac{-h+\sqrt{h^2-4mk}}{2m} & \text{for } 2\sqrt{mk} < h, \text{ Case II overdamped.} \end{cases}$$

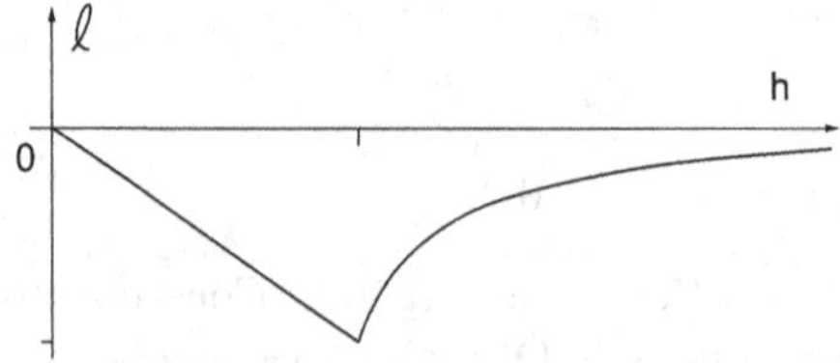

Fig. 10.10 Coefficient of decay ℓ is minimal at $h = 2\sqrt{mk}$

We next study how ℓ varies as the friction constant h changes while m and k remain fixed. Properties of the coefficient of decay ℓ:

(a) $\ell(h)$ is a continuous function for $0 \le h$.
This is true because at the point $h = 2\sqrt{mk}$ where Case I joins Case II, the two formulas for ℓ furnish the same value.

(b) $\ell(h)$ is a decreasing function of h for $0 \le h < 2\sqrt{mk}$.
This is true because for $0 \le h < 2\sqrt{mk}$, the derivative of ℓ is $-\dfrac{1}{2m} < 0$.

(c) $\ell(h)$ is an increasing function of h for $2\sqrt{mk} < h$.
This is true because for $h > 2\sqrt{mk}$, the derivative of ℓ is positive. To see this, note in $\ell'(h)$ that since $h > \sqrt{h^2 - 4mk}$, the fraction in the large parentheses below is greater than 1:

$$\ell'(h) = \frac{1}{2m}\left(-1 + \frac{h}{\sqrt{h^2 - 4mk}}\right).$$

(d) $\ell(h)$ reaches its minimum value at the *critical damping* $h = 2\sqrt{mk}$.
This is a consequence of the first three items.

Note that the function $\ell(h)$ is continuous and its absolute value is largest at $h = 2\sqrt{mk}$. It is not differentiable at $h = 2\sqrt{mk}$, as can be seen from Fig. 10.10. As the graph indicates, $\ell(h)$ tends to zero as h tends to infinity. Knowing the value of h that maximizes $|\ell|$ is important. For example, in an automobile bouncing after hitting a pothole, the springs provide a restoring force and the shock absorbers provide frictional damping. In fact, the shock is absorbed by the springs; the role of the shock absorbers is to dissipate the energy resulting from a sudden displacement.

10.1f Linear Systems Driven by an External Force

Next, we study the motion of particles under the influence of a restoring frictional force and a *driving force* f_d presented as a known function of time. This is a frequently occurring situation; examples of it are

(a) the motion of the eardrum driven by pressure pulses in the air,
(b) the motion of a magnetic diaphragm under an electromagnetic force,
(c) the motion of air in the resonating cavity of a violin under the force exerted by a vibrating violin string, and
(d) the motion of a building under the force exerted by wind or tremors in the Earth.

Of course, these examples are much more complicated than the case of a single particle that we shall investigate.

Newton's law of motion governing a single particle says that

$$mx'' = f_{\mathrm{re}}(x) + f_{\mathrm{fr}}(v) + f_d(t).$$

We shall discuss the case in which the restoring force and the frictional force are linear functions of their arguments and the driving force is the simple periodic function

$$f_d(t) = F\cos(qt),$$

where F is a positive constant. Substituting these forces into Newton's law, we get the equation

$$mx'' + hx' + kx = F\cos(qt). \tag{10.15}$$

See Fig. 10.11.

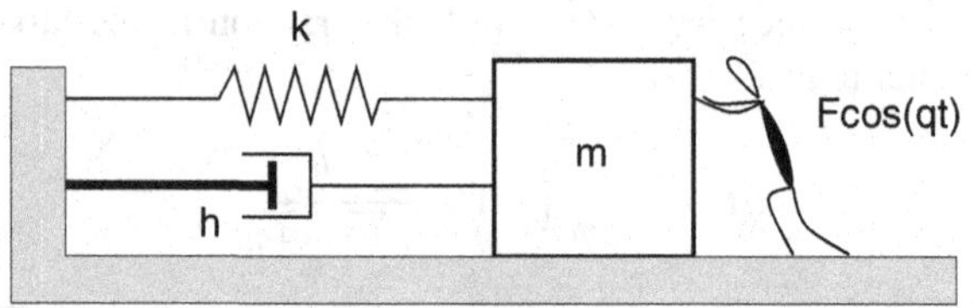

Fig. 10.11 Forces are applied to a mass m with position $x(t)$ in Eq. (10.15): the linear spring restoring force $-kx$, the linear frictional force $-hx'$, and the applied force $F\cos(qt)$

We begin by establishing a simple relation between any two solutions of this equation. Let x_0 be another solution:

$$mx_0'' + hx_0' + kx_0 = F\cos(qt).$$

Subtracting from Eq. (10.15), we get

$$m(x - x_0)'' + h(x - x_0)' + k(x - x_0) = 0, \tag{10.16}$$

i.e., the difference of *any* two solutions of Eq. (10.15) is a solution of Eq. (10.16). But this is the equation governing the motion of particles subject only to a restoring force and a frictional force. In the previous section, we showed that all solutions of Eq. (10.16) tend to zero as t tends to infinity (see Figs. 10.8 and 10.9). This shows that for large t, *any two solutions of* Eq. (10.15) *differ by very little*. Thus we may study the large-time behavior of any one solution.

We shall find a solution of Eq. (10.15) by the following trick. We look for complex-valued solutions z of the complex equation

$$mz'' + hz' + kz = Fe^{iqt}. \tag{10.17}$$

We ask you in Problem 10.16 to verify that if z is a complex-valued solution of Eq. (10.17), then $x = \operatorname{Re} z$ is a real solution of Eq. (10.15). The advantage of z is the ease with which we can calculate with exponentials.

We take z of the same form as the driving force, because it is reasonable to guess that the mass oscillates at the same frequency at which it is pushed:

$$z(t) = Ae^{iqt}.$$

Then by Theorem 9.5,

$$z' = Aiqe^{iqt} \quad \text{and} \quad z'' = -Aq^2e^{iqt}.$$

Substituting these into Eq. (10.17), we get, after division by e^{iqt},

$$A(-mq^2 + ihq + k) = F.$$

Solving for A from this equation, we get that

$$z(t) = \frac{F}{-mq^2 + ihq + k}e^{iqt}$$

is a solution to Eq. (10.17). The real part x of z is a solution of the real part of the complex equation, which is Eq. (10.15), the equation we originally wanted to solve.

The Response Curve. The absolute value of the complex solution $z(t)$ is

$$\frac{F}{|-mq^2 + ihq + k|}$$

for all t. This is the maximum of the absolute value of its real part $x(t)$, reached for those values of t for which z is real. This maximum is called the *amplitude* of the vibration. Furthermore, F is the maximum of the absolute value of the imposed force; it is called the amplitude of the force. The ratio of the two amplitudes is

$$R(q) = \frac{\max|x|}{F} = \frac{1}{|-mq^2 + ihq + k|}.$$

In many ways, the most interesting question is this: for what value of q is $R(q)$ the largest? Clearly, $R(q)$ tends to zero as q tends to infinity, so $R(q)$ has a maximum. We shall calculate the value of the maximum. It occurs at the same frequency q at which the reciprocal of $R(q)$ is minimized:

$$\frac{1}{(R(q))^2} = |-mq^2 + ihq + k|^2 = (k - mq^2)^2 + h^2q^2.$$

The derivative of this with respect to q is

$$4mq(mq^2 - k) + 2h^2q = 2q(2m^2q^2 - 2mk + h^2),$$

and it is zero at $q = 0$. To find other possible zeros, we set the remaining factor equal to zero: $2m^2q^2 - 2mk + h^2 = 0$. After rearrangement, we get

$$q^2 = \frac{2mk - h^2}{2m^2} = \frac{k}{m} - \frac{h^2}{2m^2}.$$

If the quantity on the right is negative, which is the overdamped Case II, $h > \sqrt{2mk}$, the equation cannot be satisfied, and we conclude that the maximum of R is reached at $q = 0$. If, however, the quantity on the right is positive, in Case I, then

$$q_r = \sqrt{\frac{k}{m} - \frac{h^2}{2m^2}}$$

is a possible candidate for the value for which $R(q)$ achieves its maximum. A direct calculation shows that

$$R(q_r) = \frac{1}{h}\frac{1}{\sqrt{\frac{k}{m} - \frac{h^2}{4m^2}}}.$$

In Problem 10.19, we ask you to show that $R(q_r)$ is greater than $R(0) = \frac{1}{k}$. So for $h < \sqrt{2mk}$, the graph of $R(q)$ looks qualitatively like the example in Fig. 10.12. The graph of R is called the *response curve* of the vibrating system.

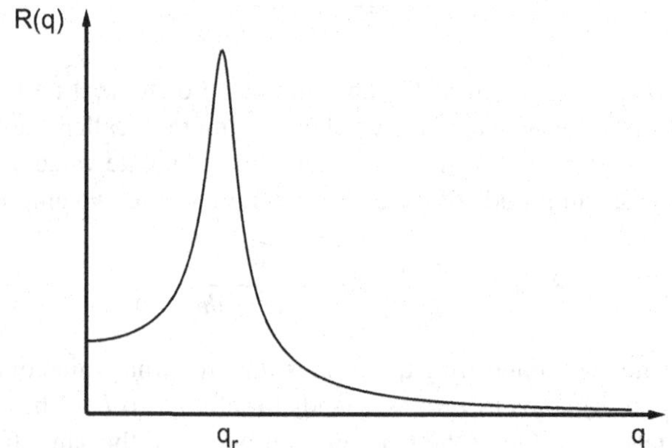

Fig. 10.12 Graph of the response function $R(q) = \dfrac{1}{\sqrt{(-q^2+1)^2 + \left(\frac{1}{5}q\right)^2}}$ for the damped equation $x'' + \frac{1}{5}x' + x = F\cos(qt)$. The maximum is $R(0.989\ldots) = 5.025\ldots$

The significance of the maximum at q_r is that *among all driving forces of the form $F\cos qt$, the one with $q = q_r$ causes the motion with the largest amplitude.* This phenomenon is called *resonance*, and $\frac{q_r}{2\pi}$ is called the *resonant frequency*. Resonance is particularly striking if friction is small, i.e., if h is small, for then $R(q_r)$ is so large that at the resonant frequency, even a low-amplitude driving force will cause a motion of large amplitude. A known dramatic example of this kind of resonance is the shattering of a wine glass by a musical note pitched at the resonant frequency of the glass.

We conclude this section with a summary:
For motion under a restoring force without friction:

(a) Total energy is conserved.
(b) All motions are periodic.
(c) All motions with relatively small amplitude have approximately the same period.
For motion under a restoring force with friction:
(d) Total energy is decreasing.
(e) All motion decays to zero at an exponential rate.
(f) There is a critical value of the coefficient of friction that maximizes the rate at which solutions decay to zero.
We have proved (e) and (f) only for a linear restoring force and linear friction.
For motions under a linear restoring force, linear friction, and a sinusoidal driving force:
(g) All motions tend toward a sinusoidal motion with the same frequency as the driving force.
(h) If friction is not too large, there is a resonant frequency.

Problems

10.1. Which of the following differential equations are examples of the model (10.2) that we developed for vibrations of a mechanical system? Be sure to check the required properties of the frictional and restoring forces.

(a) $2x'' - x = 0$
(b) $x'' + x' + x + x^3 = 0$
(c) $x'' + x' = 0$
(d) $x'' - x^2 = 0$
(e) $x'' - 0.07x' - 3x = 0$

10.2. Verify that since we assumed that the restoring force $f_{\text{re}}(x)$ is an odd function, the potential energy $p(x) = -\int_0^x f_{\text{re}}(y)\,dy$ is an even function.

10.3. Solve $x'' + x' = 0$, which has no restoring force, by trying a combination of exponential solutions $x(t) = \mathrm{e}^{rt}$ for the two cases

(a) $x(0) = 5,\ x'(0) = 7,$
(b) $x(0) = 5,\ x'(0) = -7.$

Do the solutions have limits as t tends to infinity?

10.4. Find an equation $mx'' + hx' + kx = 0$ if the roots to $mr^2 + hr + k = 0$ are $r_\pm = -\frac{1}{10} \pm \mathrm{i}$.

10.5. Find exponential solutions e^{rt} of $2x'' + 7x' + 3x = 0$.

10.6. Find trigonometric solutions of $2y'' + 3y = 0$.

10.7. As indicated by the graph in Fig. 10.10 of the coefficient of decay ℓ, some solutions $x(t)$ of $mx'' + hx' + kx = 0$ decay toward zero very slowly if h is either very small or very large. Sketch typical solutions for both cases.

10.8. Find a complex exponential solution $z(t)$ of the equation $z'' + 4z' + 5z = 0$ and verify that the real part $x(t) = \mathrm{Re}\, z(t)$ is a solution of $x'' + 4x' + 5x = 0$.

10.9. Let $x_1(t)$ and $x_2(t)$ be real-valued functions that are solutions of the nth-order differential equation

$$A_n x^{(n)}(t) + \cdots + A_2 x''(t) + A_1 x'(t) + A_0 x(t) = 0,$$

where the A_i are real constants.

(a) Show that if c is any real constant, then $cx_1(t)$ is a solution.
(b) Show that $y(t) = x_1(t) + x_2(t)$ is a solution.

Combining these observations, we observe that $c_1 x_1(t) + c_2 x_2(t)$ is a solution whenever x_1 and x_2 are solutions and the c's are constant. This is an example of the *linearity* of this differential equation.

10.10. Suppose that the coefficients $A_0, A_1, \ldots, A_n$ in the differential equation of Problem 10.9 are functions of t. Are the assertions made there still valid? Are the assertions true if instead, the equation is modified to

$$A_n x^{(n)}(t) + \cdots + A_2 x''(t) + A_1 x'(t) + A_0 x(t) = \cos t\ ?$$

10.11. Suppose $x_1(t) = p_1(t) + iq_1(t)$ and $x_2(t) = p_2(t) + iq_2(t)$ are complex-valued solutions of

$$mx'' + hx' + kx = 0$$

and that c_1 and c_2 are complex numbers. Show that $c_1 x_1(t) + c_2 x_2(t)$ is a solution.

10.12. The function $x(t) = \mathrm{e}^{-bt}\cos(wt)$ represents a motion under a linear restoring force and linear friction.

(a) Show that the interval between successive times when $x(t) = 0$ has length $\dfrac{\pi}{w}$.

(b) Show that the time interval between successive local maxima is $\dfrac{2\pi}{w}$.

10.13. Consider the equation of motion $mx'' + hx' + kx = 0$, and suppose that h has the critical value $2\sqrt{mk}$.

(a) Show that the only solution of the form e^{rt} has $r = -\sqrt{\dfrac{k}{m}}$.

(b) Show that $t\mathrm{e}^{-\sqrt{\frac{k}{m}}t}$ is a solution.

10.14. Find all solutions $x(t)$ of the equation of motion $x'' + x' + x = 0$.

10.15. Find a complex exponential solution $z(t)$ of the equation $z''+z'+6z=52e^{6it}$, and verify that the real part $x(t)=\operatorname{Re} z(t)$ is a solution of $x''+x'+6x=52\cos(6t)$.

10.16. Show that if $z(t)$ is any solution of $mz''+hz'+kz=Fe^{iqt}$, then the real part $x(t)=\operatorname{Re} z(t)$ is a solution of $mx''+hx'+kx=F\cos(qt)$.

10.17. Find a solution $x_1(t)$ of the equation $x''+x'+x=\cos t$. Verify that you may add any solution of $y''+y'+y=0$ to your solution to get another solution $x_2=y+x_1$ of $x''+x'+x=\cos t$.

10.18. A heavy motor runs at 1800 rpm, causing the floor to vibrate with small vertical displacements $y(t)=A\cos(\omega t)$. Find ω if t is measured in minutes.

10.19. Prove that in the damped case $h<\sqrt{2mk}$, the response maximum $R(q_r)$ is greater than $R(0)=\frac{1}{k}$, i.e., that $\frac{1}{h}\frac{1}{\sqrt{\frac{k}{m}-\frac{h^2}{4m^2}}}>\frac{1}{k}$.

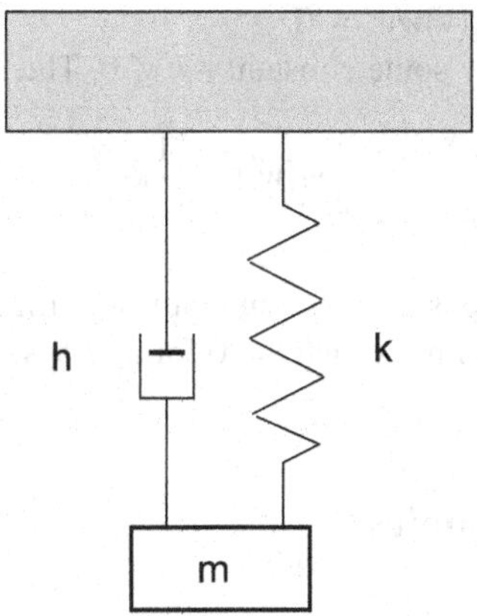

Fig. 10.13 A spring and mass with friction. Before gravity is applied, $y=0$ is the equilibrium. $f_{\text{re}}(y)=-ky$ $f_{\text{fr}}(v)=-hv$. See Problem 10.20

10.20. Newton's equation of motion for a particle at the end of a vertical spring (see Fig. 10.13) under the influence of the restoring force of the spring, friction, and the applied force of gravity is

$$my''+hy'+ky=mg.$$

Here the displacement y is measured as positive downward, and m, h, k, and g are positive constants.

(a) Show that the difference of any two solutions solves the equation for the case of no gravity, $mx'' + hx' + kx = 0$.
(b) Find a constant solution y.
(c) Show that every solution is of the form $y(t) = \frac{gm}{k} + x(t)$, where x solves the case of no gravity.
(d) Show that as t tends to infinity, every solution $y(t)$ tends to the constant solution.

10.21. Justify the following items, which prove the uniqueness theorem, Theorem 10.1, stated at the end of Sect. 10.1a.

(a) If $mx'' - f_{\text{fr}}(x') - f_{\text{re}}(x) = 0$ and $my'' - f_{\text{fr}}(y') - f_{\text{re}}(y) = 0$, denote by w the difference $w(t) = y(t) - x(t)$. Then

$$mw'' - \left(f_{\text{fr}}(w' + x') - f_{\text{fr}}(x')\right) - \left(f_{\text{re}}(w + x) - f_{\text{re}}(x)\right) = 0.$$

(b) For each t, there is u between $x'(t)$ and $y'(t)$ and v between $x(t)$ and $y(t)$ such that

$$mw'' - f'_{\text{fr}}(u)w' - f'_{\text{re}}(v)w = 0.$$

(c) Therefore, $mw''w' - f'_{\text{re}}(v)ww' \le 0$.
(d) f'_{re} is bounded above by some constant $-k \le 0$. Therefore, $mw''w' + kww' \le 0$, and

$$\frac{1}{2}m(w')^2 + \frac{1}{2}kw^2$$

is nonincreasing.
(e) A nonnegative nonincreasing function that is 0 at time s must be 0 for all times $t > s$. Explain why this implies $w(t) = 0$ for all $t > s$.

10.2 Population Dynamics

In this section, calculus is used to study the evolution of populations—animal, vegetable, or mineral. About half the material is devoted to formulating the laws governing population changes in the form of differential equations, and the other half to studying their solutions. Only in the simplest cases can this be accomplished by obtaining explicit formulas for solutions. When explicit solutions are not available, relevant qualitative and quantitative properties of solutions can nevertheless be deduced directly from the equations, as our examples will show. Using numerical methods that extend those we mention in Sect. 10.4, one can generate extremely accurate approximations to any specific solution of a differential equation. These may lead to the answers we seek or suggest trends that once perceived, can often be deduced logically from the differential equations.

Theoretical population models have become more and more useful in such diverse fields as the study of epidemics and the distribution of inherited traits. Yet the most important application, the one about which the public needs to be informed in order to make intelligent decisions, is to demography, the study of human populations. Indeed, as Alexander Pope put it, "The proper study of Mankind is Man."

In Sect. 10.2a, we develop a theory of differential equations needed for the study of population growth. In Sect. 10.2b, we describe the dynamics of a population consisting of a single species, and in Sect. 10.2c, the dynamics of a population consisting of two species.

10.2a The Differential Equation $\frac{\mathrm{d}N}{\mathrm{d}t} = R(N)$

In this section, we analyze the type of differential equations that govern both population growth and chemical reactions, but without reference to these applications. We consider the equation

$$\frac{\mathrm{d}N}{\mathrm{d}t} = R(N), \tag{10.18}$$

where $R(N)$ is a known rate depending on N. In Sect. 3.3, we solved one equation of this type, $\frac{\mathrm{d}N}{\mathrm{d}t} = kN$. We saw that the solutions $N(t) = N(0)\mathrm{e}^{kt}$ are exponential functions, including the constant function $N(t) = 0$. In Fig. 10.14, we plot the solutions of $\frac{\mathrm{d}N}{\mathrm{d}t} = -N$ for five different initial conditions $N_0 = N(0) = 3, 1, 0, -1, -2$. We know that as t tends to infinity, each solution shown, $N(t) = N(0)\mathrm{e}^{-t}$, regardless of the initial condition, tends to the constant solution $N = 0$. In the context of populations, $N_0 < 0$ may not make sense, but the differential equation has solutions for those initial conditions, so we include them in our analysis.

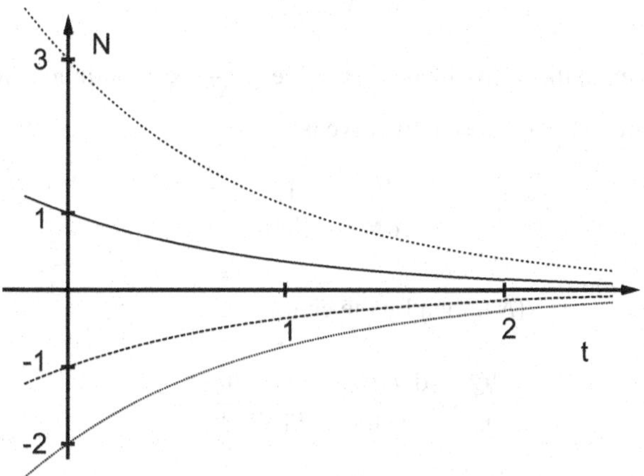

Fig. 10.14 Graphs of five solutions to $\frac{\mathrm{d}N}{\mathrm{d}t} = -N$

The equation $\frac{dN}{dt} = 2N - N^2 = N(2-N)$ is another example of an equation of the form $\frac{dN}{dt} = R(N)$. In Sect. 10.2b, you can see that we will find explicit solution formulas for this equation, some of which we have graphed on the right side of Fig. 10.16.

Both of these differential equations are statements about the relative growth rate $\frac{\frac{dN}{dt}}{N}$ of the population. In the first equation, the relative growth rate $\frac{\frac{dN}{dt}}{N} = -1$ is constant. In the second, the relative growth rate $\frac{\frac{dN}{dt}}{N} = 2 - N$ decreases as N increases from 0 to 2, perhaps due to a lack of resources.

A third equation of the form (10.18) is $\frac{dN}{dt} = -N(N-1)(N-2)$. Solutions of this equation are plotted in Fig. 10.15. A fourth such equation is given in Example 10.7, whose solutions are plotted in Fig. 10.22.

You may have begun to perceive a pattern. It appears that the constant solutions, i.e., the places where $R(N) = 0$, play a key role in describing the long-term behavior of the other solutions. Our first task is to determine conditions under which solutions exist and are determined uniquely by the initial condition. Then we will show how the zeros of $R(N)$ are related to the long-term behavior of solutions.

To begin, assume that $R(N)$ is a continuous function of N, different from zero. Then we can divide both sides of

$$\frac{dN}{dt} = R(N)$$

by $R(N)$, obtaining

$$\frac{1}{R(N)} \frac{dN}{dt} = 1.$$

By the fundamental theorem of calculus, since $\frac{1}{R(N)}$ is a continuous function of N, there is a function $Q(N)$ whose derivative is

$$\frac{dQ}{dN} = \frac{1}{R(N)}. \tag{10.19}$$

If $N(t)$ satisfies $\frac{dN}{dt} = R(N)$, then by the chain rule,

$$\frac{dQ}{dt} = \frac{dQ}{dN}\frac{dN}{dt} = \frac{1}{R(N)}\frac{dN}{dt} = 1.$$

A function with constant derivative is linear, so

$$Q(N(t)) = t + c, \quad c \text{ a constant.}$$

It follows from $\frac{dQ}{dN} = \frac{1}{R(N)}$ that $\frac{dQ}{dN}$ is not zero, and from the continuity of R that $\frac{dQ}{dN}$ does not change sign. Therefore, $Q(N)$ is strictly monotonic and hence invertible. This means that $Q(N(t)) = t + c$ can be solved for

$$N(t) = Q^{-1}(t+c).$$

The constant c can be related to the initial value $N(0) = N_0$ by setting $t = 0$:

$$N(0) = Q^{-1}(c) \quad \text{or} \quad Q(N_0) = c.$$

With this determination of c, the function

$$N(t) = Q^{-1}(t + Q(N_0)) \tag{10.20}$$

is the solution of $\frac{dN}{dt} = R(N)$ with initial value N_0. Thus we have proved the following theorem.

Theorem 10.2. Existence. *If $R(N)$ is a continuous function of N that is never 0, then the differential equation*

$$\frac{dN}{dt} = R(N), \quad \text{with} \quad N(0) = N_0,$$

has a unique solution on a possibly infinite t-interval (r,s). If one of the endpoints r or s is finite, the solution $N(t)$ approaches plus or minus infinity as t approaches the endpoint.

It is instructive to look at the example $\frac{dN}{dt} = N^2 + 1$, with initial value $N(0) = 0$. We divide this equation by $N^2 + 1$ and get $\frac{1}{N^2+1}\frac{dN}{dt} = 1$. The left side is the derivative of $\tan^{-1} N$, so integration gives $\tan^{-1} N = t + c$. Since we specified that $N(0) = 0$, it follows that $c = 0$, and so $N(t) = \tan t$, defined on the interval $\left(-\frac{\pi}{2}, \frac{\pi}{2}\right)$. As t approaches the left or right endpoint of this interval, $N(t)$ tends to minus or plus infinity.

We now turn to the more interesting case that $R(N)$ vanishes at some points. The derivation of formula (10.20) for the solution of the initial value problem shows more than what is stated in Theorem 10.2. It shows that even if $R(N)$ vanishes at some points, if it is not zero for $R(N_0)$, then the method used to solve the initial value problem yields a solution on a short time interval $(-d,d)$.

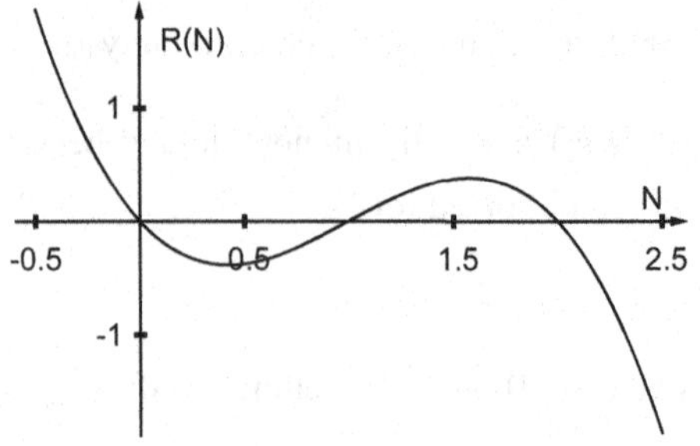

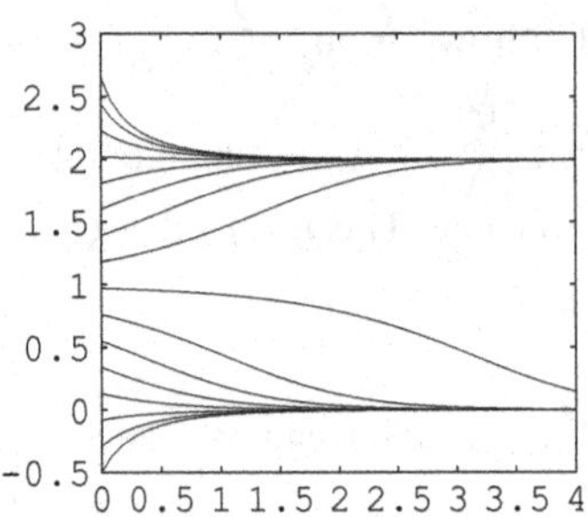

Fig. 10.15 *Left*: graph of the right-hand side $R(N) = -N(N-1)(N-2)$. *Right*: graphs of some computed solutions to $\frac{\mathrm{d}N}{\mathrm{d}t} = -N(N-1)(N-2)$

Theorem 10.3. *Suppose that the function $R(N)$ is differentiable and $N(t)$ is a solution of*

$$\frac{\mathrm{d}N}{\mathrm{d}t} = R(N).$$

Suppose that $N(0)$ is not a zero of the function $R(N)$, that is, $R(N(0)) \neq 0$.

Then $N(t)$ is not a zero of $R(N)$ for any value of t.

Proof. We shall argue indirectly. Suppose, to the contrary, that at some s, $N(s)$ is a zero of R(N). Denote this zero by Z:

$$N(s) = Z, \qquad R(Z) = 0.$$

We shall show that then $R(N(t))$ is zero for all t.

Since $R(Z) = 0$, we have $R(N) = R(N) - R(Z)$. Using the mean value theorem, we obtain $R(N) = k(N-Z)$, where k is the value of the derivative of the function R at some point between N and Z. Since Z is a constant, we can rewrite the differential equation governing N as

$$\frac{\mathrm{d}(N-Z)}{\mathrm{d}t} = R(N) = k(N-Z).$$

Denote the function $N(t) - Z$ by $M(t)$ and write the differential equation for $N - Z$ as $\frac{\mathrm{d}M}{\mathrm{d}t} = kM$. Multiply this equation by $2M$. We get $2M\frac{\mathrm{d}M}{\mathrm{d}t} = 2kM^2$. Denote the function M^2 by P and rewrite the equation above as

$$\frac{\mathrm{d}P}{\mathrm{d}t} = kP.$$

Denote by m an upper bound for the function $k(N)$. We deduce from this differential equation the inequality

$$\frac{dP}{dt} \le mP.$$

We write this inequality as $\frac{dP}{dt} - mP \le 0$, and multiply it by e^{-mt}. We get

$$e^{-mt}\frac{dP}{dt} - me^{-mt}P \le 0.$$

The left side is the derivative of $e^{-mt}P$, and since it is nonpositive, $e^{-mt}P$ is a nonincreasing function of t. The function P was defined as M^2, and M as $N - Z$. Since Z is the value of the function N at s, $M(s)$ is zero, and so is $P(s)$. Since the function P is a square, its values are nonnegative, and so are the values of $e^{-mt}P(t)$. We have shown before that $e^{-mt}P(t)$ is a nonincreasing function of t. But since $e^{-mt}P(t)$ is zero at s, it follows that $e^{-mt}P(t)$, and therefore $P(t)$, is zero for t greater than s. Using a similar argument but with a lower bound for the function $k(N)$, we show similarly that $P(t)$ is zero for all t less than s.

Since $P(t)$ is the square of $M(t)$ and $M(t)$ is $N(t) - Z$, this proves that $N(t) = Z$ for all t. But this contradicts the assumption that $N(0)$ is not a zero of $R(N)$. Since we got into this contradiction by denying Theorem 10.3, this proves the theorem.

□

Next we see how the zeros of $R(N)$ are related to the long-term behavior of the solutions of $\frac{dN}{dt} = R(N)$. From Theorem 10.3, we shall deduce the following property of solutions of Eq. (10.18).

Theorem 10.4. *Denote by $N(t)$ a solution of $\frac{dN}{dt} = R(N)$, and its value $N(0)$ by N_0. Suppose that $R(N)$ is differentiable, its derivative bounded, and assume that $R(N)$ is positive for N large negative and that it is negative for N large positive.*

(a) If $R(N_0)$ is negative, then the solution $N(t)$ decreases as t increases, and as t tends to infinity, $N(t)$ tends to the largest zero of $R(N)$ that is less than N_0.
(b) Similarly, if $R(N_0)$ is positive, then $N(t)$ is an increasing function of t, and as t tends to infinity, $N(t)$ tends to the smallest zero of $R(N)$ that is larger than N_0.

Before we write the proof of Theorem 10.4, let us see what the theorem tells us about our two examples:

- $\frac{dN}{dt} = N(2-N)$. If $0 < N_0 < 2$, then $R(N_0)$ is positive. According to Theorem 10.4, as t tends to infinity, the solution $N(t)$ increases to $N = 2$. If $2 < N_0$, $R(N_0)$ is negative, and the solution decreases to 2, the largest zero of $R(N)$ that is less than N_0.
- $\frac{dN}{dt} = -N(N-1)(N-2)$. The graph of $R(N)$ on the left side of Fig. 10.15 will help us tell where $R(N_0)$ is positive or negative. If $N_0 < 0$, $R(N_0)$ is positive, so

according to Theorem 10.4, $N(t)$ increases to $N = 0$. If $0 < N_0 < 1$, then $R(N_0)$ is negative, so $N(t)$ decreases to $N = 0$. If $1 < N_0 < 2$, then $R(N_0)$ is positive, so $N(t)$ increases to the smallest zero that is larger than N_0, i.e., $N = 2$. If $2 < N_0$, then $R(N_0)$ is negative, and so $N(t)$ decreases to $N = 2$ as t tends to infinity. This agrees with the approximate solutions computed in Fig. 10.15.

Now we will prove the theorem that makes so much qualitative information about the solutions readily available.

Proof. We prove part (a); assume that $R(N_0)$ is negative. Since $R(N)$ is positive for N large negative, it has a zero less than N_0. Denote by M the largest zero of $R(N)$ less than N_0. When $R(N_0)$ is negative, then according to $\frac{dN}{dt} = R(N)$, the derivative of $N(t)$ is negative at $t = 0$ and remains negative as long as $N(t)$ is greater than M, because $R(N)$ is negative for all values of N between M and N_0. It follows that $N(t)$ is a decreasing function of t and keeps decreasing as long as $N(t)$ is greater than M. According to Theorem 10.3, $N(t)$ is not equal to a zero of $R(N)$. Therefore, $N(t)$ is greater than M for all positive t.

We show now that as t tends to infinity, $N(t)$ tends to M. We again argue indirectly and assume to the contrary that for all t, $N(t)$ is greater than $M + p$, p some positive number. The function $R(N)$ is negative on the interval $[M + p, N_0]$. Denote by m the maximum of $R(N)$ on this interval; m is a negative number. We apply the mean value theorem to the function $N(t)$:

$$\frac{N(t) - N(0)}{t} = \frac{dN}{dt}(c),$$

where c is some number between 0 and t. Since $N(t)$ is a solution of the differential equation, this gives

$$\frac{N(t) - N(0)}{t} = \frac{dN}{dt}(c) = R(N(c)) \le m.$$

We deduce that $N(t) \le N(0) + mt$ for all positive t. Since m is negative, this would imply that $N(t)$ tends to minus infinity as t tends to infinity. This is contrary to our previous demonstration that $N(t)$ is greater than M for all t. Therefore, our assumption that $N(t)$ is greater than $M + p$ for all t must be false.

This completes the proof in the case that $R(N_0)$ is negative. The proof for $R(N_0)$ positive is analogous. □

Remark. Take the case that the zeros M of $R(N)$ are *simple* in the sense that $\frac{dR}{dN}(M)$ is not zero. Theorem 10.4 can be expressed in this way: *The zeros of $R(N)$ where the derivative $\frac{dR}{dN}$ is negative attract solutions of $\frac{dN}{dt} = R(N)$.* Here is why: the proof shows that solutions less than a root M will increase toward M where R is positive, and solutions greater than M decrease toward M when R is negative. But if R is positive below M and negative above, then $\frac{dR}{dN}(M)$ must be either negative or 0.

The zeros of $R(N)$ are called equilibrium solutions. A zero that attracts solutions which that nearby it is called a *stable* equilibrium. A zero that repels some near solutions, such as $N = 1$ in Fig. 10.15, is called unstable.

10.2b Growth and Fluctuation of Population

The Arithmetic of Population and Development

> The rate of population growth is itself impeding efforts at social and economic development. Take the case, for example, of a developing country which has achieved an annual increase in its gross national product of five percent–a very respectable effort indeed, and one which few countries have been able to maintain on a continuing basis. Its population is increasing at 3 percent annually. Thus, its per capita income is increasing by 2 percent each year, and will take 35 years to double, say from \$100 per year to \$200. In the meantime, its population will have almost tripled, so that greatly increased numbers of people will be living at what is still only a subsistence level. Reduction of the rate of population growth is not a *sufficient* condition for social and economic development–other means, such as industrialization, must proceed along with such reduction–but it is clear that it is a *necessary* condition without which the development process is seriously handicapped.
>
> Dr. John Maier
> Director for Health Sciences
> The Rockefeller Foundation

In this section we shall study the growth of population of a single species and of several species living in a shared environment. The growth rate of a population is related to the birth and death rates. The basic equation governing the growth in time t of a single population of size $N(t)$ is

$$\frac{\mathrm{d}N}{\mathrm{d}t} = B - D,$$

where B is the *birth rate* and D is the *death rate* for the total population. What do B and D depend on? They certainly depend on the age distribution within the population; a population with a high percentage of old members will have a higher death rate and lower birth rate than a population of the same size that has a low percentage of old members. Yet in this section we shall disregard this dependence of birth and death rates on age distribution. The results we shall derive are quantitatively relevant in situations in which the age distribution turns out to change fairly little over time.

If we assume, in addition, that the basic biological functions of the individuals in the population are unaffected by the population size, then it follows that *both birth rate and death rate are proportional to the population size*. The mathematical expression of this idea is

$$B = cN, \quad D = dN,$$

where c, d are constants. Substituting this into the differential equation leads to

$$\frac{\mathrm{d}N}{\mathrm{d}t} = aN,$$

where $a = c - d$. The solution of this equation is

$$N(t) = N_0 e^{at},$$

where $N_0 = N(0)$ is the initial population size. For positive a this is the celebrated—and lamented—Malthusian law of population explosion.

The Verhulst Model. If the population grows beyond a certain size, the sheer size of the population will *depress* the birth rate and *increase* the death rate. We summarize this as

$$\frac{\mathrm{d}N}{\mathrm{d}t} = aN - \text{effect of overpopulation.}$$

How can we quantify the effect of overpopulation? Let us assume that the *effect of overpopulation is proportional to the number of encounters between members of the population* and that these encounters are *by chance*, i.e., are due to individuals bumping into each other without premeditation. For each individual, the number of encounters is proportional to the population size. In Chap. 11, we find that probabilities of independent events need to be multiplied, and therefore, the total number of such encounters is proportional to the *square* of the population. So the effect of overpopulation is to depress the rate of population growth by bN^2, for b some positive number. The resulting growth equation is

$$\frac{\mathrm{d}N}{\mathrm{d}t} = aN - bN^2, \quad a,b > 0. \tag{10.21}$$

This equation was introduced into the theory of population growth by Verhulst. It is a special instance of the equation $\frac{\mathrm{d}N}{\mathrm{d}t} = R(N)$, discussed in Sect. 10.2a.

Example 10.4. Consider $\frac{\mathrm{d}N}{\mathrm{d}t} = 2N - N^2 = N(2-N)$ in Fig. 10.16. Note that when the population N is between 0 and 2, it must increase, because the rate of change $N(2-N)$ is positive then.

Now we find a solution formula for the Verhulst model (10.21). Suppose that the right side of Eq. (10.21) is not zero. Then dividing by $aN - bN^2$, we get

$$1 = \frac{1}{aN - bN^2}\frac{\mathrm{d}N}{\mathrm{d}t}.$$

We write the right side as a derivative:

$$1 = \frac{1}{\frac{a}{N} - b}\frac{1}{N^2}\frac{\mathrm{d}N}{\mathrm{d}t} = \frac{1}{\frac{a}{N} - b}\frac{\mathrm{d}}{\mathrm{d}t}\left(-\frac{1}{N}\right) = -\frac{1}{a}\frac{\mathrm{d}}{\mathrm{d}t}\left(\log\left(\frac{a}{N} - b\right)\right).$$

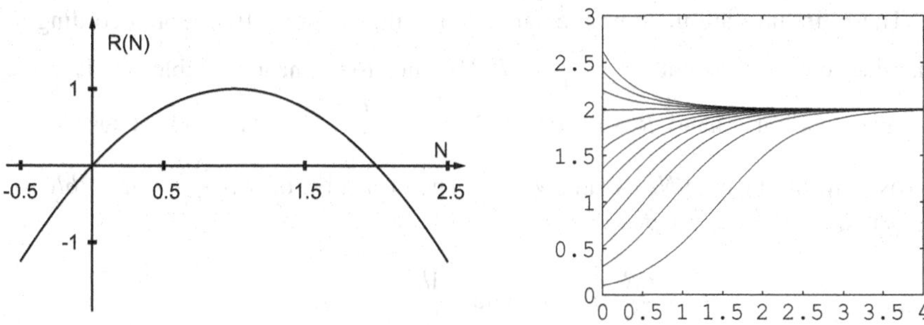

Fig. 10.16 *Left*: graph of the right-hand side $R(N) = N(2-N)$ in Example 10.4. *Right*: graphs of some computed solutions to $\frac{dN}{dt} = N(2-N)$

Integrating yields $\log\left(\frac{a}{N} - b\right) = c - at$ for some number c. If N_0 denotes the initial value of N, then $\log\left(\frac{a}{N_0} - b\right) = c$. Therefore,

$$\log\left(\frac{a}{N} - b\right) = \log\left(\frac{a}{N_0} - b\right) - at.$$

Combining the logarithms gives

$$\log\left(\frac{\frac{a}{N} - b}{\frac{a}{N_0} - b}\right) = -at.$$

Applying the exponential function gives $\frac{\frac{a}{N} - b}{\frac{a}{N_0} - b} = e^{-at}$, which can be solved for $N(t)$ as

$$N(t) = \frac{\frac{a}{b}N_0}{N_0 - \left(N_0 - \frac{a}{b}\right)e^{-at}}.$$

Many interesting properties of $N(t)$ can be deduced from this formula by inspection.

Theorem 10.5. *Assume that the initial value $N(0) = N_0$ is positive in the Verhulst model*

$$\frac{dN}{dt} = aN - bN^2.$$

(a) If $N_0 > \frac{a}{b}$, then $N(t) > \frac{a}{b}$ for all t and decreases as time increases.
(b) If $N_0 = \frac{a}{b}$, then $N(t) = \frac{a}{b}$ for all t.
(c) If $N_0 < \frac{a}{b}$, then $N(t) < \frac{a}{b}$ for all t and increases as time increases.

In all cases, $N(t)$ tends to $\frac{a}{b}$ as t tends to infinity.

These findings are in complete agreement with Theorem 10.4. For according to that theorem, every solution of $\frac{\mathrm{d}N}{\mathrm{d}t} = R(N)$ tends to the nearest stable steady state. For the equation at hand, $R(N) = aN - bN^2 = bN\left(\frac{a}{b} - N\right)$. The steady states are the zeros of R, in this case $N = 0$ and $N = \frac{a}{b}$. The derivative of R is $\frac{\mathrm{d}R}{\mathrm{d}N} = a - 2bN$, so its values at the zeros of R are

$$\frac{\mathrm{d}R}{\mathrm{d}N}(0) = a \quad \text{and} \quad \frac{\mathrm{d}R}{\mathrm{d}N}\left(\frac{a}{b}\right) = -a.$$

Since a is positive, we conclude that both zeros are simple, and that $\frac{\mathrm{d}R}{\mathrm{d}N}$ is positive at $N = 0$, negative at $N = \frac{a}{b}$. Therefore, the zero 0 is unstable and the zero $\frac{a}{b}$ is stable, and all solutions with initial value $N_0 > 0$ tend to the stable steady state $\frac{a}{b}$ as t tends to infinity. This is exactly what we found by studying the explicit formula for all solutions. It is gratifying that properties of solutions can be deduced directly from the differential equation that they satisfy without help from an explicit formula for solutions. Indeed, there are very few differential equations whose solutions can be described by an explicit formula.

The result we have just obtained, that all solutions of the Verhulst model (10.21) tend to $\frac{a}{b}$ as t tends to infinity, has great demographic significance, for it predicts the eventual steady state of any population that can reasonably be said to be governed by an equation of that form.

An Extinction Model. We now return to the basic equation $\frac{\mathrm{d}N}{\mathrm{d}t} = R(N)$ of population growth and again we assume that the *death rate is proportional to the population size*. This amounts to assuming that death is due to "natural" causes, and not due to one member of the population eating the food needed by another member, or due to one member eating another. On the other hand, we *challenge the assumption that birth rate is proportional to population size*. This assumption holds for extremely primitive organisms, such as amoebas, which reproduce by dividing. It is also true of well-organized species, such as human beings, who seek out a partner and proceed to produce a biologically or socially determined number of offspring. But there are important classes of organisms whose reproductive sophistication falls between those of the amoeba and humans, who need a partner for reproduction but must rely on chance encounters for meeting a mate. The expected number of encounters is proportional to the *product* of the numbers of males and females. If these are equally distributed in the population, the number of encounters—and so the birth rate—is proportional to N^2. The death rate, on the other hand, is proportional to the population size N. Since the rate of population growth is the difference between birth rate and death rate, the equation governing the growth of such populations is

$$\frac{\mathrm{d}N}{\mathrm{d}t} = bN^2 - aN, \; a, b > 0.$$

This equation is of the form $\frac{\mathrm{d}N}{\mathrm{d}t} = R(N)$ with

$$R(N) = bN^2 - aN = bN\left(N - \frac{a}{b}\right), \quad a, b > 0.$$

This function has two zeros, 0 and $\frac{a}{b}$. The derivative is $\frac{\mathrm{d}R}{\mathrm{d}N} = 2bN - a$, so its values at the zeros of R are

$$\frac{\mathrm{d}R}{\mathrm{d}N}(0) = -a \quad \text{and} \quad \frac{\mathrm{d}R}{\mathrm{d}N}\left(\frac{a}{b}\right) = a.$$

Since a is positive, it follows that both zeros are simple, and that $R'(N)$ is negative at $N = 0$, positive at $N = \frac{a}{b}$. Therefore, the zero 0 is stable, and the zero $\frac{a}{b}$ is unstable; *all solutions with initial value* $N_0 < \frac{a}{b}$ tend to 0 as t tends to infinity.

Example 10.5. The case $\frac{\mathrm{d}N}{\mathrm{d}t} = N^2 - 2N$ can be viewed in Fig. 10.16 by time reversal, where we imagine t increasing from right to left along the horizontal axis. We ask you to explore this idea in Problem 10.24.

This stability of 0 is the stability of death; what we have discovered by our analysis is a very interesting and highly significant threshold effect. *Once the population size* N_0 *drops below the critical size* $\frac{a}{b}$, *the population tends to extinction.* This notion of a critical size is important for the preservation of a species. A species is classified *endangered* if its current size is perilously close to its critical size.

10.2c Two Species

We now turn to a situation involving *two* species, where one species feeds on nourishment whose supply is ample, and the other species feeds on the first species. We denote the population sizes of the two species by N and P, N denoting the *prey*, P denoting the *predators*. Both N and P are functions of t, and their growth is governed by differential equations of the form $\frac{\mathrm{d}N}{\mathrm{d}t} = B - D$. The initial task is to choose suitable functions B and D describing the birth rates and death rates of each species.

We assume that the two species encounter each other by chance, at a rate proportional to the product of the size of the two populations. If we assume that the principal cause of death among the first species is due to being eaten by a member of the second species, then the death rate for N is proportional to the product NP. We assume that the birth rate for the predator is proportional to the population size P, and that the portion of the young that survive is proportional to the available food supply N. Thus the effective birth rate is proportional to NP. Finally, we assume that the birth rate of the prey and the death rate of the predator are proportional to the size of their respective populations. So the equations governing the growth of these species are called the Lotka-Volterra equation, Table 10.1.

Table 10.1 The Lotka-Volterra equation

Species	Growth rate		Birth rate		Death rate
Prey	$\frac{dN}{dt}$	$=$	aN	$-$	bNP
Predator	$\frac{dP}{dt}$	$=$	hNP	$-$	cP

In the Lotka-Volterra equations, a, b, c, and h are all positive constants. These equations were first set down and analyzed, independently, by Volterra and by Lotka. Lotka's work on this and other population models is described in his book *Elements of Physical Biology*, originally published in 1925 and reprinted by Dover, New York, in 1956. The work of Volterra, inspired by the fluctuations in the size and composition of the catch of fish in the Adriatic, appeared in *Cahier Scientifique*, vol. VII, Gauthier-Villars, Paris, 1931, under the romantic title "Leçons sur la théorie mathématique de la lutte pour la vie" (Lessons on the mathematical theory of the fight for survival). It is reprinted in his collected works published by Accademia dei Lincei, Rome. We first give an example in which there is no predation.

Example 10.6. Consider the case in which there is no interaction between the predator and prey, so $b = h = 0$ and $a = 2,\ c = 3$. Then the system reads

Species	Growth rate		Birth rate		Death rate
Prey	$\frac{dN}{dt}$	$=$	$2N$		
Predator	$\frac{dP}{dt}$	$=$		$-$	$3P$

The solutions are exponential:

$$N(t) = N_0 e^{2t}, \qquad P(t) = P_0 e^{-3t}.$$

Note that by properties of exponents, $(e^{2t})^{-3/2} = e^{-3t}$, so

$$\frac{P(t)}{P_0} = \left(\frac{N(t)}{N_0}\right)^{-3/2}.$$

We plot two such relations in the (N,P)-plane in Fig. 10.17.

Next we consider the general case that all of a, b, c, and h are positive. The first order of business is to show that these laws of growth, and knowledge of the initial population size, are sufficient to determine the size of both populations at all future times. We formulate this as a uniqueness theorem.

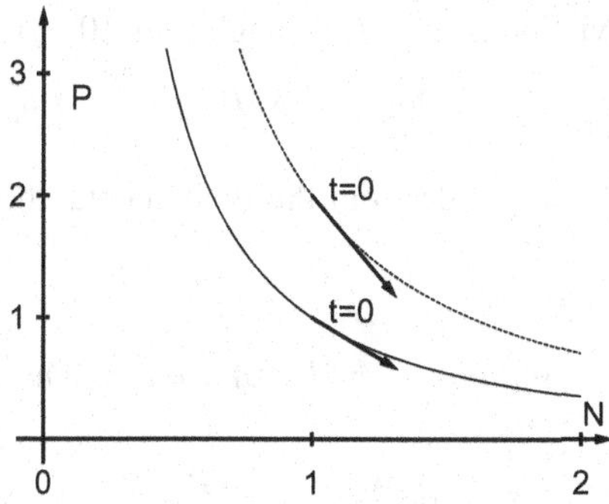

Fig. 10.17 The (N,P)-plane in a case of no predation, $\frac{dN}{dt} = 2N$, $\frac{dP}{dt} = -3P$. Two time histories are shown, one starting from $(N_0,P_0) = (1,1)$ and the other from $(N_0,P_0) = (1,2)$. See Example 10.6

Theorem 10.6. Uniqueness. *A solution of the Lotka-Volterra equations (Table 10.1) is uniquely determined for all time by the specification of the initial values N_0, P_0 of N and P. That is, if N,P and n,p are solutions with the same initial values, then $N(t) = n(t)$ and $P(t) = p(t)$ for all t.*

We ask you to fill in steps of a proof of this theorem in Problem 10.26. We omit a proof of existence of solutions, and instead investigate the properties of solutions. We take the case that both species are present, i.e., $P > 0$, $N > 0$. We divide the Lotka–Volterra equations by N and P respectively, obtaining

$$\frac{1}{N}\frac{dN}{dt} = a - bP,$$
$$\frac{1}{P}\frac{dP}{dt} = hN - c.$$

It follows that $P = \frac{a}{b}$ and $N = \frac{c}{h}$ are steady-state solutions. That is, if the initial values are $N_0 = \frac{c}{h}$ and $P_0 = \frac{a}{b}$, then $P = \frac{a}{b}$ and $N = \frac{c}{h}$ for all t. To study the non-steady-state solutions, we multiply the first equation by $hN - c$, the second equation by $bP - a$, and add. The sum of the right sides is 0, so we get the relation

$$\left(h - \frac{c}{N}\right)\frac{dN}{dt} + \left(b - \frac{a}{P}\right)\frac{dP}{dt} = 0.$$

Using the chain rule, we rewrite this relation as

$$\frac{d}{dt}\left(hN - c\log N + bP - a\log P\right) = 0.$$

We introduce the abbreviations H and K through (Fig. 10.18)

$$H(N) = hN - c\log N, \quad K(P) = bP - a\log P.$$

Then $\dfrac{\mathrm{d}(H+K)}{\mathrm{d}t} = 0$. We conclude from the fundamental theorem of calculus the following.

Theorem 10.7. *For any solution of the Lotka-Volterra equations*

$$\frac{\mathrm{d}N}{\mathrm{d}t} = aN - bNP,$$

$$\frac{\mathrm{d}P}{\mathrm{d}t} = hNP - cP,$$

the quantity

$$H(N) + K(P) = hN - c\log N + bP - a\log P$$

is independent of t.

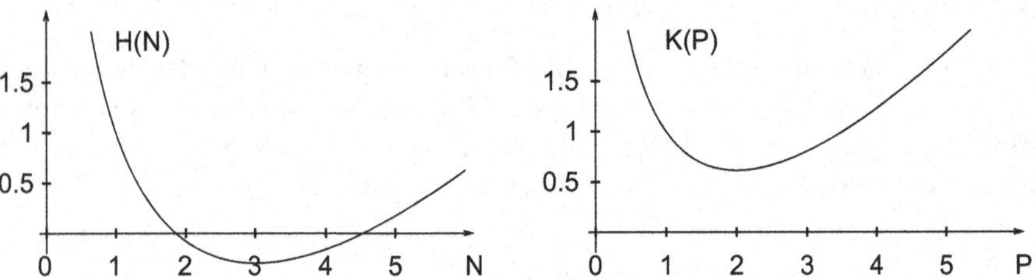

Fig. 10.18 Graphs of the functions $H(N) = N - 3\log N$ and $K(P) = P - 2\log P$ for the system $\dfrac{\mathrm{d}N}{\mathrm{d}t} = 2N - NP$, $\dfrac{\mathrm{d}P}{\mathrm{d}t} = -3P + NP$

The constancy of the sum $H + K$ is strongly reminiscent of the law of conservation of energy in mechanics and can be used, like the law of conservation of energy, to gain qualitative and quantitative information about solutions. For that purpose, we note the following properties of functions the H and K. Inspection of their definitions shows that the functions $H(N)$ and $K(P)$ tend to infinity as N and P tend to zero or to infinity. Since the functions $H(N)$ and $K(P)$ are continuous, they have minimum values, which we now locate. The derivatives of these functions are

$$\frac{\mathrm{d}H}{\mathrm{d}N} = h - \frac{c}{N}, \qquad \frac{\mathrm{d}K}{\mathrm{d}P} = b - \frac{a}{P},$$

and these are zero at

$$N_m = \frac{c}{h} \quad \text{and} \quad P_m = \frac{a}{b}.$$

Notice that these are the steady-state values for the Lotka–Volterra equations.

Theorem 10.8. *Consider species N and P that satisfy the Lotka-Volterra equations (Table 10.1)*

(a) Neither species can become extinct, i.e., there is a positive lower bound for each population, throughout the whole time history.
(b) Neither species can proliferate ad infinitum, i.e., there is an upper bound for each population throughout its time history.
(c) The steady state is neutrally stable in the following sense: if the initial state N_0, P_0 *is near the steady state, then* $N(t)$, $P(t)$ *stays near the steady state throughout the whole time history.*

Proof. All three results follow from the conservation law: The sum $H(N)+K(P)$ tends to infinity if either N or P tends to 0 or infinity. Since this is incompatible with the constancy of $H+K$, we conclude that neither $N(t)$ nor $P(t)$ can approach 0 or infinity. This proves parts (a) and (b).

For part (c), we specify the meaning of "near the steady state" by describing sets G_s of points as follows. The minimum of $H(N)$ is reached at N_m, and the minimum of $K(P)$ is reached at P_m. Let s be a small positive number. Denote by G_s the set of points in the (N,P)-plane where $H(N)+K(P)$ is less than $H(N_m)+K(P_m)+s$. For s small, G_s is a small region around the point (N_m,P_m); see Fig. 10.19. Choose $(N(0),P(0))$ in G_s. Then since $H(N)+K(P)$ has the same value for all t, $H(N)+K(P)$ is less than $H(N_m)+K(P_m)+s$ for all t. This shows that $(N(t),P(t))$ remains in G_s for all t. This completes the proof of part (c). □

The next result, Theorem 10.9, is both interesting and surprising. It is suggested by some computed solutions that are shown in Fig. 10.20.

Theorem 10.9. *Every time history is periodic, i.e., for every solution* $N(t),P(t)$ *of the Lotka-Volterra equations*

$$\frac{dN}{dt} = aN - bNP,$$
$$\frac{dP}{dt} = hNP - cP,$$

there is a time T such that

$$N(T) = N(0), \qquad P(T) = P(0).$$

The number T is called the *period* of this particular time history. Different time histories have different periods. It is instructive to picture the time histories

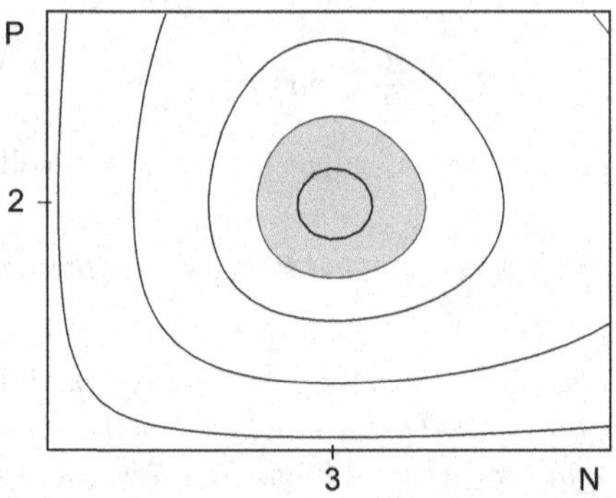

Fig. 10.19 Stability: solutions starting in the region G_s remain in that region, as shown in the proof of Theorem 10.8. The gray region here is $G_{1/10}$, where $N - 3\log N + P - 2\log P$ exceeds its minimum by less than $\frac{1}{10}$

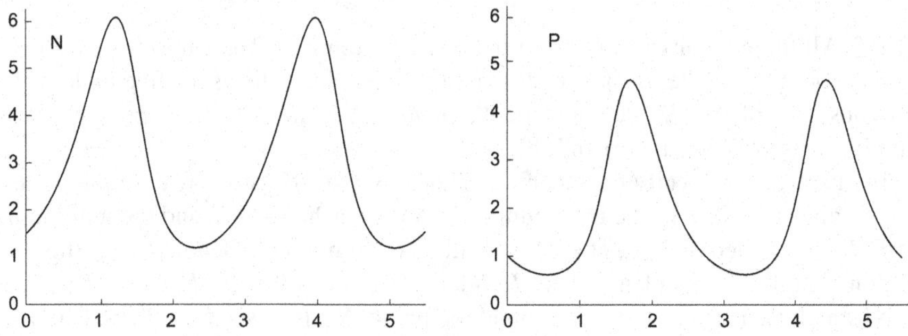

Fig. 10.20 Computed graphs of $N(t)$ and $P(t)$ for the system $\frac{dN}{dt} = 2N - NP$, $\frac{dP}{dt} = -3P + NP$ with $N(0) = 1.5$, $P(0) = 1$. Two cycles are shown

$N(t)$, $P(t)$ graphed in Fig. 10.20 as curves in the (P,N)-plane, in Fig. 10.21. Periodicity means that these curves close. Compare to Fig. 10.17, where the species do not interact.

Proof. We write the Lotka–Volterra equations in terms of the steady-state values as

$$\frac{dN}{dt} = aN - bNP = bN\left(\frac{a}{b} - P\right) = bN(P_m - P)$$

$$\frac{dP}{dt} = hNP - cP = hP\left(N - \frac{c}{h}\right) = hP(N - N_m)$$

and conclude from the monotonicity criterion that N and P are increasing or decreasing functions of t, depending on whether the right sides are positive or negative. Therefore,

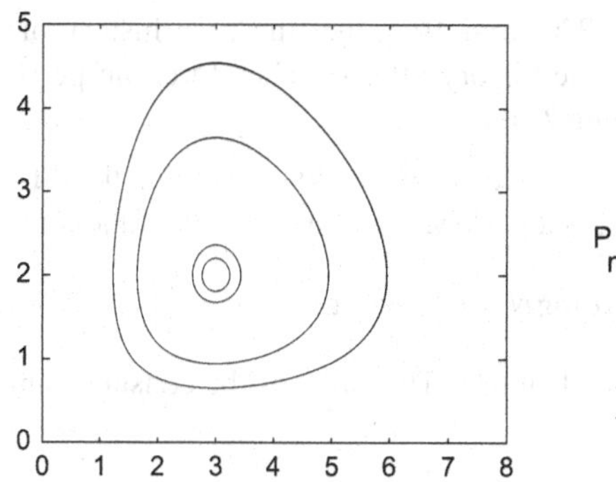

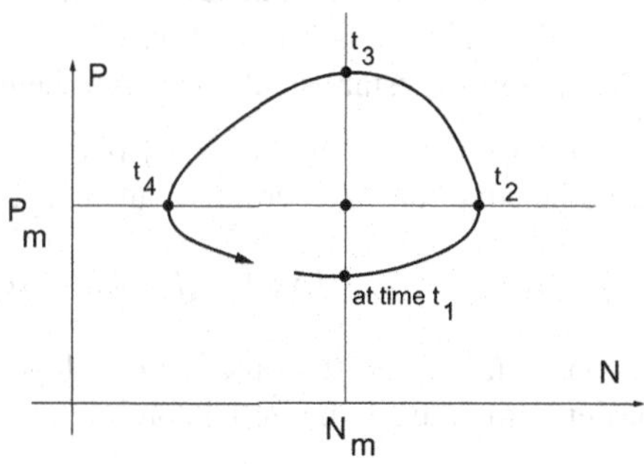

Fig. 10.21 *Left*: four computed time histories folded into the (N,P)-plane, for the same system as in Fig. 10.20. *Right*: a sketch showing the notation used in the proof of Theorem 10.9

$$N(t)\begin{cases}\text{increases when} & P < P_m,\\ \text{decreases when} & P > P_m,\end{cases}$$

$$P(t)\begin{cases}\text{decreases when} & N < N_m,\\ \text{increases when} & N > N_m.\end{cases}$$

We shall use these relations to trace qualitatively the time histories of $N(t)$ and $P(t)$. See Fig. 10.21. The initial values N_0, P_0 may be chosen arbitrarily. For sake of definiteness, we choose $N_0 < N_m$, $P_0 < P_m$. Then N starts to increase, P to decrease, until time t_1, when N reaches N_m. At time t_1, P starts to increase, and N continues to increase, until time t_2, when P reaches P_m. Then N starts to decrease, and P continues to increase, until time t_3, when N reaches N_m. Then P starts to decrease, and N continues to decrease, until time t_4, when P reaches P_m. Then N starts to increase, and P continues to decrease, until time t_5, when N reaches N_m.

We claim that at time t_5, the value of P is the same as at time t_1. To convince you of this, we appeal to the conservation law. Denoting the values of N and P at t_1 and t_5 by subscripts 1 and 5, we conclude from the conservation law that

$$H(N_5) + K(P_5) = H(N_1) + K(P_1).$$

The times t_1 and t_5 were chosen so that both N_1 and N_5 are equal to N_m. It follows then from the conservation law that

$$K(P_5) = K(P_1).$$

The value of $K(P)$ decreases for $P < P_m$. Recall that each value of a decreasing function occurs only once. Since t_1 and t_5 were chosen so that both P_1 and P_5 are less than P_m, it follows therefore from $K(P_5) = K(P_1)$ that

$$P_5 = P_1.$$

It follows from the uniqueness theorem, Theorem 10.6, that the time history of $N(t)$, $P(t)$ after t_5 is a repetition of its time history after t_1. Therefore, the periodicity claim is established, the period being $T = t_5 - t_1$. □

The closed curves in the (N,P)-plane can be determined without solving the differential equations. According to the conservation law, on each curve, the function

$$H(N) + K(P) = hN - c\log N + bP - a\log P$$

is constant, because the function is independent of t. The value of the constant can be determined from the initial condition

$$H(N_0) + K(P_0) = \text{constant}.$$

As remarked before, each solution is periodic, but different solutions have different periods. It is quite remarkable that the following quantities are the same for all solutions.

Theorem 10.10. *The average values of P and N over a period are the same for all solutions of the Lotka-Volterra equations (Table 10.1), and they equal their steady-state values* $P_m = \frac{a}{b}$ *and* $N_M = \frac{c}{h}$. *That is,*

$$\frac{1}{T}\int_0^T N(t)\,\mathrm{d}t = N_m, \quad \frac{1}{T}\int_0^T P(t)\,\mathrm{d}t = P_m,$$

where T is the period of N and P.

Proof. Write the Lotka–Volterra equations as

$$\frac{1}{N}\frac{\mathrm{d}N}{\mathrm{d}t} = a - bP, \quad \frac{1}{P}\frac{\mathrm{d}P}{\mathrm{d}t} = hN - c.$$

We integrate both equations from 0 to T, where T is the period of the solution in question. Using the chain rule, we get

$$\log N(T) - \log N(0) = \int_0^T \frac{1}{N}\frac{\mathrm{d}N}{\mathrm{d}t}\,\mathrm{d}t = \int_0^T (a - bP)\,\mathrm{d}t,$$

$$\log P(T) - \log P(0) = \int_0^T \frac{1}{P}\frac{\mathrm{d}P}{\mathrm{d}t}\,\mathrm{d}t = \int_0^T (hN - c)\,\mathrm{d}t.$$

Since T is the period of N and P, the left sides are zero. So we obtain the relations

$$0 = aT - b\int_0^T P(t)\,\mathrm{d}t, \qquad 0 = h\int_0^T N(t)\,\mathrm{d}t - cT.$$

Dividing the first equation by bT, the second by hT, we get

$$\frac{1}{T}\int_0^T P(t)\,\mathrm{d}t = \frac{a}{b}, \quad \text{and} \quad \frac{1}{T}\int_0^T N(t)\,\mathrm{d}t = \frac{c}{h}.$$

The expressions on the left are the average values of P and N over a period, while those on the right are their steady-state values. This concludes the proof. □

This result contains several interesting features; we mention two. The constants a, b, c, h that determine the steady state have nothing to do with the initial values P_0, N_0 of our populations. So it follows that the *average values of P and N are independent of the initial values*. Thus, if we were to increase the initial population N_0, for example by stocking a lake with fish, this would not affect the average size of $N(t)$ over a period, but would only lead to different oscillations in the size of $N(t)$. We ask you in Problem 10.28 to explore cases in which stocking the lake may result in either larger or smaller oscillations in $N(t)$.

For another application, suppose we introduce fishing into the model. Assuming that the catch of predator and prey is proportional to the number of each, fishing diminishes each population at a rate proportional to the size of that population. Denoting by f the constant of proportionality, we have the following modification of the equations:

$$\frac{dN}{dt} = aN - bNP - fN, \qquad \frac{dP}{dt} = -cP + hNP - fP.$$

We may write these equations in the form

$$\begin{aligned}\frac{dN}{dt} &= (a-f)N - bPN,\\ \frac{dP}{dt} &= -(c+f)P + hNP,\end{aligned}$$

and observe that they differ from the original system only in that the coefficient a of N in the first has been replaced by $a-f$, and the coefficient $-c$ of P in the second has been replaced by $-(c+f)$. According to Theorem 10.10, the average values of P and N are $\dfrac{a-f}{b}$ and $\dfrac{c+f}{h}$, respectively. In other words, *increased fishing depresses the average population of predators, but increases the average population of edible fish.* During the First World War, the Italian fishing industry reported a marked increase in the ratio of sharks to edible fish. Since less fishing was done during that war than before, this observation is consistent with Volterra's surprising result.

In more complicated models, numerical computations are indispensable. They not only provide numerical answers that cannot be found in any other way, but often reveal patterns of behavior amenable to mathematical analysis. For example, the computed solutions in Fig. 10.20 suggested that solutions are periodic, and we proved that they are.

To conclude, we point out simplifications that were made in the models presented in this section:

- We have neglected to take into account the *age distribution* of the population. Since birth rate and death rate are sensitive to this, our models are deficient and

would not describe correctly population changes accompanied by shifts in the population in and out of childbearing age. This phenomenon is particularly important in demography, the study of human populations.
- We have assumed that the population is *homogeneously distributed* in its environment. In many cases this is not so; the population distribution changes from location to location.

In problems such as the geographic spread of epidemics and the invasion of the territory belonging to one species by another, the interesting phenomenon is precisely the change in population as a function of time and location. Population sizes that depend on age and location as well as time are prototypes of functions of several variables. The calculus of functions of several variables is the natural language for the formulation of laws governing the growth of such populations.

You probably noticed throughout this section that we have treated population size as a differentiable function of t, whereas in fact, population changes by whole numbers, and so is not even a continuous function of t. Our defense is that these models are just models, i.e., approximations to reality, where some less-essential features are sacrificed for the sake of simplicity. The point we are making is that the *continuous is sometimes simpler than the discrete, since it allows us to use the powerful notions and tools of calculus.* Analogous simplifications are made in dealing with matter, e.g., in applying calculus to such physical quantities as pressure or density as functions of time or space. After all, according to the atomic theory of matter, these functions, too, change discontinuously.

Problems

10.22. Take the case of Eq. (10.19) where $Q(N) = N^{4/3}$ and $N_0 = 1000$.

(a) Solve $Q(N) = t + c$ for N as a function of t.
(b) Evaluate c.
(c) Find $R(N)$ and verify that your answer $N(t)$ is a solution to $\frac{dN}{dt} = R(N)$.

10.23. Show that the differential equation

$$\frac{dN}{dt} = \sqrt{N}$$

is satisfied by both functions $N(t) = 0$ and $N(t) = \frac{1}{4}t^2$ for $t \geq 0$. Since both functions are 0 at $t = 0$, does this contradict Theorem 10.3, according to which two solutions with the same initial value agree for all t?

10.24. Verify that the change of variables $n(t) = N(-t)$ converts the Verhulst model $\frac{dN}{dt} = 2N - N^2$ of Example 10.4 into the extinction model $\frac{dn}{dt} = n^2 - 2n$ of Example 10.5.

10.25. Consider the differential equation

$$\frac{dN}{dt} = N^2 - N$$

for values $0 < N < 1$. Derive a formula for the solutions. Use your formula to verify that if the initial value N_0 is between 0 and 1, then $N(t)$ tends to 0 as t tends to infinity.

10.26. Let p and n be functions of t that satisfy the following differential equations:

$$n' = f(p), \quad p' = g(n),$$

where the prime denotes differentiation with respect to t, and f and g are differentiable functions.

(a) Let n_1, p_1 and n_2, p_2 be two pairs of solutions. Show that the differences

$$n_1 - n_2 = m, \quad p_1 - p_2 = q$$

satisfy the inequalities

$$|m'| \leq k|q|, \quad |q'| \leq k|m|,$$

where k is an upper bound for the absolute value of the derivatives of the functions f and g.

(b) Deduce that

$$mm' + qq' \leq 2k|m||q| \leq k(m^2 + q^2).$$

(c) Define $E = \frac{1}{2}m^2 + \frac{1}{2}q^2$. Prove that $E' \leq 2kE$.

(d) Deduce that $e^{-2kt}E$ is a nonincreasing function of t. Deduce from this that if $E(0) = 0$, then $E(t) = 0$ for all $t > 0$. Show that this implies that two solutions n_1, p_1 and n_2, p_2 that are equal at $t = 0$ are equal forever after.

10.27. Consider the relation between numbers N and P given by the equation

$$H(N) + K(P) = \text{constant}, \tag{10.22}$$

where H and K are convex functions, and suppose that $K(P)$ is a decreasing function of P for P less than some number P_m, and an increasing function for $P > P_m$.

(a) Show that solutions of Eq. (10.22) where $P > P_m$ can be described by expressing P as a function of N. Show that solutions where $P < P_m$ can be described similarly. Denote these functions by $P_+(N)$ and $P_-(N)$.

(b) Let $P(N)$ be either of the functions $P_+(N)$, $P_-(N)$. Show by differentiating Eq. (10.22) twice that

$$\frac{dH}{dN} + \frac{dK}{dP}\frac{dP}{dN} = 0$$

and

$$\frac{d^2H}{dN^2} + \frac{d^2K}{dP^2}\left(\frac{dP}{dN}\right)^2 + \frac{dK}{dP}\frac{d^2P}{dN^2} = 0.$$

(c) Use the result of part (b) to express the second derivative as

$$\frac{d^2P}{dN^2} = -\frac{\frac{d^2H}{dN^2} + \frac{d^2K}{dP^2}\left(\frac{dP}{dN}\right)^2}{\frac{dK}{dP}}.$$

Deduce from this formula and the information given about H and K that $P_+(N)$ is a concave function and $P_-(N)$ is convex.

Remark. This confirms that the oval shapes computed in Fig. 10.21 are qualitatively correct.

10.28. What is the common point contained in all the ovals in Fig. 10.21?

10.3 Chemical Reactions

We give an elementary introduction to the theory of chemical reactions. This subject is of enormous interest to chemical engineers and to theoretical chemists. It also plays a central role in two topics that have recently been at the center of public controversy: emission by automobile engines and the deleterious effect on ozone of the accumulation of fluorocarbon compounds in the stratosphere.

In high-school chemistry, we studied the concept of a *chemical reaction*: it is the formation of one or several compounds called the *products* of the reaction out of one or several compounds or elements called *reactants*. Here is a familiar example:

$$2H_2 + O_2 \to 2H_2O$$

In words: two molecules of hydrogen and one molecule of oxygen form two molecules of water. Another example is

$$H_2 + I_2 \to 2HI.$$

In words: one molecule of hydrogen and one molecule of iodine form two molecules of hydrogen iodide.

A chemical reaction may require energy or may release energy in the form of heat; the technical terms are *endothermic* and *exothermic*. Familiar examples of reactions that release energy are the burning of coal or oil, and, more spectacularly, the burning of an explosive. In fact, the whole purpose of these chemical reactions is to garner the energy they release; the products of these reactions are uninteresting. In fact, they can be a severe nuisance, namely pollution. On the other hand, in the chemical industry, the desired commodity is the end product of the reactions or of a series of reactions.

The above description of chemical reactions deals with the phenomenon entirely in terms of its initial and final states. In this section, we shall study *time histories* of chemical reactions. This branch of chemistry is called *reaction kinetics*. An understanding of kinetics is essential in the chemical industry, because many reactions necessary in certain production processes must be set up so that they occur in the right order within specified time intervals. Similarly, the kinetics of burning must be understood in order to know what the end products are, for when released into the atmosphere, these chemicals affect global warming. The effect of fluorocarbons on depletion of ozone in the stratosphere must be judged by computing the rates at which various reactions involving these molecules occur. Last but not least, reaction kinetics is a valuable experimental tool for studying the structure of molecules.

In this section we shall describe the kinetics of fairly simple reactions, in particular those in which both reactants and products appear as gases. Furthermore, we shall assume that all components are homogeneously distributed in the vessel in which the reaction takes place. That is, we assume that the concentration, temperature, and pressure of all components at any given time are the same at all points in the vessel.

The *concentration* of a reactant measures the number of molecules of that reactant present per unit volume. Note that if two components in a vessel have the same concentration, then that vessel contains the same number of molecules of each component.

In what follows, we shall denote different molecules as well as atoms, ions, and radicals that play important roles in chemical reactions by different capital letters such as A, B, C, and we shall denote their concentrations by the corresponding lowercase letters such as a, b, c. (In the chemical literature, the concentration of molecule A would be denoted by [A].) These concentrations change with time. The *rates* at which they change, i.e., the derivatives of the concentrations with respect to time, are called the *reaction rates*. A basic principle of reaction kinetics says that the reaction rates are completely determined by pressure, temperature, and the concentrations of all components present. Mathematically, this can be expressed by specifying the rates as functions of pressure, temperature, and concentrations; then the laws of reaction kinetics take the form of differential equations:

$$\frac{\mathrm{d}a}{\mathrm{d}t} = f(a,b;T,p), \quad \frac{\mathrm{d}b}{\mathrm{d}t} = g(a,b;T,p),$$

where f, g are functions specific to each particular reaction. In the simple reactions considered here, we suppress the dependence of f, g, on the temperature T and the pressure p. The determination of these functions is the task of the theorist and experimenter. We shall start with some theoretical observations; of course, the last word belongs to the experimentalist.

The products of a chemical reaction are built out of the same basic components as the reactants, i.e., the same nuclei and the same number of electrons, but the components are now arranged differently. In other words, the chemical reaction is the process by which the rearrangement of the basic components occurs. One can think of this process of rearrangement as a continuous distortion, starting with the original component configuration and ending up with the final one. There is an *energy* associated with each transient configuration; the initial and the final states are *stable*, which means that energy is at a local minimum in those configurations. It follows that during a continuous distortion of one state into the other, energy increases until it reaches a peak and then decreases as the final configuration is reached. There are many paths along which this distortion can take place; the reaction is channeled mainly along the path where the peak value is minimum. The difference between this minimum peak value of energy and the energy of the initial configuration is called the *activation energy*. It is an energy barrier that has to be surmounted for the reaction to take place.

This description of a chemical reaction as rearrangement in one step is an oversimplification; it is applicable to only a minority of cases, called *elementary reactions*. In the great majority of cases, the reaction is *complex*, meaning that it takes place in a number of stages that lead to the formation of a number of intermediate states. The intermediate states—atoms, free radicals, and activated states—disappear when the reaction is completed. The transitions from the initial state to an intermediate state, from one intermediate state to another, and from an intermediate state to the final state are all elementary reactions. So a complex reaction may be thought of as a network of elementary reactions.

We now study the rate of an elementary reaction of form

$$A_2 + B_2 \to 2AB,$$

where one A_2 molecule consisting of two A atoms and one B_2 molecule consisting of two B atoms combine to form two molecules of the compound AB. The reaction takes place only if the two molecules collide and are energetic enough. The kinetic energies of the molecules in a vessel are not uniform, but are distributed according to a Maxwellian probability distribution (see Chap. 11). Therefore, some molecules always have sufficient kinetic energy to react when they collide, to supply the activation energy needed for the reaction. The frequency with which this happens is proportional to the *product* of the concentrations of A_2 and B_2 molecules, i.e., is equal to

$$kab, \qquad k \text{ a positive number.}$$

Here a and b denote the concentrations of A_2 and B_2, and k is the *rate constant*. This is called the *law of mass action*. Denote the concentration of the reaction product AB at time t by $x(t)$. By the law of mass action, x satisfies the differential equation

$$\frac{\mathrm{d}x}{\mathrm{d}t} = kab.$$

Denote by a_0 and b_0 the initial concentrations of A_2 and B_2. Since each molecule of A_2 and B_2 make two molecules of AB, the concentrations at time t are

$$a(t) = a_0 - \frac{x(t)}{2}, \qquad b(t) = b_0 - \frac{x(t)}{2}.$$

Substituting this into the differential equation yields

$$\frac{\mathrm{d}x}{\mathrm{d}t} = k\left(a_0 - \frac{x}{2}\right)\left(b_0 - \frac{x}{2}\right).$$

This equation is of the form of our population model $\frac{\mathrm{d}N}{\mathrm{d}t} = R(N)$ in Eq. (10.18), with x in place of N:

$$\frac{\mathrm{d}x}{\mathrm{d}t} = R(x), \qquad R(x) = k\left(a_0 - \frac{x}{2}\right)\left(b_0 - \frac{x}{2}\right).$$

If the initial concentration of AB is zero, then

$$x(0) = x_0 = 0.$$

Since $R(0) = ka_0b_0$ is positive, the solution $x(t)$ with initial value $x(0) = 0$ starts to increase. According to Theorem 10.4, this solution tends to the zero of $R(x)$ to the right of $x = 0$ that is nearest to $x = 0$. The zeros of $R(x)$ are $x = 2a_0$ and $x = 2b_0$. The one nearest to zero is the smaller of the two. We denote it by x_∞:

$$x_\infty = \min\{2a_0, 2b_0\}.$$

It follows, then, that as t tends to infinity, $x(t)$ tends to x_∞. Observe that the quantity x_∞ is the largest amount of AB that can be made out of the given amounts a_0 and b_0 of A_2 and B_2. Therefore, our result shows that as t tends to infinity, one or the other of the reactants gets completely used up.

Example 10.7. Consider

$$\frac{\mathrm{d}x}{\mathrm{d}t} = (1-x)(3-x).$$

The smaller root is $x_\infty = 1$. Some computed solutions are plotted in Fig. 10.22. We see that solutions starting between 0 and 3 tend to 1.

Our second observation is that although $x(t)$ tends to x_∞, $x(t)$ never reaches x_∞. So strictly speaking, the reaction goes on forever. However, when the difference between $x(t)$ and x_∞ is so small that it makes no practical difference, the reaction is practically over. We show how to estimate the time required for the practical completion of the reaction, using a linear rate instead of the quadratic rate $R(x)$, as follows.

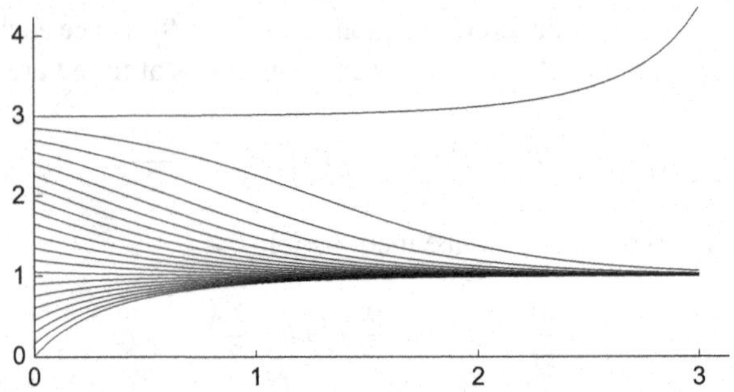

Fig. 10.22 Graphs of several solutions to $x' = R(x) = (1-x)(3-x)$ of Example 10.7. The function R is graphed in Fig. 10.23

The function $R(x) = k\left(a_0 - \frac{x}{2}\right)\left(b_0 - \frac{x}{2}\right)$ is quadratic. Therefore, its graph is a parabola. The second derivative of $R(x)$ is $\frac{1}{2}k$, a positive quantity. Therefore, as explained in Sect. 4.2b, its graph is a convex curve. This means that the points of the curve lie above its tangent lines. In particular, they lie above the tangent line at the point x_∞, as illustrated in Fig. 10.23. Denote the slope of the tangent line by $n = \frac{\mathrm{d}R}{\mathrm{d}x}(x_\infty)$. So we deduce that

$$R(x) > n(x - x_\infty) \qquad \text{when } x \neq x_\infty.$$

Since x_∞ is the smaller zero of $R(x)$, we deduce from Fig. 10.23 that n is negative. We set this inequality into the rate equation and get

$$\frac{\mathrm{d}x}{\mathrm{d}t} = k\left(a_0 - \frac{x}{2}\right)\left(b_0 - \frac{x}{2}\right) \geq n(x - x_\infty).$$

Since x_∞ is a constant, the relation $\frac{\mathrm{d}x}{\mathrm{d}t} \geq n(x - x_\infty)$ can be further simplified if we write $\frac{\mathrm{d}(x - x_\infty)}{\mathrm{d}t} \geq n(x - x_\infty)$: denote the difference $x_\infty - x$ by y. Then the inequality can be expressed as

$$0 \geq \frac{\mathrm{d}y}{\mathrm{d}t} - ny.$$

Keep in mind that we want to estimate how long it takes for $x(t)$ to approach x_∞, i.e., for $y(t)$ to approach 0. We multiply this inequality by e^{-nt}:

$$0 \geq \mathrm{e}^{-nt}\frac{\mathrm{d}y}{\mathrm{d}t} - \mathrm{e}^{-nt}ny.$$

We recognize the function on the right as the derivative of $y(t)\mathrm{e}^{-nt}$,

$$0 \geq \frac{\mathrm{d}}{\mathrm{d}t}\left(y(t)\mathrm{e}^{-nt}\right).$$

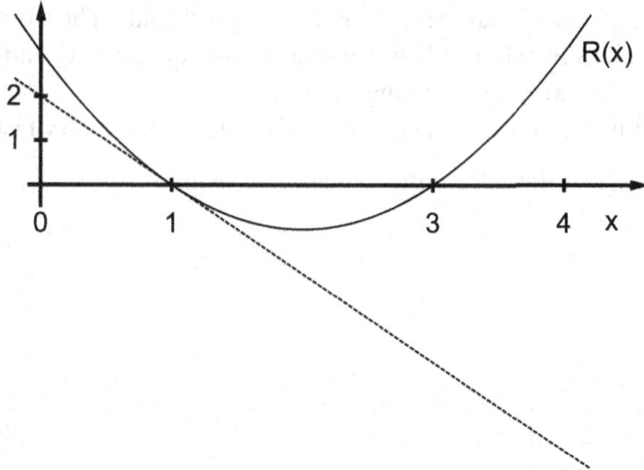

Fig. 10.23 The graph of the convex function $R(x) = (1-x)(3-x)$ lies above its tangent line at $x_\infty = 1$: $(1-x)(3-x) \geq 2-2x$. See Example 10.7

Since the derivative is nonpositive, the function $y(t)\mathrm{e}^{-nt}$ is nonincreasing. Therefore, for t positive, $y(t)\mathrm{e}^{-nt} \leq y(0)$. Multiplying both sides by e^{nt} gives

$$y(t) \leq y(0)\mathrm{e}^{nt}.$$

Since n is a negative number, this shows that $y(t)$ tends to zero at an exponential rate.

Example 10.8. For the equation in Example 10.7, we have

$$\frac{\mathrm{d}x}{\mathrm{d}t} = (1-x)(3-x) \geq 2-2x,$$

$x_\infty = 1,\ n = -2$, and $y(t) = 1-x(t)$ approaches 0 at the rate

$$|1-x(t)| \leq |1-x(0)|\mathrm{e}^{-2t}.$$

We now calculate the decay rate n for any quadratic reaction rate. We have

$$R(x) = \frac{k}{4}x^2 - \frac{k}{2}(a_0+b_0)x + ka_0b_0.$$

Differentiate:

$$\frac{\mathrm{d}R}{\mathrm{d}x} = \frac{k}{2}x - \frac{k}{2}(a_0+b_0).$$

Suppose a_0 is less than b_0. Then $x_\infty = 2a_0$, and so

$$n = \frac{\mathrm{d}R}{\mathrm{d}x}(2a_0) = ka_0 - \frac{k}{2}(a_0+b_0) = \frac{k}{2}(a_0-b_0).$$

Notice that when a_0 and b_0 are nearly equal, n is very small. Therefore, in this case $x(t)$ approaches x_∞ rather slowly. When a_0 and b_0 are equal, $n=0$, and our argument tells us nothing about $x(t)$ approaching x_∞.

We show that in case a_0 equals b_0, $x(t)$ tends to x_∞, but not very fast. In this case, $R(x)=\frac{k}{4}(x-x_\infty)^2$, so the differential equation says that

$$\frac{dx}{dt}=\frac{k}{4}(x-x_\infty)^2.$$

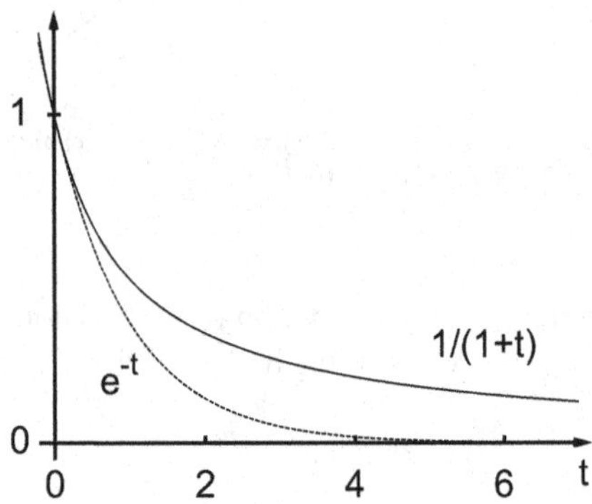

Fig. 10.24 Graphs of the functions $\frac{1}{1+t}$ and e^{-t} to illustrate differences in completion rates typical for reactions starting with equal or unequal concentrations of A_2 and B_2

Introducing as before $y=x_\infty-x>0$ as a new variable, we can rewrite the differential equation as

$$\frac{dy}{dt}=-\frac{k}{4}y^2.$$

Divide both sides by y^2. The resulting equation

$$\frac{1}{y^2}\frac{dy}{dt}=-\frac{k}{4}$$

can be written as

$$\frac{d}{dt}\left(\frac{1}{y}\right)=\frac{k}{4}.$$

Integrate from 0 to t; it follows that $\frac{1}{y(t)}=\frac{1}{y(0)}+\frac{k}{4}t$. Taking reciprocals, we get

$$y(t)=\frac{y(0)}{1+\frac{1}{4}ky(0)t}.$$

Since k and $y(0)$ are positive, $y(t)$ is defined for all $t \geq 0$, and tends to 0 as t tends to infinity. This proves that $x(t) = x_\infty - y(t)$ tends to x_∞, but at a very slow rate. See Fig. 10.24.

There is a good chemical reason why the reaction proceeds to completion much more slowly when the ingredients a_0 and b_0 are so perfectly balanced that they get used up simultaneously. If there is a shortage of both kinds of molecules, a collision leading to a reaction is much less likely than when there is a scarcity of only one kind of molecule but an ample supply of the other.

Complex Reactions. We consider a typical complex reaction such as the spontaneous decomposition of some molecule A, for example N_2H_4. The decomposition occurs in two stages; the first stage is the formation of a population of *activated molecules B* followed by the spontaneous splitting of the activated molecules. The mechanism for the formation of activated molecules B is through collision of two sufficiently energetic A molecules. See Fig. 10.25. The number of these collisions per unit time in a unit volume is proportional to a^2, the square of the concentration of A. There is also a reverse process of *deactivation*, due to collisions of activated and nonactivated molecules; the number of these per unit time in a unit volume is proportional to the product ab of the concentrations of A and B. There is, finally, a spontaneous decomposition of B molecules into the end products C. The number of these decompositions per unit time in a unit volume is proportional to the concentration of B. If we denote the rate constant of the formation of activated molecules by k, that of the reverse process by r, and that of the spontaneous decomposition by d, we get the following rate equations:

$$\frac{da}{dt} = -ka^2 + rab,$$

$$\frac{db}{dt} = ka^2 - rab - db, \tag{10.23}$$

and $\frac{dc}{dt} = db$, which may be used after $b(t)$ is determined. It can be shown that $a(t)$ and $b(t)$ tend to zero as t tends to infinity, but the argument goes beyond the scope of this chapter.

Finally, we again call attention to the striking similarity between the differential equations governing the evolution of concentrations of chemical compounds during reaction and the laws governing the evolution of animal species interacting with each other. This illustrates the universality of mathematical ideas.

Problems

10.29. Consider the differential equations

(a) $\frac{dy}{dt} = -y^2$,

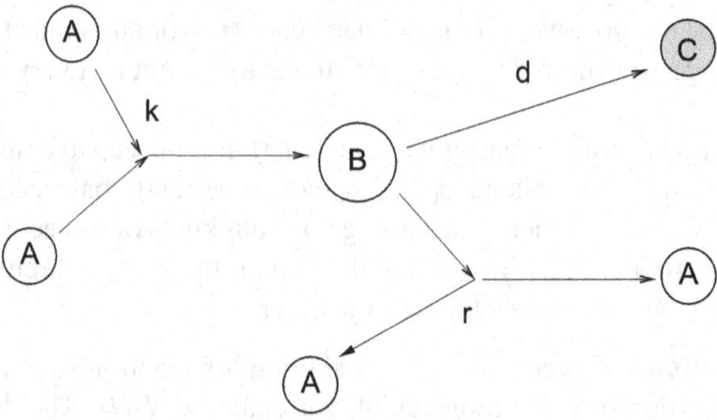

Fig. 10.25 An illustration for a complex decomposition of A to C. Here k is the rate of formation of activated B from two A, r the reverse rate, and d the rate of spontaneous decomposition of B to C

(b) $\dfrac{\mathrm{d}y}{\mathrm{d}t} = -y.$

Take $y(0)$ positive. For which equation does $y(t)$ tend to zero faster as t tends to infinity?

10.30. Show that if $a(t)$ and $b(t)$ are positive functions that satisfy the differential equations (10.23), than $a+b$ is a decreasing function of t.

10.31. In Eq. (10.23) let $p(t) = rb(t) - ka(t)$, so that

$$\frac{\mathrm{d}a}{\mathrm{d}t} = ap, \quad \frac{\mathrm{d}b}{\mathrm{d}t} = -ap - db.$$

Divide the positive quadrant of the (a,b)-plane into two parts along the ray $rb = ka$. Show that on the side where $p(t) < 0$, $a(t)$ is a decreasing function, and that on the side where $p(t) > 0$, $b(t)$ is decreasing.

10.4 Numerical Solution of Differential Equations

In Sect. 10.2a we showed how integration and inverting a function can be used to find solutions of an equation linking a function N to its first derivative. Such methods no longer work for finding solutions of equations involving functions and their higher derivatives, or for systems of equations relating the derivatives of several functions. Such equations can be solved only by numerical methods. In this section, we show how this is done with a very simple example.

The basis of most numerical methods for finding approximate solutions of a differential equation

$$\frac{\mathrm{d}N}{\mathrm{d}t} = N' = R(N,t), \qquad N(0) = N_0,$$

is to replace the derivative $\frac{dN}{dt}$ in the differential equation by a difference quotient

$$\frac{N(t+h)-N(t)}{h}.$$

Instead of the differential equation, we solve the *difference* equation

$$\frac{N(t+h)-N(t)}{h} = R(N(t),t), \qquad N(0) = N_0,$$

which we rearrange as

$$N(t+h) = N(t) + hR(N(t),t).$$

This is called Euler's method for approximating solutions.

We denote solutions of the difference equation by $N_h(t)$, defined for values of t that are integer multiples nh of h. One of the results of the theory of approximations by difference equations is that $N_h(nh)$ tends to $N(t)$ as h tends to zero and nh tends to t. The proof goes beyond the scope of this book. However, we will show that the method converges in the case $N' = N$. In the process, we will encounter some familiar sequences from Chap. 1.

The Equation $N' = N$. The solution of the differential equation

$$N'(t) = N(t), \qquad \text{with initial condition } N(0) = 1, \tag{10.24}$$

is our old friend $N(t) = e^t$. Denote by $e_h(t)$ a function that satisfies the equation obtained by replacing the derivative on the left in Eq. (10.24) by a difference quotient:

$$\frac{e_h(t+h)-e_h(t)}{h} = e_h(t), \qquad e_h(0) = 1. \tag{10.25}$$

For h small, the difference quotient doesn't differ too much from the derivative, so it is reasonable to expect that the solution e_h of Eq. (10.25) does not differ too much from the solution e^t of Eq. (10.24). From Eq. (10.25), we can express $e_h(t+h)$ in terms of $e_h(t)$ as follows:

$$e_h(t+h) = (1+h)e_h(t). \tag{10.26}$$

Set $t = 0$ in Eq. (10.26). Since $e_h(0) = 1$, we get $e_h(h) = 1+h$. Set $t = h$ in Eq. (10.26), and we get $e_h(2h) = (1+h)e_h(h) = (1+h)^2$. Continuing in this fashion, we get for any positive integer k, $e_h(kh) = (1+h)^k$. When $h = \frac{1}{n}$ and $t = kh$, $k = nt$, we have (Fig. 10.26)

$$e_h(t) = e_{1/n}\left(k\frac{1}{n}\right) = \left(1+\frac{1}{n}\right)^k = \left(1+\frac{1}{n}\right)^{nt} = \left(\left(1+\frac{1}{n}\right)^n\right)^t.$$

Recall that $\left(1+\frac{1}{n}\right)^n$ is the number we called e_n in Sect. 1.4. So we have

$$e_{1/n}(t) = (e_n)^t$$

for every positive rational number t.

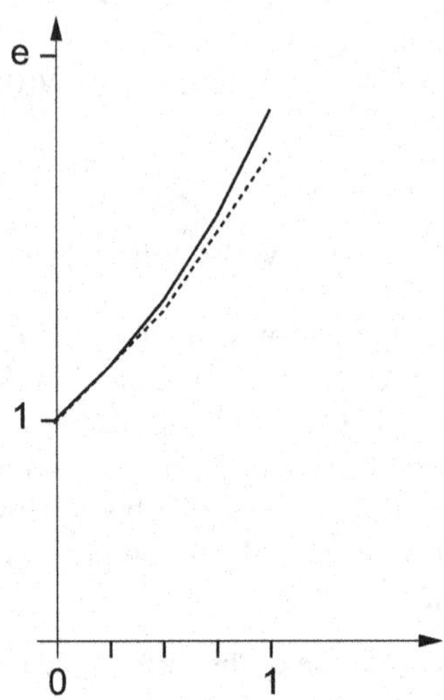

Fig. 10.26 Approximations $e_h(kh)$ using Eq. (10.26) with $h = \dfrac{1}{n}$. Values are connected by segments. *Dotted line*: $n = 2$ and $h = 0.5$. *Solid line*: $n = 4$ and $h = 0.25$. The highest points are $\left(1+\dfrac{1}{2}\right)^2$ and $\left(1+\dfrac{1}{4}\right)^4$

We investigate now another way of replacing the differential equation (10.24) by a difference equation. We denote this approximate solution by $f_h(t)$. As before, we replace the derivative by the same difference quotient, but we set this equal to $f_h(t+h)$:

$$\frac{f_h(t+h) - f_h(t)}{h} = f_h(t+h), \quad f_h(0) = 1. \tag{10.27}$$

This equation can be used to express $f_h(t+h)$ in terms of $f_h(t)$:

$$f_h(t+h) = \frac{1}{1-h} f_h(t).$$

Since $f_h(0) = 1$, we deduce from this equation that $f_h(h) = \dfrac{1}{1-h}$. Arguing as before, where $t = kh$, we deduce that for every positive integer k, $f_h(kh) = \left(\dfrac{1}{1-h}\right)^k$.

Set $h = \dfrac{1}{n+1}$. Then

$$f_h(kh) = \left(\frac{1}{1-\frac{1}{n+1}}\right)^k = \left(\frac{n+1}{n}\right)^k = \left(1+\frac{1}{n}\right)^{t(n+1)}.$$

Recall that $\left(1+\frac{1}{n}\right)^{n+1}$ is the number we called f_n in Sect. 1.4. So $f_{1/(n+1)}(t) = (f_n)^t$.

Next we use calculus to compare the solutions to the difference equations with the exact solution of the differential equation (10.24), for the case $t = 1$.

Theorem 10.11. *For every positive integer n,*

$$\left(1+\frac{1}{n}\right)^n < \mathrm{e} < \left(1+\frac{1}{n}\right)^{n+1}. \tag{10.28}$$

Proof. We use the mean value theorem to express the difference quotient

$$\frac{\mathrm{e}^{t+h}-\mathrm{e}^t}{h} = \mathrm{e}^c,$$

where c lies between t and $t+h$. Since e^t is an increasing function, we get that if $h > 0$, then

$$\mathrm{e}^t < \mathrm{e}^c = \frac{\mathrm{e}^{t+h}-\mathrm{e}^t}{h} < \mathrm{e}^{t+h}.$$

Multiply by h and rearrange terms to get

$$(1+h)\mathrm{e}^t < \mathrm{e}^{t+h} < \frac{1}{1-h}\mathrm{e}^t. \tag{10.29}$$

Choosing first $t = 0$ and then $t = h$, we deduce from the inequality on the left in Eq. (10.29) that

$$1+h < \mathrm{e}^h, \quad (1+h)\mathrm{e}^h < \mathrm{e}^{2h}.$$

Multiply the first of these by $(1+h)$ and use the second to see that $(1+h)^2 < \mathrm{e}^{2h}$. Similarly, for every positive integer n, $(1+h)^n < \mathrm{e}^{nh}$. Choosing $h = \dfrac{1}{n}$, we get

$$\left(1+\frac{1}{n}\right)^n < \mathrm{e}. \tag{10.30}$$

Using similarly the inequality on the right in Eq. (10.29), we get $\mathrm{e}^h < \dfrac{1}{1-h}$. Then

$$\mathrm{e}^{2h} < \frac{1}{1-h}\mathrm{e}^h < \left(\frac{1}{1-h}\right)^2,$$

and so forth. Taking $n+1$ steps and $h=\frac{1}{n+1}$ gives

$$e=e^{(n+1)\frac{1}{n+1}}<\left(\frac{1}{1-\frac{1}{n+1}}\right)^{n+1}=\left(1+\frac{1}{n}\right)^{n+1}.$$

This completes the proof. □

In Sect. 1.4 we proved inequality (10.28) using the A-G inequality; here we have given an entirely different proof.

From Theorem 10.11, we can easily deduce that both $\left(1+\frac{1}{n}\right)^n$ and $\left(1+\frac{1}{n}\right)^{n+1}$ tend to e as n tends to infinity. Take their difference:

$$\left(1+\frac{1}{n}\right)^{n+1}-\left(1+\frac{1}{n}\right)^{n}=\left(1+\frac{1}{n}\right)^{n}\frac{1}{n}<\frac{e}{n}.$$

The inequality we have just proved shows that the difference tends to zero. Since e lies between these two numbers, it follows that their difference from e also tends to zero. This proves the convergence of both difference schemes (10.25) and (10.27).

The Rate of Convergence. In Sect. 1.4, we saw that the convergence of e_n and f_n to the limit e is very slow. For example, with $n=1000$ we had $e_{1000}=2.717\ldots$ and $f_{1000}=2.719\ldots$, only two correct digits after the decimal point. Now we shed some light on why these approximations to e are so crude. Both e_n and f_n are derived from one-sided approximations to the derivative. We saw in Sect. 4.4 that for a twice differentiable function g, the error in using the asymmetric difference quotient

$$g'(t)-\frac{g(t+h)-g(t)}{h}$$

tends to zero with h, while the error in using the symmetric difference quotient

$$g'(t)-\frac{g(t+h)-g(t-h)}{2h}$$

is equal to sh, where s tends to zero with h, which gives a better approximation. We can take advantage of this observation to improve the approximate solution to $y'=y$. Use the equation

$$\frac{g(t+h)-g(t-h)}{2h}=\frac{g(t+h)+g(t-h)}{2},$$

in which we use the symmetric difference quotient on the left-hand side to approximate y', and on the right hand-side we average the values of g at $t+h$ and $t-h$ as an approximation of $y(t)$. Solving for $g(t+h)$, we obtain $g(t+h)=\frac{1+h}{1-h}g(t-h)$. Replace t by $t+h$ to get

$$g(t+2h)=\frac{1+h}{1-h}g(t).$$

Taking $g(0) = 1$, this gives

$$g(2h) = \frac{1+h}{1-h}, \quad g(4h) = \frac{1+h}{1-h}g(2h) = \left(\frac{1+h}{1-h}\right)^2, \quad \dots \quad g(2nh) = \left(\frac{1+h}{1-h}\right)^n.$$

Take $h = \dfrac{1}{2n}$. We obtain

$$g(1) = \left(\frac{1+\frac{1}{2n}}{1-\frac{1}{2n}}\right)^n.$$

For $n = 10$ and 20, this gives the estimates

$$\left(\frac{1.05}{0.95}\right)^{10} = 2.7205\ldots \quad \text{and} \quad \left(\frac{1.025}{0.975}\right)^{20} = 2.7188\ldots,$$

much closer to e than the numbers e_n and f_n.

Problems

10.32. Use Euler's numerical method with $h = 0.1$ to approximate the solution to

$$\frac{dy}{dt} = -1-t, \quad y(0) = 1$$

for several steps, sufficient to estimate the time t at which y becomes 0. Compare to the exact solution.

10.33. Verify that for any differential equation $y' = f(t)$ with $y(0) = 0$, Euler's numerical method with n subdivisions gives exactly the approximate integral

$$y_n = I_{\text{left}}(f, [0, nh]).$$

10.34. Consider the differential equation $y' = a - y$, where a is a constant.

(a) Verify that the constant function $y(t) = a$ is a solution.

(b) Suppose y is a solution, and for some interval of t we have $y(t) > a$. Is y increasing or decreasing?

(c) Consider two numerical methods. For the first, we use Euler's method to produce a sequence y_n according to

$$y_{n+1} = y_n + h(a - y_n).$$

For the second, we use a method similar to that in Eq. (10.27) to produce a sequence Y_n according to

$$Y_{n+1} = Y_n + h(a - Y_{n+1}).$$

Show that if some y_n is equal to a, then $y_{n+1} = a$, and if some Y_n is equal to a, then $Y_{n+1} = a$.

(d) Show that if some Y_n is greater than a, then $Y_{n+1} > a$.

(e) Find a value of h such that the sequence y_n alternates between numbers less than a and greater than a.

Chapter 11
Probability

Abstract Probability is the branch of mathematics that deals with events whose individual outcomes are unpredictable, but whose outcomes on average are predictable. In this chapter we shall describe the rules of probability. We shall apply these rules to specific situations. As you will see, the notions and methods of calculus play an extremely important part in these applications. In particular, the logarithmic and exponential functions are ubiquitous. For these reasons, this chapter has been included in this book.

The origins of calculus lie in Newtonian mechanics, of which a brief preview was given in Sect. 10.1, where we considered the motion of a particle under the combination of a restoring force and friction. We saw that once the force acting on a particle is ascertained and the initial position and velocity of the particle specified, the whole future course of the particle is predictable. Such a predictable motion is called *deterministic*. In fact, every system of particles moving according to Newton's laws by ascertainable forces describes a predictable path. On the other hand, when the forces acting on a particle cannot be ascertained exactly, or even approximately, or when its initial position and velocity are not under our control or even our power to observe, then the path of the object is far from being predictable. Many—one is tempted to say almost all—motions observed in everyday life are of this kind. Typical examples are the wafting of smoke, drifting clouds in the sky, dice thrown, cards shuffled and dealt. Such unpredictable motion is called *nondeterministic* or *random*.

Even though the outcome of a single throw of a die is unpredictable, the average outcome in the long run is quite predictable, at least if the die is the standard kind: each number will appear in about one-sixth of a large number of throws. Similarly, if we repeatedly shuffle and deal out the top card of a deck of 52 cards, each card will appear about $1/52$ times the number of deals. With certain types of cloud formations, experience may indicate rain in three out of five cases on average.

P.D. Lax and M.S. Terrell, *Calculus With Applications*, Undergraduate Texts in Mathematics, DOI 10.1007/978-1-4614-7946-8_11,

11.1 Discrete Probability

We shall consider some simple, almost simplistic, experiments such as the tossing of a die, the shuffling of a deck of cards and dealing the top card, and the tossing of a coin. A more realistic example is the performance of a physical experiment. The two stages of an experiment are setting it up and observing its outcome. In many cases, such as meteorology, geology, oceanography, the setting up of the experiment is beyond our power; we can merely observe what has been set up by nature.

We shall deal with experiments that are *repeatable* and *nondeterministic*. Repeatable means that it can be set up repeatedly any number of times. Nondeterministic means that any single performance of the experiment may result in a variety of *outcomes*. In the simple examples mentioned at the beginning of this section, the possible outcomes are respectively a whole number between 1 and 6, any one of 52 cards, heads or tails. In this section, we shall deal with experiments that like the examples above, have *a finite number of possible outcomes*. We denote the number of possible outcomes by n, and shall number them from 1 to n.

Finally, we assume that the outcome of the experiment, unpredictable in any individual instance, is *predictable on average*. By this we mean the following: Suppose we could repeat the experiment as many times as we wished. Denote by S_j the number of instances among the first N experiments in which the jth outcome was observed to take place. Then the frequency $\frac{S_j}{N}$ with which the jth outcome has been observed to occur tends to a limit as N tends to infinity. We call this limit the *probability of the jth outcome* and denote it by p_j:

$$p_j = \lim_{N\to\infty} \frac{S_j}{N}. \tag{11.1}$$

These probabilities have the following properties:

(a) Each probability p_j is a real number between 0 and 1:

$$0 \le p_j \le 1.$$

(b) The sum of all probabilities equals 1:

$$p_1 + p_2 + \cdots + p_n = 1.$$

Both these properties follow from Eq. (11.1), for $\frac{S_j}{N}$ lies between 0 and 1, and therefore so does its limit p_j. This proves the first assertion. On the other hand, there are altogether n possible outcomes, so that each of the first N outcomes of the sequence of experiments performed falls into one of these n cases. Since S_j is the number of instances among the first N when the jth outcome was observed, it follows that

$$S_1 + S_2 + \cdots + S_n = N.$$

Dividing by N, we get

$$\frac{S_1}{N} + \frac{S_2}{N} + \cdots + \frac{S_n}{N} = 1.$$

Now let N tend to infinity. The limit of $\frac{S_1}{N}$ is p_1, that of $\frac{S_2}{N}$ is p_2, etc., so in the limit, we see that $p_1 + p_2 + \cdots + p_n = 1$, as asserted.

Sometimes, in fact very often, we are not interested in all the details of the outcome of an experiment, but merely in a particular aspect of it. For example, in drawing a card we may be interested only in the suit to which it belongs, and in throwing a die, we may be interested only in whether the outcome is even or odd. An occurrence such as drawing a spade or throwing an even number is called an *event*. In general, *we define an event E as any collection of possible outcomes*. Thus drawing a spade is the collective name for the outcomes of drawing the deuce of spades, the three of spades, etc., all the way up to drawing the ace of spades. Similarly, an even throw of a die is the collective name for throwing a two, a four or a six.

We define the probability $p(E)$ of an event E similarly to the way we defined the probability of an outcome:

$$p(E) = \lim_{N\to\infty} \frac{S(E)}{N},$$

where $S(E)$ is the number of instances among the first N performances of the experiment when the event E took place. It is easy to show that this limit exists. In fact, it is easy to give a formula for $p(E)$. For by definition, the event E takes place whenever the outcome belongs to the collection of the possible outcomes that make up the event E. Therefore, $S(E)$, the number of instances in which E has occurred, is the sum of all S_j for those j that make up E:

$$S(E) = \sum_{j \text{ in } E} S_j.$$

Divide by N:

$$\frac{S(E)}{N} = \sum_{j \text{ in } E} \frac{S_j}{N}.$$

This relation says that $\frac{S(E)}{N}$ is the sum of the frequencies $\frac{S_j}{N}$, where j is in E. We deduce that in the limit as N tends to infinity,

$$p(E) = \sum_{j \text{ in } E} p_j. \tag{11.2}$$

Two events E_1 and E_2 are called *disjoint* if both cannot take place simultaneously. That is, the set of outcomes that constitute the event E_1 and the set of outcomes that constitute the event E_2 have nothing in common. Here are some examples of disjoint events:

Example 11.1. If the experiment consists in drawing one card, let E_1 be the event of drawing a spade, and E_2 the event of drawing a heart:

$$E_1 = \{2\spadesuit, 3\spadesuit, \ldots, 10\spadesuit, J\spadesuit, Q\spadesuit, K\spadesuit, A\spadesuit\}, \quad E_2 = \{2\heartsuit, 3\heartsuit, \ldots, A\heartsuit\}.$$

Each event contains 13 outcomes, and they have no outcome in common; E_1 and E_2 are disjoint events.

Example 11.2. Suppose the experiment is to roll one die. Let E_1 be the event of throwing an even number, and E_2 the event of throwing a 3. Then E_1 consists of the outcomes 2, 4, and 6, while E_2 consists of outcome 3 only. These are disjoint.

We define the *union* of two events E_1 and E_2, denoted by $E_1 \cup E_2$, as the event of either E_1 or E_2 (or both) taking place. That is, the outcomes that constitute $E_1 \cup E_2$ are the outcomes that constitute E_1 combined with the outcomes that constitute E_2.

Example 11.3. In the card experiment, Example 11.1, $E_1 \cup E_2$ consists of half the deck: all the spades and all the hearts. In the die experiment, Example 11.2, $E_1 \cup E_2$ consists of outcomes 2, 3, 4, and 6.

The following observation is as important as it is simple: *The probability of the union of two disjoint events* is the sum of the probabilities of each event:

$$p(E_1 \cup E_2) = p(E_1) + p(E_2).$$

This is called the *addition rule for disjoint events.* This result follows from formula (11.2) for the probability of an event, for by definition of union,

$$p(E_1 \cup E_2) = \sum_{j \text{ in } E_1 \text{ or } E_2} p_j.$$

On the other hand, disjointness means that an outcome j may belong either to E_1 or to E_2 but not to both. Therefore,

$$p(E_1 \cup E_2) = \sum_{j \text{ in } E_1 \text{ or } E_2} p_j = \sum_{j \text{ in } E_1} p_j + \sum_{j \text{ in } E_2} p_j = p(E_1) + p(E_2),$$

as asserted.

Next we turn to another important idea in probability, the *independence* of two experiments. Take two experiments such as (1) throwing a die and (2) shuffling a deck and dealing the top card. Our common sense plus everything we know about the laws of nature tells us that these experiments are totally independent of each other in the sense that the outcome of one cannot possibly influence the other, nor is the outcome of both under the influence of a common cause. We state now, precisely in the language of probability theory, an important consequence of independence.

Given two experiments, we can compound them into a single *combined experiment* simply by performing them simultaneously. Let E be any event in the framework of one of the experiments, F any event in the framework of the other. The combined event of both E and F taking place will be denoted by $E \cap F$.

Example 11.4. For instance, if E is the event an even throw, and F is the event drawing a spade, then $E \cap F$ is the event of an even throw *and* drawing a spade.

We claim that *if the experiments are independent, then the probability of the combined event $E \cap F$ is the product of the separate probabilities of the events E and F*:

$$p(E \cap F) = p(E)p(F). \tag{11.3}$$

We refer to this relation as the *product rule* for independent experiments.

We now show how to deduce the product rule. Imagine the combined experiment repeated as many times as we wish. We look at the first N experiments of this sequence. Among the first N, count the number of times E has occurred, F has occurred, and $E \cap F$ has occurred. We denote these numbers by $S(E)$, $S(F)$, and $S(E \cap F)$. By definition of the probability of an event,

$$p(E) = \lim_{N\to\infty} \frac{S(E)}{N},$$

$$p(F) = \lim_{N\to\infty} \frac{S(F)}{N},$$

$$p(E \cap F) = \lim_{N\to\infty} \frac{S(E \cap F)}{N}.$$

Suppose that we single out from the sequence of combined experiments the *subsequence* of those in which E occurred. The frequency of occurrence of F in this subsequence is $\dfrac{S(E \cap F)}{S(E)}$. If the two events E and F are truly independent, the frequency with which F occurs in this subsequence should be the same as the frequency with which F occurs in the original sequence. Therefore,

$$\lim_{N\to\infty} \frac{S(E \cap F)}{S(E)} = \lim_{N\to\infty} \frac{S(F)}{N} = p(F).$$

Now we write the frequency $S(E \cap F)/N$ as the product

$$\frac{S(E \cap F)}{N} = \frac{S(E \cap F)}{S(E)} \frac{S(E)}{N}.$$

Then

$$\lim_{N\to\infty} \frac{S(E \cap F)}{N} = \lim_{N\to\infty} \frac{S(E \cap F)}{S(E)} \cdot \lim_{N\to\infty} \frac{S(E)}{N}.$$

Therefore, $p(E \cap F) = p(E)p(F)$.

Suppose that one experiment has m outcomes numbered $1, 2, \ldots, j, \ldots, m$, and the other has n outcomes numbered $1, 2, \ldots, k, \ldots, n$. Denote their respective probabilities by $p_1, \ldots, p_m$ and $q_1, \ldots, q_n$. The combined experiment then has mn possible outcomes, namely all pairs of outcomes (j, k). If the experiments are independent, then the product rule tells us that the outcome (j,k) of the combined experiment has probability

$$p_j q_k.$$

This formula plays a very important role in probability theory. We now give an illustration of its use.

Suppose that both of the two experiments we have been discussing are the tossing of a die. Then the combined experiment is the tossing of a pair of dice. Each experiment has six possible outcomes, with probability $\frac{1}{6}$. There are 36 outcomes for the combined experiment, which we can list from $(1,1)$ to $(6,6)$. According to the product rule for independent events, $p(E \cap F) = p(E)p(F)$, so each combined outcome has probability $\frac{1}{36}$. We now ask the following question: What is the probability of the event of tossing a 7? There are six ways of tossing a 7:

$$(1,6),\ (2,5),\ (3,4),\ (4,3),\ (5,2),\ (6,1).$$

The probability of tossing a 7 is the sum of the probabilities of these six outcomes that constitute the event. That sum is

$$\frac{1}{36}+\frac{1}{36}+\frac{1}{36}+\frac{1}{36}+\frac{1}{36}+\frac{1}{36}=\frac{1}{6}.$$

Similarly, we can calculate the probability of tossing any number between 2 and 12. We ask you in Problem 11.6 to go through the calculations of determining the probabilities that the numbers $2,3,\ldots,12$ will be thrown. The results are in Table 11.1.

Table 11.1 Probabilities for the sum of two independent dice

Throw	2	3	4	5	6	7	8	9	10	11	12
Probability	$\frac{1}{36}$	$\frac{1}{18}$	$\frac{1}{12}$	$\frac{1}{9}$	$\frac{5}{36}$	$\frac{1}{6}$	$\frac{5}{36}$	$\frac{1}{9}$	$\frac{1}{12}$	$\frac{1}{18}$	$\frac{1}{36}$

Numerical Outcome. We now turn to another important concept of probability, the *numerical outcome* of an experiment. In physical experiments designed to measure the value of a single physical quantity, the numerical outcome is simply the measured value of the quantity in question. For the simple example of throwing a pair of dice, the numerical outcome might be the sum of the face values of each die. For the experiment of dealing a bridge hand, the numerical outcome might be the point count of the bridge hand. In general, the *numerical outcome* of an experiment *means the assignment of a real number x_j to each of the possible outcomes,* $j = 1,2,\ldots,\ n$.

Note that different outcomes may be assigned the same number, as in the case of the dice; the numerical outcome 7 is assigned to the six different outcomes $(1,6),(2,5),(3,4),(4,3),(5,2),(6,1)$.

Expectation. We show now that in a random experiment with n possible outcomes of probability p_j and numerical outcome x_j $(j = 1,2\ldots n)$, the average numerical outcome, called the *mean* of x, or *expectation* of x, denoted by $\overline{x}$ or $E(x)$, is given by the formula

$$\overline{x} = E(x) = p_1 x_1 + \cdots + p_n x_n. \tag{11.4}$$

To prove this, denote as before by S_j the number of instances among the first N in which the jth outcome was observed. The average numerical outcome among the first N is therefore

$$\frac{S_1 x_1 + S_2 x_2 + \cdots + S_n x_n}{N}.$$

We rewrite this as

$$\frac{S_1}{N} x_1 + \frac{S_2}{N} + \cdots + \frac{S_n}{N} x_n.$$

By hypothesis, each of the ratios $\frac{S_j}{N}$ tends to the limit p_j. It follows that the average numerical outcome tends to $\overline{x}$, as asserted.

We now give an example of formula (11.4) for the average numerical outcome. Take the experiment of throwing a pair of dice. We classify the outcomes as throwing a 2, 3, ..., up to 12. We take these numbers to be the numerical outcomes of the experiment. The probability of each outcome is given in Table 11.1. We get the following value for the average numerical outcome of a throw of a pair of dice:

$$\overline{x} = \frac{1}{36}2 + \frac{1}{18}3 + \frac{1}{12}4 + \frac{1}{9}5 + \frac{5}{36}6 + \frac{1}{6}7 + \frac{5}{36}8 + \frac{1}{9}9 + \frac{1}{12}10 + \frac{1}{18}11 + \frac{1}{36}12 = 7.$$

Variance. We have shown that if we perform a random experiment with numerical outcomes many times, the average of the numerical outcomes will be very close to the mean, given by Eq. (11.4). A natural question is this: by how much do the numerical outcomes differ on average from the mean? The average difference is

$$\sum_{i=1}^{n} (x_i - \overline{x}) p_i = \sum_{i=1}^{n} p_i x_i - \Big(\sum_{i=1}^{n} p_i\Big)\overline{x} = \overline{x} - \overline{x} = 0,$$

not very informative. It turns out that a related concept, the *variance*, has much better mathematical properties.

Definition 11.1. The *variance*, denoted by V, is the expected value of the square of the difference of the outcome and its expected value:

$$V = \overline{(x - \overline{x})^2} = E\big((x - E(x))^2\big).$$

We show how to express the variance in terms of the numerical outcomes and their probabilities. The numerical outcome x_j differs from the mean $\overline{x}$ by $x_j - \overline{x}$. Its square is $(x_j - \overline{x})^2$, which is equal to

$$x_j^2 - 2x_j\overline{x} + (\overline{x})^2. \tag{11.5}$$

Denote as before by S_j the number of times the jth outcome occurred among the first N events. The expected value of the quantity in Eq. (11.5) is

$$\frac{S_1x_1^2+\cdots+S_nx_n^2}{N}-2\frac{S_1x_1+\cdots+S_nx_n}{N}\overline{x}+(\overline{x})^2.$$

As N tends to infinity, $\dfrac{S_j}{N}$ tends to p_j. Therefore, the expected value above tends to

$$V=p_1x_1^2+\cdots+p_nx_n^2-2(p_1x_1+\cdots+p_nx_n)\overline{x}+(\overline{x})^2=\overline{x^2}-(\overline{x})^2.$$

We denote the expected value of the square of the outcome by $\overline{x^2}=E(x^2)$. This leads to an alternative way to calculate the variance:

$$V=E\big((x-E(x))^2\big)=E(x^2)-\big(E(x)\big)^2. \tag{11.6}$$

Definition 11.2. The square root of the variance is called the *standard deviation.*

The Binomial Distribution. Suppose a random experiment has two possible outcomes A and B, with probabilities p and q respectively, where $p+q=1$. For example, think of a coin toss, or an experiment with two outcomes, A success and B failure. Choose any positive integer N and repeat the experiment N times. Assume that the repeated experiments are independent of each other. This new combined experiment has outcomes that are a string of successes and failures. For example, if $N=5$, then *ABAAB* is one possible outcome, and another is *BABBB*. If we let x be the numerical outcome that this repeated experiment results in A occurring exactly x times, we see that x has possible values

$$x=0,\ 1,\ ,\ldots,N.$$

The probability that the outcome A occurs exactly k times and the outcome B exactly $N-k$ times is given by the expression

$$b_k(N)=\binom{N}{k}p^kq^{N-k}.$$

For since the outcomes of the experiments are independent of each other, the probability of a particular sequence of (k) A's and $(N-k)$ B's is p^kq^{N-k}. Since there are exactly $\binom{N}{k}$ arrangements of (k) A's and $(N-k)$ B's, this proves the result. We make the following definition.

Definition 11.3. The probabilities

$$b_k(N) = \binom{N}{k} p^k q^{N-k} \tag{11.7}$$

are called the *binomial* distribution. We call $b_k(N)$ the probability of k successes in N independent trials, where p is the probability of success in one trial, and $q = 1 - p$ it the probability of failure.

The sum of the probabilities of all possible outcomes is

$$\sum_{k=0}^{N} \binom{N}{k} p^k q^{N-k} = \sum_{k=0}^{N} \binom{N}{k} p^k (1-p)^{N-k}.$$

According to the binomial theorem, this sum is equal to $(p+1-p)^N = 1^N = 1$.

Example 11.5. Suppose a fair coin is tossed 10 times, and x is the number of heads. What is the probability of getting 7 heads and 3 tails? "Fair" means that the probabilities of heads or tails on one toss are each $\frac{1}{2}$. We take $N = 10$. There are $\binom{10}{7} = 120$ ways for 7 heads to turn up out of 10 tosses. The probability of each is $\left(\frac{1}{2}\right)^{10}$. Therefore,

$$p(x=7) = \binom{10}{7} \left(\frac{1}{2}\right)^7 \left(\frac{1}{2}\right)^3$$

$$= \frac{(10)(9)(8)}{3!} \left(\frac{1}{2}\right)^{10} = (120)\frac{1}{1024} = 0.1171875$$

We calculate now the expected value of the number occurrences of outcome A when there are N independent trials. Let $(x = k)$ be the numerical outcome that there are exactly k occurrences of A, and let $p(x = k)$ be its probability. We saw above that $p(x = k) = \binom{N}{k} p^k (1-p)^{N-k}$ for each possible value of $x = 0, 1, 2, \ldots, k, \ldots, N$. By definition of expectation,

$$E(x) = \sum_{k=0}^{N} kp(x=k) = \sum_{k=0}^{N} k \binom{N}{k} p^k q^{N-k} = \sum_{k=1}^{N} k \binom{N}{k} p^k q^{N-k}.$$

Using the formula for the binomial coefficients, we can write the formula for the expected value as

$$E(x) = \sum_{k=1}^{N} \frac{kN!}{k!(N-k)!} p^k q^{N-k} = Np \sum_{k=1}^{N} \frac{(N-1)!}{(k-1)!(N-k)!} p^{k-1} q^{N-k}.$$

Using the binomial theorem, we can rewrite the last sum as

$$Np(p+q)^{N-1} = Np.$$

Thus we have proved that for the binomial distribution, the expected number of successes, $E(x)$, is Np.

Note that since the probability of the outcome A in a single trial is p, it is reasonable to expect the outcome A to occur Np times after the experiment has been performed N times.

The Poisson Distribution. Suppose you know that each week, a large number of vehicles pass through a busy intersection and there are on average u accidents. Let us assume that the probability of a vehicle having an accident is independent of the occurrence of previous accidents. We use a binomial distribution to determine the probability of k accidents in a week:

$$b_k(N) = \binom{N}{k} p^k(1-p)^{N-k} = \frac{N(N-1)\dots(N-k+1)}{k!} p^k(1-p)^{N-k}$$

$$= \left(1-\frac{1}{N}\right)\cdots\left(1-\frac{k-1}{N}\right)\frac{N^k p^k(1-p)^{N-k}}{k!}.$$

Setting $p = \frac{u}{N}$, we can rewrite this as

$$= \left[\frac{\left(1-\frac{1}{N}\right)\cdots\left(1-\frac{k-1}{N}\right)}{(1-p)^k}\right]\frac{u^k}{k!}\left(1-\frac{u}{N}\right)^N.$$

As N tends to infinity, p tends to 0, and the factor in brackets tends to 1, because both the numerator and the denominator tend to 1. As shown in Sect. 1.4, the third factor tends to e^{-u}. Therefore,

$$\lim_{N\to\infty,\ u=Np} b_k(N) = \frac{u^k}{k!}\mathrm{e}^{-u}.$$

This gives us an estimate for $b_k(N)$ when N is large, p is small, and $Np = u$.

Definition 11.4. The *Poisson* distribution is the set of probabilities

$$p_k(u) = \frac{u^k}{k!}\mathrm{e}^{-u}, \tag{11.8}$$

where u is a parameter. The number p_k is the probability of k favorable outcomes, $k = 0, 1, 2, \dots$.

Note that the sum of the $p_k(u)$ equals 1:

$$\sum_{k=0}^{\infty} p_k(u) = \mathrm{e}^{-u} \sum_{k=0}^{\infty} \frac{u^k}{k!} = \mathrm{e}^{-u}\mathrm{e}^{u} = 1.$$

Here we have used the expression of the exponential function given by its Taylor series. The Poisson process is an example of discrete probability with infinitely many possible outcomes.

Next, we show that the combination of two Poisson processes is again a Poisson process. Denote by $p_k(u)$ and $p_k(v)$ the probability of k favorable outcomes in these processes, where p_k is given by formula (11.8). We claim that the probability of k favorable outcomes when both experiments are performed is $p_k(u+v)$, assuming that the experiments are independent.

Proof. There will be k favorable outcomes for the combined experiment if the first experiment has j favorable outcomes and the second experiment has $(k-j)$. If the experiments are independent, the probability of such a combined outcome is the product of the probabilities,

$$p_j(u)p_{k-j}(v),$$

so the probability of the combined experiment to have k favorable outcomes is the sum

$$\sum_j p_j(u)p_{k-j}(v) = \sum_j \frac{u^j}{j!}\mathrm{e}^{-u}\frac{v^{k-j}}{(k-j)!}\mathrm{e}^{-v}.$$

We rewrite this sum as

$$\frac{1}{k!}\mathrm{e}^{-(u+v)} \sum_j \frac{k!}{j!(k-j)!} u^j v^{k-j}.$$

The sum in this formula is the binomial expression for $(u+v)^k$. Therefore, the probability of k favorable outcomes for the combined experiment is

$$\frac{1}{k!}(u+v)^k e^{-(u+v)},$$

which is the Poisson distribution $p_k(u+v)$. □

Problems

11.1. Calculate the variance of the outcome when one die is rolled.

11.2. Find the probability of getting exactly three heads in six independent tosses of a fair coin.

11.3. Let E be an event consisting of a certain collection of outcomes of an experiment. We may call these outcomes *favorable* from the point of view of the event that interests us. The collection of all unfavorable outcomes, i.e., those that do not belong to E, is called the event *complementary* to E. Denote by E' the complementary event. Prove that

$$p(E) + p(E') = 1.$$

11.4. We have said that the probability of the outcome (j,k) of the combination of two independent experiments is $p_j q_k$ when the outcomes of one experiment have probabilities p_j and the other q_k, where $j = 1, 2, \ldots, n$ and $k = 1, 2, \ldots, m$. Show that the sum of all these probabilities is 1.

11.5. An event E is *included* in the event F if whenever E takes place, F also takes place. Another way of expressing this relationship is to say that the outcomes that constitute E form a *subset* of the outcomes that make up F. The assertion "the event E is included in the event F" is expressed in symbols by $E \subset F$. For example, the event E of drawing a spade is included in the event F of drawing a black card.

Show that if $E \subset F$, then

$$p(E) \le p(F).$$

11.6. Verify the probabilities shown for two independent dice in Table 11.1.

11.7. Let $E_1, E_2, \ldots, E_m$ be a collection of m events that are *disjoint* in the sense that no outcome can belong to more than one event. Denote the union of the events E_j by

$$E = E_1 \cup E_2 \cup \cdots \cup E_m.$$

Show that the *additive* rule holds:

$$p(E) = p(E_1) + \cdots + p(E_m).$$

11.8. Show that the variance of the number of successes in the binomial distribution is $Np(1-p)$.

11.9. Let x represent the number of successes in a Poisson distribution (11.8). Show that the expected value

$$E(x) = \sum_{k=0}^{\infty} k p_k(u)$$

is equal to u.

11.2 Information Theory: How Interesting Is Interesting?

It is a universal human experience that some information is dull, some interesting. Man bites dog is news, dog bites man is not. In this section, we describe a way of assigning a quantitative measure to the value of a piece of information.

"Interesting," in this discussion, shall mean the degree of surprise at being informed that a certain event E, whose occurrence is subject to chance, has occurred.

An event is a collection of possible outcomes of an experiment. The frequency with which the event E occurs in a large number of performances of the experiment is its probability $p(E)$. We assume in this theory that the information gained on learning that an event has occurred depends only on the probability p of the event. We denote by $f(p)$ the information thus gained. In other words, we could think of $f(p)$ as a measure of the element of surprise generated by the event that has occurred.

What properties does this function f have? We claim that the following four are mandatory:

(a) $f(p)$ increases as p decreases.
(b) $f(1) = 0$.
(c) $f(p)$ tends to infinity as p tends to 0.
(d) $f(pq) = f(p) + f(q)$.

Property (a) expresses the fact that the occurrence of a less probable event is more surprising than the occurrence of a more probable one and therefore carries more information. Property (b) says that the occurrence of an event that is a near certainty imparts almost no new information, while property (c) says that the occurrence of a rare event is of great interest and furnishes a great deal of new information.

Property (d) expresses a property of independent events. Suppose that two events E and F are *independent*. Since such events are totally unrelated, being informed that both of them have occurred conveys no more information than learning that each has occurred separately, i.e., the information gained on learning that both have occurred is the *sum* of the information gained by learning of each occurrence separately. Denote by p and q the probabilities of events E and F, respectively. According to the product rule (11.3), the probability of the combined event $E \cap F$ is the product pq. It is not hard to show that the only continuous function that satisfies property (d) is a constant multiple of $\log p$. So we conclude that

$$f(p) = k \log p. \tag{11.9}$$

What is the value of this constant? According to property (a), $f(p)$ increases with decreasing p. Since $\log p$ increases with increasing p, we conclude that the constant must be *negative*. What about its magnitude? There is no way of deciding that without first adopting an arbitrary unit of information. For convenience, we choose the constant to be -1, and so define

$$f(p) = -\log p.$$

We ask you to verify properties (b) and (c) in Problem 11.10.

Now consider an experiment with n possible outcomes having probabilities $p_1, p_2, \ldots, p_n$. If in a single performance of the experiment, the jth outcome occurs, we have gained information in the amount $-\log p_j$. We now ask the following question: If we perform the experiment repeatedly many times, what is the *average information gain*? The answer to this question is contained in formula (11.4) concerning the average numerical outcome of a series of experiments. According to that formula, if the jth numerical outcome is x_j, and the average numerical outcome is

$p_1x_1 + \cdots + p_nx_n$. In our case, the numerical outcome, the information gained in the jth outcome, is

$$x_j = -\log p_j.$$

So the average information gain I is $I = -(p_1 \log p_1 + p_2 \log p_2 + \cdots + p_n \log p_n)$. To indicate the dependence of I on the probabilities, we write

$$I = I(p_1, \ldots, p_n) = -p_1 \log p_1 - p_2 \log p_2 - \cdots - p_n \log p_n. \tag{11.10}$$

This definition of information is due to the physicist Léo Szilárd. It was introduced in the mathematical literature by Claude Shannon.

Let us look at the simplest case that there are only two possible outcomes, with probabilities p and $1-p$. We can write the formula for information gain as follows:

$$I = -p \log p + (p-1) \log(1-p).$$

How does I depend on p? To study how I changes with p, we use the methods of calculus: we differentiate I with respect to p and get

$$\frac{\mathrm{d}I}{\mathrm{d}p} = -\log p - 1 + \log(1-p) + 1 = -\log p + \log(1-p).$$

Using the functional equation of the logarithm function, we can rewrite this as

$$\frac{\mathrm{d}I}{\mathrm{d}p} = \log\left(\frac{1-p}{p}\right).$$

We know that $\log x$ is positive for $x > 1$ and negative for $x < 1$. Also,

$$\frac{(1-p)}{p} \begin{cases} > 1 \text{ for } 0 < p < \frac{1}{2}, \\ < 1 \text{ for } \frac{1}{2} < p < 1. \end{cases}$$

Therefore,

$$\frac{\mathrm{d}I}{\mathrm{d}p} \begin{cases} > 0 \text{ for } 0 < p < \frac{1}{2}, \\ < 0 \text{ for } \frac{1}{2} < p < 1. \end{cases}$$

It follows that $I(p)$ is an increasing function of p from 0 to $\frac{1}{2}$, and a decreasing function as p goes from $\frac{1}{2}$ to 1. Therefore, *the largest value of I occurs when $p = \frac{1}{2}$*. In words: the most information that can be gained on average from an experiment with two possible outcomes occurs when the probabilities of the two outcomes are equal.

We now extend this result to experiments with n possible outcomes.

Theorem 11.1. *The function*

$$I(p_1,\ldots,p_n) = -p_1\log p_1 - p_2\log p_2 - \cdots - p_n\log p_n,$$

defined for positive numbers with $p_1+p_2+\cdots+p_n = 1$, *is largest when*

$$p_1 = p_2 = \cdots = p_n = \frac{1}{n}.$$

Proof. We have to show that

$$I(p_1,\ldots,p_n) < I\left(\frac{1}{n},\ldots,\frac{1}{n}\right)$$

unless all the p_j are equal to $\frac{1}{n}$. In order to apply the methods of calculus to proving this inequality, we consider the following functions $r_j(s)$:

$$r_j(s) = sp_j + (1-s)\frac{1}{n}, \qquad j = 1,\ldots,n.$$

These functions are designed so that at $s = 0$, the value of each r_j is $\frac{1}{n}$, and at $s = 1$, the value of r_j is p_j:

$$r_j(0) = \frac{1}{n}, \qquad r_j(1) = p_j, \qquad j = 1,\ldots,n.$$

So, if we define the function $J(s) = I(r_1(s),\ldots,r_n(s))$, then

$$J(0) = I\left(\frac{1}{n},\ldots,\frac{1}{n}\right), \qquad J(1) = I(p_1,\ldots,p_n).$$

Therefore, the inequality to be proved can be expressed simply as $J(1) < J(0)$. We shall prove this by showing that $J(s)$ is a decreasing function of s. We use the monotonicity criterion to demonstrate the decreasing character of $J(s)$, by verifying that its derivative is negative. To calculate the derivative of $J(s)$, we need to know the derivative of each r_j with respect to s. This is easily calculated:

$$\frac{\mathrm{d}r_j(s)}{\mathrm{d}s} = p_j - \frac{1}{n}. \tag{11.11}$$

Note that the derivative of each r_j is constant, since each r_j is a linear function of s. Using the definition of I, we have

$$J(s) = -r_1\log r_1 - \cdots - r_n\log r_n.$$

We calculate the derivative of J using the chain rule and Eq. (11.11):

$$\begin{aligned}\frac{\mathrm{d}J}{\mathrm{d}s} &= -(1+\log r_1)\frac{\mathrm{d}r_1}{\mathrm{d}s} - \cdots - (1+\log r_n)\frac{\mathrm{d}r_n}{\mathrm{d}s} \\ &= -(1+\log r_1)\left(p_1 - \frac{1}{n}\right) - \cdots - (1+\log r_n)\left(p_n - \frac{1}{n}\right). \end{aligned} \tag{11.12}$$

Since $r_j(0) = \dfrac{1}{n}$, we get with $s = 0$ that

$$\frac{\mathrm{d}J}{\mathrm{d}s}(0) = -\left(1+\log\left(\tfrac{1}{n}\right)\right)\left(p_1 - \frac{1}{n}\right) - \cdots - \left(1+\log\left(\tfrac{1}{n}\right)\right)\left(p_n - \frac{1}{n}\right).$$

Since the sum of the p_j is 1, this gives

$$\frac{\mathrm{d}J}{\mathrm{d}s}(0) = -\left(1+\log\left(\tfrac{1}{n}\right)\right)\left(1 - n\frac{1}{n}\right) = 0.$$

Switching to the J' notation, we have $J'(0) = 0$. We claim that for all positive values of s,

$$J'(s) < 0. \tag{11.13}$$

If we can show this, our proof that J is a decreasing function is complete. To verify Eq. (11.13), we shall show that $J'(s)$ itself is a decreasing function of s. Since $J'(0) = 0$, then J' will be negative for all positive s.

To show that J' is decreasing, we apply the monotonicity criterion once more, this time to J', and show that J'' is negative. We compute J'' by differentiating Eq. (11.12) and using Eq. (11.11):

$$\begin{aligned} J'' &= -\frac{1}{r_1}r_1'\left(p_1 - \frac{1}{n}\right) - \cdots - \frac{1}{r_n}r_n'\left(p_n - \frac{1}{n}\right) \\ &= -\frac{1}{r_1}\left(p_1 - \frac{1}{n}\right)^2 - \cdots - \frac{1}{r_n}\left(p_n - \frac{1}{n}\right)^2. \end{aligned}$$

Each term in this sum is negative or zero. Since not all p_j are equal to $\dfrac{1}{n}$, at least some terms are negative. This proves $J'' < 0$, and completes the proof of our theorem. □

Problems

11.10. Verify that the function $f(p) = -k\log p$ has properties (b) and (c) that we listed at the outset of Sect. 11.2.

11.11. Suppose that an experiment has three possible outcomes, with probabilities p, q, and r, with

$$p+q+r=1.$$

Suppose that we simplify the description of our experiment by lumping the last two cases together, i.e., we look on the experiment as having two possible outcomes, one with probability p, the other with probability $1-p$. The average information gain in looking at the full description of the experiment is

$$-p\log p - q\log q - r\log r.$$

In looking at the simplified description, the average information gain is

$$-p\log p - (1-p)\log(1-p).$$

Prove that the average information gain from the full experiment is greater than that obtained from its simplified description. The result is to be expected: if we lump data together, we lose information.

11.12. Let $p_1, \ldots, p_n$ be the probabilities of the n possible outcomes of an experiment, and $q_1, \ldots, q_m$ the probabilities of the outcomes of another experiment. Suppose that the experiments are *independent*, i.e., if we combine the two experiments, the probability of the first experiment having the jth outcome and the second experiment having the kth outcome is the product

$$r_{jk} = p_j q_k.$$

Show that in this case, the average information gain from the combined experiment is the *sum* of the average information gains in the performance of each experiment separately:

$$I(r_{11},\ldots,r_{mn}) = I(p_1,\ldots,p_n) + I(q_1,\ldots,q_m).$$

11.13. Suppose an experiment can have n possible outcomes, the jth having probability p_j, $j=1,\ldots,n$. The information gained from this experiment is on average

$$-p_1\log p_1 - \cdots - p_n\log p_n.$$

Suppose we simplify the description of the experiment by lumping the last $n-1$ outcomes together as failures of the first case. The average information gain from this description is

$$-p_1\log p_1 - (1-p_1)\log(1-p_1).$$

Prove that we gain on average more information from the full description than from the simplified description.

11.3 Continuous Probability

The probability theory developed in Sect. 11.1 deals with experiments that have finitely many possible numerical outcomes. This is a good model for experiments such as tossing a coin (labeling the numerical outcome 0 or 1) or throwing a die, but it is artificial for experiments such as making a physical measurement with an apparatus subject to random disturbances that can be reduced but not totally eliminated. *Every real number is a possible numerical outcome* of such an experiment. This section is devoted to developing a probability theory for such situations. The experiments we study are, just like the previous ones, repeatable and nondeterministic but predictable on average.

By "predictable on average" we mean this: Repeat the experiment as many times as we wish and denote by $S(x)$ the number of instances among the first N performances for which the numerical outcome was less than x. Then the frequency $\frac{S(x)}{N}$ with which this event occurs tends to a limit as N tends to infinity. This limit is the *probability that the outcome is less than x*, and is denoted by $P(x)$:

$$P(x) = \lim_{N\to\infty} \frac{S(x)}{N}.$$

The probability $P(x)$ has the following properties:

(i) Each probability lies between 0 and 1:

$$0 \le P(x) \le 1.$$

(ii) $P(x)$ is a nondecreasing function of x.

Properties (i) and (ii) are consequences of the definition, for the number $S(x)$ lies between 0 and N, so that the ratio $\frac{S(x)}{N}$ lies between 0 and 1; but then so does the limit $P(x)$. Secondly, $S(x)$ is a nondecreasing function of x, so that the ratio $\frac{S(x)}{N}$ is a nondecreasing function of x; then so is the limit $P(x)$. We shall assume two further properties of $P(x)$:

(iii) $P(x)$ tends to 0 as x tends to minus infinity.
(iv) $P(x)$ tends to 1 as x tends to infinity.

Property (iii) says that the probability of a very large negative outcome is very small. Property (iv) implies that very large positive outcomes are very improbable, as we ask you to explain in Problem 11.14. As in Sect. 11.1, we shall be interested in collections of outcomes, which we call *events*.

Example 11.6. Examples of events are:

(a) The outcome is less than x.

(b) The outcome lies in the interval I.
(c) The outcome lies in a given collection of intervals.

The probability of an event E, which we denote by $P(E)$, is defined as in Sect. 11.1, as the limit of the frequencies:

$$\lim_{N\to\infty} \frac{S(E)}{N} = P(E),$$

where $S(E)$ the number of times the event E took place among the first N of an infinite sequence of performances of an experiment. The argument presented in Sect. 11.1 can be used in the present context to show the additive rules for disjoint events: Suppose E and F are two events that have probabilities $P(E)$ and $P(F)$, and suppose that they are *disjoint* in the sense that one event precludes the other. That is, no outcome can belong to both E and F. In this case, the union $E \cup F$ of the events, consisting of all outcomes either in E or in F, also has a probability that is the sum of the probabilities of E and F:

$$P(E \cup F) = P(E) + P(F).$$

We apply this to the events

$$E: \text{ the outcome } x < a$$

and

$$F: \text{ the outcome } a \le x < b.$$

The union of these two is

$$E \cup F: \text{ the outcome } x < b.$$

Then

$$P(E) = P(a), \quad P(E \cup F) = P(b).$$

We conclude that

$$P(F) = P(b) - P(a)$$

is the probability of an outcome less than b but greater than or equal to a.

We now make the following assumption:

(v) $P(x)$ is a continuously differentiable function.

This assumption holds in many important cases and allows us to use the methods of calculus. We denote the derivative of P by p:

$$\frac{\mathrm{d}P(x)}{\mathrm{d}x} = p(x)$$

The function $p(x)$ is called the *probability density*. According to the mean value theorem, for every a and b, there is a number c lying between a and b such that

$$P(b) - P(a) = p(c)(b-a). \tag{11.14}$$

According to the fundamental theorem of calculus,

$$P(b) - P(a) = \int_a^b p(x)\,dx. \tag{11.15}$$

Since by assumption (iii), $P(a)$ tends to 0 as a tends to minus infinity, we conclude that

$$P(b) = \int_{-\infty}^{b} p(x)\,dx.$$

Since by assumption (iv), $P(b)$ tends to 1 as b tends to infinity, we conclude that

$$1 = \int_{-\infty}^{\infty} p(x)\,dx.$$

This is the continuous analogue of the basic fact that $p_1 + p_2 + \cdots + p_n = 1$ in discrete probability. According to property (ii), $P(x)$ is a nondecreasing function of x. Since the derivative of a nondecreasing function is nowhere negative, we conclude that $p(x)$ is nonnegative for all x:

$$0 \le p(x).$$

We now define the *expectation*, or mean, $\overline{x}$ of an experiment analogously to the discrete case. Imagine the experiment performed as many times as we wish, and denote the sequence of outcomes by

$$a_1, a_2, \ldots, a_N, \ldots.$$

Theorem 11.2. *If an experiment is predictable on average, and if the outcomes are restricted to lie in a finite interval, then*

$$\overline{x} = \lim_{N\to\infty} \frac{a_1 + \cdots + a_N}{N}$$

exists and is equal to

$$\overline{x} = \int_{-\infty}^{\infty} x p(x)\,dx. \tag{11.16}$$

The assumption that the outcomes lie in a finite interval is a realistic one if one thinks of the experiment as a measurement. After all, every measuring apparatus has a finite range. However, there are probability densities of great theoretical interest, such as the ones we shall discuss in Sect. 11.4, that are positive for all real x. Theorem 11.2 remains true for these experiments, too, under the additional assumption that the improper integral defining $\overline{x}$ exists.

Proof. Divide the interval I in which all outcomes lie into n subintervals $I_1, \dots, I_n$. Denote the endpoints by

$$e_0 < e_1 < \cdots < e_n.$$

The probability P_j of an outcome lying in interval I_j is the difference of the values of P at the endpoints of I_j. According to formula (11.14), this difference is equal to

$$P_j = P(e_j) - P(e_{j-1}) = p(x_j)(e_j - e_{j-1}), \tag{11.17}$$

where x_j is a point in I_j guaranteed by the mean value theorem, and $(e_j - e_{j-1})$ denotes the length of I_j. We now simplify the original experiment by recording merely the intervals I_j in which the outcome falls, and calling the numerical outcome in this case x_j, the point in I_j that appears in formula (11.17). The actual outcome of the full experiment and the numerical outcome of the simplified experiment always belong to the same subinterval of the subdivision we have taken. Therefore, *these two outcomes differ by at most w, the length of largest of the subintervals I_j.*

Now consider the sequence of outcomes a_1, $a_2, \dots$ of the original experiment. Denote the corresponding outcomes of the simplified experiment by b_1, $b_2, \dots$. The simplified experiment has a finite number of outcomes. For such discrete experiments, we have shown in Sect. 11.1 that the average of the numerical outcomes tends to a limit, called the expectation. We denote it by $\overline{x}_n$:

$$\overline{x}_n = \lim_{N\to\infty} \frac{b_1 + \cdots + b_N}{N}, \tag{11.18}$$

where n is the number of subintervals of I. The expectation $\overline{x}_n$ of the simplified experiment can be calculated by formula (11.4):

$$\overline{x}_n = P_1 x_1 + \cdots + P_n x_n. \tag{11.19}$$

By Eq. (11.17), this is

$$= p(x_1)x_1(e_1 - e_0) + \cdots + p(x_n)x_n(e_n - e_{n-1}).$$

We recognize this as an approximating sum for the integral of $xp(x)$ over I. If the subdivision is fine enough, *the approximating sum $\overline{x}_n$ differs very little from the value of the integral*

$$\int_{e_0}^{e_n} xp(x)\,\mathrm{d}x. \tag{11.20}$$

We recall that the outcomes of the simplified experiment and the full experiment differ by less than w, the length of the largest subinterval I_j. Therefore, the expectation of the simplified experiment tends to the expectation of the full experiment as the lengths of the subintervals tend to zero. This proves that the expectation of the full experiment is given by the integral (11.20). Since $p(x)$ is zero outside the interval I, the integrals (11.20) and (11.16) are equal. This concludes the proof of Theorem 11.2. □

We now give some examples of expectation.

Example 11.7. Let A be a positive number, and define $p(x)$ by

$$p(x) = \begin{cases} 0 & \text{for } x < 0, \\ 1/A & \text{for } 0 \le x < A, \\ 0 & \text{for } A \le x. \end{cases}$$

This is intended to mean that the numerical outcome x is equally likely to occur anywhere in $[0, A]$. This choice of p satisfies $\int_{-\infty}^{\infty} p(x)\,\mathrm{d}x = \int_0^A \frac{\mathrm{d}x}{A} = 1$. We now compute the expected value

$$\overline{x} = \int_{-\infty}^{\infty} xp(x)\,\mathrm{d}x = \int_0^A \frac{x}{A}\,\mathrm{d}x = \left[\frac{x^2}{2A}\right]_0^A = \frac{A}{2}.$$

Example 11.8. Let A be a positive number and set

$$p(x) = \begin{cases} 0 & \text{for } x < 0, \\ A\mathrm{e}^{-Ax} & \text{for } 0 \le x. \end{cases}$$

Let us check that p satisfies $\int_{-\infty}^{\infty} p(x)\,\mathrm{d}x = 1$. Using the fundamental theorem of calculus, we have

$$\int_{-\infty}^{\infty} p(x)\,\mathrm{d}x = \int_0^{\infty} A\mathrm{e}^{-Ax}\,\mathrm{d}x = -\mathrm{e}^{-Ax}\big|_0^{\infty} = 1.$$

We now compute $\overline{x}$. Using integration by parts and then the fundamental theorem, we have

$$\overline{x} = \int_{-\infty}^{\infty} xp(x)\,\mathrm{d}x = \int_0^{\infty} xA\mathrm{e}^{-Ax}\,\mathrm{d}x = \int_0^{\infty} \mathrm{e}^{-Ax}\,\mathrm{d}x = \left[\frac{-\mathrm{e}^{-Ax}}{A}\right]_0^{\infty} = \frac{1}{A}.$$

Example 11.9. Assume that $p(x)$ is an even function. Then $xp(x)$ is an odd function, and so

$$\overline{x} = \int_{-\infty}^{\infty} xp(x)\,\mathrm{d}x = 0.$$

Let $f(x)$ be any function of x. We define the *expected value of* f with respect to the probability density $p(x)$ as

$$\overline{f} = \int_{-\infty}^{\infty} f(x)p(x)\,\mathrm{d}x.$$

One can show, analogously to the foregoing discussion, that if $a_1,\ldots,a_N,\ldots$ is a sequence of outcomes, then

$$\lim_{N\to\infty}\frac{f(a_1)+\cdots+f(a_N)}{N}=\overline{f}.$$

Independence. We now turn to the important concept of *independence*. The intuitive notion is the same as in the discrete models discussed in Sect. 11.1: two experiments are independent if the outcome of either has no influence on the other, nor are they both influenced by a common cause. We analyze the consequences of independence the same way we did previously, by constructing a *combined experiment* consisting of performing both experiments.

We first analyze the case that the outcome of the first experiment may be any real number, but the second experiment can have only a finite number of outcomes. As before, we denote by $P(a)$ the probability that the numerical outcome of the first experiment is less than a. The second experiment has n possible numerical outcomes $a_1,\ldots,a_n$, which occur with probabilities Q_1, Q_2, $\ldots$, Q_n. We define the *numerical outcome* of the combined experiment to be the *sum* of the separate numerical outcomes of the two experiments that constitute it.

We now derive a useful and important formula for the probability that the numerical outcome of the combined experiment is less than x. We denote this event by $E(x)$, and denote its probability by $U(x)$. We shall show that

$$U(x)=Q_1P(x-a_1)+\cdots+Q_nP(x-a_n). \tag{11.21}$$

Proof. The numerical outcome of the second experiment is one of the n numbers a_j. The numerical outcome of the combined experiment is then less than x if and only if the outcome of the first experiment is less than $x-a_j$. We denote this event by $E_j(x)$. Thus the event $E(x)$ is the union

$$E(x)=E_1(x)\cup\cdots\cup E_n(x).$$

The events $E_j(x)$ are disjoint, that is, an outcome cannot belong to two distinct events $E_j(x)$ and $E_k(x)$. It follows then from the addition rule for disjoint events that the probability of their union $E(x)$ is the sum of the probabilities of the events $E_j(x)$.

Since the two experiments are independent, the probability of $E_j(x)$ is given by the product of the probabilities of the two experiments,

$$Q_jP(x-a_j).$$

The sum of the probabilities of the $E_j(x)$ is $U(x)$, the probability of $E(x)$. This completes the proof of Eq. (11.21). □

We now turn to the situation in which both experiments can have any real number as outcome. We denote by $P(a)$ and $Q(a)$ the probabilities that the outcome is less than a in each of the two experiments, respectively.

We shall prove the following analogue of formula (11.21): Suppose that $Q(x)$ is continuously differentiable, and denote its derivative by $q(x)$. Then $U(x)$, the probability that the outcome of the combined experiment is less than x, is given by

$$U(x) = \int_{-\infty}^{\infty} q(a)P(x-a)\,da. \tag{11.22}$$

The proof deduces Eq. (11.22) from Eq. (11.21). We assume that the outcome of the second experiment always lies in some finite interval I. We subdivide I into a finite number n of subintervals $I_j = [e_{j-1}, e_j]$. Let us denote by Q_j the probability that the outcome of the experiment Q lies in I_j. According to the mean value theorem,

$$Q_j = Q(e_j) - Q(e_{j-1}) = q(a_j)(e_j - e_{j-1}), \tag{11.23}$$

where a_j is some point in I_j.

We *discretize* the second experiment by lumping together all outcomes that lie in the interval I_j and *redefine* the numerical outcome in that case to be a_j, the number guaranteed to exist by the mean value theorem in Eq. (11.23). The probability of the outcome a_j is then the probability that the outcome lies in the interval I_j, i.e., it is Q_j.

Substitute for each Q_j the expressions given in Eq. (11.23). According to formula (11.21), the probability that the outcome of the discretized experiment is less than x is

$$U_n(x) = q(a_1)P(x-a_1)(e_1 - e_0) + \cdots + q(a_n)P(x-a_n)(e_n - e_{n-1}).$$

The sum on the right is an approximating sum for the integral

$$\int_{-\infty}^{\infty} q(a)P(x-a)\,da.$$

This function was denoted by $U(x)$ in formula (11.22). Since approximating sums tend to the integral as the subdivision is made finer and finer, we conclude that for every x, $U_n(x)$ tends to $U(x)$. This proves our contention.

Now suppose that $P(x)$ is continuously differentiable, and denote its derivative by $p(x)$. It follows from Theorem 7.8 that $U(x)$ as defined by Eq. (11.22) is differentiable, and its derivative, which we denote by $u(x)$, can be obtained by differentiating the integrand with respect to x:

$$u(x) = \int_{-\infty}^{\infty} q(a)p(x-a)\,da. \tag{11.24}$$

We summarize what we have proved:

Theorem 11.3. *Consider two experiments whose outcomes lie in some finite interval and have probability densities p and q respectively. Suppose the experiments are independent. In the combined experiment consisting of performing both experiments, define the outcome of the combined experiment to be the sum of the outcomes of the individual experiments. Then the combined experiment has probability density* $u(x)$ *given by* $u(x) = \int_{-\infty}^{\infty} q(a)p(x-a)\,\mathrm{d}a$.

The restriction of the outcomes of the experiments to a finite interval is too confining for many important applications. Fortunately, the theorem, although not our proof, holds under more general conditions.

Definition 11.5. The function u defined by $u(x) = \int_{-\infty}^{\infty} q(a)p(x-a)\,\mathrm{d}a$ is called the *convolution* of the functions q and p. This relation is denoted by

$$u = q * p. \tag{11.25}$$

Example 11.10. Consider the following example of evaluating the convolution of two functions, where A and B are positive numbers.

$$p(a) = \begin{cases} 0 & \text{for } a < 0, \\ \mathrm{e}^{-Aa} & \text{for } 0 \le a, \end{cases} \qquad q(a) = \begin{cases} 0 & \text{for } a < 0, \\ \mathrm{e}^{-Ba} & \text{for } 0 \le a. \end{cases}$$

Substitute these definitions of p and q into the definition of the convolution:

$$u(x) = (p * q)(x) = \int_{-\infty}^{\infty} p(a)q(x-a)\,\mathrm{d}a.$$

Both $p(t)$ and $q(t)$ were defined to be zero for $t < 0$. It follows from this that the first factor in the integrand, $p(a)$, is zero for a negative. If $x < 0$, the second factor, $q(x-a)$, is zero for a positive. So for x negative, the integrand is zero for all values of a, and therefore so is the integral. This shows that $u(x) = 0$ for $x < 0$. For $x > 0$, the same analysis shows that the integrand is nonzero only in the range $0 \le a \le x$. So for $x > 0$,

$$u(x) = \int_0^x \mathrm{e}^{-Aa-B(x-a)}\,\mathrm{d}a$$

$$= \mathrm{e}^{-Bx}\int_0^x \mathrm{e}^{(B-A)a}\,\mathrm{d}a = \left[\mathrm{e}^{-Bx}\frac{\mathrm{e}^{(B-A)a}}{B-A}\right]_{a=0}^{x} = \frac{1}{B-A}(\mathrm{e}^{-Ax} - \mathrm{e}^{-Bx}).$$

Convolution is an important operation among functions, with many uses. We now state and prove some of its basic properties without any reference to probability.

Theorem 11.4. *Let $q_1(x)$, $q_2(x)$, and $p(x)$ be continuous functions defined for all real numbers x, and assume that the functions are zero outside a finite interval.*

(a) Convolution is distributive: $(q_1+q_2)*p = q_1*p+q_2*p$.
(b) Let k be any constant. Then $(kq)*p = k(q*p)$.
(c) Convolution is commutative: $q*p = p*q$.

Proof. The first result follows from the additivity of integrals:

$$(q_1+q_2)*p(x) = \int_{-\infty}^{\infty} (q_1(a)+q_2(a))\,p(x-a)\,\mathrm{d}a$$

$$= \int_{-\infty}^{\infty} q_1(a)p(x-a)\,\mathrm{d}a + \int_{-\infty}^{\infty} q_2(a)p(x-a)\,\mathrm{d}a = q_1*p(x)+q_2*p(x).$$

The second result follows from

$$(kq)*p(x) = \int_{-\infty}^{\infty} kq(a)p(x-a)\,\mathrm{d}a = k\int_{-\infty}^{\infty} q(a)p(x-a)\,\mathrm{d}a = k(q*p)(x).$$

The third result follows if we make the change of variable $b = x-a$:

$$q*p(x) = \int_{-\infty}^{\infty} q(a)p(x-a)\,\mathrm{d}a = \int_{-\infty}^{\infty} q(x-b)p(b)\,\mathrm{d}b = p*q(x).$$

□

The following result is another basic property of convolution.

Theorem 11.5. *Suppose p and q are continuous functions, both zero outside some finite interval. Denote their convolution by u:*

$$u = p*q.$$

Then the integral of the convolution is the product of the integrals of the factors:

$$\int_{-\infty}^{\infty} u(x)\,\mathrm{d}x = \int_{-\infty}^{\infty} p(x)\,\mathrm{d}x \int_{-\infty}^{\infty} q(a)\,\mathrm{d}a. \tag{11.26}$$

Proof. By definition of the convolution $u = p*q$,

$$u(x) = \int_{-\infty}^{\infty} p(x-a)q(a)\,\mathrm{d}a. \tag{11.27}$$

Suppose that the function $p(a)$ is zero outside the interval $I = [-b,b]$, so that $\int_{-\infty}^{\infty} p(x)\,\mathrm{d}x = \int_{-b}^{b} p(x)\,\mathrm{d}x$, and $q(a)$ is zero outside the interval J. It follows that $u(x)$ is zero when x lies outside the interval $I \cup J$.

Approximate the integral (11.27) by the sum

$$u_n(x) = \sum_{j=1}^{n} p(x - a_j)q(a_j)(a_{j+1} - a_j), \tag{11.28}$$

where the numbers $a_1, \ldots, a_n$ are n equally spaced points in the interval J of integration. It follows from the definition of integral as the limit of approximate sums that $u_n(x)$ tends to $u(x)$, uniformly for all x in the interval $I \cup J$. It follows that the integral of $u_n(x)$ with respect to x over $I \cup J$ tends to the integral of $u(x)$. It follows from formula (11.28) that the integral of $u_n(x)$ over $I \cup J$ is

$$\int_{-b}^{b} p(x)\,\mathrm{d}x \sum_{j=1}^{n} q(a_j)(a_{j+1} - a_j).$$

The limit of this sum as n tends to infinity is the integral of q. This concludes the proof of Eq. (11.26) in Theorem 11.5. □

The numerical outcome of the combination of two experiments was defined as the *sum* of the numerical outcomes of its two constituents. We now give some realistic examples to illustrate why this definition is of interest.

Suppose the outcomes of the two experiments represent *income* from two entirely different sources. Their sum is then the total income; its probability distribution is of considerable interest.

Here is another example: Suppose the two outcomes represent amounts of water entering a reservoir in a given period from two different sources. Their sum represents the total inflow, again an object of considerable interest.

Problems

11.14. We have said that the assumption $P(x)$ tends to 1 as x tends to infinity means that very large positive outcomes x are improbable. Justify that statement.

11.15. Define p by

$$p(x) = \begin{cases} 0 & \text{for } x < 0, \\ \frac{2}{A}\left(1 - \frac{x}{A}\right) & \text{for } 0 \le x \le A, \\ 0 & \text{for } A < x. \end{cases}$$

(a) Show that $\int_{-\infty}^{\infty} p(x)\,dx = 1$.

(b) Calculate the expected value of x, i.e., find $\overline{x} = \int_{-\infty}^{\infty} xp(x)\,dx$.

(c) Calculate the expected value $\overline{x^2} = \int_{-\infty}^{\infty} x^2 p(x)\,dx$.

(d) Give a definition of standard deviation and calculate it for this case.

11.16. Define p by

$$p(x) = k|x|e^{-kx^2}, \quad k > 0.$$

Show that p is a probability density, i.e.,

$$\int_{-\infty}^{\infty} p(x)\,dx = 1.$$

11.17. Let A and B be two positive numbers. Define p and q by

$$p(t) = \begin{cases} 0 & \text{for } t < 0, \\ \frac{1}{A} & \text{for } 0 \le t \le A, \\ 0 & \text{for } A < t, \end{cases} \quad q(t) = \begin{cases} 0 & \text{for } t < 0, \\ \frac{1}{B} & \text{for } 0 \le t \le B, \\ 0 & \text{for } B < t. \end{cases}$$

(a) Show that p and q are probability densities, i.e., that they satisfy

$$\int_{-\infty}^{\infty} p(t)\,dt = 1, \qquad \int_{-\infty}^{\infty} q(t)\,dt = 1.$$

(b) Let u denote the convolution of p and q. Show that $u(x) = 0$ for $x < 0$ and for $x > A + B$.

(c) Verify that $u(x)$ is constant if $B < x < A$.

(d) Determine all values of $u(x)$ for the case $B < A$.

11.18. The purpose of this problem is to give an alternative proof of Theorem 11.5. Let p and q be a pair of functions, both zero outside some finite interval J. Let u be the convolution of p and q.

(a) Let h be a small number. Show that the sum

$$\sum_i p(ih)q(x - ih)h \tag{11.29}$$

is an approximating sum to the integral defining $u(x)$.

(b) Show that

$$\sum_j u(jh)h \tag{11.30}$$

is an approximating sum to the integral $\int_{-\infty}^{\infty} u(x)\,dx$.

(c) Substitute the approximations (11.29) for $u(x)$ into Eq. (11.30) with $x = jh$. Show that the result is the *double sum* $\sum_{i,j} p(ih)q((j-i)h)h^2$.

(d) Denote $j - i$ by ℓ and rewrite the above double sum as $\sum_{i,\ell} p(ih)q(\ell h)h^2$.

(e) Show that this double sum can be written as the product of two single sums:

$$\left(\sum_i p(ih)h\right)\left(\sum_\ell q(\ell h)h\right).$$

(f) Show that the single sums are approximations to the integrals

$$\int_{-\infty}^{\infty} p(x)\,\mathrm{d}x \quad \text{and} \quad \int_{-\infty}^{\infty} q(x)\,\mathrm{d}x.$$

(g) Show that as h tends to zero, you obtain the identity in Theorem 11.5.

11.19. Define

$$|u|_1 = \int_{-\infty}^{\infty} |u(x)|\,\mathrm{d}x$$

as a quantity that measures the size of functions $u(x)$ that are defined for all x and are zero outside a finite interval.

(a) Evaluate $|u|_1$ if $u(x) = 5$ on $[a,b]$, and zero outside of $[a,b]$.
(b) Verify the properties $|cu|_1 = |c||u|_1$ when c is constant, and $|u+v|_1 \le |u|_1 + |v|_1$.
(c) Prove for convolution that $|u*v|_1 \le |u|_1|v|_1$.

11.4 The Law of Errors

In this section, we shall analyze a particular experiment. The experiment consists in dropping pellets from a fixed point at a certain height onto a horizontal plane. If the hand that releases the pellet were perfectly still and if there were no air currents diverting the pellet on its downward path, then we could predict with certainty that the pellet would end up directly below the point where it was released. But even the steadiest hand trembles a little, and even on the stillest day, minute air currents buffet the pellet in its downward flight, in a random fashion. These effects become magnified and very noticeable if the pellets are dropped from a great height, say the tenth floor of a building. Under such circumstances, the experiment appears to be nondeterministic, i.e., it is impossible to predict where each pellet is going to land.[1]

Although it is impossible to predict where any particular pellet would fall, the outcome can be predicted very well on average. That is, let G be any region such as a square, rectangle, triangle, or circle. Denote by $S(G)$ the number of instances

[1] G.I. Taylor (1886–1975), a famous British applied mathematician, described the following experience during the First World War: Taylor was working on a project to develop aerial darts; his task was to record the patterns created when a large number of darts were dropped from an airplane. This he did by putting a piece of paper under each dart where it had fallen in the field. These papers were to be photographed from the air. He had just finished this tedious task when a cavalry officer came by on horseback and demanded to know what Taylor was doing. Taylor explained the dart project, whereupon the officer exclaimed, "And you chaps managed to hit all those bits of paper? Good show!"

among the first N in a sequence of experiments in which the pellet landed in G. Then the frequencies $\frac{S(G)}{N}$ tend to a limit, called the probability of landing in G and denoted by $C(G)$:

$$\lim_{N\to\infty} \frac{S(G)}{N} = C(G).$$

In this section, we shall investigate the nature of this probability.

Suppose that the region G is a very small one. Then we expect the probability of landing in G to be nearly proportional to the area $A(G)$ of G. We can express this surmise more precisely as follows: Let g be any point in the plane. Then there is a number $c = c(g)$, called the *probability density* at g, such that for any region G containing g

$$C(G) = \big(c(g) + \text{small}\big)A(G),$$

where "small" means a quantity that tends to zero as G shrinks to the point g.

What can we say about the probability density $c(g)$? It depends on how close g is to the bullseye, i.e., the point directly underneath where the pellet is released. The closer g is, the greater the probability of a hit near g. In particular, the maximum value of c is achieved when g is the bullseye. We now adopt the following two hypotheses about the way in which the uncontrolled tremors of the hand and the unpredictable gusts of wind influence the distribution of hits and misses:

(i) $c(g)$ depends only on the distance of g from the bullseye, and not on the direction in which g lies.
(ii) Let x and y be perpendicular directions. Displacement of pellets in the x-direction is independent of their displacement in the y-direction.

Example 11.11. A special case illustrating hypothesis (ii) consists of two half-spaces bounded by lines through the origin in perpendicular directions. The probability of the pellet falling in either half-plane is $\frac{1}{2}$. The probability that the pellet falls in the quarter-plane that is the intersection of the two half-planes is $\frac{1}{4}$, and this is equal to $(\frac{1}{2})(\frac{1}{2})$.

To express these hypotheses in a mathematical form, we introduce a Cartesian coordinate system with the origin, naturally, at the bullseye. We denote by (a,d) the coordinates of the point g, as in Fig. 11.1. We denote by $P(a)$ the probability that the pellet falls in the half-plane

$$x < a.$$

The probability that the pellet falls in the strip $a \le x < b$ is

$$P(b) - P(a).$$

Assume that $P(a)$ has a continuous derivative for all a. We denote it by $p(a)$. According to the mean value theorem, the difference

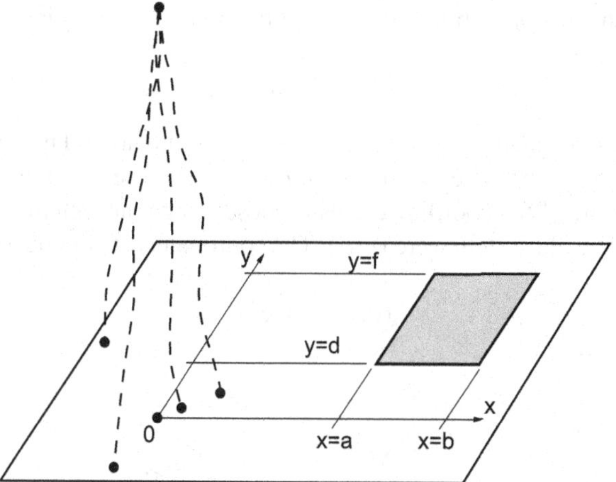

Fig. 11.1 Pellets dropped directly over the origin might land in the *shaded rectangle*

$$P(b) - P(a) \quad \text{is equal to} \quad p(a_1)(b-a),$$

for some number a_1 between a and b.

What is the probability that the pellet falls in the rectangle

$$a \le x < b, \quad d \le y < f?$$

This event occurs when the pellet falls in the strip $a \le x < b$ and the strip $d \le y < f$. According to hypothesis (ii), these two events are independent, and therefore, according to the *product rule*, the probability of the combined event is the product of the probabilities of the two separate events whose simultaneous occurrence constitutes the combined event. Thus the probability of a pellet falling in the rectangle is the product

$$p(a_1)(b-a)p(d_1)(f-d).$$

Since the product $(b-a)(f-d)$ is the area A of the rectangle, we can rewrite this as

$$p(a_1)p(d_1)A.$$

Now consider a sequence of rectangles that tend to the point $g = (a,d)$ by letting b tend to a and f tend to d. Since a_1 lies between a and b and d_1 lies between d and f, and since p is a continuous function, it follows that $p(a_1)$ tends to $p(a)$ and $p(d_1)$ tends to $p(d)$. Thus, in this case, we can express the probability that the pellet lands in the rectangle as

$$\big(p(a)p(d) + \text{small}\big)A.$$

We conclude that the probability density c at the point $g = (a,d)$ is

$$c(g) = p(a)p(d). \tag{11.31}$$

Next, we exploit the symmetry of the experimental setup around the bullseye by introducing another coordinate system, as in Fig. 11.2, whose origin is still the bullseye but where one of the coordinate axes is chosen to go through the point g whose coordinates in the old system were (a,d). The coordinates of g in the new system are

$$\left(0, \sqrt{a^2+d^2}\right).$$

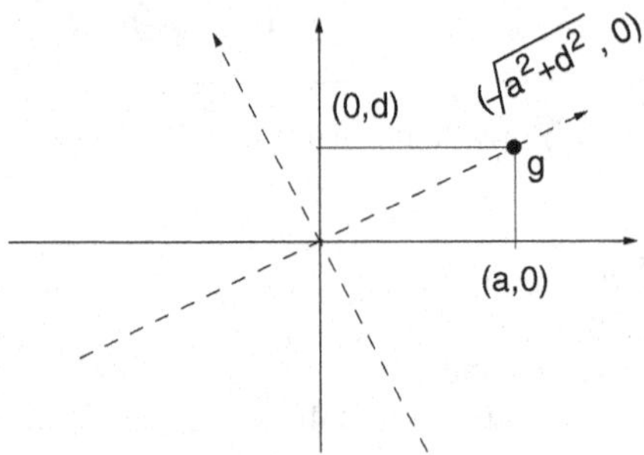

Fig. 11.2 A rotated coordinate system

According to hypothesis (i), we can apply relation (11.31) in any coordinate system. Then c in the new coordinate system is

$$c(g) = p(0)p\left(\sqrt{a^2+d^2}\right). \tag{11.32}$$

Since the value of $c(g)$ expressed in two different coordinate systems as in Eqs. (11.31) and (11.32) are equal, we conclude that

$$p(a)p(d) = p(0)p\left(\sqrt{a^2+d^2}\right). \tag{11.33}$$

This is a functional equation for $p(x)$. It can be solved by the trick of writing the function $p(x)$ in terms of another function $f(x) = \dfrac{p(\sqrt{x})}{p(0)}$. Set $x = a^2$ and $y = d^2$ in Eq. (11.33), which gives

$$f(x)f(y) = f(x+y).$$

This, at last, is the familiar functional equation satisfied by exponential functions and only by them, as explained in Sect. 2.5c. So we conclude that $f(x) = \mathrm{e}^{Kx}$. Using the relation $f(x) = \dfrac{p(\sqrt{x})}{p(0)}$, we deduce that $p(a) = p(0)\mathrm{e}^{Ka^2}$. We claim that the constant K is negative. For as a tends to infinity, the probability density $p(a)$ tends to zero, and this is the case only if K is negative. To put this into evidence, we rename K as $-k$, and rewrite

$$p(x) = p(0)\mathrm{e}^{-kx^2}. \tag{11.34}$$

Since p is a probability density, it satisfies $\int_{-\infty}^{\infty} p(x)\,\mathrm{d}x = 1$. Substituting Eq. (11.34) into this relation gives

$$p(0)\int_{-\infty}^{\infty} \mathrm{e}^{-kx^2}\,\mathrm{d}x = 1. \tag{11.35}$$

Introduce $y = \sqrt{2k}x$ as a new variable of integration. We get

$$\int_{-\infty}^{\infty} \mathrm{e}^{-kx^2}\,\mathrm{d}x = \frac{1}{\sqrt{2k}}\int_{-\infty}^{\infty} \mathrm{e}^{-\frac{y^2}{2}}\,\mathrm{d}y. \tag{11.36}$$

It follows from Eq. (7.10) that $\int_{-\infty}^{\infty} \mathrm{e}^{-\frac{y^2}{2}}\,\mathrm{d}y = \sqrt{2\pi}$. Therefore,

$$\int_{-\infty}^{\infty} \mathrm{e}^{-kx^2}\,\mathrm{d}x = \sqrt{\frac{\pi}{k}}.$$

Setting this in Eq. (11.35) gives $p(0) = \sqrt{\frac{k}{\pi}}$. Therefore, using Eq. (11.34), we get

$$p(x) = \sqrt{\frac{k}{\pi}}\mathrm{e}^{-kx^2}. \tag{11.37}$$

Substituting this into $c(g) = p(a)p(d)$, we deduce

$$c(x,y) = \frac{k}{\pi}\mathrm{e}^{-k(x^2+y^2)}. \tag{11.38}$$

The derivation of the law of errors presented above is due to the physicist James Clerk Maxwell (1831–1879), who made profound investigations of the significance of probability densities of the form (11.37) and (11.38) in physics. For this reason, such densities in physics are called *Maxwellian*. Even before Maxwell, Carl Friedrich Gauss (1777–1855) investigated probabilities of the form (11.37). Mathematicians call such densities Gaussian. Another name for probabilities of this form is *normal*.

In Fig. 11.3, we see the shape of the normal distributions $p(x)$ for three different values: $k = 0.5$, $k = 1$, $k = 2$. These graphs indicate that the larger the value of k, the greater the concentration of the probability near the bullseye. The rest of this section is about some of the basic properties of normal distributions.

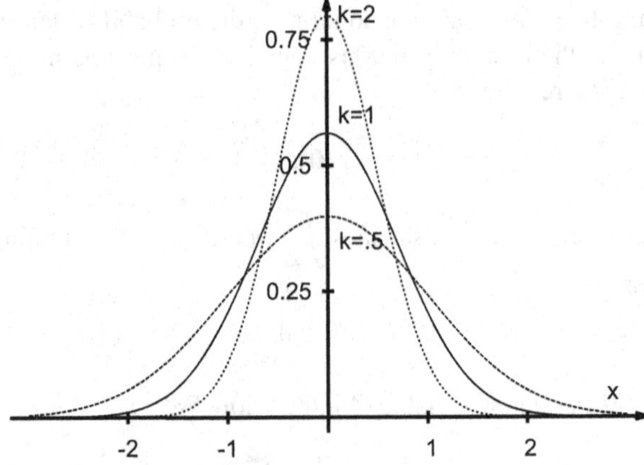

Fig. 11.3 Normal distributions with $k = 0.5$, $k = 1$, and $k = 2$

Theorem 11.6. *The convolution of two normal distributions is normal.*

Proof. We denote the two normal distributions by

$$p(x) = \sqrt{\frac{k}{\pi}}\mathrm{e}^{-kx^2}, \qquad \text{and} \quad q(x) = \sqrt{\frac{m}{\pi}}\mathrm{e}^{-mx^2}. \tag{11.39}$$

Their convolution is

$$(q*p)(x) = \int_{-\infty}^{\infty} q(a)p(x-a)\,\mathrm{d}a$$

$$= \frac{\sqrt{mk}}{\pi}\int_{-\infty}^{\infty} \mathrm{e}^{-ma^2-k(x-a)^2}\,\mathrm{d}a = \frac{\sqrt{mk}}{\pi}\mathrm{e}^{-kx^2}\int_{-\infty}^{\infty}\mathrm{e}^{-\left((m+k)a^2-2akx\right)}\,\mathrm{d}a.$$

To evaluate the integral, we complete the exponent under the integral sign to a perfect square:

$$(m+k)a^2 - 2akx = (m+k)\left(a - \frac{kx}{m+k}\right)^2 - \frac{k^2}{m+k}x^2.$$

Setting this into the integral above, we get, using $a - \dfrac{kx}{m+k} = b$ as the new variable of integration,

$$(q*p)(x) = \frac{\sqrt{mk}}{\pi}\mathrm{e}^{-k+\frac{k^2}{m+k}x^2}\int_{-\infty}^{\infty}\mathrm{e}^{-(m+k)b^2}\,\mathrm{d}b.$$

The integral is of the same form as the integral (11.36), with $(m+k)$ in place of k. Therefore, the value of the integral is $\sqrt{\dfrac{\pi}{m+k}}$. This gives

$$(q*p)(x) = \sqrt{\frac{1}{\pi}}\sqrt{\frac{mk}{m+k}}\mathrm{e}^{-(k-\frac{k^2}{m+k})x^2} = \sqrt{\frac{1}{\pi}}\sqrt{\frac{mk}{m+k}}\mathrm{e}^{-\frac{km}{m+k}x^2}.$$

We summarize: With p and q given by Eq. (11.39),

$$q*p = \sqrt{\frac{\ell}{\pi}}\mathrm{e}^{-\ell x^2}, \qquad \text{where } \ell = \frac{km}{k+m}. \tag{11.40}$$

□

We turn next to the continuous analogue of Theorem 11.1 for discrete probability:

Theorem 11.7. *Among all probability densities $q(x)$ that satisfy*

$$\int_{-\infty}^{\infty} x^2 q(x)\,\mathrm{d}x = \frac{1}{2k},$$

the quantity $I(q) = -\int_{-\infty}^{\infty} q(x)\log q(x)\,\mathrm{d}x$ is largest for the Gaussian, i.e., when $q = p$ given by $p(x) = \sqrt{\frac{k}{\pi}}\mathrm{e}^{-kx^2}$.

Remarks.

(i) This result is a continuous analogue of Theorem 11.1, which asserts that among all probability distributions for n events, $-\sum_1^n p_j \log p_j$ is largest when all the p_j are equal. Our proof is similar to the proof given in the discrete case.

(ii) The functional $I(q)$ is the *entropy* of $q(x)$, an important quantity.

(iii) Implicit in the statement of the theorem is that

$$\int_{-\infty}^{\infty} x^2 p(x)\,\mathrm{d}x = \frac{1}{2k} \qquad \text{when} \quad p(x) = \sqrt{\frac{k}{\pi}}\mathrm{e}^{-kx^2},$$

which we ask you to derive in Problem 11.20.

Proof. We construct the following one-parameter family of probability densities $r(s)$ in the interval $0 \le s \le 1$:

$$r(s) = sq + (1-s)p. \tag{11.41}$$

This function is designed so that $r(0) = p$ and $r(1) = q$. To show that $I(p) \ge I(q)$, it suffices to verify that $I(r(s))$, which we abbreviate as $F(s)$, is a decreasing function of s. According to the monotonicity criterion, the decreasing character of $F(s)$

can be shown by verifying that the derivative of $F(s)$ is negative. To this end, we calculate the derivative of $F(s) = -\int_{-\infty}^{\infty} r(s)\log r(s)\,dx$ using the differentiation theorem for integrals, Theorem 7.8. Differentiate $r(s)\log r(s)$ with respect to s; since $\frac{dr}{ds} = p - q$, we get

$$\frac{d}{ds}(r(s)\log r(s)) = (1+\log r(s))\frac{dr}{ds} = (1+\log r(s))(p-q).$$

Thus

$$\frac{dF(s)}{ds} = -\int_{-\infty}^{\infty} (1+\log r(s))(p(x)-q(x))\,dx. \tag{11.42}$$

For $s=0$, we have $r(0) = p(x)$ and $\log p(x) = \log\sqrt{\frac{k}{\pi}} - kx^2$. So if we set $s=0$ in the derivative, we get

$$\frac{dF}{ds}(0) = -\int_{-\infty}^{\infty}\left(1+\log\sqrt{\frac{k}{\pi}} - kx^2\right)(p(x)-q(x))\,dx$$

$$= -\left(1+\log\sqrt{\frac{k}{\pi}}\right)\int_{-\infty}^{\infty}(p(x)-q(x))\,dx - k\int_{-\infty}^{\infty}x^2(p(x)-q(x))\,dx.$$

Since both p and q are probability densities, $\int_{-\infty}^{\infty} p(x)\,dx = \int_{-\infty}^{\infty} q(x)\,dx = 1$. Furthermore, $\int_{-\infty}^{\infty} x^2 p(x)\,dx = \int_{-\infty}^{\infty} x^2 q(x)\,dx = \frac{1}{2k}$. Therefore, $\int_{-\infty}^{\infty}(p(x)-q(x))\,dx$ and $\int_{-\infty}^{\infty} x^2(p(x)-q(x))\,dx$ are both zero, and $\frac{dF}{ds}(0) = 0$. To show that $\frac{dF}{ds}(s) < 0$ for all s between 0 and 1, it suffices to show that $\frac{d^2F}{ds^2} < 0$. We now calculate the second derivative of F by again applying the differentiation theorem for integrals to Eq. (11.42). We get that

$$\frac{d^2F(s)}{ds^2} = \int_{-\infty}^{\infty}(p(x)-q(x))\frac{1}{r(s)}\frac{dr}{ds}\,dx = \int_{-\infty}^{\infty} -\frac{(p(x)-q(x))^2}{r(s)}\,dx.$$

This last integral is negative unless q and p are identical; hence the second derivative of F is negative. □

Remark. In our proof we have applied the differentiation theorem for integrals to improper integrals over the infinite interval $(-\infty,\infty)$, whereas this differentiation theorem was proved only for proper integrals. To get around this difficulty, we assume that $q(x)$ equals $p(x)$ outside a sufficiently large interval (a,b) and derive the inequality $I(q) \le I(p)$ for this subclass of q. From this, we can deduce the inequality for any q by approximating q by a sequence of q's belonging to the subclass. We omit the details of this step in the proof.

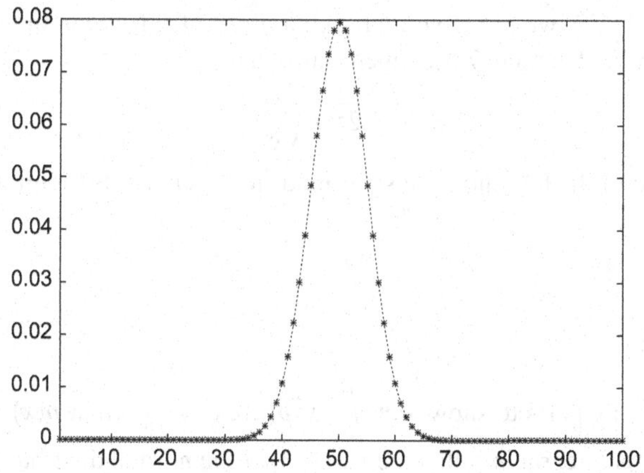

Fig. 11.4 The binomial distribution $b_k(100)$ and a normal distribution

The Limit of the Binomial Distribution. We have defined the binomial distribution as $b_k(n) = \binom{n}{k} p^k q^{n-k}$. To simplify the discussion, we take $p = q = \frac{1}{2}$. In this case,

$$b_k(n) = 2^{-n}\binom{n}{k}.$$

We have plotted these probabilities in Fig. 11.4 for $n = 100$ together with the function $\frac{1}{10}\sqrt{\frac{2}{\pi}}\mathrm{e}^{-2y^2}$, which is a multiple of the normal distribution. Note that the points b_k lie (nearly) on the graph of the normal distribution. The figures suggest that for n large, binomial distributions tend toward normal distributions. The precise statement is the following theorem.

Theorem 11.8. *The binomial distribution*

$$b_k(n) = 2^{-n}\binom{n}{k}$$

is approximately normal in this sense: set $y = \dfrac{k - \frac{1}{2}n}{\sqrt{n}}$. *Then*

$$b_k(n) \sim \frac{1}{\sqrt{n}}\sqrt{\frac{2}{\pi}}\mathrm{e}^{-2y^2},$$

where $\sim$ *means asymptotic as* n *and* k *tend to infinity with* y *fixed.*

In Problem 11.23, we guide you through a proof of this theorem. It is based on Stirling's formula (Theorem 7.5), which states that

$$m! \sim \sqrt{2\pi m}\left(\frac{m}{\mathrm{e}}\right)^m,$$

that is, the ratio of the left and right sides tends to 1 as m tends to infinity.

Problems

11.20. Integrate by parts to show that $\int_{-\infty}^{\infty} x^2 p(x)\,\mathrm{d}x = \frac{1}{2k}$ when $p(x) = \sqrt{\frac{k}{\pi}}\mathrm{e}^{-kx^2}$. Explain why this integral is called the variance of the normal distribution.

11.21. The purpose of this problem is to evaluate the integral $\int_{-\infty}^{\infty} \mathrm{e}^{-y^2}\,\mathrm{d}y$ numerically. Since the interval of integration is infinite, we truncate it by considering for large N, the approximate integral

$$I_{\text{mid}}\left(\mathrm{e}^{-y^2}, \left[-(N+\frac{1}{2})h, (N+\frac{1}{2})h\right]\right) = h\sum_{n=-N}^{N} \mathrm{e}^{-(nh)^2}$$

with subintervals of length h.

(a) Prove that $\sum_{n=K}^{\infty} \mathrm{e}^{-(nh)^2} \le \sum_{n=K}^{\infty} \mathrm{e}^{-Knh^2} = \frac{\mathrm{e}^{-K^2h^2}}{1-\mathrm{e}^{-Kh^2}}$. Use this to show that for h=1, the sum $\sum_{n=4}^{\infty} \mathrm{e}^{-(nh)^2}$ is less than 10^{-6}.

(b) Evaluate $I_{\text{mid}} \approx 1.77263\ldots$ numerically using just the sum from -3 to 3 with $h = 1$.

Remark. The value of the integral is $\sqrt{\pi} = 1.77245\ldots$. Thus we see that the midpoint rule gives an astonishingly good approximation to the value of the integral, even when we divide the interval of integration into subintervals of length $h = 1$, a rather crude subdivision.

11.22. Let

$$p(x,t) = \frac{1}{\sqrt{4\pi t}}\mathrm{e}^{-\frac{x^2}{4t}}.$$

(a) Determine the derivative of p with respect to x. Denote it by p_x.
(b) Determine the second derivative of p with respect to x. Denote it by p_{xx}.
(c) Determine the derivative of p with respect to t. Denote it by p_t.
(d) Verify that $p_t = p_{xx}$.
(e) In one application, p has an interpretation as the temperature of a metal rod, which varies with position and time. Suppose $p(x,t)$ is graphed as a function of x. In an interval of x where the graph is convex, will the temperature increase or decrease with time, according to part (d)?

11.23. Verify the following steps, which prove Theorem 11.8.

(a) Use Stirling's formula for each factorial in $2^{-n}\binom{n}{k}$ to show that

$$2^{-n}\binom{n}{k} \sim \frac{1}{\sqrt{2\pi}}\sqrt{\frac{n}{k(n-k)}}\frac{n^n}{2^n k^k (n-k)^{n-k}}.$$

(b) Substitute $k = \frac{n}{2} + \sqrt{n}y$ and rearrange the previous expression to show that

$$2^{-n}\binom{n}{k} \sim \frac{1}{\sqrt{2\pi}}\sqrt{\frac{n}{\frac{n^2}{4} - ny^2}}\frac{1}{\left(1-\frac{4y^2}{n}\right)^{\frac{n}{2}}\left(\frac{n}{2}+\sqrt{n}y\right)^{\sqrt{n}y}\left(\frac{n}{2}-\sqrt{n}y\right)^{-\sqrt{n}y}}.$$

(c) We showed in Sect. 2.6 that $\left(1+\frac{x}{m}\right)^m$ tends to e^x as m tends to infinity. Use this to show that the right side in (b) is asymptotic to

$$\frac{1}{\sqrt{2\pi}}\sqrt{\frac{n}{\frac{n^2}{4} - ny^2}}\frac{1}{\mathrm{e}^{-2y^2}\mathrm{e}^{2y^2}\mathrm{e}^{2y^2}}.$$

(d) Show that the last expression is asymptotic to $\dfrac{1}{\sqrt{n}}\sqrt{\dfrac{2}{\pi}}\mathrm{e}^{-2y^2}$.

Answers to Selected Problems

Chapter 1

1.1

(a) $[-1,7]$
(b) $[-52,-48]$
(c) $y<6$ or $y>8$
(d) $|3-x|=|x-3|$ so same as (a)

1.3

(a) $|x|\le 3$
(b) $-3\le x\le 3$

1.5 $\frac{2+4+8}{3}=\frac{14}{3}$, $(2\cdot 4\cdot 8)^{1/3}=(2^6)^{1/3}=2^2=4<\frac{14}{3}$.

1.7

(a) $(\sqrt{x}-\sqrt{y})(\sqrt{x}+\sqrt{y})=x-y$ and $0<\dfrac{1}{\sqrt{x}+\sqrt{y}}\le\dfrac{1}{4}$, so $\sqrt{x}-\sqrt{y}\le\frac{1}{4}(x-y)$.
(b) Since $x>y$, $|\sqrt{x}-\sqrt{y}|=\sqrt{x}-\sqrt{y}\le\frac{1}{4}(0.02)=0.005$

1.9 $|x|<m$ means $-m<x<m$. So if $|b-a|<\varepsilon$, then:

(a) At least one of $0\le b-a$ and $0\le-(b-a)$ is true. The upper bound ε comes from the assumption.
(b) $-\varepsilon<b-a<\varepsilon$ is a restatement of $|b-a|<\varepsilon$
(c) add a to both sides in (b)
(d) $|a-b|=|b-a|$, so $-\varepsilon<a-b<\varepsilon$ is a restatement of $|b-a|<\varepsilon$
(e) add b to both sides in (d)

1.11 5/3

1.13

(a) $(1(1)(x))^{1/3}\le\frac{1+1+x}{3}$
(b) $(1\cdots(1)(x))^{1/n}\le\frac{1+\cdots+1+x}{n}=\frac{x+n-1}{n}$
(c) $\frac{2n-1}{n}=2-\frac{1}{n}<2$

P.D. Lax and M.S. Terrell, *Calculus With Applications*, Undergraduate Texts in Mathematics,
DOI 10.1007/978-1-4614-7946-8,

1.15

(a) Since $1 \le 2 \le \cdots \le n$, we get $n! < n^n$. Taking the nth root gives $(n!)^{1/n} \le n$.
(b) By the A-G inequality, $(1 \cdot 2 \cdot 3 \cdots n)^{1/n} \le \frac{1+2+3+\cdots+n}{n}$, so $(n!)^{1/n} \le \frac{\frac{1}{2}n(n+1)}{n} = \frac{n+1}{2}$.

1.16

(a) We have $|ab - a_0b_0| = |ab - ab_0 + ab_0 - a_0b_0|$. By the triangle inequality,

$$|ab - a_0b_0| = |ab - ab_0 + ab_0 - a_0b_0| \le |ab - ab_0| + |ab_0 - a_0b_0|$$

Recall that $|ab| = |a||b|$. Then $|ab - a_0b_0| \le |a||b - b_0| + |b_0||a - a_0|$.
(b) $|a| \le 10$ and $|b_0| \le 10.001$, so $|ab - a_0b_0| \le 10(0.001) + 10.001(0.001)$

1.19 $m = 1$ because $\sqrt{3} = 1.732\cdots = 1.7 + (0.032\cdots)$ and $(0.032\cdots) < 10^{-1}$.

1.21

(a) a_n is part of the area of the 1 by 1 square, so $a_n < 1$
(b) S has an upper bound of 1, therefore has a least upper bound, the area of the quarter-circle of radius 1, which is $\frac{\pi}{4}$.

1.27 $s_1 = 1,\ s_2 = (1/2)(s_1 + 3/s_1) = 2,\ s_3 = (1/2)(2 + 3/2) = 7/4 = 1.75$, $s_4 = (1/2)(7/4 + 12/7) = 97/56 = 1.7321\ldots$ If you start with $s_1 = 2$ instead, it just shifts the sequence, since 2 already occurred as s_2.

1.29 If $s > \sqrt{2}$, then $\frac{1}{s} < \frac{1}{\sqrt{2}}$. Multiply by 2 to get $\frac{2}{s} < \frac{2}{\sqrt{2}} = \sqrt{2}$.

1.31 Suppose $s \ge \sqrt{2} + q$ for some number q. Then

$$2 + p > s^2 \ge 2 + 2\sqrt{2}q + q^2 \ge 2 + 2\sqrt{2}q.$$

This is possible only if $p > 2\sqrt{2}q$. Therefore, taking $q = \frac{p}{2^{3/2}}$, we get $s < \sqrt{2} + q$.

1.35

(a) a_n is arbitrarily close to a when n is sufficiently large, so in particular, there is N such that a_n is within 1 of a when $n > N$.
(b) $|a_n| = |a + (a_n - a)| \le |a| + |a_n - a| < |a| + 1$
(c) use the definition of α

1.37 $a_1 = s_1 = \frac{1}{3}$, $a_2 = s_2 - a_1 = \frac{2}{4} - \frac{1}{3} = \frac{1}{6}$, and the sum is the limit of the s_n, which is 1.

1.39 $\dfrac{1}{1 - \frac{5}{7}}$

1.41 $\left|\dfrac{(n+1)a_{n+1}}{na_n}\right| = \dfrac{n+1}{n}\left|\dfrac{a_{n+1}}{a_n}\right|$ has the same limit as $\left|\dfrac{a_{n+1}}{a_n}\right|$.
The sum $\sum_{n=0}^{\infty}(-1)^n n^5 a_n$ also converges absolutely by the ratio test if $\sum_{n=0}^{\infty} a_n$ does so.

1.47

(a) Converges absolutely by the limit comparison theorem; compare with $\sum \left(\frac{2}{3}\right)^n$.
(b) Diverges by the comparison theorem; compare with the harmonic series
(c) Converges by the alternating series theorem
(d) Diverges because the nth term does not tend to 0
(e) Diverges by the limit comparison theorem; compare with the harmonic series
(f) Converges by the ratio test

1.49

(a) The series of absolute values is a convergent geometric series, so the series converges absolutely
(b) Converges by comparing with the geometric series $\sum(10)^{-n}$
(c) For any number b, the series converges absolutely by the ratio test
(d) Converges by comparing with the geometric series $\sum \frac{2}{3^n}$
(e) Sum of two convergent series is convergent
(f) Diverges because the term does not tend to 0

1.51 The sequence converges to x, so it must be Cauchy by the theorem that every convergent sequence is Cauchy. To see this case specifically, a_n is within 10^{-n} of x, so if n and m are both greater than N, then $|a_n - a_m| = |a_n - x + x - a_m| \le |a_n - x| + |x - a_m| < (2)10^{-N}$.

1.53

(a) $\left(1+\frac{1}{n-1}\right)^n = e_{n-1}\left(1+\frac{1}{n-1}\right) < (3)(2)$.
(b) This is $\left(\frac{n}{n-1}\right)^n < 6$, or $n^n < 6(n-1)^n$. Therefore $n^{n-1} < \frac{6}{n}(n-1)^n \le (n-1)^n$ if $n \ge 6$.
Take roots to get $n^{1/n} < (n-1)^{1/(n-1)}$.
(c) If $n^{1/n}$ were less than 1, its powers, such as n, would be less than 1. Therefore, we have a decreasing sequence bounded below by 1, which then has a limit $r \ge 1$.
(d) $(2n)^{1/(2n)} = 2^{1/(2n)}\sqrt{n^{1/n}}$ tends to $r = \sqrt{r}$. So $r = 1$.

Chapter 2

2.1

(a) Not bounded, not bounded away from 0
(b) Not bounded, bounded away from 0
(c) Bounded, not bounded away from 0
(d) Not bounded, not bounded away from 0

2.3

(a) Cancel common factors
(b) f is defined except at 0 and -3; g is defined except at -3. h is defined for all numbers.
(c) The graph of h is a line, that of g is the line with one point deleted, and that of f is the line with two points deleted.

2.5 51,116.80

2.7 The change in radius is $\frac{1}{2\pi}$ times the change in circumference, about 3 meters.

2.9

(a) -3
(b) -6
(c) -2

2.13 Numerator is polynomial, continuous on $[-20, 120]$, has a maximum value M and a minimum value m. Denominator is $x^2 + 2 \ge 2$ and has maximum $(120)^2 + 2$. Therefore,

$$\frac{m}{(120)^2+2} \le f(x) \le \frac{M}{2}.$$

So f is bounded.

2.15 It appears to be approximately $(-0.4, 0.4)$.

2.17 No. The truncation of $x = 9.a_1a_2a_3a_4a_5a_6a_7a_8a_9\dots$ is $y = 9.a_1a_2a_3a_4a_5a_6a_7a_8$. The difference is $x - y = 0.00000000a_9 \dots < 0.000000010 = 10^{-8}$. Then

$$x^2 - y^2 = (x+y)(x-y) < (20)(10^{-8}) = 2 \times 10^{-7}.$$

In fact, if we take an example with a_9 as large as possible, then

$$(9.000000009)^2 - 9^2 = 0.000000162\dots > 10^{-7}.$$

2.18 f has a minimum value on each closed interval contained in (a,b).

Take an expanding sequence of closed intervals, such as $I_n = \left[a + \frac{b-a}{2n}, b - \frac{1}{n}\right]$.

It might happen that the minimum value of f on I_n decreases with n. If so, there must be locations x_n at which the minimums occur, for which x_n tends to a or to b. This is a contradiction, because the values $f(x_n)$ must tend to infinity.

2.20 (a) 3×10^{-m} (b) $(1/3) \times 10^{-7}$ (c) all x

2.23 We suppose the bottle only has one volume for a given height, $V = f(H)$, and only one height for a given volume, $H = g(V)$. Then $H = g(V) = g(f(H)) = (g \circ f)(H)$ and $V = f(H) = f(g(V)) = (f \circ g)(V)$. So f and g are inverses.

2.25 There are two, x and x^{-1}.

2.27

(a) $k \circ k$
(b) $g \circ k$
(c) $k \circ g$

2.29 Let $f(x) = \sqrt{x^2+1} - \sqrt[3]{x^5+2}$. Then f is continuous because polynomials and roots are continuous and composites of continuous functions are continuous. We have $f(0) = 1 - \sqrt[3]{2} < 0$ and $f(-1) = \sqrt{2} - a > 0$. By the intermediate value theorem, $f(x) = 0$ for some number x in $[-1, 0]$.

2.31

(a) If $a < b$ then $f(a) < f(b)$, so $f(f(a)) < f(f(b))$.
(b) Let $f(x) = -x$, for example. Then f is decreasing, but $f(f(x)) = x$ is increasing.

2.33

(a) By the continuity of f
(b) By the limit of g
(c) Combine parts (a) and (b)
(d) restates part (c)

2.35 Since $x^2 + y^2 = 1$ on the unit circle, $\cos^2 s + \sin^2 s = 1$. This identity does not hold for the first pair, but does for the second.

2.37 The graph has to be very wide if you use the same scale on both axes!

2.39 Maximum height 1.2 reached at $3t = \frac{\pi}{2}$ and $3t = \frac{\pi}{2} + 2\pi$. The t difference is $2\pi/3$.

2.43

(a) $\sin(\tan^{-1}(z))$ is the y-coordinate of the point on the unit circle whose radius points toward $(1,z)$. In order to rescale $(1,z)$ back to the unit circle, multiply by some number c: (c, cz) is on the unit circle if $c^2 + c^2z^2 = 1$, so $c = \frac{1}{\sqrt{1+z^2}}$. Therefore, $\sin(\tan^{-1}(z)) = y = cz = \frac{z}{\sqrt{1+z^2}}$.

(b) $\cos(\sin^{-1}(y))$ is the x-coordinate of the point whose vertical coordinate is y. By the Pythagorean theorem, it is $\sqrt{1-y^2}$.

2.45 $g(x+y) = f(c(x+y)) = f(cx+cy) = f(cx)f(cy) = g(x)g(y)$.

2.47 $p(0) = 800$, $p(d) = 1600 = 800(1.023)^d$ gives $2 = d\log(1.023)$ so $d = 87.95\ldots$.

2.51 $f(1/2) = 3f(0) = 3 = m$, so $f(1) = 3f(1/2) = 9 = ma$, so $a = 3$.

2.53 $P(N) = P(1)P(N-1) = P(1)P(1)P(N-2) = \cdots = P(1)\cdots P(1) = P(1)^N$. So the sequence is $1 + P(1) + (P(1))^2 + \cdots + (P(1))^N$.

2.55 $e > 2$, so $e^{10} > 2^{10} = 1024 > 1000 = e^{\log 1000}$. But e^x is increasing; therefore $\log 1000 < 10$. Then $\log(1\,000\,000) = \log(1000) + \log(1000) < 20$

2.57 With $\frac{e^x}{x^2} > 1$ for large x, you get $e^x > x^2 = e^{\log(x^2)}$. So $x > \log(x^2)$, or $\sqrt{y} > \log y$.

2.59 $\left|1 + x + x^2 + x^3 + x^4 - \frac{1}{1-x}\right| = \left|\frac{x^5}{1-x}\right|$. If $-\frac{1}{2} \le x \le \frac{1}{2}$, the numerator does not exceed 1/32, and the denominator is at least 1/2, so the error $\le 1/16$.

2.63 Since $\sum_{n=0}^{\infty} a_n(x-2)^n$ converges at $x = 4$, it must converge at every x that is closer to 2, i.e., $|x-2| < 2$. So the radius of convergence is at least 2, and f is continuous at least on $(0,4)$.

2.65 We have said that the series in question, which is centered at a, converges for $m < x < M$. We have also said that when a power series centered at a converges at any particular number x, then it converges for every number closer to a. So it converges on intervals *symmetric* about a. Therefore $M - a = a - m$.

2.67

(a) You need to argue that since $p_n^{1/n}$ tends to ℓ, it is eventually smaller than every number such as r that is greater than ℓ. Then compare with a geometric series.
(b) Is similar to part (a)
(c) If $p_n^{1/n} = |a_n|^{1/n}|x|$ tends to $L|x|$. If $L|x| < 1$, the series converges by part (a). If $L|x| > 1$, it diverges by part (b). Therefore $1/L$ is the radius of convergence.

2.69 $\sum_{n=0}^{\infty} nx^n$ converges for $|x| < 1$ by the ratio test. If $\{n^{1/n}\}$ tends to a limit $r > 1$, take x between $1/r$ and 1, so that $rx > 1$. Then the nth root $(nx^n)^{1/n} = n^{1/n}x = \frac{n^{1/n}}{r}(rx)$ tends to $rx > 1$. According to the root test, the series diverges, contradiction.

2.71

(a) $\frac{1}{1+t^2}$, for $|t| < 1$
(b) $\frac{1}{1-x} - 1 - x - x^2$, for $|x| < 1$
(c) We don't have a name for this one.
(d) $\frac{1}{1-\frac{1}{2}t} + \frac{1}{1-3t^2}$, for $|t| < 1/\sqrt{3}$

2.73 Take the list of $n+1$ numbers $1 + \frac{x}{n}$ (n times) and 1. The A-G inequality gives

$$\left(1 + \frac{x}{n}\right)^{n/(n+1)} < \frac{n\left(1 + \frac{x}{n}\right) + 1}{n+1} = 1 + \frac{x}{n+1}.$$

The $(n+1)$st power gives the result.

2.75 $e_n(-x)$ will do.

Chapter 3

3.1 A linear function is its own tangent.

3.3 $\ell(x) = 5(x-2)+6 = f(2)+f'(2)(x-2)$.
Use the properties $\ell(2) = f(2) = 6$ and $\ell(2) = 5 = f'(2)$

3.5 (c) $\dfrac{f(a+h)-f(a)}{h} = \dfrac{1}{h}(a^3+3a^2h+3ah^2+h^3-a^3)$ tends to $3a^2 = 3$.
Tangent line $y = -1+3(x+1)$.

3.7 $y = -x - \frac{1}{4}$ is tangent at both $-\frac{1}{2}$ and $\frac{1}{2}$.

3.9 Average rate of change: $\dfrac{T(a+h)-T(a)}{h}$. If $T'(a)$ is positive, $T(x)$ is locally increasing at a; hence it will be hotter to the right. If it is cooler to the left of a, $T'(a)$ should be positive. If the temperature is constant, $T(a+h)-T(a) = 0$, so $T'(a) = 0$.

3.11 Since $f'(3) = 5$, $g'(3) = 6$, I would say that g is more sensitive to change near 3.

3.15 $(f(h)-f(0))/h = h^{-1/3}$ does not have a limit at h tends to 0. The one-sided derivative does not exist, so f is not differentiable on $[0,1]$.

3.17

(a) $\frac{1}{2}(x^3+1)^{-1/2}(3x^2)$
(b) $3\left(x+\frac{1}{x}\right)^2\left(1-\frac{1}{x^2}\right)$
(c) $\frac{1}{2}(1+\sqrt{x})^{-1/2}\left(\frac{1}{2}x^{-1/2}\right)$
(d) 1

3.19

(a) Positions $f(0) = 0$, $f(2) = -6$
(b) Velocities $f'(0) = 1$, $f'(2) = -11$
(c) Direction of motion right, left, assuming axis is drawn positive to the right.

3.21 Only in parts (a), (b), and (c) do higher derivatives vanish.

(a) $f(x) = x^3$, $f'(x) = 3x^2$, $f''(x) = 6x$, $f'''(x) = 6$, 0, 0, 0
(b) t^3+5t^2, $3t^2+10t$, $6t+10$, 6, 0, 0, 0
(c) r^6, $6r^5$, $(6)(5)r^4$, $(6)(5)(4)r^3$, $(6)(5)(4)(3)r^2$, $(6)(5)(4)(3)(2)r$, $6!$
(d) x^{-1}, $-x^{-2}$, $2x^{-3}$, $3!x^{-4}$, $-4!x^{-5}$, $5!x^{-6}$, $-6!x^{-7}$
(e) $t^{-3}+t^3$, $-3t^{-4}+3t^2$, $12t^{-5}+6t$, $-60t^{-6}+6$, $360t^{-7}$, $-2520x^{-8}$, $40160x^{-9}$

3.23

(a) $(f^2)' = 2ff' = 2(1+t+t^2)(1+2t)$
(b) 0
(c) $6(5^4)(4)$

3.25

(a) $2GmMr^{-3}$
(b) $2GmMr^{-3}r'(t) = 2GmM(2000000+1000t)^{-3}(1000)$
(c) $\dfrac{dF}{dt} = \dfrac{dF}{dr}\dfrac{dr}{dt} = 2GmMr^{-3}\dfrac{dr}{dt}$

3.27 $V(t) = \frac{4}{3}\pi(r(t))^3$, and for some constant k,
$V'(t) = 4\pi(r(t))^2 r'(t) = k \cdot 4\pi(r(t))^2$. So $r'(t) = k$.

3.29 $\dfrac{\mathrm{d}P}{\mathrm{d}t} = \dfrac{7}{5}k\rho^{2/5}\dfrac{\mathrm{d}\rho}{\mathrm{d}t}$.

3.31 $f(1) = 1+2+3+1 = 7$ and $f'(1) = 3+4+3 = 10$,
so $g'(7) = (f^{-1})'(7) = 1/f'(1) = 1/10$.

3.33 $(x^a)^b = x^{ab} = x$ if $ab = 1$. If $1/p + 1/q = 1$ then multiply by pq, giving $q + p = pq$, or $(p-1)(q-1) = 1$.

3.35

(a) $f(x+h) = k + f(x) = k + y$, so $x + h = g(k+y)$
(b) f strictly monotonic means that when $h \neq 0$, $f(x+h) \neq f(x)$
(c) The algebra is correct as long as no denominator is 0, and we have shown that none is, provided $k \neq 0$. The left-hand side tends to $g'(y)$ as k tends to 0. But as k tends to 0, $h = g(k+y) - g(y)$ tends to 0 due to the continuity of g (By Theorem 2.9). Therefore, the right-hand side tends to $1/f'(x)$.

3.37 $(-1/10)^n \mathrm{e}^{-t/10}$

3.39 $2x + 0 + 0 + 2^x \log 2 + \mathrm{e}^x + \mathrm{e}x^{\mathrm{e}-1}$

3.41

(a) $1/x$
(b) $2/x$
(c) 0
(d) $-\mathrm{e}^x \mathrm{e}^{-\mathrm{e}^x}$
(e) $\dfrac{1-\mathrm{e}^{-x}}{1+\mathrm{e}^{-x}}$

3.43 $y = x - 1$

3.45 The rate is 1.5 times the size, $p'(t) = 1.5p(t)$. The solutions are $p(t) = c\mathrm{e}^{1.5t}$.
Then $p(1) = 100 = c\mathrm{e}^{1.5}$, and $p(3) = c\mathrm{e}^{4.5} = 100\mathrm{e}^{-1.5}\mathrm{e}^{4.5}$.

3.47 $(\log(fg))' = \dfrac{(fg)'}{fg} = (\log f + \log g)' = \dfrac{f'}{f} + \dfrac{g'}{g}$. Multiply by fg to get $(fg)' = f'g + fg'$.

3.49 $\left(\dfrac{1}{2}\log(x^2+1) + \dfrac{1}{3}\log(x^4-1) - \dfrac{1}{5}\log(x^2-1)\right)' = \dfrac{1}{2}\dfrac{2x}{x^2+1} + \dfrac{1}{3}\dfrac{4x^3}{x^4-1} - \dfrac{1}{5}\dfrac{2x}{x^2-1}$

3.53

(a) $\cot x$
(b) $\dfrac{\mathrm{e}^{\tan^{-1}(x)}}{x^2+1}$
(c) $\dfrac{10x}{25x^4+1}$
(d) $\dfrac{2\mathrm{e}^{2x}}{\mathrm{e}^{2x}+1}$
(e) $\mathrm{e}^{(\log x)(\cos x)}\left(\dfrac{\cos x}{x} - \sin x \log x\right)$

3.55

(a) $(\sec x)' = -(\cos x)^{-2}(-\sin x)$
(b) $(\csc x)' = -(\sin x)^{-2}(\cos x)$
(c) $(\cot x)' = \dfrac{(\sin x)(-\sin x)-(\cos x)(\cos x)}{\sin^2 x}$

3.57 $y(x) = u\cos x + v\sin x$, $y(0) = -2 = u$, $y'(0) = 3 = v$, so $y(x) = -2\cos x + 3\sin x$.

3.59

(a) For $y > 1$, sketch a triangle with legs 1 and $\sqrt{y^2-1}$ to see that $\cos(\sec^{-1} y) = 1/y$. The derivative gives $-\sin(\sec^{-1} y)(\sec^{-1} y)' = -y^{-2}$. Then

$$(\sec^{-1} y)' = \frac{1}{\sin(\sec^{-1} y)}\frac{1}{y^2} = \frac{1}{\sqrt{1-y^{-2}}}\frac{1}{y^2} = \frac{1}{y\sqrt{y^2-1}}.$$

For $y < -1$, sketch a graph of cosine and secant to see the symmetry $\sec^{-1} y = \pi - \sec^{-1}(-y)$. Then by the chain rule, $(\sec^{-1})'(y) = (\sec^{-1})'(-y)$. So $(\sec^{-1} y)' = \dfrac{1}{|y|\sqrt{y^2-1}}$.

(b) $(\cos^{-1} x)' = -(\sin^{-1} x)' = -1/\sqrt{1-x^2}$.
(c) $(\csc^{-1} x)' = -(\sec^{-1} x)' = -1/(|x|\sqrt{x^2-1}$.
(d) $\left(\tan^{-1}\dfrac{1}{x}\right)' = -(\tan^{-1} x)' = -\dfrac{1}{1+x^2}$. By the chain rule, $\dfrac{1}{1+(x^{-1})^2}\dfrac{-1}{x^2} = -\dfrac{1}{1+x^2}$.

3.61 $\sinh' x = \frac{1}{2}(e^x - e^{-x})' = \frac{1}{2}(e^x + e^{-x}) = \cosh x$ and similarly for $\cosh'$.

3.63 $\cosh^2 x - \sinh^2 x = \frac{1}{4}\left(2e^x e^{-x}) - 2(-e^x e^{-x})\right) = 1$

3.65

(a) $0 + 0^5 + \sin 0 = 3(1^2) - 3$.
(b) Apply the chain rule to $y(x) + y(x)^5 + \sin(y(x)) = 3x^2 - 3$.
(c) $\dfrac{dy}{dx} = \dfrac{6x}{1+5y^4+\cos y} = 6/2$.
(d) tangent line is $y = 3(x-1)$.
(e) $y(1.01) \approx 3(.01)$

3.67 $y(x) = u\cosh x + v\sinh x,\ y(0) = -1 = u,\ y'(0) = 3 = v,\ y(x) = -\cosh x + 3\sinh x$

3.69 Take $x = 0$ to find $u = 0$. Then evaluate the derivative at 0 to find $0 = v$.

3.71

(a) This is a statement of the binomial theorem applied to $(x+h)^n$.
(b) Apply the triangle inequality to the right-hand side in part (a). In each term, the binomial coefficient is positive, and $|x^{n-k}h^k| = |x|^{n-k}|h|^k$. Then replace $|h|$ by the equal number $\dfrac{|h|}{H}H$, and you get

$$\left|\binom{n}{2}x^{n-2}h^2 + \binom{n}{3}x^{n-3}h^3 + \cdots + h^n\right| \le \left|\binom{n}{2}x^{n-2}h^2\right| + \cdots + |h^n|$$

$$\le \binom{n}{2}|x|^{n-2}|h|^2 + \cdots + |h|^n = \binom{n}{2}|x|^{n-2}\frac{|h|^2}{H^2}H^2 + \cdots + \frac{|h|^n}{H^n}H^n$$

(c) Factor out $\dfrac{|h|^2}{H^2}$

(d) Recognize that the factor $\left(\binom{n}{2}|x|^{n-2}H^2+\cdots+H^n\right)$ consists of all but two (positive) terms of the binomial expansion

$$(|x|+H)^n = |x|^n + h|x|^{n-1}H + \left(\binom{n}{2}|x|^{n-2}H^2 + \binom{n}{3}|x|^{n-3}H^3 + \cdots + H^n\right)$$

and is therefore less than $(|x|+H)^n$.

3.73 Each side is $\left(\dfrac{1}{1-x}\right)^2$ for $|x|<1$.

Chapter 4

4.1 $0.4 \le f'(c) = \dfrac{f(2.1)-6}{0.1} \le 0.5$ gives $6.04 \le f(2.1) \le 6.05$

4.3 $h(x) = \frac{2}{3}\sin(3x) + \frac{3}{2}\cos(2x) + c$, $h(x) = \frac{2}{3}\sin(3x) + \frac{3}{2}\cos(2x) - \frac{3}{2}$.

4.5 $f'(x) = (1-x^2)/(1+x^2)$. f increases on $(-1,1)$, decreases on $(-\infty,-1)$ and on $(1,\infty)$. In $[-10,10]$, the minimum has to be either $f(-1) = -1/2$ or $f(10) = 10/101$, so $-1/2$. In $[-10,10]$, the maximum has to be either $f(-10) = -10/101$ or $f(1) = 1/2$, so $1/2$.

4.7

(a) For a rectangle x wide, area $A(x) = x(16-2x)/2 = 8x - x^2$ defined on $[0,8]$. $A'(x) = 8-2x$ is 0 when $x = 4$, $A(0) = A(8) = 0$, so $A(4) = 16$ is the maximum.
(b) Now $A(x) = x(16-2x) = 16x - 2x^2$ defined on $[0,16]$. $A'(x) = 16-4x$ is 0 when $x = 4$, $A(0) = A(16) = 0$, so $A(4) = 32$ is the maximum.

4.9

$$E'(m) = 2(y_1 - mx_1)(-x_1) + \cdots + 2(y_n - mx_n)(-x_n) = -2\sum_{i=1}^{n} x_i y_i + 2\Big(\sum_{i=1}^{n} x_i^2\Big)m$$

is 0 when $m = \dfrac{\Sigma x_i y_i}{\Sigma x_i^2}$. This gives a minimum because $E'(m) < 0$ when $m < \dfrac{\Sigma x_i y_i}{\Sigma x_i^2}$ and $E'(m) > 0$ when $m > \dfrac{\Sigma x_i y_i}{\Sigma x_i^2}$.

4.11 $x'(3/2) = 0$ and $x''(t)$ is negative for all t, and therefore $x(3/2) = 9/4$ is the maximum.

4.13 Let $c(x) = x - x^3$. Then $c'(x) = 1 - 3x^2$ is 0 when $x = \sqrt{1/3}$, and $c''(x) = -6x$ is negative for all $x > 0$. So $c(\sqrt{1/3}) = \sqrt{1/3}(2/3) = 0.384\ldots$ is the largest amount.

4.17 Let $f(x) = h(x) - g(x)$. Then we are given that $f'(x) \ge 0$ for $x > 0$ and $f(0) = 0$. Thus f is nondecreasing for $x > 0$, so if $x > 0$, $f(x) \ge f(0) = 0$. So $h(x) - g(x) \ge 0$, so $h(x) \ge g(x)$.

4.21 Since $e^0 = 1$, $\lim\limits_{x\to 0} \dfrac{e^x - 1}{x}$ is the derivative of e^x at 0, namely 1. The reciprocal $\dfrac{x}{e^x - 1}$ therefore also tends to 1.

4.25 The linear approximation theorem gives $f(t) = 0 + 3t + \frac{1}{2}f''(c)t^2$ for some c between 0 and t. So $3t + 4.9t^2 \le f(t) \le 3t + 4.905t^2$.

4.27 $f'(x) = 6x^2 - 6x + 12 = 6(x-2)(x+1)$ is negative on $(-1,2)$, $f'(-1) = f'(2) = 0$, and positive otherwise; $f(-1) = -17$ is a local minimum and $f(2) = 28$ is a local maximum because of the concavity: $f''(x) = 12x - 6$ is negative when $x < \frac{1}{2}$, so f is concave there, convex when $x > \frac{1}{2}$.

4.29 Tangent is below graph, because the function is convex.

4.31 Set $g(x) = \mathrm{e}^{-1/x}$. Then $g(x) > 0$, $g'(x) = x^{-2}g(x)$, and
$g''(x) = (-2x^{-3} + x^{-4})g(x) = (1-2x)x^{-4}g(x)$. So g is convex on $(0, \frac{1}{2})$.

4.33 Yes. $(\mathrm{e}^f)' = f'\mathrm{e}^f$, $(\mathrm{e}^f)'' = f''\mathrm{e}^f + (f')^2\mathrm{e}^f$. Therefore if $f'' > 0$, then $(\mathrm{e}^f)'' > 0$.

4.35 Write $h = \frac{1}{2}(b-a)$ and $c = \frac{1}{2}(a+b)$. Then $c = a+h = b-h$, and linear approximation gives

$$f(a) = f(c) + f'(c)h + \frac{1}{2}f''(c_1)h^2, \quad f(b) = f(c) - f'(c)h + \frac{1}{2}f''(c_2)h^2,$$

for some c_1 and c_2 in $[a,b]$. Average these to get $\frac{1}{2}(f(a)+f(b)) = f(c) + \frac{1}{4}(f''(c_1) + f''(c_2))h^2$. The last term has $|\frac{1}{4}(f''(c_1) + f''(c_2))h^2| \le \frac{1}{4}2Mh^2 = \frac{M}{8}(b-a)^2$.

4.39 $f' > 0$ on $[-5,-1.8]$ and $[0.5,5]$ $f' < 0$ on $[-1.8,0.5]$
$f'' > 0$ on $[-1.8,2.5]$ $f'' < 0$ on $[-5,-1.8]$ and $[2.5,5]$

4.41

(a) See Fig. 11.5.
(b) $f'(x) = -x\mathrm{e}^{-\frac{x^2}{2}}$ is positive when x is negative, so f is increasing on $(-\infty,0)$ and decreasing on $(0,\infty)$. $f''(x) = (-1+x^2)\mathrm{e}^{-\frac{x^2}{2}}$ is negative when $-1 < x < 1$, so f is concave on $(-1,1)$ and convex on $(-\infty,-1)$ and on $(1,\infty)$. The only critical point is for the maximum at $x = 0$.
(c) $g'(x) = x^{-2}\mathrm{e}^{-1/x}$ is positive and g is increasing on $(0,\infty)$.
$g''(x) = (-2x^{-3} + x^{-4})\mathrm{e}^{-1/x} = x^{-4}(1-2x)\mathrm{e}^{-1/x}$ is negative on $(0,1/2)$ and positive on $(1/2,\infty)$ so g is convex on $(0,1/2)$ and concave on $(1/2,\infty)$.

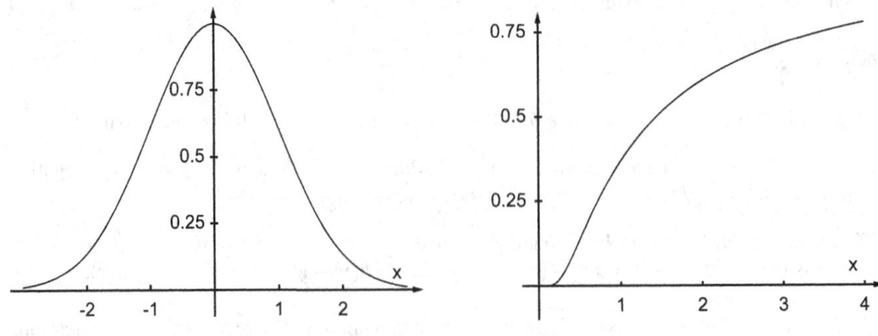

Fig. 11.5 *Left*: the graph of $f(x) = \mathrm{e}^{-\frac{x^2}{2}}$. *Right*: the graph of $g(x) = \mathrm{e}^{-1/x}$. See Problem 4.41

4.43 $\cos x = 1 - \dfrac{x^2}{2} + \dfrac{x^4}{4!} - \dfrac{x^6}{6!} + \cdots$ converges for all x.

4.45 $t_3 = t_4$ because $\sin''''(0) = 0$. It is better to use t_4 to take advantage of the 5! in the remainder. This gives $|\sin x - t_3(x)| = |\sin x - t_4(x)| \le \frac{x^5}{120}$.

4.47 $\cosh' x = \sinh x$, $\sinh' x = \cosh x$, give Taylor polynomials $t_n(x) = 1 + \frac{x^2}{2!} + \frac{x^4}{4!} + \alpha\frac{x^n}{n!}$, where $\alpha = 1$ when n is even and $\alpha = 0$ when n is odd. The remainder is $\cosh^{(n+1)}(c)\frac{x^{n+1}}{(n+1)!}$ for some c between 0 and x. By definition, $\sinh b$ and $\cosh b$ are each less than e^b. Therefore, for x in $[-b,b]$,

$$|t_n(x) - \cosh x| \le e^b \frac{b^{n+1}}{(n+1)!},$$

and this tends to 0 as b tends to infinity. The convergence is uniform on $[-b,b]$.

4.49 Let $f(x) = \cos x$. Then $f(\pi/3) = \frac{1}{2}$, $f'(\pi/3) = -\frac{\sqrt{3}}{2}$, $f''(\pi/3) = -\frac{1}{2}$, $f'''(\pi/3) = \frac{\sqrt{3}}{2}, \dots$, and $\cos x = \frac{1}{2} - \frac{\sqrt{3}}{2}\left(x - \frac{\pi}{3}\right) - \frac{1}{2}\frac{1}{2!}\left(x - \frac{\pi}{3}\right)^2 + \frac{\sqrt{3}}{2}\frac{1}{3!}\left(x - \frac{\pi}{3}\right)^3 - \frac{1}{2}\frac{1}{4!}\left(x - \frac{\pi}{3}\right)^4 + \cdots$

4.51

$$\sqrt{1+y} = 1 + \frac{1}{2}y + \frac{\frac{1}{2}\left(\frac{1}{2}-1\right)}{2!}y^2 + \frac{\frac{1}{2}\left(\frac{1}{2}-1\right)\left(\frac{1}{2}-2\right)}{3!}y^3$$
$$+ \frac{\frac{1}{2}\left(\frac{1}{2}-1\right)\left(\frac{1}{2}-2\right)\left(\frac{1}{2}-3\right)}{4!}y^4 + \cdots = 1 + \frac{1}{2}y - \frac{1}{8}y^2 + \frac{1}{16}y^3 - \frac{5}{128}y^4 + \cdots$$

4.53 $\sqrt{x} = 1 + \frac{1}{2}(x-1) - \frac{1}{8}(x-1)^2 + \frac{1}{16}(x-1)^3 - \frac{15}{16}c^{-7/2}\frac{(x-1)^4}{4!}$. For x in $[1, 1+d]$, $c \ge 1$ and $|x-1| \le d$ give $|\sqrt{x} - t_3(x)| \le \frac{15}{16}\frac{d^4}{4!} = \frac{5d^4}{128}$. This is less than

(a) 0.1 when $d \le ((0.1)128/5)^{1/4} = 1.26\dots$
(b) 0.01 when $d \le ((0.01)128/5)^{1/4} = 0.71\dots$
(c) 0.001 when $d \le ((0.001)128/5)^{1/4} = 0.4$

4.55 $t_6(0.7854) = 1 - 0.308427 + 0.0158545 - 0.000325996 = 0.70710\dots$
This is nearly $\cos\left(\frac{\pi}{4}\right) = \frac{1}{\sqrt{2}}$.

4.57

(a) $g(0) = f(a)$, $g(-a) = f(0) + af'(0) + \frac{a^2}{2}f''(0) + \cdots + \frac{a^n}{n!}f^{(n)}(0)$
(b) $g'(x) = f'(x+a) - f'(x+a) - xf''(x+a) + xf''(x+a) - \cdots + \frac{(-1)^n x^{n-1}}{(n-1)!}f^{(n)}(x+a)$.
The last term would involve $f^{(n+1)}$, but this is zero because f has degree n. All terms cancel, and $g'(x) = 0$ for all x.
(c) Since $g'(x) = 0$ for all x, $g(x)$ is constant.
(d) Since g is constant, $g(0) = g(-a)$, giving $f(a) = f(0) + af'(0) + \cdots + \frac{a^n}{n!}f^{(n)}(0)$.

4.59 $\frac{(x+h)^2 - x^2}{h} = 2x + h$, $\frac{(x+h)^2 - (x-h)^2}{2h} = 2x$, $\frac{(x+h)^3 - x^3}{h} = 3x^2 + 3xh + h^2$, $\frac{(x+h)^3 - (x-h)^3}{2h} = 3x^2 + h^2$.
$\frac{(10.1)^2 - 10^2}{0.1} = 20.1 = (\text{derivative}) + 0.1$, $\frac{(10.1)^2 - (9.9)^2}{0.2} = 20 = (\text{derivative})$,
$\frac{(10.1)^3 - 10^3}{0.1} = 33.01 = (\text{derivative}) + 3.01$, $\frac{(10.1)^3 - (9.9)^3}{.2} = 30.01 = (\text{derivative}) + 0.01$.

4.63 (a) $f'(x) = 2x\sin\left(\frac{1}{x}\right) - \cos\left(\frac{1}{x}\right)$ when $x \neq 0$.

(b) $\dfrac{h^2 \sin\left(\frac{1}{h}\right) - 0}{h} = h\sin\left(\frac{1}{h}\right)$ tends to 0 by the squeeze theorem as h tends to 0, because $|\sin| \le 1$.

(c) $f'\left(\frac{1}{n\pi}\right) = -\cos(n\pi)$ has no limit as n tends to infinity, so $f'(x)$ has no limit as x tends to 0.

Chapter 5

5.3 $y(t) = -4.9t^2 + 10t + 0,\ y(1) = -4.9 + 10t = 5.1,\ y'(1) = -9.8 + 10 = 0.2,$
$y(2) = -4.9(4) + 10(2) = 0.4,\ y'(2) = -9.8(2) + 10 = -9.6,$

5.5 $my'' = g - f_{\text{up}}$.

5.7 Functions with the same derivative on an interval differ by a constant. It is a consequence of the mean value theorem.

5.9

(a) With $x = y^6$, $f(x) = 1 + x^{1/3} - x^{1/2}$ becomes $g(y) = 1 + y^2 - y^3$. Since $g(1) > 0$ and $g(2) < 0$, there is a root y between 1 and 2. Starting from $y_1 = 1$ and iterating

$$y_{\text{new}} = y - \frac{1 + y^2 - y^3}{2y - 3y^2}$$

produces 2, 1.625, 1.4858, 1.4660, 1.4656, 1.4656. So $x = (1.4656)^6 = 9.9093$.

(b) Experiments give $f(-1) = -3,\ f(0) = 1,\ f(2) = -3,\ f(3) = 1$. Since these are $- + - +$, there are three real roots, one in each interval $[-1,0],\ [0,2]$, and $[2,3]$.

(c) $f(x) = \dfrac{x}{x^2+1} + 1 - \sqrt{x}$ has derivative $f'(x) = \dfrac{1-x^2}{1+x^2} - \dfrac{1}{2\sqrt{x}}$. The derivative is negative on $[1,\infty)$, so there can be only one zero at most; since $f(1) = 1/2$ and $f(3) = 0.3 + 1 - \sqrt{3} < 0$, the root is in $[1,3]$.

5.11 There are two statements to be explained, that this finds the maximum, and the statement about a zero of f' in (b).

In (b), there is a zero of f' because f is a continuous function on the closed subinterval $[x_{j-1}, x_{j+1}]$ for which the endpoints do *not* give the maximum.

Reason for large N, and why this finds the maximum, is that only two things could go wrong: (1) If there is more than one zero of f' in a subinterval, then Newton could converge to the wrong one, (2) there might be a maximum at one of the x_j.

Take N so large that f is well approximated by a quadratic function in each subinterval. This solves (1), and replacing N by $N+1$ solves (2).

5.13

(a) Take $z_1 = 1$, then $z_2 = z_1 - \frac{1}{2}(z_1^2 - 2) = 1.5,\ z_3 = 1.375,\ z_4 = 1.42969,\ z_5 = 1.40768,$ $z_6 = 1.4169$ and in general, $z_{n+1} - \sqrt{2} = (z_n - \sqrt{2})\left(1 - \frac{1}{2}(z_n + \sqrt{2})\right)$; since the second factor is negative, the z_n alternate greater than and less than $\sqrt{2}$.

(b) Take $z_1 = 1$, then $z_2 = z_1 - \frac{1}{3}(z_1^2 - 2) = 1.33333,\ z_3 = 1.40741,\ z_4 = 1.41381,\ z_5 = 1.41381,$ $z_6 = 1.41419$ and in general, $z_{n+1} - \sqrt{2} = (z_n - \sqrt{2})\left(1 - \frac{1}{3}(z_n + \sqrt{2})\right)$; since the second factor is positive, the $z_n - \sqrt{2}$ are all of the same sign.

5.17

(a) Q is fixed, so the cost of producing q in plant 1 plus the cost of producing $Q - q$ in plant 2 is $C_1(q) + C_2(Q - q)$.

(b) If C has a minimum at q, then $C'(q) = 0$ gives $C_1'(q) = C_2'(Q-q)$.
(c) $2aq = 2b(Q-q)$, $q = \frac{b}{a+b}Q = \frac{1.2a}{a+1.2a}Q = 0.545Q$.

5.19 Marginal cost is $C'(q) = akq^{k-1}$, and average cost is $\frac{C(q)}{q} = aq^{k-1} + \frac{b}{q}$. These are equal if there is a q for which $a(k-1)q^k = b$. Since a and b are positive, this is possible as long as $k > 1$.

Chapter 6

6.1

(a) Four parts: $1 < 2 < 3 < 4 < 5$.

$$(1)(1)+(2)(1)+(3)(1)+(4)(1) \le R(x,[1,5]) \le (2)(1)+(3)(1)+(4)(1)+(5)(1)$$

gives $10 \le R(x,[1,5]) \le 14$.
(b) Eight parts: $1 < 1.5 < 2 < 2.5 < 3 < 3.5 < 4 < 4.5 < 5$.

$$(1)(0.5)+(1.5)(0.5)+(2)(0.5)+(2.5)(0.5)+(3)(0.5)+(3.5)(0.5)+(4)(0.5)+(4.5)(0.5)$$
$$\le R(x,[1,5]) \le$$
$$(1.5)(0.5)+(2)(0.5)+(2.5)(0.5)+(3)(0.5)+(3.5)(0.5)+(4)(0.5)+(4.5)(0.5)+(5)(0.5)$$

gives $11 \le R(x,[1,5]) \le 13$.

6.3

(a) See Fig. 11.6.
(b) Since f is negative only on $(-1,1)$, we know that $A(x^2-1,[-3,-2])$ is certainly positive, and $A(x^2-1,[-1,0])$ is certainly negative. The other two would require some extra effort.
(c) $I_{\text{upper}} = ((-3)^2-1+(-2)^2-1+(-1)^2-1+1^2-1+2^2-1)1 = 14$.
$I_{\text{lower}} = ((-2)^2-1+(-1)^2-1+(0)^2-1+0^2-1+1^2-1)1 = 1$.

6.5

(a) $I_{\text{approx}}(f,[1,3]) = f(1.2)(0.5)+f(2)(0.5)+f(2.5)(1) = 12.57$
(b) $I_{\text{approx}}(\sin,[0,\pi]) = \left(\sin(0)+\sin(\frac{\pi}{4})+\sin(\frac{\pi}{2})+\sin(\frac{3\pi}{4})\right)\frac{\pi}{4} = 1.896$

6.7 (a) The area of the large rectangle is e, so the area of the shaded region is 1.
(b) If you flip the picture over the diagonal, the shaded region is the region under the graph $\log t$ from $t = 1$ to $t = e$, so the integral is 1.

6.9

(a) left: $(130+75+65+63+61)(3)/15 = 78.8$
right: $(75+65+63+61+60)(3)/15 = 64.8$

(b) left: $(130(1)+120(1)+90(2)+70(2)+65(9))/15$
right: $(120(1)+90(1)+70(2)+65(2)+60(9))/15$

6.11 The graph of f_k is the graph of f stretched horizontally by a factor of k. If we have a set of rectangles approximating the "area under the curve" f with total area A and error ε, stretching them by a factor of k will produce a set of rectangles approximating the "area under the curve" f_k with total area kA and error $k\varepsilon$. Since we can make ε arbitrarily small using narrower rectangles, we can also make $k\varepsilon$ arbitrarily small. The area of the new region is kA.

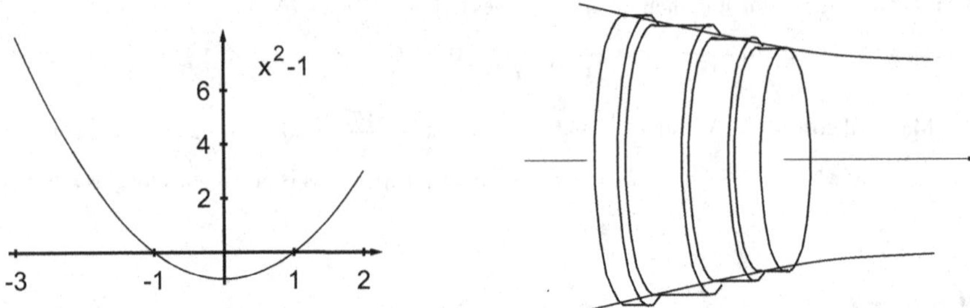

Fig. 11.6 *Left*: The graph for Problem 6.3. *Right*: A stack of cylinders approximates volume in Problem 6.31

6.13 Each approximate integral for f using $a_0 \le t_1 \le a_1 < \cdots < a_n$ becomes an approximate integral for f_- using $a_n < \cdots < a_1 < t_1 < a_0$.

6.15
(a) x^3
(b) $x^3 e^{-x}$
(c) $s^6 e^{-s^2}(2s)$
(d) $2\cos\left(\frac{\pi}{2}\right) = 0$

6.17
(a) $\tan^{-1}\left(\frac{\pi}{4}\right) = 1$
(b) $\left[\frac{1}{5}x^5 + \frac{4}{3}x^3 + 4x\right]_0^1$
(c) $\left[4\sqrt{x} - \frac{2}{3}x^{3/2}\right]_1^4$
(d) $\left[2t - 4t^{-1} + 4t^{-2}\right]_{-2}^{-1} = 7$
(e) $\left[s^2 + \log(s+1)\right]_2^6 = 32 + \log 7 - \log 3$

6.19 By the fundamental theorem, $F'(x) = \dfrac{1}{\sqrt{1-x^2}}$. Then the chain rule applied to $t = F(\sin t)$ gives $1 = F'(\sin t)(\sin t)' = \dfrac{1}{\sqrt{1-(\sin t)^2}}(\sin t)'$ or $(\sin t)' = \sqrt{1-(\sin t)^2}$.

6.21 This is the chain rule combined with the fundamental theorem.

6.23
(a) $3t^2(1+t^3)^3 = \left(\frac{1}{4}(1+t^3)^4\right)'$, so this is an example of the fundamental theorem.
(b) $a(t) = \left(v(t)\right)'$, so this is an example of the fundamental theorem

6.25 $k = \dfrac{2000}{0.004} = 500000$. $W = \displaystyle\int_0^{0.004} kx\,dx = k\frac{1}{2}x^2\Big|_0^{0.004} = 4$ joule

6.27 If the pump starts at $t = 0$, the volume drained in T minutes is $\displaystyle\int_0^T (2t+10)\,dt = T^2 + 10T$. This is 200 when $T = 10$ minutes.

6.29 $I_{\text{left}} = 1.237 > \displaystyle\int_1^2 \sqrt{1+x^{-2}}\,dx > I_{\text{right}} = 1.207$ because the integrand is decreasing.

6.31 See Fig. 11.6.

Chapter 7

7.1

(a) $\int_0^1 t^2(\mathrm{e}^t)'\,\mathrm{d}t = [t^2\mathrm{e}^t]_0^1 - \int_0^1 2t(\mathrm{e}^t)'\,\mathrm{d}t = [t^2\mathrm{e}^t - 2t\mathrm{e}^t]_0^1 + 2\int_0^1 \mathrm{e}^t\,\mathrm{d}t = \mathrm{e} - 2\mathrm{e} + 2(\mathrm{e}-1) = \mathrm{e}-1$

(b) $\frac{\pi}{2} - 1$

(c) $\frac{\pi^2}{4} - 2$. Integrate twice by parts.

(d) $\int_0^1 x^3(1+x^2)^{1/2}\,\mathrm{d}x = \int_0^1 \frac{x^2}{2}\left(\frac{2}{3}(1+x^2)^{3/2}\right)'\mathrm{d}x = [\frac{x^2}{2}\frac{2}{3}(1+x^2)^{3/2}]_0^1 - \frac{1}{3}\int_0^1 2x(1+x^2)^{3/2}\,\mathrm{d}x$
$= [\frac{x^2}{2}\frac{2}{3}(1+x^2)^{3/2}]_0^1 - \frac{1}{3}[\frac{2}{5}(1+x^2)^{5/2}]_0^1 = \frac{1}{3}2^{3/2} - \frac{2}{15}(2^{5/2}-1) = \frac{2}{15}(1+\sqrt{2})$

7.3

(a) Integrate by parts, differentiating $\tan^{-1}x$. Answer $\frac{1}{4}\pi - \frac{1}{2}$.
(b) $u\sin u$

7.5 Set $f = f_1 - f_2$. Then $f'' - vf = 0$, $f(a) = 0$, and $f(b) = 0$. Integrate by parts

$$0 \le \int_a^b v(t)f(t)f(t)\,\mathrm{d}t = \int_a^b f''(t)f(t)\,\mathrm{d}t = -\int_a^b f'(t)f'(t)\,\mathrm{d}t \le 0.$$

Therefore $\int_a^b v(t)(f(t))^2\,\mathrm{d}t = 0$. Since $v > 0$, f must be identically 0 on $[a,b]$. Therefore $f_1 = f_2$.

7.7

(a) $\int_0^{\pi/2} \sin^2 t\,\mathrm{d}t = \int_0^{\pi/2} \frac{1}{2}(1-\cos(2t))\,\mathrm{d}t = \frac{\pi}{4}$

(b) $\int_0^{\pi/2} \sin^3 t\,\mathrm{d}t = \int_0^1 (1-u^2)\,\mathrm{d}u = \frac{2}{3}$ (let $u = \cos t$)

7.9

(a) $\mathrm{e}^{-1/x} + C$
(b) $\int x^{-1}\mathrm{e}^{-1/x}\,\mathrm{d}x = x\mathrm{e}^{-1/x} - \int \mathrm{e}^{-1/x}\,\mathrm{d}x.$
(c) $(x^{-1}+1)\mathrm{e}^{-1/x} + C$
(d) $\mathrm{e}^{-1/x} - (x^2+2x+1)\mathrm{e}^{-x} + C$

7.11

(a) $\sin x - x\cos x + C$
(b) $K_m(x) = -x^m\cos x + \int mx^{m-1}\cos x\,\mathrm{d}x = -x^m\cos x + mx^{m-1}\sin x - m(m-1)K_{m-2}(x)$
(c) $K_0(x) = C - \cos x$, $K_2(x) = -x^2\cos x + 2x\sin x - 2K_0(x)$,
$K_4(x) = -x^4\cos x + 4x^3\sin x - 12K_2(x)$. Then $\int_0^\pi x^4\sin x\,\mathrm{d}x = K_4(x)\Big|_0^\pi = \pi^4 - 12\pi^2 + 48.$
(d) $K_3(x) = -x^3\cos x + 3x^2\sin x - 6K_1(x)$.

7.15

(a) Let $u = t^2+1$, then $\int_0^1 \frac{t}{t^2+1}\,\mathrm{d}t = \int_1^2 \frac{1}{2u}\,\mathrm{d}u = \frac{\log 2}{2}.$
(b) $\frac{1}{4}$
(c) Let $t = \tan u$, then $\int_0^1 \frac{1}{(t^2+1)^2}\,\mathrm{d}t = \int_0^{\pi/4} \frac{\sec^2 u}{\tan^2 u + 1}\,\mathrm{d}u = \int_0^{\pi/4} \cos^2 u\,\mathrm{d}u$
$= \int_0^{\pi/4} \frac{1}{2}(1+\cos 2u)\,\mathrm{d}u = [\frac{1}{2}u + \frac{1}{4}\sin 2u]_0^{\pi/4} = \frac{\pi}{8} + \frac{1}{4}.$

(d) $\int_{-1}^{1} x^2 e^{x^3}\,dx = \frac{1}{3}\int_{-1}^{1} (e^{x^3})'\,dx = \frac{1}{3}(e - e^{-1})$.

(e) $\int_{-1}^{1} \frac{2t+3}{t^2+9}\,dt = 2\tan^{-1}(\frac{1}{3})$ (rewrite the integrand as $\frac{2t}{t^2+9} + \frac{3}{t^2+9}$. Let $u = t^2+9$ on the left and $v = \frac{t}{3}$ on the right)

(f) With $t = \sqrt{2}\sinh u$ we have $2+t^2 = 2+2\sinh^2 u = 2+2(\cosh^2 u - 1) = 2\cosh^2 u$. So

$$\int_0^1 \sqrt{2+t^2}\,dt = \int_0^b \sqrt{2}\cosh u \sqrt{2}\cosh u\,du,$$

where $1 = \sqrt{2}\sinh b$. But $\cosh^2 u = \frac{1}{4}(e^{2u} + 2 + e^{-2u})$, so the integral is equal to

$$\frac{2}{4}\left[\frac{e^{2u}}{2} + 2u + \frac{e^{-2u}}{-2}\right]_0^b = \frac{e^{2b}}{4} + b - \frac{e^{-2b}}{4}.$$

Then since $b = \sinh^{-1}\frac{1}{\sqrt{2}} = \log\left(\frac{1}{\sqrt{2}} + \sqrt{1+\frac{1}{2}}\right)$, the integral is

$$\frac{1}{4}\left(\frac{1}{\sqrt{2}} + \sqrt{\frac{3}{2}}\right)^2 + \log\left(\frac{1}{\sqrt{2}} + \sqrt{1+\frac{1}{2}}\right) - \frac{1}{4}\left(\frac{1}{\sqrt{2}} + \sqrt{\frac{3}{2}}\right)^{-2} = 1.5245043\ldots.$$

7.17 With $x = t^2$, $0 \le t \le 1$, and change of variables gives $\int_0^1 \sqrt{1+\sqrt{x}}\,dx = \int_0^1 \sqrt{1+t}\,2t\,dt$. Then integrate by parts to get

$$\left[\frac{2}{3}(1+t)^{3/2}\,2t\right]_0^1 - \int_0^1 \frac{2}{3}(1+t)^{3/2}\,2\,dt = \frac{4}{3}2^{3/2} - \frac{4}{15}\left[(1+t)^{5/2}\right]_0^1 = \frac{4}{3}2^{3/2} - \frac{4}{15}2^{7/2}$$

7.19 f has an antiderivative F. Then

$$\int_a^b f(g(t))|g'(t)|\,dt = -\int_a^b F'(g(t))g'(t)\,dt = -F\circ g\Big|_a^b = -\int_{g(a)}^{g(b)} F'(u)\,du = \int_{g(b)}^{g(a)} F'(u)\,du$$

7.21

(a) Set $u = x - r$. Then $u = a$ when $x = a+r$, and $u = b$ when $x = b+r$, so

$$\int_a^b f(u)\,du = \int_{a+r}^{b+r} f(x-r)\,dx.$$

(b) Set $u = -x$. Then $u = -a$ when $x = a$, and $u = -b$ when $x = b$, so

$$\int_a^b f(u)\,du = \int_{-a}^{-b} f(-x)(-1)\,dx = \int_{-b}^{-a} f(-x)\,dx.$$

7.23

(a) Yes: $\frac{na_n}{nb_n} = \frac{a_n}{b_n}$ tends to 1. Yes: $\frac{na_n}{\sqrt{1+n^2}b_n} = \frac{n}{\sqrt{1+n^2}}\frac{a_n}{b_n}$ tends to 1.

(b) $\lim_{n\to\infty}(\log a_n - \log b_n) = \log 1 = 0$.

7.25

(a) $\sum_{n=1}^{\infty} \frac{1}{n^2} \leq 1 + \int_1^{\infty} x^{-2} dx = -x^{-1}\Big|_1^{\infty} = 1$. The series converges.

(b) $\sum_{n=1}^{\infty} \frac{1}{n^{1.2}} \leq 1 + \int_1^{\infty} x^{-1.2} dx = \frac{x^{-.2}}{-.2}\Big|_1^{\infty}$. The series converges.

(c) $\sum_{n=2}^{\infty} \frac{1}{n \log n} \geq \int_2^{\infty} \frac{1}{x \log x} dx = \log(\log x)\Big|_2^{\infty}$. This limit does not exist; the series diverges.

(d) $\sum_{n=1}^{\infty} \frac{1}{n^{.9}} \geq \int_1^{\infty} x^{-.9} dx = 10x^{.1}\Big|_1^{\infty} = \infty$. This limit does not exist; the series diverges.

7.27

(a)

$$\int_1^b \frac{p_0 + \cdots + p_{n-2}x^{n-2}}{q_0 + \cdots + q_n x^n}\,\mathrm{d}x = \int_{1/b}^1 \frac{p_0 + \cdots + p_{n-2}z^{-(n-2)}}{q_0 + \cdots + q_n z^{-n}} \left| -z^{-2} \right| \frac{z^n}{z^n}\,\mathrm{d}z$$

$$= \int_{1/b}^1 \frac{p_0 z^{n-2} + \cdots + p_{n-2}}{q_0 z^{n+2} + \cdots + q_n}\,\mathrm{d}z$$

tends to a proper integral on $[0,1]$, that is, it is proper and remains proper as b tends to infinity because the denominator is never 0 in $[0,1]$. The result: $\int_1^{\infty} f(x)\,\mathrm{d}x = \int_0^1 f(z^{-1})z^{-2}\,\mathrm{d}z$.

(b) $f(x) = \frac{1}{1+x^3}$ has denominator of degree $3 = n \geq 2$, numerator of degree $0 \leq 3-2$, and $1+x^3 \neq 0$ for $x \geq 1$, so part (a) applies. Then $x = z^{-1}$ gives $f(x) = \frac{1}{1+z^{-3}}$, $\mathrm{d}x = -z^{-2}\,\mathrm{d}z$, so $\int_1^{\infty} \frac{1}{1+x^3}\,\mathrm{d}x = -\int_1^0 \frac{1}{1+z^{-3}} z^{-2}\,\mathrm{d}z = \int_0^1 \frac{z}{z^3+1}\,\mathrm{d}z$.

7.29 $\lim_{b\to\infty} \int_1^b \frac{\sin x}{x}\,\mathrm{d}x = \lim_{b\to\infty} \left(-\frac{1}{x}\cos x\Big|_1^b - \int_1^b \frac{\cos x}{x^2}\,\mathrm{d}x \right)$ The limit of the first term is $\cos 1$, and the integral of $\frac{\cos x}{x^2}$ converges by comparison with $\int_1^{\infty} \frac{1}{x^2}\,\mathrm{d}x$, which converges. For the integral $\lim_{b\to\infty} \int_1^b \frac{|\sin x|}{x}\,\mathrm{d}x$, we have

(a) ignore integral from 1 to π,
(b) in each subinterval $\frac{1}{x} \geq \frac{1}{k\pi}$,
(c) each integral is equal to $\int_0^{\pi} \sin x\,\mathrm{d}x = 2$, and
(d) the integral has been shown to be larger than each partial sum of the divergent harmonic series.

7.31 $\int_s^b \frac{1}{x \log x}\,\mathrm{d}x = [\log(\log x)]_s^b$ tends to infinity with b because $\log b$ tends to infinity.

7.33 An antiderivative for $\frac{1}{y-y^2} = \frac{1}{y} + \frac{1}{1-y}$ is $\log|y| - \log|1-y|$. Therefore,

$$\int_2^b \frac{1}{y-y^2}\,\mathrm{d}y = \log\left|\frac{b}{1-b}\right| - \log\left|\frac{2}{1-2}\right|.$$

Since $\frac{b}{1-b}$ tends to -1 as b tends to infinity, the integral tends to $-\log 2$.

7.35 Denote the integrals by I_1, I_2 respectively. Then

$$I_1 = \int_0^\infty \sin(at)e^{-pt}\,\mathrm{d}t = -\frac{1}{p}\sin(at)e^{-pt}\Big|_0^\infty + \frac{1}{p}\int_0^\infty a\cos(at)e^{-pt}\,\mathrm{d}t = \frac{a}{p}I_2$$

$$I_1 = \int_0^\infty \sin(at)e^{-pt}\,\mathrm{d}t = -\frac{1}{a}\cos(at)e^{-pt}\Big|_0^\infty - \frac{p}{a}\int_0^\infty \cos(at)e^{-pt}\,\mathrm{d}t = \frac{1}{a} - \frac{p}{a}I_2.$$

Then solve for I_1 and I_2 from these relations.

7.37 Since $(x^n)' = nx^{n-1}$ is true for real positive n, integration by parts gives

$$\int_0^b x^n \mathrm{e}^{-x}\,\mathrm{d}x = \left[-\mathrm{e}^{-x}x^n\right]_0^b - \int_0^b (-\mathrm{e}^{-x})nx^{n-1}\,\mathrm{d}x.$$

As b tends to infinity, this becomes $n! = 0 + n\big((n-1)!\big)$.

7.39 Change $x = y^2$ in $\left(\frac{1}{2}\right)! = \int_0^\infty x^{1/2}\mathrm{e}^{-x}\,\mathrm{d}x = \int_0^\infty y\mathrm{e}^{-y^2}2y\,\mathrm{d}y = 2\int_0^\infty y^2\mathrm{e}^{-y^2}\,\mathrm{d}y = \frac{1}{2}\sqrt{\pi}.$

7.41 (a) $10 + \frac{1000}{3} + 10t$ (b) 10 (c) 10 (d) They are equal.

Chapter 8

8.1

(a) $I_{\text{left}}(x^3,[1,2]) = 1^3(2-1) = 1,\ (1^3+(1.5)^3)(2-1)/2 = 2.1875,$
$(1^3+(1.25)^3+(1.5)^3+(1.75)^3)(2-1)/4 = 2.92188$
$I_{\text{right}}(x^3),[1,2] = 2^3(2-1) = 8,\ ((1.5)^3+2^3)(2-1)/2 = 5.6875,$
$((1.25)^3+(1.5)^3+(1.75)^3+2^3)(2-1)/4 = 4.67188$
$I_{\text{mid}}(x^3),[1,2] = (1.5)^3(2-1) = 3.375,\ ((1.25)^3+(1.75)^3)(2-1)/2 = 3.35625,$
$((1.125)^3+(1.375)^3+(1.625)^3+(1.875)^3)(2-1)/4 = 3.72656$

(b) pseudocode to compute $I_{\text{mid}}(f,[a,b])$ with n subdivisions.

```
function iapprox = Imid(a,b,n)
 h = (b-a)/n;               [ width of subinterval ]
 x = a+h/2;                 [ midpoint of 1st subinterval ]
 iapprox = f(x);
for k = 2 up to n
 x = x+h;                   [ move to next midpoint ]
 iapprox = iapprox+f(x);    [ add value of f there ]
endfor
iapprox = iapprox*h;        [ multiply by width h last ]

function y = f(x)
 y = sqrt(1-x*x);
```

$I_{\text{left}}\left(\sqrt{1-x^2}, \left[0, \frac{1}{\sqrt{2}}\right]\right) = 0.70711, 0.68427, 0.66600$

$I_{\text{right}}\left(\sqrt{1-x^2}, \left[0, \frac{1}{\sqrt{2}}\right]\right) = 0.5, 0.58072, 0.61422;$

$I_{\text{mid}}\left(\sqrt{1-x^2}, \left[0, \frac{1}{\sqrt{2}}\right]\right) = 0.66144, 0.66722, 0.66399$

(c) $I_{\text{left}}\left(\frac{1}{1+x^2}, [0,1]\right) = 1.00000, 0.90000, 0.84529$
$I_{\text{right}}\left(\frac{1}{1+x^2}, [0,1]\right) = 0.5, 0.65000, 0.72079;\ \ I_{\text{mid}}\left(\frac{1}{1+x^2}, [0,1]\right) = 0.80000, 0.79059, 0.78670$

8.3

(a) $n=1:0.447214,\ n=5:0.415298,\ n=10:0.414483,\ n=100:0.414216$
actual value $\sqrt{2}-1\approx 0.414214$
(b) $n=1:0.8,\ n=5:0.78623,\ n=10:0.78561,\ n=100:0.78540$
actual value $\tan^{-1}1=\frac{\pi}{4}\approx 0.78540$
(c) $n=1:0.5,\ n=5:0.65449,\ n=10:0.66350,\ n=100:0.66663$
actual value $\frac{2}{3}\approx 0.66666$

8.5

(a) The graph of a convex function lies above each tangent line, in particular the tangent line at the midpoint of each subinterval $[c,d]$. Since then $f\geq\ell$ for that linear function ℓ on the subinterval, $\int_c^d f(x)\,dx\geq\int_c^d \ell(x)\,dx$. But the integral of ℓ is the midpoint rule for f.

(b) The graph of a convex function lies above each tangent line, in particular the secant line on each subinterval $[c,d]$. Since then $f\leq\ell$ for that linear secant function ℓ on the subinterval, $\int_c^d f(x)\,dx\leq\int_c^d \ell(x)\,dx$. But the integral of ℓ is the trapezoidal rule for f.

8.7

(a) The midpoint rule with one interval $[-h,h]$ gives $2hf(0)$.
(b) The derivatives for Taylor are
$K(h)-K(-h)$; at $h=0$ it is 0.
$K'(h)+K'(-h)=f(0)-f(h)+\big(f(0)-f(-h)\big)$; at $h=0$ it is 0.
$K''(h)-K''(-h)=-f'(h)+f'(-h)$; at $h=0$ it is 0.
$K'''(h)+K'''(-h)=-f''(h)-f''(-h)$.

(c) In the last step recognize that the subinterval width is $2h=(b-a)/n$, and use the triangle inequality, $|-f''(c_2)-f''(-c_2)|\leq 2M_2$.

8.9 $(1/2)((1/4)^2+(3/4)^2)+(1/24)(1/2)^2(4-0)=0.2002\ldots$

8.11 $n=100:0.7853575,\quad n=1000:0.7853968,\quad \frac{\pi}{4}=0.7853981$. The error term depends on the maximum value of $|f^{(4)}(x)|$ on the interval, which is unbounded as $x\to 1$.

Chapter 9

9.1

(a) $\sqrt{2^2+3^2}=\sqrt{13},\ \sqrt{4^2+(-1)^2}=\sqrt{17}$
(b) $2-3\mathrm{i},\ 4+\mathrm{i}$
(c) $\frac{1}{2+3\mathrm{i}}=\frac{1}{2+3\mathrm{i}}\frac{2-3\mathrm{i}}{2-3\mathrm{i}}=\frac{2-3\mathrm{i}}{13}=\frac{2}{13}-\frac{3}{13}\mathrm{i}$, and the reciprocal of the conjugate is the conjugate of the reciprocal: $\frac{2}{13}+\frac{3}{13}\mathrm{i}$.
$\frac{1}{4-\mathrm{i}}=\frac{1}{4-\mathrm{i}}\frac{4+\mathrm{i}}{4+\mathrm{i}}=\frac{4+\mathrm{i}}{17}=\frac{4}{17}+\frac{1}{17}\mathrm{i}$, and $\frac{4}{17}-\frac{1}{17}\mathrm{i}$
(d) $(2+3\mathrm{i})+(2-3\mathrm{i})=4=2(2),\ (4-\mathrm{i})+(4+\mathrm{i})=8=2(4)$
(e) $(2+3\mathrm{i})(2-3\mathrm{i})=(4+9)=\sqrt{13}^2,\ (4-\mathrm{i})(4+\mathrm{i})=(16+1)=\sqrt{17}^2$

9.3 $z=(4-1)+2\mathrm{i}=3+2\mathrm{i},\ \bar{z}=(3-2\mathrm{i})$

9.5 (a) 5 (b) $\sqrt{61}$ (c) $5/\sqrt{61}$ (d) 1

9.7 If $z=x+\mathrm{i}y$ then these say $y=\dfrac{x+\mathrm{i}y-(x-\mathrm{i}y)}{2\mathrm{i}}$ and $x=\dfrac{x+\mathrm{i}y+(x-\mathrm{i}y)}{2}$, which are true.

9.9 $(a^2+b^2)(c^2+d^2) = a^2c^2+a^2d^2+b^2c^2+b^2d^2$ and

$$(ac-bd)^2+(ad+bc)^2 = (ac)^2-2acbd+(bd)^2+(ad)^2+2adbc+(bc)^2$$
$$= a^2c^2+a^2d^2+b^2c^2+b^2d^2$$

Then if $z=a+bi$ and $w=c+di$, the left-hand side is $|z|^2|w|^2$ and $zw=(ac-bd)+(ad+bc)\mathrm{i}$, so the right-hand side is $|zw|^2$. Taking the square root of both sides gives $|z||w| = \pm|zw|$. However, since absolute values are only positive, $|z||w| = |zw|$.

9.11

(a) $|z_1-z_2|^2 = (z_1-z_2)\overline{(z_1-z_2)} = (z_1-z_2)(\overline{z_1}-\overline{z_2}) = z_1\overline{z_1}-z_1\overline{z_2}-z_2\overline{z_1}+z_2\overline{z_2}$
$= |z_1|^2+|z_2|^2-(z_1\overline{z_2}+\overline{z_1\overline{z_2}}) = |z_1|^2+|z_2|^2-2\mathrm{Re}\,(z_1\overline{z_2})$

(b) By the triangle inequality, $|z_1| = |z_2+(z_1-z_2)| \le |z_2|+|z_1-z_2|$.
Thus, $|z_1|-|z_2| \le |z_1-z_2|$. Similarly, $|z_2|-|z_1| \le |z_2-z_1| = |z_1-z_2|$. These combine to give us the desired inequality.

9.13 $-1=\cos(\pi)$ has cube roots $\cos\left(\frac{\pi+2k\pi}{3}\right)+\mathrm{i}\sin\left(\frac{\pi+2k\pi}{3}\right)$ for $k=0,1,2$.

These are $\frac{1}{2}+\frac{\sqrt{3}}{2}\mathrm{i}$, -1, and $\frac{1}{2}-\frac{\sqrt{3}}{2}\mathrm{i}$. These three roots are equally spaced around the unit circle.

9.15 The area formula says that triangle $(0,a,b)$ has area $A(0,a,b) = \frac{1}{2}|\mathrm{Im}\,(\overline{a}b)|$. If $a=a_1+\mathrm{i}a_2$, $b=b_1+\mathrm{i}b_2$, you obtain

$$\overline{a}b = (a_1-\mathrm{i}a_2)(b_1+\mathrm{i}b_2) = a_1b_1+a_2b_2+\mathrm{i}(a_1b_2-a_2b_1).$$

Therefore, the area is $\frac{1}{2}|a_1b_2-a_2b_1|$.

9.17 On the unit circle, $\overline{p} = \frac{1}{p}$. Therefore,

(a) $(p-1)^2\overline{p} = (p^2-2p+1)\overline{p} = p-2+\overline{p}$. This is real because $z+\overline{z}$ is twice the real part of z.
(b) When q is on the unit circle, so is $\overline{q}$. According to part (a), $(\overline{q}-1)^2q$ is real.
Then $\left((p-1)(\overline{q}-1)\right)^2\overline{p}q$ is the product of two real numbers, which is real.
(c) From the figure, and using Problem 9.16 part (b), β is the argument of $q\overline{p}$, and α is minus the argument of $(\overline{q-1})(p-1)$. But since $\left((p-1)(\overline{q}-1)\right)^2\overline{p}q$ is real, its argument, which is $-2\alpha+\beta$, must be 0.

9.19

(a) Because of the identity $(x-1)w(x) = x^5-1$, and because $w(1)=5\neq 0$, it follows that four of the five fifth roots of 1 are the roots of w.
(b) The n roots of 1 are equally spaced around the unit circle and one of them is 1.
(c) If $r^n=1$, then every integer power of r is also a root: $(r^p)^n = r^{np} = (r^n)^p = 1$. The only issue is whether these powers are *all* of the roots. [For example, powers of i^2 do not give all the fourth roots of 1.] Taking the one with smallest argument makes

$$r = \cos\left(\frac{2\pi}{n}\right)+\mathrm{i}\sin\left(\frac{2\pi}{n}\right),$$

and the powers have arguments $\frac{2\pi}{n}$, $2\frac{2\pi}{n}$, etc., so this gives all the roots. The identity $(x-1)(x^{n-1}+\cdots+x^2+x+1) = x^n-1$ and the same argument as in part (a) show that $r,\ldots,r^{n-1}$ are the roots of w and $r^n=1$ is the other root of $x^n=1$.

9.21

(a) $e^t + i\cos t$
(b) $-\dfrac{1}{(t-i)^2} - \dfrac{1}{(t+i)^2}$
(c) $ie^{t^2}2t$
(d) $i\cos t - (t+3+i)^{-2}$

9.23 Since $\overline{e^{it}} = e^{-it} = \dfrac{1}{e^{it}}$, $\cos t = \mathrm{Re}\, e^{it} = \dfrac{e^{it}+e^{-it}}{2}$ and $\sin t = \mathrm{Im}\, e^{it} = \dfrac{e^{it}-e^{-it}}{2i}$.

9.25 Define $\cosh z = \frac{1}{2}(e^z + e^{-z})$. Then $\cosh(it) = \frac{1}{2}(e^{it}+e^{-it}) = \cos t$.

9.27 $\displaystyle\int_0^b e^{ikx-x}\,dx = \frac{e^{ikb}-1}{ik-1}$ tends to $\dfrac{1}{1-ik}$. (e^{-b} tends to 0 as b tends to infinity, $|e^{ikb}| = 1$.)

Chapter 10

10.1 Only (b), and (c) if we allow f_{re} to be 0. The others do not match the descriptions of the frictional and restoring forces.

10.3 We need $r^2 + r = 0$, so $r = 0$ or -1. Then trying $x(t) = c_1 e^0 + c_2 e^{-t}$, we obtain

(a) $x(t) = 12 - 7e^{-t}$ tends to 12
(b) $x(t) = -2 + 7e^{-t}$ tends to -2

10.5 $2r^2 + 7r + 3 = 0$ gives $r = (-7 \pm \sqrt{49-24})/4 = -\frac{1}{2}, -3$. So $e^{-\frac{1}{2}t}$ and e^{-3t} are solutions.

10.7 Look at solutions e^{rt} where $mr^2 + hr + k = 0$, $r = \dfrac{-h \pm \sqrt{h^2-4km}}{2m}$. When h is small, this is roughly $r \approx -\dfrac{h}{2m} \pm i\sqrt{\dfrac{k}{m}}$, and you have a solution $x \approx e^{-\frac{h}{2m}}\cos\left(\sqrt{\dfrac{k}{m}}\right)$ with a gradually decreasing amplitude. When h is large, one value of r is a negative number close to 0 by the binomial theorem,

$$-h + \sqrt{h^2-4km} = -h + h\sqrt{1-\frac{4km}{h^2}} \approx -h + h(1-\frac{2km}{h^2}),$$

and you have a solution $x \approx e^{-(\mathrm{small})t}$, which is a gradually decreasing exponential.

10.9 We use $(cf)' = cf'$ and $(f+g)' = f' + g'$ repeatedly.

(a) Let $x = cx_1$. Then $A_n x^{(n)} + \cdots = A_n c x_1^{(n)} + \cdots = c\big(A_n x_1^{(n)} + \cdots\big) = c(0) = 0$.
(b) $A_n y^{(n)} + \cdots = \big(A_n x_1^{(n)} + \cdots\big) + \big(A_n x_2^{(n)} + \cdots\big) = 0 + 0 = 0$.

10.11 $m(c_1x_1 + c_2x_2)'' + h(c_1x_1 + c_2x_2)' + k(c_1x_1 + c_2x_2)$
$= m(c_1x_1'' + c_2x_2'') + h(c_1x_1' + c_2x_2') + k(c_1x_1 + c_2x_2)$
$= c_1(mx_1'' + hx_1' + kx_1) + c_2(mx_2'' + hx_2' + kx_2) = 0 + 0 = 0$.

10.13

(a) $mr^2 + 2\sqrt{mk}r + k = (\sqrt{m}r + \sqrt{k})^2$, so $r = -\sqrt{k/m}$
(b) $x' = (1+rt)e^{rt}$, $x'' = (2r + r^2t)e^{rt}$. Then

$$mx'' + 2\sqrt{mk}x' + kx = \left(2mr + mr^2t + 2\sqrt{km}(1+rt) + kt\right)e^{rt} = (-2\sqrt{mk} + 0t + 2\sqrt{km})e^{rt} = 0$$

10.15 Suppose $z(t) = ae^{6it}$. Then $z'' + z' + 6z - 52e^{6it} = \big(a(-36+6i+6) - 52\big)e^{6it}$ is zero if $a = \dfrac{52}{-30+6i} = \dfrac{52}{6}\dfrac{-5-i}{26} = \dfrac{-5-i}{3}$. Then the real part of $z(t) = \dfrac{-5-i}{3}e^{6it}$ is $x(t) = -\frac{5}{3}\cos(6t) + \frac{1}{3}\sin(6t)$.

10.17 Try $x = \operatorname{Re} z$, where $z = ae^{it}$ solves $z'' + z' + z = e^{it}$.
We need $(-1+i+1)ae^{it} = e^{it}$, so $a = -i$. Then $z = i\cos t + \sin t$ and $x_1(t) = \sin t$.
With $x_2 = y + x_1$, you have $x_2'' + x_2' + x_2 = y'' + y' + y + x_1'' + x_1' + x_1 = 0 + \cos t$.

10.19 The inequality holds if its square holds: $\dfrac{1}{h^2}\dfrac{1}{\frac{k}{m} - \frac{h^2}{4m^2}} > \dfrac{1}{k^2}$,

if the reciprocals $k^2 > h^2\left(\dfrac{k}{m} - \dfrac{h^2}{4m^2}\right)$, if $4m^2$ times it $4m^2k^2 > h^2(4mk - h^2)$.
But that is true because since $h < \sqrt{2mk}$, we have $h^2 < 2mk$, $0 > 4m^2k^2 - 2mkh^2$, $4m^2k^2 > 8m^2k^2 - 2mkh^2 = 2mk(4mk - h^2) > h^2(4mk - h^2)$.

10.21

(a) Use $w'' = y'' - x''$ and subtract the differential equations.
(b) This is due to the mean value theorem.
(c) Using (b), $mw''w' - f_{\text{re}}'(v)ww' = f_{\text{fr}}'(u)(w')^2 \le 0$ because f_{fr} is decreasing.
(d) Due to the law of decrease of energy, $x(t), y(t)$ are both bounded with values in some interval $[-M, M]$. Let k be the bound of $|f_{\text{re}}'|$ in $[-M, M]$. Because v is between x and y, $v \in [-M, M]$, so $f_{\text{re}}'(v)$ is bounded above by $-k$. Therefore,

$$mw''w' + kww' \le mw''w' - f_{\text{re}}'(v)ww' \le 0,$$

so its antiderivative $\frac{1}{2}m(w')^2 + \frac{1}{2}kw^2$ is nonincreasing.
(e) The function $\frac{1}{2}m(w')^2 + \frac{1}{2}kw^2$ is nonincreasing, nonnegative, and is zero when $t = s$. Therefore it is zero for all $t > s$. Therefore w is zero for all $t > s$.

10.23 For $N(t) = 0$, $N' = 0 = \sqrt{0}$, so that is a solution. For $N(t) = \frac{1}{4}t^2$, if $t \ge 0$ then $\sqrt{N} = \frac{1}{2}t$. Then $N' = \frac{2}{4}t = \sqrt{N}$, so that is a solution. (Note: $\frac{1}{4}t^2$ is not a solution when $t < 0$ because the derivative has the wrong sign.)

There is no contradiction: the existence theorem does not apply to $N' = \sqrt{N}$ with $N(0) = 0$ because the function $\sqrt{N}$ is not defined in an interval containing $N(0)$.

10.25 $N' = N^2 - N$, $\dfrac{1}{N^2 - N}N' = 1$, $\dfrac{-N^{-2}}{N^{-1} - 1}N' = 1$, $\log(N^{-1} - 1) = t + c$,
$N^{-1} - 1 = e^{t+c}$, $N_0^{-1} - 1 = e^c$. So $N(t) = \dfrac{1}{1 + e^{t+c}} = \dfrac{1}{1 + (N_0^{-1} - 1)e^t} = \dfrac{N_0}{N_0 + (1 - N_0)e^t}$. With N_0 between 0 and 1, the denominator is more than 1 and tends to infinity, so $N(t)$ is less than N_0 and tends to 0.

10.27

(a) Where $P > P_m$, K is increasing, so has an inverse. Then $P = K^{-1}(c - H(N))$ defines one function P_+. Similarly for P_- with values less than P_m.
(b) is by the chain rule
(c) For P_+, K is increasing, so in the formula for $\dfrac{d^2P_+}{dN^2}$, the denominator is positive. All numbers in the numerator are positive because H and K are convex. Therefore $\dfrac{d^2P_+}{dN^2}$ is negative. The denominator reverses sign for P_-.

10.29

(a) If $y(0)$ is positive, then $y(t)$, being continuous, will remain positive for some interval of time. Divide by $y(t)$ to get $-y^{-2}y' = 1$. Then we integrate from 0 to t to obtain $y^{-1}(t) - y^{-1}(0) = t$. Rearrange to

$$y(t) = \frac{1}{t + y^{-1}(0)} = \frac{y(0)}{y(0)t + 1}.$$

Note that in fact, $y(t)$ remains positive for all t.

(b) We know this equation; the answer is $y(t) = y(0)e^{-t}$.

For the second equation, y tends exponentially to zero, faster than the $1/t$ rate for the first one.

10.31 $\frac{da}{dt} = ap$, which is negative when $a > 0$ and $p < 0$.

$\frac{db}{dt} = -ap - db$, which is negative when $a > 0$, $b > 0$, and $p > 0$.

10.33 Using subintervals of length h, and $y_{n+1} = y_n + hf(nh)$, we obtain
$y_1 = y_0 + f(0)h = I_{\text{left}}(f,[0,h])$, $y_2 = y_1 + f(h)h = (f(0) + f(h))h = I_{\text{left}}(f,[0,2h])$
$y_3 = y_2 + f(2h)h = (f(0) + f(h) + f(2h))h = I_{\text{left}}(f,[0,3h])$ and so forth.

Chapter 11

11.1 The expected value of the die roll is $\frac{1}{6}(1+2+3+4+5+6) = 3.5$. The expected value of the squared difference between 3.5 and the value rolled is: $\frac{1}{6}(2.5^2 + 1.5^2 + 0.5^2 + 0.5^2 + 1.5^2 + 2.5^2) = \frac{35}{12}$ This is the variance.

11.3 Let $S(E)$ be the number of instances among the first N experiments when E occurred. Then $S(E') = N - S(E)$. So $p(E) + p(E') = \lim_{N\to\infty}\left(\frac{S(E)}{N} + \frac{S(E')}{N}\right) = \lim 1 = 1$.

11.5 Each time an outcome of E occurs, it is also an outcome of F, so $S(E) \le S(F)$. Then $P(E) = \lim_{n\to\infty} \frac{S(E)}{N} \le \lim_{n\to\infty} \frac{S(F)}{N} = P(F)$.

11.7 Since each outcome that occurs in an experiment can belong to only one of the events, the count $S(E_1 \cup \cdots \cup E_m)$ increases by exactly one each time an event occurs in some E_j, and only that one $S(E_j)$ increases by one. So $S(E_1 \cup \cdots \cup E_m) = S(E_1) + \cdots + S(E_m)$.

11.9 $\sum_{k=0}^{\infty} k\frac{u^k}{k!}e^{-u} = \sum_{k=1}^{\infty} u\frac{u^{k-1}}{(k-1)!}e^{-u} = u\sum_{k=0}^{\infty} \frac{u^k}{k!}e^{-u} = ue^u e^{-u} = u$

11.11 We need to show that

$$q\log q + r\log r < (1-p)\log(1-p) = (q+r)\log(q+r) = q\log(q+r) + r\log(q+r).$$

But this is true because log is increasing, that is, $\log q < \log(q+r)$ and $\log r < \log(q+r)$.

11.13 We need to show that $p_2\log p_2 + \cdots + p_n\log p_n < (1-p_1)\log(1-p_1)$

$$= (p_2 + \cdots + p_n)\log(p_2 + \cdots + p_n) = p_2\log(p_2 + \cdots + p_n) + \cdots + p_n\log(p_2 + \cdots + p_n),$$

which is true since log is increasing.

11.15

(a) $\int_{-\infty}^{\infty} p(x)\,dx = \int_0^A \frac{2}{A}\left(1-\frac{x}{A}\right)dx = \frac{2}{A}\left(A - \frac{A^2}{2A}\right) = 1$

(b) $\overline{x} = \int_{-\infty}^{\infty} xp(x)\,dx = \int_0^A \frac{2}{A}\left(x - \frac{x^2}{A}\right)dx = \frac{2}{A}\left(\frac{A^2}{2} - \frac{A^3}{3A}\right) = \frac{A}{3}$

(c) $\overline{x^2} = \int_{-\infty}^{\infty} x^2 p(x)\,dx = \int_0^A \frac{2}{A}\left(x^2 - \frac{x^3}{A}\right)dx = \frac{2}{A}\left(\frac{A^3}{3} - \frac{A^4}{4A}\right) = \frac{A^2}{6}$

(d) Using $\sqrt{\overline{x^2} - (\overline{x})^2}$ we get $\sqrt{\frac{A^2}{6} - \left(\frac{A}{3}\right)^2} = \frac{1}{18}A$.

11.17

(a) $\int_0^A \frac{1}{A}\,dx = 1$, similarly for p.

(b) $u(x) = \int_{-\infty}^{\infty} p(t)q(x-t)\,dt = \int_0^A p(t)q(x-t)\,dt$. In this integral, $-A \le -t \le 0$. If $x < 0$, then $x - t < 0$, so $q(x-t) = 0$. So $u(x) = 0$ when $x < 0$. If $x > A+B$, then $x - t > A+B-A = B$, so again $q(x-t) = 0$, showing that $u(x) = 0$ when $x > A+B$.

(c) If you graph $q(x-t)$ as a function of t, you see that the convolution integral gives the area of a rectangle whose size is constant when $B < x < A$. The width is B, height $\frac{1}{AB}$, so $u = 1/A$.

(d) $u(x) = \frac{x}{AB}$ on $[0,B]$, $\frac{1}{A}$ on $[B,A]$, and $\frac{1}{A} - \frac{x}{AB}(x-A)$ in $[A, A+B]$.

11.19

(a) $|w|_1 = \int_a^b 5\,dx = 5(b-a)$.

(b) $|cu|_1 = \int_{-\infty}^{\infty} |cu(x)|\,dx = \int_{-\infty}^{\infty} |c||u(x)|\,dx = |c||u|_1$,

$|u+v|_1 = \int_{-\infty}^{\infty} |u(x)+v(x)|\,dx \le \int_{-\infty}^{\infty} \left(|u(x)| + |v(x)|\right)dx = |u|_1 + |v|_1$.

(c)

$$|u*v|_1 = \int_{-\infty}^{\infty} |u*v(x)|\,dx = \int_{-\infty}^{\infty} \left| \int_{-\infty}^{\infty} u(y)v(x-y)\,dy \right| dx$$
$$\le \int_{-\infty}^{\infty}\int_{-\infty}^{\infty} |u(y)v(x-y)|\,dy\,dx = \int_{-\infty}^{\infty}\int_{-\infty}^{\infty} |u(y)||v(x-y)|\,dy\,dx$$

This inner integral is the convolution of $|u|$ and $|v|$. The integral of a convolution is the product of the integrals. Therefore, we get

$$|u*v|_1 = \left(\int_{-\infty}^{\infty} |u(x)|\,dx\right)\left(\int_{-\infty}^{\infty} |v(x)|\,dx\right) = |u|_1|v|_1.$$

11.21

(a) $n^2 > kn$, then use a geometric series times $e^{-K^2h^2}$. Then $\sum_4^{\infty} \le e^{-16}/(1-e^{-4}) = (1.1463)10^{-7}$.

(b) $1 + 2e^{-1} + 2e^{-4} + 2e^{-9} = 1.7726369797\ldots$

11.23

(a) Use the fact that if $p_n \sim q_n$ and $r_n \sim s_n$, then $p_n r_n \sim q_n s_n$. Then

$$2^{-n}\binom{n}{k} = 2^{-n}\frac{n!}{k!(n-k)!} \sim 2^{-n}\frac{\sqrt{2\pi n}\left(\frac{n}{e}\right)^n}{\sqrt{2\pi k}\left(\frac{k}{e}\right)^k \sqrt{2\pi(n-k)}\left(\frac{n-k}{e}\right)^{n-k}}$$

All the powers of e factor out as $e^{-n+k+n-k} = 1$.

(b) Part of this step is just substitution, but the hard part is

$$\frac{n^n}{2^n k^k} = \left(\frac{n}{2}\right)^{n/2}\left(\frac{n}{2}\right)^{n/2}\frac{1}{k^k} = \left(\frac{n}{2}\right)^{n/2}\frac{1}{\left(\frac{2}{n}\right)^{n/2}\left(\frac{n}{2}+\sqrt{n}y\right)^{\frac{n}{2}}\left(\frac{n}{2}+\sqrt{n}y\right)^{\sqrt{n}y}}$$

$$= \left(\frac{n}{2}\right)^{n/2}\frac{1}{\left(1+\frac{2y}{\sqrt{n}}\right)^{\frac{n}{2}}\left(\frac{n}{2}+\sqrt{n}y\right)^{\sqrt{n}y}}$$

Then handle the $(n-k)^{n-k}$ factor similarly:

$$\frac{n^n}{2^n k^k (n-k)^{n-k}} = \frac{1}{\left(1+\frac{2y}{\sqrt{n}}\right)^{\frac{n}{2}}\left(\frac{n}{2}+\sqrt{n}y\right)^{\sqrt{n}y}}\left(\frac{n}{2}\right)^{n/2}\frac{1}{\left(\frac{n}{2}-\sqrt{n}y\right)^{\frac{n}{2}}\left(\frac{n}{2}-\sqrt{n}y\right)^{-\sqrt{n}y}}$$

$$= \frac{1}{\left(1+\frac{2y}{\sqrt{n}}\right)^{\frac{n}{2}}\left(\frac{n}{2}+\sqrt{n}y\right)^{\sqrt{n}y}}\frac{1}{\left(1-\frac{2y}{\sqrt{n}}\right)^{\frac{n}{2}}\left(\frac{n}{2}-\sqrt{n}y\right)^{-\sqrt{n}y}}$$

and combine the $n/2$ powers to get $\dfrac{1}{\left(1-\frac{4y^2}{n}\right)^{\frac{n}{2}}\left(\frac{n}{2}+\sqrt{n}y\right)^{\sqrt{n}y}\left(\frac{n}{2}-\sqrt{n}y\right)^{-\sqrt{n}y}}$.

(c) Two of the factors in the denominator,

$$\left(\frac{n}{2}+\sqrt{n}y\right)^{\sqrt{n}y}\left(\frac{n}{2}-\sqrt{n}y\right)^{-\sqrt{n}y} = \left(\frac{\frac{n}{2}+\sqrt{n}y}{\frac{n}{2}-\sqrt{n}y}\right)^{\sqrt{n}y} = \left(\frac{1+\frac{2y}{\sqrt{n}}}{1+\frac{2y}{\sqrt{n}}}\right)^{\sqrt{n}y}$$

which tends to $\dfrac{(e^{2y})^y}{(e^{-2y})^y}$. The factor $\left(1-\dfrac{4y^2}{n}\right)^{\frac{n}{2}}$ tends to $\left(e^{-4y^2}\right)^{1/2}$.

(d) The coefficient, $\dfrac{n}{\frac{n^2}{4}-ny^2} = \dfrac{4}{n-4y^2} \sim \dfrac{4}{n}$, giving $\dfrac{1}{\sqrt{2\pi}}\sqrt{\dfrac{n}{\frac{n^2}{4}-ny^2}} \sim \dfrac{1}{\sqrt{n}}\sqrt{\dfrac{2}{\pi}}$.

Index

P.D. Lax and M.S. Terrell, *Calculus With Applications*, Undergraduate Texts in Mathematics, DOI 10.1007/978-1-4614-7946-8,

Printed in the United States
By Bookmasters